T0201429

Global Climate Change and Terrestrial Invertebrates

Global Climate Change and Terrestrial Invertebrates

Edited by
Scott N. Johnson
T. Hefin Jones

WILEY Blackwell

Library of Congress Cataloging-in-Publication Data applied for:

ISBN: 9781119070900

A catalogue record for this book is available from the British Library.

Wiley also publishes its books in a variety of electronic formats. Some content that appears in print may not be available in electronic books.

Cover image: © Gettyimages/aureliano1704

Set in 10/12pt Warnock by SPi Global, Chennai, India
Printed in Singapore by C.O.S. Printers Pte Ltd

1   2017

Contents

List of Contributors *xiii*
Preface *xvii*

1   Introduction to Global Climate Change and Terrestrial Invertebrates *1*
Scott N. Johnson and T. Hefin Jones
1.1   Background *1*
1.2   Predictions for Climate and Atmospheric Change *2*
1.3   General Mechanisms for Climate Change Impacts on Invertebrates *2*
1.3.1   Direct Impacts on Physiology, Performance and Behaviour *3*
1.3.2   Indirect Impacts on Habitats, Resources and Interacting Organisms *3*
1.4   Themes of the Book *4*
1.4.1   Methods for Studying Invertebrates and Global Climate Change *4*
1.4.2   Friends and Foes: Ecosystem Service Providers and Vectors of Disease *4*
1.4.3   Multi-Trophic Interactions and Invertebrate Communities *5*
1.4.4   Evolution, Intervention and Emerging Perspectives *6*
Acknowledgements *7*
References *7*

Part I   Methods for Studying Invertebrates and Climate Change *9*

2   Using Historical Data for Studying Range Changes *11*
Georgina Palmer and Jane K. Hill
Summary *11*
2.1   Introduction *11*
2.2   Review of Historical Data Sets on Species' Distributions *13*
2.3   Methods for Using Historical Data to Estimate Species' Range Changes *15*
2.3.1   Measuring Changes in Distribution Size *16*
2.3.2   Measuring Change in the Location of Species Ranges *16*
2.3.3   An Invertebrate Example: Quantifying Range Shift by the Comma Butterfly *Polygonia c-album* in Britain *17*
2.4   Challenges and Biases in Historical Data *19*
2.4.1   Taxonomic Bias *19*

2.4.2   Spatial and Temporal Biases *20*

2.4.3   Accounting for Temporal and Spatial Biases *21*

2.5   New Ways of Analysing Data and Future Perspectives *23*

       Acknowledgements *24*

       References *24*

3   **Experimental Approaches for Assessing Invertebrate Responses to Global Change Factors** *30*

       Richard L. Lindroth and Kenneth F. Raffa

       Summary *30*

3.1   Introduction *30*

3.2   Experimental Scale: Reductionist, Holistic and Integrated Approaches *32*

3.3   Experimental Design: Statistical Concerns *33*

3.4   Experimental Endpoints: Match Metrics to Systems *35*

3.5   Experimental Systems: Manipulations From Bottle to Field *36*

3.5.1   Indoor Closed Systems *36*

3.5.2   Outdoor Closed Systems *38*

3.5.3   Outdoor Open Systems *39*

3.6   Team Science: the Human Dimension *40*

3.6.1   Personnel *41*

3.6.2   Guiding Principles *41*

3.6.3   Operation and Communication *41*

3.7   Conclusions *41*

       Acknowledgements *42*

       References *42*

4   **Transplant Experiments – a Powerful Method to Study Climate Change Impacts** *46*

       Sabine S. Nooten and Nigel R. Andrew

       Summary *46*

4.1   Global Climate Change *46*

4.2   Climate Change Impacts on Species *47*

4.3   Climate Change Impacts on Communities *48*

4.4   Common Approaches to Study Climate Change Impacts *48*

4.5   Transplant Experiments – a Powerful Tool to Study Climate Change *49*

4.5.1   Can Species Adapt to a Warmer Climate? *50*

4.5.2   The Potential of Range Shifts *50*

4.5.3   Changes in the Timing of Events *51*

4.5.4   Shifts in Species Interactions *52*

4.5.5   Disentangling Genotypic and Phenotypic Responses *54*

4.5.6   Shifts in Communities *54*

4.6   Transplant Experiment Trends Using Network Analysis *57*

4.7   What's Missing in Our Current Approaches? Next Steps for Implementing Transplant Experiments *60*

       Acknowledgements *62*

       References *62*

Part II   Friends and Foes: Ecosystem Service Providers and Vectors of Disease *69*

5   Insect Pollinators and Climate Change *71*
         Jessica R. K. Forrest
         Summary *71*
5.1   Introduction *71*
5.2   The Pattern: Pollinator Populations and Climate Change *72*
5.2.1   Phenology *72*
5.2.2   Range Shifts *75*
5.2.3   Declining Populations *75*
5.3   The Process: Direct Effects of Climate Change *76*
5.3.1   Warmer Growing-Season Temperatures *76*
5.3.2   Warmer Winters and Reductions in Snowpack *79*
5.4   The Process: Indirect Effects of Climate Change *81*
5.4.1   Interactions with Food Plants *81*
5.4.2   Interactions with Natural Enemies *82*
5.5   Synthesis, and the View Ahead *83*
         Acknowledgements *84*
         References *84*

6   Climate Change Effects on Biological Control in Grasslands *92*
         Philippa J. Gerard and Alison J. Popay
         Summary *92*
6.1   Introduction *92*
6.2   Changes in Plant Biodiversity *94*
6.3   Multitrophic Interactions and Food Webs *94*
6.3.1   Warming and Predator Behaviour *97*
6.3.2   Herbage Productivity and Quality *98*
6.3.3   Plant Defence Compounds *98*
6.3.4   Fungal Endophytes *100*
6.3.5   Changes in Plant Phenology *101*
6.4   Greater Exposure to Extreme Events *102*
6.4.1   Changes in Precipitation *102*
6.4.2   Drought Effects *103*
6.5   Range Changes *103*
6.6   Greater Exposure to Pest Outbreaks *104*
6.7   Non-Target Impacts *104*
6.8   Conclusion *105*
         Acknowledgements *105*
         References *105*

7   Climate Change and Arthropod Ectoparasites and Vectors of Veterinary Importance *111*
         Hannah Rose Vineer, Lauren Ellse and Richard Wall
         Summary *111*
7.1   Introduction *111*
7.2   Parasite–Host Interactions *113*

7.3   Evidence of the Impacts of Climate on Ectoparasites and Vectors   *114*

7.4   Impact of Human Behaviour and Husbandry on Ectoparasitism   *116*

7.5   Farmer Intervention as a Density-Dependent Process   *118*

7.6   Predicting Future Impacts of Climate Change on Ectoparasites and Vectors   *118*
        Acknowledgements   *123*
        References   *123*

8   Climate Change and the Biology of Insect Vectors of Human Pathogens   *126*
        Luis Fernando Chaves
        Summary   *126*

8.1   Introduction   *126*

8.2   Interaction with Pathogens   *129*

8.3   Physiology, Development and Phenology   *131*

8.4   Population Dynamics, Life History and Interactions with Other Vector Species   *132*

8.5   Case Study of Forecasts for Vector Distribution Under Climate Change: The Altitudinal
        Range of *Aedes albopictus* and *Aedes japonicus* in Nagasaki, Japan   *134*

8.6   Vector Ecology and Evolution in Changing Environments   *138*
        Acknowledgements   *139*
        References   *140*

**9   Climate and Atmospheric Change Impacts on Aphids as Vectors of Plant
        Diseases**   *148*
        James M. W. Ryalls and Richard Harrington
        Summary   *148*

9.1   The Disease Pyramid   *148*

9.1.1   Aphids   *149*

9.1.2   Host-Plants   *152*

9.1.3   Viruses   *154*

9.2   Interactions with the Pyramid   *155*

9.2.1   Aphid–Host-Plant Interactions   *155*

9.2.2   Host-Plant–Virus Interactions   *158*

9.2.3   Virus–Aphid Interactions   *160*

9.2.4   Aphid–Host-Plant–Virus Interactions   *162*

9.3   Conclusions and Future Perspectives   *162*
        Acknowledgements   *163*
        References   *164*

Part III   Multi-Trophic Interactions and Invertebrate Communities   *177*

10   Global Change, Herbivores and Their Natural Enemies   *179*
        William T. Hentley and Ruth N. Wade
        Summary   *179*

10.1   Introduction   *180*

10.2   Global Climate Change and Insect Herbivores   *181*

10.3   Global Climate Change and Natural Enemies of Insect Herbivores   *185*

10.3.1 Elevated Atmospheric CO_2 *185*

10.3.1.1 Prey Location   *185*

10.3.1.2 Prey Quality   *186*

10.3.2   Temperature Change   *186*
10.3.3   Reduction in Mean Precipitation   *188*
10.3.4   Extreme Events   *190*
10.3.5   Ozone and UV-B   *190*
10.4   Multiple Abiotic Factors   *191*
10.5   Conclusions   *192*
         Acknowledgements   *193*
         References   *193*

11   Climate Change in the Underworld: Impacts for Soil-Dwelling Invertebrates   *201*
         Ivan Hiltpold, Scott N. Johnson, Renée-Claire Le Bayon and Uffe N. Nielsen
         Summary   *201*
11.1   Introduction   *201*
11.1.1   Soil Community Responses to Climate Change   *202*
11.1.2   Scope of the Chapter   *202*
11.2   Effect of Climate Change on Nematodes: Omnipresent Soil Invertebrates   *203*
11.2.1   Nematode Responses to eCO_2 *203*
11.2.2   Nematode Responses to Warming   *205*
11.2.3   Nematode Responses to Altered Precipitation Regimes   *206*
11.2.4   Ecosystem Level Effects of Nematode Responses to Climate Change   *207*
11.3   Effect of Climate Change on Insect Root Herbivores, the Grazers of the Dark   *207*
11.3.1   Insect Root Herbivore Responses to eCO_2 *208*
11.3.2   Insect Root Herbivore Responses to Warming   *210*
11.3.3   Insect Root Herbivore Responses to Altered Precipitation   *210*
11.3.4   Soil-Dwelling Insects as Modifiers of Climate Change Effects   *211*
11.4   Effect of Climate Change on Earthworms: the Crawling Engineers of Soil   *212*
11.4.1   Earthworm Responses to eCO_2 *212*
11.4.2   Earthworm Responses to Warming and Altered Precipitation   *214*
11.4.3   Climate Change Modification of Earthworm–Plant–Microbe Interactions   *214*
11.4.4   Influence of Climate Change on Earthworms in Belowground Food Webs   *215*
11.4.5   Influence of Climate Change on Earthworm Colonization of New Habitats   *215*
11.5   Conclusions and Future Perspectives   *216*
         Acknowledgements   *217*
         References   *218*

12   Impacts of Atmospheric and Precipitation Change on Aboveground-Belowground Invertebrate Interactions   *229*
         Scott N. Johnson, James M.W. Ryalls and Joanna T. Staley
         Summary   *229*
12.1   Introduction   *229*
12.1.1   Interactions Between Shoot and Root Herbivores   *231*
12.1.2   Interactions Between Herbivores and Non-Herbivorous Invertebrates   *232*
12.1.2.1 Detritivore–Shoot Herbivore Interactions   *232*
12.1.2.2 Root Herbivore–Pollinator Interactions   *232*
12.2   Atmospheric Change – Elevated Carbon Dioxide Concentrations   *233*
12.2.1   Impacts of $e[CO_2]$ on Interactions Mediated by Plant Trait Modification   *233*
12.2.2   Impacts of $e[CO_2]$ and Warming on Interactions Mediated by Plant Trait Modification   *234*

12.2.3   Impacts of Aboveground Herbivores on Belowground Invertebrates via Deposition Pathways *234*

12.3   Altered Patterns of Precipitation *236*

12.3.1   Precipitation Effects on the Outcome of Above–Belowground Interactions *236*

12.3.1.1 Case Study – Impacts of Simulated Precipitation Changes on Aboveground–Belowground Interactions in the Brassicaceae *237*

12.3.2   Aboveground–Belowground Interactions in Mixed Plant Communities Under Altered Precipitation Scenarios *239*

12.3.3   Altered Precipitation Impacts on Decomposer–Herbivore Interactions *240*

12.3.4   Impacts of Increased Unpredictability and Variability of Precipitation Events on the Frequency of Above–Belowground Interactions *240*

12.4   Conclusions and Future Directions *242*

12.4.1   Redressing the Belowground Knowledge Gap *243*

12.4.2   Testing Multiple Environmental Factors *243*

12.4.3   New Study Systems *244*

12.4.4   Closing Remarks *245*

     Acknowledgements *245*

     References *245*

13   **Forest Invertebrate Communities and Atmospheric Change** *252*

     Sarah L. Facey and Andrew N. Gherlenda

     Summary *252*

13.1   Why Are Forest Invertebrate Communities Important? *253*

13.2   Atmospheric Change and Invertebrates *253*

13.3   Responses of Forest Invertebrates to Elevated Carbon Dioxide Concentrations *254*

13.3.1   Herbivores *254*

13.3.2   Natural Enemies *259*

13.3.3   Community-Level Responses *259*

13.4   Responses of Forest Invertebrates to Elevated Ozone Concentrations *263*

13.4.1   Herbivores *263*

13.4.2   Natural Enemies *264*

13.4.3   Community-Level Studies *265*

13.5   Interactions Between Carbon Dioxide and Ozone *265*

13.6   Conclusions and Future Directions *267*

     Acknowledgements *268*

     References *268*

14   **Climate Change and Freshwater Invertebrates: Their Role in Reciprocal Freshwater–Terrestrial Resource Fluxes** *274*

     Micael Jonsson and Cristina Canhoto

     Summary *274*

14.1   Introduction *274*

14.2   Climate-Change Effects on Riparian and Shoreline Vegetation *275*

14.3   Climate-Change Effects on Runoff of Dissolved Organic Matter *277*

14.4   Climate Change Effects on Basal Freshwater Resources Via Modified Terrestrial Inputs *278*

14.5   Effects of Altered Terrestrial Resource Fluxes on Freshwater Invertebrates *279*

14.6   Direct Effects of Warming on Freshwater Invertebrates *280*

14.7   Impacts of Altered Freshwater Invertebrate Emergence on Terrestrial Ecosystems *282*

14.8   Conclusions and Research Directions   284
14.8.1   Effects of Simultaneous Changes in Resource Quality and Temperature on Freshwater
          Invertebrate Secondary Production   284
14.8.2   Effects of Changed Resource Quality and Temperature on the Size Structure of Freshwater
          Invertebrate Communities   284
14.8.3   Effects of Changed Resource Quality on Elemental Composition (i.e., Stoichiometry,
          Autochthony versus Allochthony, and PUFA Content) of Freshwater Invertebrates   284
14.8.4   Effects of Changed Freshwater Invertebrate Community Composition and Secondary
          Production on Freshwater Insect Emergence   285
14.8.5   Effects of Changed Quality (i.e., Size Structure and Elemental Composition) of Emergent
          Freshwater Insects on Terrestrial Food Webs   285
14.8.6   Effects of Climate Change on Landscape-Scale Cycling of Matter Across the
          Freshwater–Terrestrial Interface   285
          Acknowledgements   286
          References   286

15   **Climatic Impacts on Invertebrates as Food for Vertebrates**   295
          Robert J. Thomas, James O. Vafidis and Renata J. Medeiros
          Summary   295
15.1   Introduction   295
15.2   Changes in the Abundance of Vertebrates   296
15.2.1   Variation in Demography and Population Size   296
15.2.2   Local Extinctions   299
15.2.3   Global Extinctions   299
15.3   Changes in the Distribution of Vertebrates   300
15.3.1   Geographical Range Shifts   300
15.3.2   Altitudinal Range Shifts   301
15.3.3   Depth Range Shifts   302
15.3.4   Food-Mediated Mechanisms and Trophic Consequences of Range Shifts   302
15.4   Changes in Phenology of Vertebrates, and Their Invertebrate Prey   303
15.4.1   Consequences of Phenological Changes for Trophic Relationships   303
15.4.2   Phenological Mismatches in Marine Ecosystems   303
15.4.3   Phenological Mismatches in Terrestrial Ecosystems   304
15.4.3.1 Behaviour and Ecology of the Vertebrates   305
15.4.3.2 Habitat Differences in Prey Phenology   306
15.5   Conclusions   307
15.6   Postscript: Beyond the Year 2100   308
          Acknowledgements   308
          References   308

          Part IV   Evolution, Intervention and Emerging Perspectives   317

16   **Evolutionary Responses of Invertebrates to Global Climate Change: the Role of
          Life-History Trade-Offs and Multidecadal Climate Shifts**   319
          *Jofre Carnicer, Chris Wheat, Maria Vives, Andreu Ubach, Cristina Domingo, Sören Nylin, Constantí
          Stefanescu, Roger Vila, Christer Wiklund and Josep Peñuelas*
          Summary   319
16.1   Introduction   319

16.2 Fundamental Trade-Offs Mediating Invertebrate Evolutionary Responses to Global Warming *327*
16.2.1 Background *327*
16.2.2 Mechanisms Underpinning Trade-Offs *328*
16.2.2.1 Endocrine Hormone-Signalling Pathway – Antagonistic Pleiotropy Trade-Off Hypothesis *330*
16.2.2.2 The Thermal Stability – Kinetic Efficiency Trade-Off Hypothesis *330*
16.2.2.3 Resource-Allocation Trade-Off Hypothesis *331*
16.2.2.4 Enzymatic-Multifunctionality (Moonlighting) Hypothesis *331*
16.2.2.5 Respiratory Water Loss – Total Gas Exchange Hypothesis *332*
16.2.2.6 Water-Loss Trade-Off Hypotheses *332*
16.3 The Roles of Multi-Annual Extreme Droughts and Multidecadal Shifts in Drought Regimens in Driving Large-Scale Responses of Insect Populations *333*
16.4 Conclusions and New Research Directions *337*
Acknowledgements *339*
References *339*

17 Conservation of Insects in the Face of Global Climate Change *349*
Paula Arribas, Pedro Abellán, Josefa Velasco, Andrés Millán and David Sánchez-Fernández
Summary *349*
17.1 Introduction *349*
17.1.1 Insect Biodiversity *349*
17.1.2 Insect Biodiversity and Climate Change: the Research Landscape *350*
17.2 Vulnerability Drivers of Insect Species Under Climate Change *352*
17.3 Assessment of Insect Species Vulnerability to Climate Change *353*
17.4 Management Strategies for Insect Conservation Under Climate Change *355*
17.5 Protected Areas and Climate Change *357*
17.6 Perspectives on Insect Conservation Facing Climate Change *359*
Acknowledgements *360*
References *361*

18 Emerging Issues and Future Perspectives for Global Climate Change and Invertebrates *368*
Scott N. Johnson and T. Hefin Jones
18.1 Preamble *368*
18.2 Multiple Organisms, Asynchrony and Adaptation in Climate Change Studies *368*
18.3 Multiple Climatic Factors in Research *369*
18.4 Research Into Extreme Climatic Events *371*
18.5 Climate change and Invertebrate Biosecurity *372*
18.6 Concluding Remarks *374*
References *374*

Species Index *379*

Subject Index *385*

List of Contributors

Pedro Abellán
Department of Biology
Queens College
City University of New York
Flushing NY11367
USA

Nigel R. Andrew
Centre for Behavioural and Physiological
Ecology, Zoology
University of New England
Armidale
NSW 2351
Australia

Paula Arribas
Department of Life Sciences
Natural History Museum
London SW7 5BD
UK

and

Department of Life Sciences
Imperial College London
Ascot SL5 7PY
UK

Cristina Canhoto
Centre of Functional Ecology
Department of Life Sciences
University of Coimbra
3000-456 Coimbra
Portugal

Jofre Carnicer
GELIFES, Conservation Ecology Group
9747 AG
Groningen
The Netherlands

and

CREAF, Cerdanyola del Vallès 08193
Spain

and

Department of Ecology
University of Barcelona
08028
Barcelona
Spain

Luis Fernando Chaves
Nagasaki University Institute of Tropical
Medicine (NEKKEN), Sakamoto 1-12-4
Nagasaki
Japan

and

Programa de Investigación en Enfermedades
Tropicales (PIET)
Escuela de Medicina Veterinaria
Universidad Nacional
Apartado Postal 304-3000
Heredia
Costa Rica

Cristina Domingo
Department of Geography
Autonomous University of Barcelona
Spain

and

CREAF, Cerdanyola del Vallès 08193
Spain

Lauren Ellse
School of Biological Sciences
Life Sciences Building
University of Bristol
Bristol BS8 1TH
UK

Sarah L. Facey
Hawkesbury Institute for the Environment
Western Sydney University
NSW 2751
Australia

Jessica R. K. Forrest
Department of Biology
University of Ottawa
Ottawa
ON K1N 6N5
Canada

Philippa J. Gerard
AgResearch
Ruakura Research Centre
Private Bag 3123
Hamilton
New Zealand

Andrew N. Gherlenda
Hawkesbury Institute for the Environment
Western Sydney University
NSW 2751
Australia

Richard Harrington
Rothamsted Insect Survey
Rothamsted Research
Harpenden
AL5 2JQ
UK

William T. Hentley
Department of Animal and Plant Sciences
University of Sheffield
Sheffield
UK

Jane K. Hill
Department of Biology
University of York
YO10 5DD
UK

Ivan Hiltpold
Department of Entomology and Wildlife
Ecology
University of Delaware
DE 19716
USA

Scott N. Johnson
Hawkesbury Institute for the Environment
Western Sydney University
NSW 2751
Australia

T. Hefin Jones
School of Biosciences
Cardiff University
Cardiff CF10 3AX
Wales
UK

Micael Jonsson
Department of Ecology and Environmental
Science
Umeå University
SE 901 87 Umeå
Sweden

Renée-Claire Le Bayon
Functional Ecology Laboratory
University of Neuchâtel
Switzerland

Richard L. Lindroth
Department of Entomology
University of Wisconsin-Madison
Madison
WI 53706
USA

Renata J. Medeiros
School of Biosciences
Cardiff University Cardiff CF10 3AX
Wales
UK

Andrés Millán
Department of Ecology and Hydrology
University of Murcia
Espinardo 30100
Spain

Uffe N. Nielsen
Hawkesbury Institute for the Environment
Western Sydney University
NSW 2751
Australia

Sabine S. Nooten
Hawkesbury Institute for the Environment
Western Sydney University
NSW 2751
Australia

Sören Nylin
Department of Zoology
Stockholm University
Sweden

Georgina Palmer
Department of Biology
University of York
YO10 5DD
UK

Josep Peñuelas
CREAF, Cerdanyola del Vallès 08193
Catalonia
Spain

and

CSIC, Global Ecology Unit CREAF-CSIC-UAB
Bellaterra 08193
Catalonia
Spain

Alison J. Popay
AgResearch
Ruakura Research Centre
Private Bag 3123
Hamilton
New Zealand

Kenneth F. Raffa
Department of Entomology
University of Wisconsin-Madison
Madison
WI 53706
USA

James M. W. Ryalls
Hawkesbury Institute for the Environment
Western Sydney University
NSW 2751
Australia

Hannah Rose Vineer
School of Veterinary Sciences
Life Sciences Building
University of Bristol
Bristol BS8 1TH
UK

David Sánchez-Fernández
Institute of Evolutionary Biology
CSIC-University Pompeu Fabra
Barcelona 08003
Spain

and

Institute of Environmental Sciences
University of Castilla-La Mancha
Toledo 45071
Spain

Joanna T. Staley
Centre for Ecology and Hydrology
Wallingford
UK

Constanti Stefanescu
CREAF, Cerdanyola del Vallès 08193
Spain

and

Museum of Natural Sciences of Granollers
08402, Granollers
Spain

Robert Thomas
School of Biosciences
Cardiff University
Cardiff CF10 3AX
Wales
UK

Andreu Ubach
Department of Ecology
Universitat de Barcelona
08028, Barcelona
Spain

James Vafidis
School of Biosciences
Cardiff University
Cardiff CF10 3AX
Wales
UK

Josefa Velasco
Department of Ecology and Hydrology
University of Murcia
Espinardo 30100
Spain

Roger Vila
IBE, Institute of Evolutionary Biology
08003, Barcelona
Spain

Ruth N. Wade
Department of Animal and Plant Sciences
University of Sheffield
Sheffield
UK

Richard Wall
School of Biological Sciences
Life Sciences Building
University of Bristol
Bristol BS8 1TH
UK

Chris Wheat
Department of Zoology
Stockholm University
Sweden

Christer Wiklund
Department of Zoology
Stockholm University
Sweden

Preface

The title of this book should more accurately be 'Global Climate and Atmospheric Change and Terrestrial Invertebrates' because many of the contributors consider the effects of changes in greenhouse gases, especially carbon dioxide, on invertebrates. Our students, past and present, will be bemused because for many years we've laboured the point that carbon dioxide is an atmospheric chemical and not a climatic variable. We decided to use the term climate change as a 'catch all' to include atmospheric change, not just because the title is snappier, but in most peoples' minds, climate change includes components such as greenhouse gases. Public engagement with global climate change research has increased dramatically in the last few decades, helped in part by using accessible language without getting stuck on strict definitions, so we think this is a small compromise to make.

Invertebrates account for over 95% of multicellular life on our plant and represent an unrivalled level of diversity from nematodes which are a few microns in size to the colossal squid (*Mesonychoteuthis hamiltoni*) which can reach 14 metres. While we focus on terrestrial invertebrates in this book, this still represents a massively diverse group which occupy disparate habitats, aboveground and belowground. It would be impossible to provide comprehensive coverage of all groups in a single volume. Our second *mea culpa* is therefore that we have not been able to consider some groups of invertebrates of interest to readers of this book. Our selection of topics reflected those that we considered were ripe for synthesis and could be related to one another in a single volume.

Given the importance of invertebrates to our planet we felt consideration of this group in the context of global climate change was much needed. There are many books on global climate change, some which blend diverse disciplines such as the humanities, economics and science (e.g., Bloom, 2010)[1] and others that cover broad disciplines, such as biology (e.g., Newman et al., 2011).[2] This book aims to fill a gap by taking a more in-depth examination of a crucial group of organisms that shape the world we live in. The book would have not been possible without the work of the 44 contributors throughout the globe and we are indebted to them for their efforts. We sincerely hope that this book will provide a good survey introduction to the issue of global climate change and terrestrial invertebrates.

Scott N. Johnson
Sydney, Australia

T. Hefin Jones
Cardiff, United Kingdom
October 2016

1 Bloom, A.J. (2010) Global Climate Change, Convergence of Disciplines, Sinauer Associates, MA, USA.
2 Newman, J.A., Abnand, M., Henry, H.A.L., Hunt, S. & Gedalof, Z. (2011) Climate Change Biology, CABI, Wallingford, Oxfordshire, UK.

1

Introduction to Global Climate Change and Terrestrial Invertebrates

Scott N. Johnson[1] and T. Hefin Jones[2]

[1] *Hawkesbury Institute for the Environment, Western Sydney NSW 2751, Australia*
[2] *School of Biosciences, Cardiff University, Cardiff CF10 3AX, UK*

> *"If all mankind were to disappear, the world would regenerate back to the rich state of equilibrium that existed ten thousand years ago. If insects were to vanish, the environment would collapse into chaos."*
>
> E. O. Wilson

> *"The great ecosystems are like complex tapestries – a million complicated threads, interwoven, make up the whole picture. Nature can cope with small rents in the fabric; it can even, after a time, cope with major disasters like floods, fires, and earthquakes. What nature cannot cope with is the steady undermining of its fabric by the activities of man."*
>
> Gerald Durrell

1.1 Background

'Little things that run the world' is how the biologist E.O. Wilson described invertebrates (Wilson, 1987). There is a great deal of truth in this, with invertebrates playing major roles in the functioning and processes of most terrestrial and aquatic ecosystems. In terms of human wellbeing, their influence ranges from the beneficial ecosystem services of pollinators to lethal vectors of human diseases. Invertebrate pests, for example, destroy enough food to feed 1 billion people (Birch et al., 2011) at a time when global populations are expected to exceed 9.7 billion by 2050 and 11.2 billion by 2100 (UN, 2015) and therefore represent a significant challenge to secure global food security (Gregory et al., 2009). Conversely, invertebrates provide an unrivalled array of ecosystem services; globally €153 billion per year via pollination (Gallai et al., 2009), US$417 billion annually in terms of pest control (Costanza et al., 1997). This latter figure is somewhat dated, but if it increased in line with the general trend for ecosystem services calculated by Costanza et al. (2014) for 2011 this would be closer to US$1.14 trillion per year.

Besides humankind, invertebrates shape the world around us perhaps more than any other group and their response to climate change is pivotal in future global challenges, including food security, conservation, biodiversity and human health. In this book, we synthesise the current state of knowledge about how terrestrial invertebrates will respond and adapt to predicted changes in our climate and atmosphere, and, in some cases even moderate the impacts of such changes.

1.2   Predictions for Climate and Atmospheric Change

Between September 2013 and April 2014 the Fifth Assessment Report of the Intergovernmental Panel for Climate Change (IPCC) was published (IPCC, 2014). Divided into three Working Groups (WGs) and the culmination of the work of over 800 authors, the report not only focusses on the physical science basis of current climate change (WG I), but also assesses the impacts, adaptation strategies and vulnerability related to climate change (WG II) while also covering mitigation response strategies in an integrated risk and uncertainty framework and its assessments (WG III).

The report finds that the warming of the atmosphere and ocean system is *unequivocal*. Many of the associated impacts such as sea level change (among other metrics) have occurred since 1950 at rates unprecedented in the historical record. It states that there is a clear human influence on the climate and declares that it is *extremely likely* that human influence has been the dominant cause of observed warming since 1950, with the level of confidence having increased since the Fourth IPCC Report in 2007 (IPCC, 2007). In noting the current situation the 2014 Report states that (i) it is *likely* (with medium confidence) that 1983–2013 was the warmest 30-year period for 1,400 years; (ii) it is *virtually certain* the upper ocean warmed from 1971 to 2010. This ocean warming accounts, with *high confidence*, for 90% of the energy accumulation between 1971 and 2010; (iii) it can be said with *high confidence* that the Greenland and Antarctic ice sheets have been losing mass in the last two decades and that Arctic sea ice and Northern Hemisphere spring snow cover have continued to decrease in extent; (iv) there is *high confidence* that the sea level rise since the middle of the nineteenth century has been larger than the mean sea level rise of the prior two millennia; (v) concentration of greenhouse gases in the atmosphere has increased to levels unprecedented on Earth in 800,000 years; and (vi) total radiative forcing of the Earth system, relative to 1750, is positive and the most significant driver is the increase in atmospheric concentrations of carbon dioxide (CO_2).

Relying on the Coupled Model Intercomparison Project Phase 5 (CMIP5), which is an international climate modelling community effort to coordinate climate change experiments, for much of its analysis, the Fifth Report based its predictions on CO_2 concentrations reaching 421 parts per million (ppm), 538 ppm, 670 ppm and 936 ppm by the year 2100. General conclusions drawn from this analysis were that (i) further warming will continue if emissions of greenhouse gases continue; (ii) the global surface temperature increase by the end of the twenty-first century is *likely* to exceed 1.5°C relative to the 1850 to 1900 period for most scenarios, and is *likely* to exceed 2.0°C for many scenarios; (iii) the global water cycle will change, with increases in the disparity between wet and dry regions, as well as wet and dry seasons, with some regional exceptions; (iv) the oceans will continue to warm, with heat extending to the deep ocean, affecting circulation patterns; (v) decreases are *very likely* in Arctic sea ice cover, Northern Hemisphere spring snow cover, and global glacier volume; (vi) global mean sea level will continue to rise at a rate *very likely* to exceed the rate of the past four decades; (vii) changes in climate will cause an increase in the rate of CO_2 production. Increased uptake of CO_2 by the oceans will increase the acidification of the oceans; and (viii) future surface temperatures will be largely determined by cumulative CO_2, which means climate change will continue even if CO_2 emissions are stopped. This may be a moot point, however, since 2015 saw the largest ever annual increase in atmospheric CO_2 (Le Page, 2016).

1.3   General Mechanisms for Climate Change Impacts on Invertebrates

Generally speaking, predicted changes to our climate might affect invertebrates in two ways: (i) by directly affecting invertebrate physiology, performance or behaviour, and (ii) by indirectly affecting invertebrates via changes to the habitats, resources or organisms they interact with. This is a very simplified way of categorising the impacts of global climate change on invertebrates, but it provides a convenient framework for understanding more complex processes. In this introduction, we do not comprehensively review examples of these mechanisms since they are developed in more detail in subsequent chapters but simply outline the general principles of each. Invertebrates are not just affected by climate change, but they can also moderate its effects on the ecosystem. This seems especially true for soil-dwelling ecosystem engineers (see Chapters 6 and 11) which have the capacity to mitigate the negative effects of drought on plants by changing the hydrological properties of their soil environment.

1.3.1 Direct Impacts on Physiology, Performance and Behaviour

As ectotherms, invertebrates are directly and significantly affected by temperature. Increasing temperature generally increases the rate of physiological and developmental processes to a point, whereupon further increases become detrimental. Providing other resources are not limiting, increased rates of development are likely to lead to larger populations of invertebrates and possibly an increased number of generations per year (Bale et al., 2002). This is most tangibly seen in the case of invasive invertebrates that move into warmer regions; the clover root weevil (*Sitona obsoletus*), for example, which is univoltine in the UK undergoes two generations per year since its accidental introduction to New Zealand in the mid-1990s (Goldson & Gerard, 2008). Precipitation changes also have direct impacts on invertebrates. Intense precipitation events can cause physical damage to invertebrates by disrupting flight, reducing foraging efficiency and increasing migration times (Barnett & Facey, 2016), though some invertebrates such as mosquitoes are dependent on heavy rainfall events. Conversely, drought can lead to desiccation, particularly in soft-bodied invertebrates though many have physiological and behavioural adaptations to reduced moisture (Barnett & Facey, 2016). Precipitation events will clearly have greater impacts on terrestrial invertebrates than those in aquatic habitats. Atmospheric changes are generally thought to have negligible direct impacts on invertebrates.

1.3.2 Indirect Impacts on Habitats, Resources and Interacting Organisms

Climate change can affect invertebrates indirectly via its impacts on the habitat they occupy, the resources they use or the organisms they interact with. These are enormously varied for different taxa, and can be both positive and negative. Changes in habitat complexity, for instance, could affect foraging behaviour of predatory invertebrates affecting populations of both prey and predator (Facey et al., 2014). Elevated CO_2 concentrations often increase structural complexity of habitats via changes in plant architecture (Pritchard et al., 1999), which potentially explains why web-building predatory spiders can become more abundant under elevated atmospheric CO_2 concentrations (e.g., Hamilton et al., 2012).

Every invertebrate exploits specific resources and, if the supply or nature of these resources is modified by climate change, it seems likely that the invertebrate will also be affected. When climate change affects host plant quantity or quality (e.g., nutritional value or defensive status), for example, many herbivorous insects are affected by this change in their resource (Robinson et al., 2012). Moreover, alterations in herbivore performance will change their quality as a resource for natural enemies that predate or parasitise them (Facey et al., 2014).

Where invertebrate populations are influenced by interactions with other organisms (e.g., mutualism, competition or predation), climate change has the capacity to affect indirectly invertebrates if it has direct impacts on that interacting organism. Changes in temperature and precipitation, for example, often seem to introduce asynchrony between predator and prey life-cycles, which frequently results in reductions in top-down control of the prey species (Preisser & Strong, 2004; Stireman et al., 2005). Changes in the emissions of volatile organic compounds from plants grown under elevated CO_2 concentrations can alter the foraging efficiency of parasitoids (Vuorinen et al., 2004). Changes in the phenology and range expansions of invertebrates in relation to resources they exploit or organisms they interact with (e.g., natural enemies) is another indirect means that global climate change might affect invertebrates, especially in terms of temporal mismatches (Facey et al., 2014).

1.4 Themes of the Book

The following 17 chapters can be divided into four themes; (i) methods for studying invertebrates and climate change, (ii) friends and foes: ecosystem service providers and vectors of disease, (iii) multi-trophic interactions and invertebrate communities and finally (iv) evolution, intervention and emerging perspectives. Some of these chapters use the same case studies and examples but from different perspectives.

1.4.1 Methods for Studying Invertebrates and Global Climate Change

This theme describes three mainstream approaches for understanding how invertebrates will respond to global climate change. The first of these, by Palmer and Hill (Chapter 2) considers how historical data, particularly those collected by citizen science projects, can be used to predict changes in geographical distributions of invertebrates. This includes measuring changes in distribution, abundance and changes in the location of species ranges. While these datasets often have taxonomic and spatio-temporal biases, Palmer and Hill present approaches for accounting for such biases, such as fixed effort transects. New ways for analysing existing and future datasets include combining datasets with remotely sensed satellite land cover dataset, meta-genomic DNA barcoding and new dynamic models that incorporate dispersal and evolutionary processes.

Experimental approaches for investigating the impacts of global environmental change on invertebrates are discussed by Lindroth and Raffa (Chapter 3). Such controlled experiments are important for elucidating mechanisms and disentangling interactive relationships quantitatively. These range from reductionist experiments to larger scale approaches, each with inherent strengths and weaknesses. In particular, the authors set out approaches for devising experiments that maximise statistical power and avoid pseudo-replication. In closing, they consider the importance of the often overlooked human dimension of such experiments, emphasising the need for effective team assembly, leadership, project management and communication.

In Chapter 4, Nooten and Andrew describe transplant experimental approaches. This approach involves moving species or entire communities into a new location with a novel climate, usually one that is predicted to occur for the current location. Such experiments have revealed how invertebrates may adapt to warmer climates, potential range shifts of invertebrates, changes in phenology, shifts in species interactions, genotypic and phenotypic responses and community shifts. The authors use network analysis to identify gaps in our knowledge from transplant experiments and stress the importance of understanding whether transplanted species occupy the same niche as they do in their current location.

1.4.2 Friends and Foes: Ecosystem Service Providers and Vectors of Disease

In this theme, consideration is given to how global climate change will affect key groups of invertebrates that are of economic and social importance to mankind. In the first of these, Forrest

(Chapter 5) considers how global warming will affect invertebrate pollinators. She considers how warming can cause large-bodied pollinators to overheat, deplete energy reserves, reduce adult body size and increase mortality. These can be considered direct impacts of warming. Indirect impacts include altered plant phenology and changes in floral resource production. Forrest concludes that observed pollinator declines may be partly due to global warming working in concert with habitat loss, pesticide poisoning and pathogen infection.

Chapter 6 by Gerard and Popay considers the impact of global climate change on invertebrates with biological control roles in grasslands. Grassland ecosystems are prone to invasions by exotic pests, but endemic and introduced predators and parasitoids, as well as plant defences and symbionts, play a crucial role in managing weed and pest abundance. Disruption of predator–prey interactions, particularly in terms of asynchrony between life-cycles, could lead to pest outbreaks. Nonetheless, they suggest that the manipulability of grasslands will allow climate change adaptation strategies to be implemented.

Turning to invertebrates with economic pest status, Rose Vineer, Ellse and Wall describe the likely impacts of global climate change on ectoparasites and vectors of veterinary disease in Chapter 7. They note that changes in the phenology and distribution of tick species have already changed in recent years and predicted changes in climate are likely to affect seasonal patterns of blowfly strike. While they suggest climate change may result in increased abundances of ectoparasites, and possibly disease incidence, strategic changes in animal husbandry may help mitigate these impacts.

Invertebrate vectors transmit pathogens that account for almost fifth of the burden of human infectious diseases, and these are the subject of Chapter 8 by Chaves. The chapter focuses on whether global warming could exacerbate vector-borne disease transmission and whether warming will interfere with current programmes to eliminate such diseases. The chapter presents a case study using two mosquito vectors along an altitudinal gradient in Japan to illustrate useful concepts for studying changes in vectors arising through global climate change. The chapter ends with some ideas on the evolutionary implications of climate change for invertebrate vectors and the diseases they transmit.

In the final chapter of this theme, Ryalls and Harrington look at invertebrate vectors of plant diseases focussing on one of the most important groups, aphids, which are responsible for transmitting around 40% of plant viruses. They adopt a 'disease pyramid' approach for considering how interactions between aphids, their host plants and viruses determine the overall effect of global climate change on aphids and plant disease incidence. The individual responses of aphids, plants and viruses to global climate change are likely to exacerbate each other, particularly in terms of warming and drought drivers. Ryalls and Harrington conclude that relatively few studies incorporate all these factors interactively, however, so more holistic research is needed to make accurate predictions.

1.4.3 Multi-Trophic Interactions and Invertebrate Communities

Related to Chapter 6, Hentley and Wade (Chapter 10) take an in-depth look to how climate change will affect herbivore interactions with their natural enemies. In particular, they produce a comprehensive assessment of 45 studies in this area, pointing to neutral, positive and negative outcomes for insect herbivores. Positive outcomes for herbivores occurred through various mechanisms including reduced rates of parasitism by parasitoids, longer development times for natural enemies, reduced foraging efficiency and phenological mismatches between herbivore and antagonist life cycles. Negative impacts, mostly reported for increased air temperature, arose because of increased rates of parasitism and reduced resistance to such parasites by herbivores. This chapter also discusses the possible mechanisms driving the impact of climate change on herbivore–natural enemy interactions such as altered plant-derived cues which are used by natural enemies to locate their prey.

Moving belowground, Hiltpold and colleagues look at how climate change might affect soil communities of invertebrates in Chapter 11. The authors briefly consider the few studies that have examined overall changes in soil communities, before considering responses of three key groups (nematodes, insect herbivores and earthworms) in more detail in an attempt to understand these community level outcomes. The chapter considers impacts of elevated CO_2 concentrations, elevated air temperatures and altered precipitation patterns. Unlike other chapters, increased warming is likely to have fewer direct impacts on soil-dwelling invertebrates because soils will buffer temperature variation to some extent. Likewise, soil-dwelling invertebrates are already adapted to high concentrations of CO_2, so impacts are anticipated to be entirely indirect. The authors emphasise the need for longer term studies since soil communities are likely to respond to climate change over longer periods of time than aboveground communities.

Studying linkages between above- and belowground invertebrate communities is a relatively recent development in community ecology, and is the subject of Chapter 12 by Johnson, Ryalls and Staley. In particular, they consider how changes in elevated atmospheric CO_2 and precipitation changes might affect interactions between above- and belowground invertebrates. Studies in this area are scarce, so the chapter puts forward a conceptual framework for these interactions which may be mediated by changes in plant traits, shifts in plant communities and those mediated by plant-derived organic inputs (e.g., frass and litter deposition) entering the soil. Several hypotheses for how climate change may affect these interactions are made.

In terms of broad community responses to atmospheric change, forests are amongst our best studied ecosystems thanks to manipulations in Free Air Carbon dioxide Enrichment (FACE) experiments. These studies are the subject of Chapter 13 by Facey and Gherlenda which focuses on the effects of predicted concentrations of atmospheric CO_2 and ozone on forest invertebrate communities. In general, they conclude that the former tends to increase herbivore susceptibility to attack by natural enemies, whereas the latter generally speeds up herbivore development and causes reductions in natural enemy performance. However, it is suggested that these may be short term responses and that communities may show resilience in the longer term.

While the focus of this book are terrestrial invertebrates, in Chapter 14, Jonsson and Canhoto consider the impacts of global climate change on freshwater invertebrate communities via changes in terrestrial ecosystems. They argue that the effects of climate change mediated via terrestrial ecosystems is at least as important as the direct impacts of climate change, and that these impacts likely will feedback to influence invertebrates and other consumers in terrestrial systems.

Invertebrates are a crucial source of food for vertebrates and Chapter 15 considers this with particular reference to a range of vertebrate taxa. Thomas, Vafidis and Medeiros highlight three climatic drivers for invertebrate–vertebrate interactions. The first arises because global climate change affects invertebrate abundance and therefore, food supply, driving vertebrate population change and potentially causing local extinctions. Secondly, changes in invertebrate communities are likely to induce range shifts in vertebrate populations. Thirdly, invertebrate responses to global climate change are often more rapid than vertebrate responses, potentially introducing temporal mismatches between invertebrate abundance and vertebrate phenology. Invertebrate scarcity, for example, at a time when vertebrates need to provision for their young is particularly problematic.

1.4.4 Evolution, Intervention and Emerging Perspectives

The fourth, and final, theme of the book identifies evolutionary responses to global climate change, intervention and emerging perspectives in our understanding of how invertebrates will evolve to global climate change, potential adaptation strategies and the risk of future biosecurity breaches.

Chapter 16 by Carnicer and colleagues, looks at evolutionary responses of invertebrates to global climate change from the perspective of life-history trade-offs. They consider that such trade-offs will constrain the simultaneous optimisation of correlated suites of traits to environmental change and focus on five particular trade-offs. The chapter also considers the roles of multidecadal climate dynamics and drought regime shifts in long term population responses and evolutionary responses.

Chapter 17 by Arribas and colleagues takes a forward-looking perspective and asks how understanding insect species vulnerability to global climate change might inform conservation strategies. They examine the accumulated background knowledge on vulnerability of insect species to global climate change, recent developments and methods of its assessment and the links with the management options for insect conservation including: (i) monitoring, (ii) reduction of additional threats, (iii) habitat restoration, (iv) increasing habitat connectivity, (v) expansion of reserve networks and (vi) performing assisted dispersal.

Conclusions and Future Perspectives are presented in Chapter 18. We discuss common themes and issues that have arisen in the proceeding chapters, including the need to build on single species study systems and single environmental factor experiments, while observing rigour in experimental design and analysis (see Chapter 3). We also identify the need to investigate how extreme climate change events will affect invertebrates. In closing, we argue that global climate change may severely undermine biosecurity against terrestrial invertebrates since it has the capacity to make previously unsuitable habitats or regions more suitable. Using several examples, we show how movement of exotic invertebrate species into novel environments has the capacity to cause significant harm, especially in terms of food security.

Acknowledgements

We are grateful to Philip Smith for assistance in proofreading this introduction.

References

Bale, J.S., Masters, G.J., Hodkinson, I.D., Awmack, C., Bezemer, T.M., Brown, V.K., Butterfield, J., Buse, A., Coulson, J.C., Farrar, J., Good, J.E.G., Harrington, R., Hartley, S., Jones, T.H., Lindroth, R.L., Press, M.C., Symrnioudis, I., Watt, A.D. & Whittaker, J.B. (2002) Herbivory in global climate change research: direct effects of rising temperature on insect herbivores. *Global Change Biology*, **8**, 1–16.

Barnett, K.L. & Facey, S.L. (2016) Grasslands, invertebrates, and precipitation: a review of the effects of climate change. *Frontiers in Plant Science*, 7, 1196.

Birch, A.N.E., Begg, G.S. & Squire, G.R. (2011) How agro-ecological research helps to address food security issues under new IPM and pesticide reduction policies for global crop production systems. *Journal of Experimental Botany*, **62**, 3251–3261.

Costanza, R., d'Arge, R., De Groot, R., Farber, S., Grasso, M., Hannon, B., Limburg, K., Naeem, S., O'Neill, R.V., Paruelo, J., Raskin, R.G., Sutton, P. & Van den Belt, M. (1997) The value of the world's ecosystem services and natural capital. *Nature*, **387**, 253–260.

Costanza, R., De Groot, R., Sutton, P., van der Ploeg, S., Anderson, S.J., Kubiszewski, I., Farber, S. & Turner, R.K. (2014) Changes in the global value of ecosystem services. *Global Environmental Change*, **26**, 152–158.

Facey, S.L., Ellsworth, D.S., Staley, J.T., Wright, D.J. & Johnson, S.N. (2014) Upsetting the order: how atmospheric and climate change affects predator–prey interactions. *Current Opinion in Insect Science*, **5**, 66–74.

Gallai, N., Salles, J.-M., Settele, J. & Vaissière, B.E. (2009) Economic valuation of the vulnerability of world agriculture confronted with pollinator decline. *Ecological Economics*, **68**, 810–821.

Goldson, S.L. & Gerard, P.J. (2008) Using biocontrol against root-feding pests, with particular reference to *Sitona* root weevils. *Root Feeders - an ecosystem perspective* (eds S.N. Johnson & P.J. Murray), pp. 115–133. CABI, Wallingford, UK.

Gregory, P.J., Johnson, S.N., Newton, A.C. & Ingram, J.S.I. (2009) Integrating pests and pathogens into the climate change/food security debate. *Journal of Experimental Botany*, **60**, 2827–2838.

Hamilton, J., Zangerl, A.R., Berenbaum, M.R., Sparks, J.P., Elich, L., Eisenstein, A. & DeLucia, E.H. (2012) Elevated atmospheric CO_2 alters the arthropod community in a forest understory. *Acta Oecologica*, **43**, 80–85.

IPCC (2007) Climate Change 2007: Impacts, adaptations and vulnerability. *Contribution of Working Group II to the Fourth Assessment Report of the Intergovernmental Panel on Climate Change* (eds M.L. Parry, O.F. Canziani, J.P. Palutikof, P.J. Van Der Linden & C.E. Hanson), pp. 391–431. Cambridge University Press, Cambridge, UK and New York, NY, USA.

IPCC (2014) Climate Change 2014 – Impacts, Adaptation and Vulnerability. Part A: Global and Sectoral Aspects. *Contribution of Working Group II to the Fifth Assessment Report of the Intergovernmental Panel on Climate Change* (eds C.B. Field, V.R. Baros, D.J. Dokken, K.J. Mach, M.D. Mastrandrea, T.E. Bilir, M. Chatterjee, K.L. Ebi, Y.O. Estrada, R.C. Genova, B. Girma, E.S. Kissel, A.N. Levy, S. MacCracken, P.R. Mastrandrea & L.L. White), pp. 1132. Cambridge University Press, Cambridge, UK and New York, NY, USA.

Le Page, M. (2016) Highest ever annual rise in carbon dioxide levels recorded. *New Scientist*, 8 March 2016.

Preisser, E.L. & Strong, D.R. (2004) Climate affects predator control of an herbivore outbreak. *American Naturalist*, **163**, 754–762.

Pritchard, S.G., Rogers, H.H., Prior, S.A. & Peterson, C.M. (1999) Elevated CO_2 and plant structure: a review. *Global Change Biology*, **5**, 807–837.

Robinson, E.A., Ryan, G.D. & Newman, J.A. (2012) A meta-analytical review of the effects of elevated CO_2 on plant-arthropod interactions highlights the importance of interacting environmental and biological variables. *New Phytologist*, **194**, 321–336.

Stireman, J.O. III, Dyer, L.A., Janzen, D.H., Singer, M.S., Lill, J.T., Marquis, R.J., Ricklefs, R.E., Gentry, G.L., Hallwachs, W., Coley, P.D., Barone, J.A., Greeney, H.F., Connahs, H., Barbosa, P., Morais, H.C. & Diniz, I.R. (2005) Climatic unpredictability and parasitism of caterpillars: Implications of global warming. *Proceedings of the National Academy of Sciences of the United States of America*, **102**, 17384–17387.

UN (2015) *World Population Prospects: The 2015 Revision, Volume I: Comprehensive Tables (ST/ESA/SER.A/379). Population Division of the Department of Economic and Social Affairs of the United Nations Secretariat.* New York, USA.

Vuorinen, T., Nerg, A.-M., Ibrahim, M.A., Reddy, G.V.P. & Holopainen, J.K. (2004) Emission of *Plutella xylostella*-induced compounds from cabbages grown at elevated CO_2 and orientation behavior of the natural enemies. *Plant Physiology*, **135**, 1984–1992.

Wilson, E.O. (1987) The little things that run the world (the importance and conservation of invertebrates). *Conservation Biology*, **1**, 344–346.

Part I

Methods for Studying Invertebrates and Climate Change

2

Using Historical Data for Studying Range Changes

Georgina Palmer and Jane K. Hill

Department of Biology, University of York, YO10 5DD, UK

Summary

Global climates are warming and much of our understanding about the ecological responses of species to climate comes from the analysis of historical data sets. Invertebrates are sensitive to climate changes and so historical distribution data sets from citizen science projects collected over the past few decades provide excellent opportunities to research climate change impacts. In this chapter we review the range of data sets that are available for analysis, the amount of information that is held for different taxa, as well as the types of analyses that such data have been used for. We review the different analytical methods that have been employed to quantify range changes, focussing specifically on analyses of distribution extent and change in location of species' ranges (particularly range boundary shifts). We highlight some of the problems that arise in using these data sets as a consequence of temporal and spatial biases in recorder effort, and we discuss the ways in which researchers have tried to account for these biases. We conclude by highlighting new ways in which the historical data sets might be used in future to address novel research questions, for example, in relation to identifying different components of climate to which species are responding, and the ways in which by combining distribution data with other information we might gain a better understanding of the ecological impacts of climate change.

2.1 Introduction

Global climates are changing (IPCC, 2014) and species are responding to these changes (Parmesan & Yohe, 2003), with consequences for communities and ecosystems (Chapin III et al., 2000; Walther et al., 2002; Walther, 2010). Species could potentially respond to climate changes in several ways; species may shift their distributions to track changes, they may respond and adapt to changes in situ without any range shifts, or go extinct. The evidence from analyses of historical records, and from new empirical studies, reveal that the response that is observed is likely to depend on: the taxon and type of species being studied; the location of the study within the species' geographic range; and the extent to which the climate is changing. Thus at leading-edge (cool) range margins of species' distributions, historical data sets have revealed latitudinal range expansions polewards and/or upslope expansions where the climate is improving for the species (e.g., Warren et al., 2001). In contrast, at trailing-edge (warm/dry) range margins, range retractions polewards or uphill have been observed where climatic

Global Climate Change and Terrestrial Invertebrates, First Edition. Edited by Scott N. Johnson and T. Hefin Jones.

conditions have become unfavourable for species (e.g., by becoming too warm and/or too dry for the species; e.g., Wilson et al., 2005; Franco et al., 2006), leading to local extinction of low-latitude and low-elevation populations. Taken together, these responses to climate have resulted in many species shifting their geographical ranges, with greater rates of change being observed in locations that have warmed the most (Chen et al., 2011). These responses, evident in analyses of historical data sets, mirror those seen in the geological record, revealing range shifts by species under past climate changes (e.g. for beetles: Coope, 1978; and chironomids: Brooks & Birks, 2001). Thus, some of the first evidence for human-induced climate change impacts on species and ecosystem came through the analysis of historical data sets, and re-surveys of previous studies (Parmesan et al., 1999). Assuming that the past helps to inform the future, these data sets and analyses have also been used to make projections of the likely responses of species to future climate warming (e.g., Hill et al., 2002; Oliver et al., 2015).

Some of the first studies of climate change impacts on species' distributions were on butterflies in Europe, capitalising in particular on the availability of high resolution data for Britain (Parmesan et al., 1999). Given the extensive data sets that are available for butterflies, which have been analysed to examine climate change impacts, we cite many butterfly examples in this chapter. We have also made considerable use of these UK butterfly data sets in our own research (e.g., Mair et al., 2012; Mason et al., 2015), where we are focussing on gaining a better understanding of the ecological impacts of recent climate change, and we review some of our findings in this chapter. Butterflies have been a very popular taxonomic group for study by the general public in Britain for centuries. The relatively low diversity of butterflies in Britain, but high density of recorders, has resulted in very high temporal and spatial resolution distribution data for ~60 species. There are particularly good records since the 1970s, providing a long-term record of range changes (Heath et al., 1984; Asher et al., 2001; Fox et al., 2006). These distribution data are in addition to transect data (from e.g., the UK Butterfly Monitoring Scheme, UKBMS) documenting population trends in butterflies (Pollard & Yates, 1993; Roy et al., 2001). Other countries also have good records and recording schemes for butterflies, particularly in Europe (Settele et al., 2008; Pöyry et al., 2009; Kudrna et al., 2011; Devictor et al., 2012). Throughout this chapter we use the term 'historical datasets' to refer to datasets spanning the past four to five decades, when many of the current recording schemes were initiated (e.g., 1976 for UKBMS). The number of distribution and abundance records has increased dramatically in more recent decades, as a result of increased interest in citizen science schemes.

Many butterfly species reach either a trailing-edge or leading-edge range margin to their geographical ranges in Britain providing additional potential for analysis of species' range boundary responses from these British data sets (e.g., Hickling et al., 2006; Mason et al., 2015). As with other invertebrates, butterflies are ectothermic (although some are regional heterotherms, e.g., Kingsolver & Moffat, 1982) and activities such as flight, oviposition, and larval development rates are dependent on temperature. Thus, their population dynamics are sensitive to variation in climate. Their short generation times and high mobility also make them sensitive indicators of climate change impacts, and good knowledge of their taxonomy, natural history and ecology (e.g., habitat preferences, dispersal ability, larval host plants) helps to interpret observed responses to climate change and to explore potential mechanisms driving these responses.

The ecological impacts of recent environmental changes are well documented in temperate regions, and invertebrates are often used as indicators of climate change. There are several reasons why invertebrates are studied: these taxa are often well-recorded; their taxonomy is well-resolved (particularly for butterflies) and they can often be identified easily in the field; many species are undergoing rapid declines making them important to study to understand factors responsible for the declines; population dynamics respond quickly and species are sensitive to changing climates; and species occupy a variety of habitats and have different ecologies (e.g., dispersal abilities, habitat specificities, trophic interactions) making it possible to explore the role of biotic and abiotic factors in the responses of species to environmental changes. Invertebrates span a wide range of life-history traits (e.g., body mass varies by ~1.5 orders of magnitude) providing opportunities for comparative analyses of climate change impacts among different species types. In addition, many invertebrates provide vital ecosystem services (e.g., by being pollinators, prey, predators and parasitoids, decomposers) allowing the impacts of climate change to be assessed not only in relation to species but also the ecosystem functions they support (e.g., Biesmeijer et al., 2006; Kerr et al., 2015). Thus the large historical data sets that are available for invertebrates provide an opportunity to quantify and better understand the ecological impacts of climate change on a key group of species.

Some of the first analyses of historical data sets that were carried out highlighted that species' responses to recent environmental changes have been highly heterogeneous, for example in relation to abundance trends (positive or negative), amount of change in geographic range sizes, and direction of range shifts (Warren et al., 2001; Hickling et al., 2006). More recent studies have explored reasons for such variation (e.g., Angert et al., 2011) – such studies are required so that ecologists and conservationists can understand, identify and protect species that are most vulnerable to changes in climate (Pacifici et al., 2015). Thus, the availability of historical data sets that are being regularly updated with new contemporary data provide novel research opportunities to develop new fundamental ecological understanding of the limiting factors to species' ranges, as well as helping in the development of conservation management policies to aid the protection and conservation of species under threat from climate changes. In this chapter we focus particularly on distribution data sets. We review the historical datasets available for studying range changes (section 2.2), describe methods to quantify range changes (section 2.3), and then discuss the challenges and biases of using such data (section 2.4). We conclude by outlining current knowledge gaps and new opportunities for research using these historical data sets (section 2.5).

2.2 Review of Historical Data Sets on Species' Distributions

There is a strong causal link between the amount of climate change at study sites and the responses of species' at those locations (Chen et al., 2011; IPCC, 2014). However, to understand the mechanisms driving these changes and the causes of variation among species and taxa (Chen et al., 2011; Mason et al., 2015), scientists and other stakeholders have taken advantage of the large amounts of species' data collected worldwide. For example, information on the extent of occurrence are available for some species (e.g., www.iucnredlist.org), and occurrence data are also available for many species in publically-accessible repositories such as the Global Biodiversity Information Facility (GBIF: www.gbif.org) (Boakes et al., 2010). The GBIF is a repository of ~570 million freely accessible occurrence records covering vast geographic areas and many taxonomic groups. The repository includes data from 54 participant countries and 41 international organisations (Fig. 2.1A; www.gbif.org/participation/participant-list, accessed 15/07/2015). To date, approximately 1.6 million species – including invertebrate and vertebrate animals, fungi, bacteria and plants – are included in the database, and the numbers of records are continually increasing.

We are now in the era of citizen science, which has stemmed from "easier, faster and more accessible" data capture (August et al., 2015). The advent of new technology such as cheap tracking sensors, as well as initiatives such as the data-collection apps for British butterflies (iRecord Butterflies), grasshoppers (iRecord grasshoppers), and ladybirds (iRecord Ladybirds) (August et al., 2015; www.ceh.ac.uk/citizen-science-apps, accessed 06/08/15) have all led to a dramatic increase in the amount

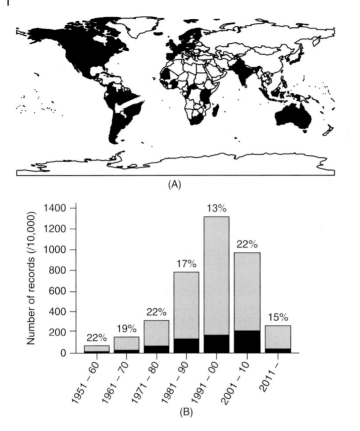

Figure 2.1 (A) Map of participant countries (black polygons) of the GBIF facility*, and (B) geo-referenced occurrence records for species in the UK, since 1950 (all 37,673,744 records = grey and black bars; Class: *Insecta* records = black bars, with the percentage of *Insecta* records displayed at the top of each bar). Bars illustrate the increase in records over time. We highlight insect data which comprise ~18% of the c.38 million geo-referenced records held in GBIF for the UK. Summary data were obtained from GBIF.org (accessed 07/08/15). *note that there are data held in the GBIF database for species that occur outside of these countries.

of data on species' distributions that are available for analysis (Pocock et al., 2015; Sutherland et al., 2015) (Fig. 2.1B; www.gbif.org/occurrence). Such data have been used in a wide variety of ways, including identifying species that are likely to invade new areas (Faulkner et al., 2014); exploring the role of species' interactions in determining large-scale distributions of species (Giannini et al., 2013); describing the relationship between beta-diversity and productivity (Andrew et al., 2012); and to predict future distributions of species under potential future climate change (Jones et al., 2013).

The three countries with the most data held in GBIF are the United States which holds ~212 million records (www.usgs.gov), Sweden which holds ~51 million records (www.gbif.se), and the UK which holds ~50 million records (www.nbn.org.uk; although ~10 million of those records are for species occurring outside of the UK). National initiatives such as the UK National Biodiversity Network (NBN) collate data from national, regional and site-based surveys, as well as from incidental sightings and old records from literature and museum/private collections. The data sets that are amassed in such databases have considerable potential to inform, particularly about large-scale patterns occurring as a result of climate change. These patterns and responses of species would otherwise be impossible to address from short-term surveys involving just a relatively small group of scientists working at small spatial scales. Examples of such uses of GBIF data include Fuller et al. (2012), who modelled the potential future distribution of the malarial vector *Anopheles albimanus* in the Caribbean and northern South America, in order to inform management efforts to limit this species' spread to those upland areas that the authors predicted would become suitable under potential future climate change. To date, over 200 hundred published studies have used data held by the UK Biological Records Centre (BRC) in the last ten years (Powney & Isaac, 2015), including many focussing on climate change topics, for example, studies describing advancement in phenology (Roy & Sparks, 2000) and polewards range shifts of species in response to climate change (e.g. Hickling et al., 2006; Chen et al., 2011; Mason et al., 2015).

Many of these data on species' distributions have come from volunteers, for example an estimated 70,000 people annually in the UK have contributed to national recording schemes (Pocock et al., 2015). In the UK, there are numerous schemes which collect data on a wide range of botanical, vertebrate, and invertebrate species groups, of which invertebrates are by far the most-studied group. For example, 75 of the 88 animal and plant recording schemes listed by the BRC are for invertebrates, ranging from well-studied butterflies and moths that we have already mentioned above, to much-understudied earthworms (Sutherland et al., 2015) (www.brc.ac.uk/recording-schemes, accessed 15/07/2015). Although there are many incidental records from previous centuries, the vast majority of dedicated recording schemes have been set up since the 1960s, but a few have been established as recently as 2006 (Barkfly Recording Scheme: http://www.brc.ac.uk/schemes/barkfly/homepage.htm) and 2007 (National Moth Recording Scheme: www.mothscount.org).

2.3 Methods for Using Historical Data to Estimate Species' Range Changes

Global meta-analyses of historical data sets have described the average range shifts of species in both terrestrial and marine ecosystems (Parmesan & Yohe, 2003; Sorte et al., 2010; Chen et al., 2011; Poloczanska et al., 2014), with average shifts in marine ecosystems being reported to be an order of magnitude higher than those in terrestrial systems. Range shifts by species may also vary between temperate and tropical ecosystems (Freeman & Freeman, 2014), although tropical data are lacking. Nonetheless, common to all of these meta-analyses is a consensus that species are shifting on average towards the poles and/or higher elevation, consistent with range shifts being driven by climate (Chen et al., 2011). In support of a causal link with climate, Sorte et al. (2010) found that 70% of the 129 studies of marine range shifts they analysed could be attributed to climate change. The fact that species vary in their rate of range shifting over time (Mair et al., 2012; Mason et al., 2015) provides further support for the important role of climate, and may explain why many studies to date generally provide little support for species' traits being important (Angert et al., 2011, but see e.g., Auer & King, 2014; Sunday et al., 2015). For example, differences in the rates of range shifts among British butterfly species have been shown to be inconsistent over time (Mair et al., 2012); in general, rates of range change were significantly greater in more recent decades (between 1995–1999 and 2005–2009) than in earlier ones (between 1970–1982 and 1995–1999). Such differences in rates of change imply that factors in addition to intrinsic species-specific sensitivities to climate are important in driving species' responses.

The majority of terrestrial data used to estimate range shifts come from the temperate zone, where species are generally considered to be temperature-limited rather than moisture-limited (Chen et al., 2011). A recent meta-analysis found that median latitudinal shifts of species were approximately

17 km per decade (Chen et al., 2011), during a period when the temperature warmed in those study sites by an average of ~0.6°C. This rate of latitudinal shift by species was much faster than previously estimated (Parmesan & Yohe, 2003), even though many species are lagging behind climate changes (Menéndez et al., 2006; Devictor et al., 2012). Studies included in the meta-analysis by Chen et al. (2011) were primarily from data-rich countries, and included data for range shifts of invertebrates from the UK (Franco et al., 2006; Hickling et al., 2006); butterflies in Finland (Pöyry et al., 2009); and intertidal invertebrates along the Chilean coast (Rivadeneira & Fernandez, 2005). Rates of range shifting have also recently been described by Mason et al. (2015), who found that the mean range shifts of nearly 1600 southerly distributed species from 21 taxonomic groups in Britain (19 invertebrate groups, plus birds and herptiles), was ~20 km per decade over the past 50 years. However, these are estimates of average range shifts for taxa; the responses of species to climate change vary greatly from species to species within taxonomic groups, with the greatest variation in range changes being observed within, rather than between, taxonomic groups (Chen et al., 2011). For example, Chen et al. (2011) found that the median range shifts of groups of British spiders, ground beetles, butterflies, and grasshoppers and allies were all in the region of 25–70 km over a 25-year period. However, within the group of spiders, for example, range shifts among species varied from an expansion of ~350 km in one species to a retraction of ~100 km in another species, over the same time period and in the same general GB study area.

2.3.1 Measuring Changes in Distribution Size

In general, two metrics of range change have been used by authors wishing to study responses of species to climate: change in the overall size of species' ranges, and change in the location of species' ranges. Many authors have defined distribution size as the number of occupied squares on gridded maps (i.e., area of occupancy), such as the British Ordnance Survey grid map (e.g., Mair et al., 2012), where the spatial resolution of the estimate of areal extent is dependent on the resolution of the underlying data. Others have quantified distribution size slightly differently, as the extent of occurrence of a species, and have measured this by calculating the area within a polygon encompassing the extremities of a species' distribution (e.g., Lyons et al., 2010). However, this assumes that all areas within a polygon are occupied by the species, which may not necessarily be the case, because species will have much more patchy distributions when defined at finer spatial scales (Jetz et al., 2008). Indeed, microhabitat heterogeneity has been shown to be important in determining the fine-scale spatial distribution of many species including plants (Maclean et al. 2015), small mammals (McCain & King, 2014), frogs (Heard et al., 2015), butterflies, ants, grasshoppers and lizards (Thomas et al., 1999).

2.3.2 Measuring Change in the Location of Species Ranges

Changes in the locations of species' ranges can be defined by calculating changes in the location of the species' range boundary, or the range core. The vector of change in the location of the centre of a species' range over time has been used to quantify changes in range core locations (e.g., Zuckerberg et al., 2009; Lyons et al., 2010; Mattila et al., 2011; Gillings et al., 2015). Using this approach, Mattilia et al. (2011) found that non-threatened butterflies in Finland had shifted their ranges further north than threatened species (30.3 km versus 7.9 km, respectively), and that non-threatened species had shifted in an east-north-east direction (73.7° deg. N) while threatened species showed no consistent angle (i.e., no consistent direction) of change. However, information regarding shifts in the core of the range may not reflect what is happening at the range boundary. For this reason, many studies calculate the change in the location of species' range boundaries, which is expected to be highly sensitive to climate changes (i.e., to deteriorating climate at species' trailing-edge boundaries, and to improving climate at leading edges).

Arguably the simplest and the most widely used method to measure boundary changes has been to calculate the north/south shift of a species' range boundary between two time periods. This has most often been done by measuring the change in mean latitude of a few (e.g., 10) most northerly or southerly occupied grid squares (e.g., Thomas & Lennon, 1999; Parmesan & Yohe, 2003; Hickling et al., 2005; Franco et al., 2006; Hickling et al., 2006; Mair et al., 2012; Mason et al., 2015). These types of analyses assume that latitudinal shifts are a good approximation of climate changes. However, while latitudinal shifts may be good correlates of temperature change in many regions, species do not necessarily shift their distributions along a north–south axis, because the interaction between changes in temperature and precipitation may result in multidirectional shifts by species (VanDerWal et al., 2013). For example, while species in montane regions have on average shifted upslope to track suitable climates, there is also evidence for down-slope shifts of some species (e.g., Konvicka et al., 2003; Hickling et al., 2006; Chen et al., 2009), possibly to take advantage of local thermal refuges (Dobrowski, 2011). Moreover, invasive species may expand in directions that primarily reflect range infilling patterns rather than climatic limitations (Simmons & Thomas, 2004). In addition, tropical species are expected to have difficulty in reaching suitable climates by latitudinal shifts given the long distances required, and so elevational shifts are expected to dominate (Colwell et al., 2008). Furthermore, focussing on data for a small subset of occupied locations at the range periphery may make estimates of range boundary locations vulnerable to issues of detectability if species distributions are patchy and abundances are low at those range margins. The availability of empty habitats for species to colonise during range shifts, and the shape of the study area may also constrain range boundary shifts and so bias estimates from these relatively simple metrics. For this reason, Gillings et al. (2015) described a new method which allows for estimation of the direction and magnitude of changes in species range boundaries which does not assume north–south shifts, by examining shifts in all directions. Using birds in Britain as exemplar species, these authors calculated changes in the locations of the range margins of species in relation to the full 360° direction over which species could potentially move (Fig. 2.2B), rather than restricting movements to either north or south. Gillings et al. (2015) calculated the mean location of the 20 occupied grid squares furthest along (i.e., closest to the range margin and furthest from the core) each of 24 axes (15° intervals; Fig. 2.2B) running along compass directions centred on the location of the species' range core. In this way, the authors were able to describe the magnitude and direction of range margin shift in multiple directions. Such information allows historical data to be used to examine if species are shifting in different directions, and thus if species may be responding to different aspect of climate and shifting along different environmental gradients. For example in Britain, species responding primarily to temperature might be expected to shift predominantly north–south, depending on which aspect of temperature they were sensitive to, whereas species sensitive to precipitation might be expected to shift along rainfall gradients that run approximately west–east in Britain. The approximately northern shifts of the range margins of British birds were driven by a combination of winter, spring and summer warming, but shifts in the location of the core range of species were not correlated with single climate variables, indicating the individualistic nature of species responses to changes in climate (Gillings et al., 2015).

2.3.3 An Invertebrate Example: Quantifying Range Shift by the Comma Butterfly *Polygonia c-album* in Britain

The choice of method to quantify changes in range boundaries can affect estimates of the direction and magnitude of range shift (Fig. 2.2). We present outputs from the same data analysed in three

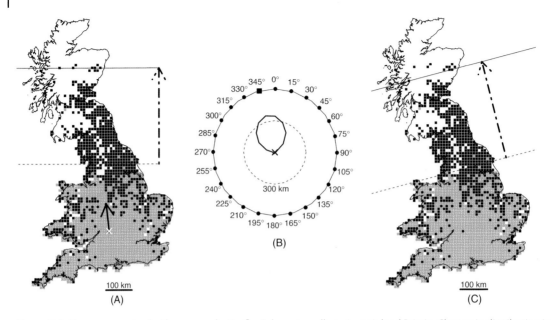

Figure 2.2 Range expansion by the comma butterfly, *Polygonia c-album*, in mainland Britain. Change in distribution is mapped between two time periods: 1970–1985 (grey squares) and 1995–2010 (grey and black squares). To take account of changes in recorder effort over time, in (A) and (C) we only plot well-recorded hectads (10 × 10 km grid squares; see Section 2.4 for definition of 'well-recorded' squares). In (A), the dot-and-dashed arrow denotes a northwards range margin shift of 345 km between the northern range margin in the first (dashed horizontal line) and second (solid horizontal line) time periods. The length and location of the solid arrow in the middle of (A) denotes the location of the range core in 1970–1985 (white cross) and the bearing and magnitude of shift of the range core over the two time periods (98 km at 353°N). In (B) the solid line denotes observed range margin change between the two time periods, along each of 24 compass directions (0 to 345°). These data demonstrate a general shift northwards, with the maximum shift of 356 km occurring along the 345° axis (black square in B). The dotted line in (B) represents a shift of 300 km between the range margins in the earlier and later time periods, along all axes, and is plotted to aid interpretation. In (C), the dot-and-dashed arrow denotes range margin shift of 356 km along the 345° axis, between the range margins in the first (dashed line) and second (solid line) time period. Distribution records were extracted on 16/04/14 from the Butterflies for the New Millennium database.

ways to illustrate this; we calculate: (i) the change in the location of the range core and (ii) changes in range margin locations using the simple north/south method and (iii) using the Gillings et al. (2015) multi-directional method, described above. By presenting these analyses, we illustrate novel ways of analysing historical data that will allow us to gain a much better understanding of species' responses. We focus on an exemplar invertebrate species, the comma butterfly *Polygonia c-album*, in Britain. This butterfly species is highly dispersive and occurs in a variety of natural and semi-natural habitats in Britain where larvae feed on *Humulus lupulus*, *Urtica dioica* and *Ulmus glabra*. The species reaches a northern range limit in Scotland, and a southern range limit in North Africa. This species has undergone rapid range expansion northwards in the UK over the past few decades (Asher et al., 2001; Braschler & Hill, 2007; Mair et al., 2012) in association with climate warming.

Our analyses of historical data for *P. c-album* revealed that the core of *P. c-album*'s range shifted 98 km over an approximately 25-year period, along a 353° bearing (i.e., approximately north; Fig. 2.2A). However, this shift in the location of the range core of *P. c-album* was much less than the shift at the range periphery. Using the simple north/south method, we estimated that the northern (i.e., leading edge) range margin of *P. c-album* shifted 345 km in Britain between the same time periods (Fig. 2.2A), consistent with this mobile species responding rapidly to a warming climate (Warren et al., 2001; Braschler & Hill, 2007). Using the multi-directional method of Gillings et al. (2015), the range margin shift was estimated to be marginally greater at 356 km along the 345° axis, in a North-North-West direction (Fig. 2.2B and C). Thus for this species, both methods to calculate changes in the range margin produced broadly similar estimates of range shift, but the simple north–south method was slightly more conservative – a similar conclusion to that reported by Gillings et al. (2015) for birds. However, the multi-directional shift method highlights that *P. c-album* has shifted north-westerly, rather than due north, and this skew is likely to be driven, in part, by the shape of the British coast line. Indeed, the directions in which the greatest proportion of British bird species extended their ranges were also in a similar direction (towards the North-East and West-North-West; Gillings et al., 2015). It would be interesting to carry out these analyses on a broader range of species with different range sizes and in different countries for which range shifts might be less constrained by the location of empty habitats and by coastlines.

Depending on the goal of the study, it may be important that different metrics of range change are calculated, describing different aspects of species' range shifts at the core and the range boundary, as evident for *P. c-album*. In addition, the multi-directional approach we describe in Fig. 2.2 (B and C) could be further developed to take account of species' habitat availability and thus examine direction and magnitude of range shifting that accounts for availability of breeding habitats. Rates of range expansion have been shown to be affected by habitat availability. For example, Hill et al. (2001) found that rates of range expansion by the speckled wood butterfly *Pararge aegeria* in two areas of Britain were up to 45% slower in locations with 24% less breeding habitat (woodland) for this butterfly species. Furthermore, a recent study by Oliver et al. (2015) found that drought-sensitive British butterflies are threatened with extinction due to climate change, but that landscape management – particularly reducing habitat fragmentation – can ameliorate negative climate impacts to improve the probability of species' persistence. Therefore, by taking account of habitat availability it may be possible to gain a better understanding of the role of climate in affecting species' observed and potential future range shifts.

2.4 Challenges and Biases in Historical Data

Our ability to detect and describe range shifts is limited by data quality and availability (Fortin et al., 2005). In particular, there are a number of issues that arise due to biases in the taxonomic, spatial and temporal coverage of historical data. In this section, we discuss each of these issues, and their implications for describing and understanding range changes.

2.4.1 Taxonomic Bias

As outlined above, there are millions of historical records available for analyses of species' distributions, but data gaps have arisen due to bias in taxonomic coverage (Parmesan, 2006). This is demonstrated by the fact that the majority of studies describing range changes of invertebrates are biased towards a small number of taxonomic groups – usually terrestrial species, and Lepidoptera, in particular. For example, a global meta-analysis of range shifts carried out by Parmesan & Yohe (2003) presented data showing that ~80% of the studies of invertebrate range shifts were of terrestrial rather than marine species. A more recent analysis of range shifts of 1573 species in Britain was biased – due to data availability – towards invertebrates (19 of 21 taxonomic groups studied), and to Lepidoptera in particular (between 40% and 60% of invertebrate species studied, depending on the time period analysed) (Mason et al., 2015). Marine invertebrates, especially open-ocean species, are often over-looked in studies of range changes (but see e.g., Rivadeneira & Fernandez, 2005; Sorte et al., 2010), despite their important roles as primary producers in those habitats. Thus more studies of a wider range of taxonomic groups are required.

Studies investigating range changes of species often exclude certain groups of species, including ubiquitous species (because there are limited opportunities for range shifting and colonising empty locations); rare species (because of data quality and methodological issues with calculating shifts); migrants (for which it is often difficult to determine breeding range); upland species (for which range shifts uphill are more likely than latitudinal shifts); species that cannot be recorded from ground-based surveys (e.g., canopy species); and species with insufficient data (e.g., Mair et al., 2014; Mason et al., 2015). Many species with insufficient data are often invertebrate species that are ecologically interesting and play important roles in ecosystem functioning, such as processing of organic material, pollination, and as prey items. Sutherland et al. (2015) have recently highlighted the need to identify and carry out surveys for understudied taxonomic groups.

However, along with recognition of the need to collect distribution data for these groups, we also require species-specific information on the taxonomy and ecology of these species (such as life history traits and behaviour; Sutherland et al., 2015), in order to interpret range shifts and to gain a better understanding of mechanisms and environmental drivers of species' responses to climate change, and the consequences for ecosystem functions. Indeed, an additional data-quality issue is the potential for misidentification of species, and therefore potentially incorrect information on the distribution of species (e.g., Johnson, 1993; Meier & Dikow, 2004). As such, many of the current recording schemes have incorporated data validation steps into their data-processing software, and there are an increasing number of online forums, apps and websites (e.g., www.ispotnature.org) which allow communities of recorders to help identify species (Sutherland et al., 2015), and reduce misidentification.

2.4.2 Spatial and Temporal Biases

In additional to taxonomic biases, there are also biases in the spatial location of records. As we previously mentioned, there is a bias towards terrestrial studies of range shifts (Parmesan, 2006), which stems from, and results in, a relative lack of information of the range shifts of marine invertebrates. However, there are also large discrepancies in the distribution of terrestrial studies and data availability worldwide (Sutherland et al., 2015); the majority of data – and therefore published studies – are from English-speaking locations, and where the gross domestic product (GDP) and security levels are high (Amano & Sutherland, 2013).

Range changes of species in tropical areas are relatively understudied (Colwell et al., 2008; Freeman & Freeman, 2014). Species in these areas are expected to exhibit different rates of change resulting from moisture- rather than temperature limitation (Chen et al., 2011), and so generalisations made from studies outside the tropics may not apply to these species – more data are required to study this. The rise of citizen science projects and new ways to record and identify species (e.g., using DNA barcoding methods) may help amass data for currently poorly studied taxa (e.g., Ji et al., 2013; Tang et al., 2015), recognising that recording species in hyper-diverse tropical regions is challenging because of problems with identification and taxonomy of many invertebrate taxa in these regions.

Historically, naturalists often focussed their data collection in easily-accessible and relatively bio-diverse areas, but more recent surveys are often randomised, increasing the representation of species in different locations and across environmental gradients (Rocchini et al., 2011). However, often there are still geographical biases in data collection for many species groups (e.g., Hill et al., 2010), arising from biases in the spatial location and activity of recorders. Such spatial biases need

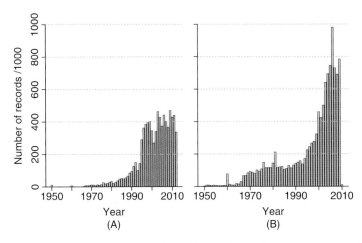

Figure 2.3 Number of (A) butterfly and (B) moth records per year since 1950 (records prior to 1950 are not shown) in the UK. Butterfly records were extracted on 16/04/14 from the Butterflies for the New Millennium database, and moth records were extracted on 08/10/13 from the National Moth Recording Scheme database.

to be taken into account when estimating range shifts, because the (historical) absence of a species in a given location may be due to under-recording, rather than its true absence in that location.

There have been huge increases in the number of records of species collected over time (e.g., Hill et al., 2010), primarily driven by the development of new technologies allowing for faster, easier and cheaper data collection (August et al., 2015). For example, the number of records of butterflies and macro-moths held by Butterfly Conservation in the UK has increased dramatically since the 1970s (Fig. 2.3), and new initiatives to engage amateur observers in data collection will further contribute to these increases in data records (e.g., www.bigbutterflycount.org, newforestcicada.info and www.harlequin-survey.org).

2.4.3 Accounting for Temporal and Spatial Biases

For reliable estimates of range changes, ideally there would be complete spatial and temporal coverage at the relevant spatial and temporal scale for the organism of interest, with no spatial or temporal biases. However, this is rarely the case, and temporal and spatial biases must therefore be taken into account (or caveats accepted) when analysing and interpreting historical data on species' ranges. In a few well-studied systems (e.g., birds in the UK), researchers can analyse data obtained through fixed-effort sampling over space and time (Gillings et al., 2015). However, in most other systems, data are collected in a sporadic way, with considerable variation in effort over both space and time.

Earlier on in this chapter we defined historical datasets as those with records spanning the past four to five decades, and as such, many of the examples of distribution changes we have discussed have been calculated over this period. However, as Hickling et al. (2006) state, there needs to be a compromise between leaving a sufficiently large gap between time periods so that range changes can take place, but not so large that substantial amounts of data are excluded. For example, in a recent meta-analysis by Poloczanska et al. (2014), the authors used only the observations from datasets which comprised at least 19 years of data to ensure sufficient time for range changes to be manifest. In general, studies using a longer span of data will be more informative than those using a shorter span of data, and so continued data collection and reporting is essential to increase the robustness of our conclusions about range changes.

Another challenge in analysing range changes using distribution data stems from the fact that recorder effort has increased dramatically in recent decades (e.g., Fig. 2.3). To account for this variation in effort, some studies group data and analyse range changes that occur between time periods rather than analysing annual data (e.g., Mason et al., 2015). Choice of time periods for grouping data may coincide with drives to collect data for taxon-specific Atlases (Isaac et al., 2014; Gillings et al., 2015), because large drops in effort often occur after these Atlas periods. By grouping data in these ways, biases associated with inter-annual variability in both recording effort and species' occurrence can be reduced. While grouping data into discrete time periods is a fairly simple way of dealing with inter-annual variation, more complex methods have been developed to account for temporal changes in recording effort. For example, increases in recording effort over time can be accounted for by sub-sampling locations from the latter time period to match the areal extent and recorder effort of the earlier time period (e.g., Warren et al., 2001); analysing data from only those locations that were visited in all time periods being studied prevents the incorrect assumption that a new sighting of a species is a colonisation when in fact no earlier record exists because the site was not visited (i.e., removing false 'absence' records). However, a single visit may not provide sufficient information to capture a true reflection of a species' presence or absence in a location, especially if the species has low detectability. As such, one can set further limits on how often each location needs to be surveyed before it is included in an analysis. Hickling et al. (2006) set several different thresholds for inclusion/exclusion of locations into ana-lyses, to investigate the consequences of being more or less strict in data selection. Being too strict might be expected to reduce statistical power if too many data are excluded, whereas not being sufficiently strict might introduce too many biases and 'noise' into analyses. Hickling et al. (2006) defined locations as either 'recorded', 'well-recorded' or 'heavily-recorded', depending on whether one species, 10% of species, or 25% of species in the taxonomic group of study were recorded at locations. This approach has since been refined (Mason et al., 2015) to take into account regional variation in species richness, by defining 'well-recorded' locations as having records of 10% of the local species pool, rather than 10% of total UK species richness (Fig. 2.2 uses these 'well-recorded' squares). We demonstrate the consequences of these different thresholds for data on UK butterflies; Fig. 2.4 shows how recorder effort declines from south to north (e.g., comparing A and C; which primarily reflects the density of recorders and centres of human populations in Britain). Fig. 2.4 also shows how refining the method to account for regional species richness rather than total species richness increases the number of locations included in analyses, particularly in northern Britain (black circles in Figs 2.4B & C; reflecting the low diversity of butterflies in these cool and wet northern and upland regions; Fig. 2.4).

Other statistical methods to measure and account for spatial and temporal biases have also been developed (see the review by Isaac et al., 2014). For example, Hill (2012) developed the '*Frescalo*' method to assess and account for bias in recording effort (available from: www.brc.ac.uk/biblio/frescalo-computer-program-analyse-your-biological-records). Using this approach, a list of 'bench-mark' species around each location are defined as those expected to also be present in the focal square, given similarities in environmental attributes, elevation, etc. (Hill, 2012; Balmer et al., 2013). Record-ing effort can then be estimated as the proportion of benchmark species found at a given location, and the observed frequencies of occurrence adjusted in order to obtain corrected occurrence trends over time (Fox et al., 2014). If all benchmark species were recorded in the focal square in one time period, but only a fraction of them were recorded in another, this indicates a reduction in recording effort in that focal square over time (Balmer et al., 2013). As such, this method can be used to deter-mine spatial patterns in recording effort, as well as changes in recording effort over time, allowing for the production of corrected occurrence trends. This method has been implemented to determine the

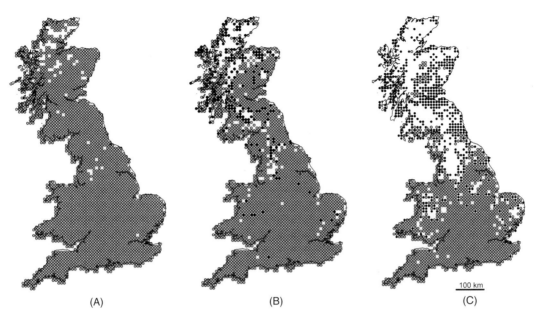

Figure 2.4 Variation in recording effort for UK butterflies, showing recorded (A), well-recorded (B) and heavily recorded (C) locations. On each map, circles represent hectads (10 × 10 km grid squares) where at least one species (A), 10% (B) or 25% (C) of the total number of species in the UK (grey circles, following Hickling et al., 2006) or in the local species pool (black and grey circles, following Mason et al., 2015) were observed in both time periods of study (1970–1985 and 1995–2010). Distribution data were extracted on 16/04/14 from the Butterflies for the New Millennium database.

temporal trends and drivers of change in the frequency of occurrence of a range of species, including British moths, as well as non-invertebrate taxa such as bryophytes and vascular plants (Fox et al., 2014; Hill & Preston, 2015).

2.5 New Ways of Analysing Data and Future Perspectives

We have focussed in this chapter on distribution data, because these types of data are most widely available for a range of invertebrate taxa. However, for a few taxa there are other types of data that, in combination with distribution data, can provide powerful new approaches for understanding range changes. In the UK, distribution data for Lepidoptera are complemented with abundance data from fixed effort transects (UKBMS data; Pollard & Yates, 1993) and light-traps (Rothamsted Insect Survey; Woiwod & Harrington, 1994). These abundance data for Lepidoptera are available from the 1960s, and provide information for the analysis of long-term population trends and phenology patterns (Hodgson et al., 2011). Other invertebrate transect schemes are also being developed (e.g., for Odonata; www.british-dragonflies.org.uk). By analysing abundance and distribution data for British butterflies, Mair et al. (2014) revealed that stable (or positive) abundance trends were a pre-requisite for range expansion, thereby helping to untangle the factors responsible for variation in species responses to climate change. Fixed effort transects are less vulnerable to recorder biases from ad hoc recording, although the locations of transects may be biased to locations with high-quality habitats and protected areas, that will not be typical of the wider countryside. This has led to the development of a new UK scheme for butterflies – the 'Wider Countryside Butterfly Survey'

(WCBS; http://butterfly-conservation.org/113/Wider-CountrysideButterflySurvey.html) that aims to monitor the abundance of widespread butterfly species across the general countryside, and hence detect population trends that are likely to be more typical of general patterns.

By combining distribution data with other data sets such as remotely-sensed satellite land cover data, it is possible to determine species' habitat associations and how habitat availability affects species' range changes. For example, in the UK, rates of range shifting for *Pararge aegeria* were much faster where there was more habitat (woodland) available for it, helping to explain some of the intra-specific variation in range expansion in this species (Hill et al., 2001). These types of analysis combining historical data with land cover data have also revealed changes in habitat associations by species under climate change (Pateman et al., 2012), revealing more factors affecting variation in species' range shifts. The launching of new satellites to complement Landsat data (Sentinels; Turner et al., 2015) will increase the temporal resolution of habitat data and the potential for new studies examining interactions between land-use change and climate. Additionally, the use of new meta-genomic DNA barcoding techniques (Ji et al., 2013) may increase the range of invertebrate taxa that can be studied, and hence help to reduce the taxonomic and spatial biases in current data sets. Combining distribution data from different species that interact (e.g., predators and prey, competing species, herbivores and host plants, parasitoids and hosts) may also provide more information on the importance of biotic interactions in range expansions (e.g., Menéndez et al., 2008; Jeffs & Lewis, 2013).

New analyses of existing historical data may be able to address whether or not species are shifting their ranges along different environmental gradients, extending the approaches of Gillings et al. (2015) to move beyond examining simple north/south shifts by species. In this chapter, we have focussed on how historical data have been used to quantify recent responses to climate change, but data can also be incorporated into models (e.g., species distribution models, SDMs) to project how distributions may change in future (e.g., Schweiger et al., 2012; Romo et al., 2014). New dynamic models that also incorporate dispersal and evolutionary processes (e.g., Rangeshifter; Bocedi et al., 2014) are likely to make rapid advances in helping us better understand and predict the ecological consequences of climate change for invertebrates.

Acknowledgements

We thank Butterfly Conservation for access to records analysed in Figs 2.2, 2.3 and 2.4. We thank the thousands of people responsible for collecting the records that we refer to in this chapter. Our work is supported by UK NERC grant NE/K00381X/1. We thank Philip Smith for proofreading this chapter.

References

Amano, T. & Sutherland, W.J. (2013) Four barriers to the global understanding of biodiversity conservation: wealth, language, geographical location and security. *Proceedings of the Royal Society B-Biological Sciences*, **280**, 2012–2649.

Andrew, M.E., Wulder, M.A., Coops, N.C. & Baillargeon, G. (2012) Beta-diversity gradients of butterflies along productivity axes. *Global Ecology and Biogeography*, **21**, 352–364.

Angert, A.L., Crozier, L.G., Rissler, L.J., Gilman, S.E., Tewksbury, J.J. & Chunco, A.J. (2011) Do species' traits predict recent shifts at expanding range edges? *Ecology Letters*, **14**, 677–689.

Asher, J., Warren, M., Fox, R., Harding, P., Jeffcoate, G. & Jeffcoate, S. (2001) *The Millennium Atlas of Butterflies in Britain and Ireland*. Oxford University Press, Oxford.

Auer, S.K. & King, D.I. (2014) Ecological and life-history traits explain recent boundary shifts in elevation and latitude of western North American songbirds. *Global Ecology and Biogeography*, **23**, 867–875.

August, T., Harvey, M., Lightfoot, P., Kilbey, D., Papadopoulos, T. & Jepson, P. (2015) Emerging technologies for biological recording. *Biological Journal of the Linnean Society*, **115**, 731–749.

Balmer, D., Gillings, S., Caffrey, B., Swann, B., Downie, I. & Fuller, R. (2013) *Bird Atlas 2007-11: the breeding and wintering birds of Britain and Ireland*. British Trust for Ornithology, Thetford.

Biesmeijer, J.C., Roberts, S.P.M., Reemer, M., Ohlemueller, R., Edwards, M., Peeters, T., Schaffers, A.P., Potts, S.G., Kleukers, R., Thomas, C.D., Settele, J. & Kunin, W.E. (2006) Parallel declines in pollinators and insect-pollinated plants in Britain and the Netherlands. *Science*, **313**, 351–354.

Boakes, E.H., McGowan, P.J.K., Fuller, R.A., Ding C.-q., Clark, N.E., O'Connor, K. & Mace, G.M. (2010) Distorted views of biodiversity: spatial and temporal bias in species occurrence data. *PLOS Biology*, **8**, e1000385.

Bocedi, G., Palmer, S.C.F., Pe'er, G., Heikkinen, R.K., Matsinos, Y.G., Watts, K. & Travis, J.M.J. (2014) RangeShifter: a platform for modelling spatial eco-evolutionary dynamics and species' responses to environmental changes. *Methods in Ecology and Evolution*, **5**, 388–396.

Braschler, B. & Hill, J.K. (2007) Role of larval host plants in the climate-driven range expansion of the butterfly *Polygonia c-album*. *Journal of Animal Ecology*, **76**, 415–423.

Brooks, S.J. & Birks, H.J.B. (2001) Chironomid-inferred air temperatures from Lateglacial and Holocene sites in north-west Europe: progress and problems. *Quaternary Science Reviews*, **20**, 1723–1741.

Chapin III, F.S., Zavaleta, E.S., Eviner, V.T., Naylor, R.L., Vitousek, P.M., Reynolds, H.L., Hooper, D.U., Lavorel, S., Sala, O.E., Hobbie, S.E., Mack, M.C. & Diaz, S. (2000) Consequences of changing biodiversity. *Nature*, **405**, 234–242.

Chen, I.-C., Hill, J.K., Ohlemueller, R., Roy, D.B. & Thomas, C.D. (2011) Rapid range shifts of species associated with high levels of climate warming. *Science*, **333**, 1024–1026.

Chen, I.-C., Shiu, H.-J., Benedick, S., Holloway, J.D., Chey, V.K., Barlow, H.S., Hill, J.K. & Thomas, C.D. (2009) Elevation increases in moth assemblages over 42 years on a tropical mountain. *Proceedings of the National Academy of Sciences of the United States of America*, **106**, 1479–1483.

Colwell, R.K., Brehm, G., Cardelús, C.L., Gilman, A.C. & Longino, J.T. (2008) Global warming, elevational range shifts, and lowland biotic attrition in the wet tropics. *Science*, **322**, 258–261.

Coope, G.R. (1978) Constancy of insect species versus inconstancy of Quaternary environments. *Diversity of Insect Faunas* (eds L.A. Mound & N. Waloff), pp. 176–187. Blackwell, Oxford.

Devictor, V., van Swaay, C., Brereton, T., Brotons, L., Chamberlain, D., Heliölä, J., Herrando, S., Julliard, R., Kuussaari, M., Lindström, Å., Reif, J., Roy, D.B., Schweiger, O., Settele, J., Stefanescu, C., Van Strien, A., Van Turnhout, C., Vermouzek, Z., WallisDeVries, M., Wynhoff, I. & Jiguet, F. (2012) Differences in the climatic debts of birds and butterflies at a continental scale. *Nature Climate Change*, **2**, 121–124.

Dobrowski, S.Z. (2011) A climatic basis for microrefugia: the influence of terrain on climate. *Global Change Biology*, **17**, 1022–1035.

Faulkner, K.T., Robertson, M.P., Rouget, M. & Wilson, J.R.U. (2014) A simple, rapid methodology for developing invasive species watch lists. *Biological Conservation*, **179**, 25–32.

Fortin, M.-J., Keitt, T.H., Maurer, B.A., Taper, M.L., Kaufman, D.M. & Blackburn, T.M. (2005) Species' geographic ranges and distributional limits: pattern analysis and statistical issues. *Oikos*, **108**, 7–17.

Fox, R., Asher, J., Brereton, T., Roy, D. & Warren, M. (2006) *The State of Butterflies in Britain and Ireland*. Pisces, Newbury.

Fox, R., Oliver, T.H., Harrower, C., Parsons, M.S., Thomas, C.D. & Roy, D.B. (2014) Long-term changes to the frequency of occurrence of British moths are consistent with opposing and synergistic effects of climate and land-use changes. *Journal of Applied Ecology*, **51**, 949–957.

Franco, A.M.A., Hill, J.K., Kitschke, C., Collingham, Y.C., Roy, D.B., Fox, R., Huntley, B. & Thomas, C.D. (2006) Impacts of climate warming and habitat loss on extinctions at species' low-latitude range boundaries. *Global Change Biology*, **12**, 1545–1553.

Freeman, B.G. & Freeman, A.M.C. (2014) Rapid upslope shifts in New Guinean birds illustrate strong distributional responses of tropical montane species to global warming. *Proceedings of the National Academy of Sciences of the United States of America*, **111**, 4490–4494.

Fuller, D.O., Ahumada, M.L., Quiñones, M.L., Herrera, S. & Beier, J.C. (2012) Near-present and future distribution of *Anopheles albimanus* in Mesoamerica and the Caribbean Basin modeled with climate and topographic data. *International Journal of Health Geographics*, **11**, 13.

Giannini, T.C., Pinto, C.E., Acosta, A.L., Taniguchi, M., Saraiva, A.M. & Alves-dos-Santos, I. (2013) Interactions at large spatial scale: The case of *Centris* bees and floral oil producing plants in South America. *Ecological Modelling*, **258**, 74–81.

Gillings, S., Balmer, D.E. & Fuller, R.J. (2015) Directionality of recent bird distribution shifts and climate change in Great Britain. *Global Change Biology*, **21**, 2155–2168.

Heard, G.W., Thomas, C.D., Hodgson, J.A., Scroggie, M.P., Ramsey, D.S.L. & Clemann, N. (2015) Refugia and connectivity sustain amphibian metapopulations afflicted by disease. *Ecology Letters*, **18**, 853–863.

Heath, J., Pollard, E. & Thomas, J.A. (1984) *Atlas of butterflies in Britain and Ireland*. Viking, London.

Hickling, R., Roy, D.B., Hill, J.K., Fox, R. & Thomas, C.D. (2006) The distributions of a wide range of taxonomic groups are expanding polewards. *Global Change Biology*, **12**, 450–455.

Hickling, R., Roy, D.B., Hill, J.K. & Thomas, C.D. (2005) A northward shift of range margins in British Odonata. *Global Change Biology*, **11**, 502–506.

Hill, J.K., Collingham, Y.C., Thomas, C.D., Blakeley, D.S., Fox, R., Moss, D. & Huntley, B. (2001) Impacts of landscape structure on butterfly range expansion. *Ecology Letters*, **4**, 313–321.

Hill, J.K., Thomas, C.D., Fox, R., Telfer, M.G., Willis, S.G., Asher, J. & Huntley, B. (2002) Responses of butterflies to twentieth century climate warming: implications for future ranges. *Proceedings of the Royal Society B-Biological Sciences*, **269**, 2163–2171.

Hill, L., Randle, Z., Fox, R. & Parsons, M. (2010) *Provisional Atlas of the UK's Larger Moths*. Butterfly Conservation, Wareham, Dorset.

Hill, M.O. (2012) Local frequency as a key to interpreting species occurrence data when recording effort is not known. *Methods in Ecology and Evolution*, **3**, 195–205.

Hill, M.O. & Preston, C.D. (2015) Disappearance of boreal plants in southern Britain: habitat loss or climate change? *Biological Journal of the Linnean Society*, **115**, 598–610.

Hodgson, J.A., Thomas, C.D., Oliver, T.H., Anderson, B.J., Brereton, T.M. & Crone, E.E. (2011) Predicting insect phenology across space and time. *Global Change Biology*, **17**, 1289–1300.

IPCC (2014) *Climate Change 2014: Impacts, Adaptation, and Vulnerability. Part A: Global and Sectoral Aspects. Contribution of Working Group II to the Fifth Assessment Report of the Intergovernmental Panel on Climate Change* (eds. C.B. Field, V.R. Barros, D.J. Dokken, K.J. Mach, M.D. Mastrandrea, T.E. Bilir, M. Chatterjee, K.L. Ebi, Y.O. Estrada, R.C. Genova, B. Girma, E.S. Kissel, A.N. Levy, S. MacCracken, P.R. Mastrandrea, & L.L. White). Cambridge University Press, Cambridge, United Kingdom and New York, NY, USA.

Isaac, N.J.B., van Strien, A.J., August, T.A., de Zeeuw, M.P. & Roy, D.B. (2014) Statistics for citizen science: extracting signals of change from noisy ecological data. *Methods in Ecology and Evolution*, **5**, 1052–1060.

Jeffs, C.T. & Lewis, O.T. (2013) Effects of climate warming on host–parasitoid interactions. *Ecological Entomology*, **38**, 209–218.

Jetz, W., Sekercioglu, C.H. & Watson, J.E.M. (2008) Ecological correlates and conservation implications of overestimating species geographic ranges. *Conservation Biology*, **22**, 110–119.

Ji, Y., Ashton, L., Pedley, S.M., Edwards, D.P., Tang, Y., Nakamura, A., Kitching, R., Dolman, P.M., Woodcock, P., Edwards, F.A., Larsen, T.H., Hsu, W.W., Benedick, S., Hamer, K.C., Wilcove, D.S., Bruce, C., Wang, X., Levi, T., Lott, M., Emerson, B.C. & Yu, D.W. (2013) Reliable, verifiable and efficient monitoring of biodiversity via metabarcoding. *Ecology Letters*, **16**, 1245–1257.

Johnson, C. (1993) *Provisional atlas of the Cryptophagidae-Atomariinae (Coleoptera) of Britian and Ireland*. Biological Records Centre, Institute of Terrestrial Ecology, Abbots Ripton, Huntingdon.

Jones, M.C., Dye, S.R., Fernandes, J.A., Froelicher, T.L., Pinnegar, J.K., Warren, R. & Cheung, W.W.L. (2013) Predicting the impact of climate change on threatened species in UK waters. *PLOS ONE*, **8**, e54216.

Kerr, J.T., Pindar, A., Galpern, P., Packer, L., Potts, S.G., Roberts, S.M., Rasmont, P., Schweiger, O., Colla, S.R., Richardson, L.L., Wagner, D.L., Gall, L.F., Sikes, D.S. & Pantoja, A. (2015) Climate change impacts on bumblebees converge across continents. *Science*, **349**, 177–180.

Kingsolver, J.G. & Moffat, R.J. (1982) Thermoregulation and the determinants of heat transfer in *Colias* butterflies. *Oecologia*, **53**, 27–33.

Konvicka, M., Maradova, M., Benes, J., Fric, Z. & Kepka, P. (2003) Uphill shifts in distribution of butterflies in the Czech Republic: effects of changing climate detected on a regional scale. *Global Ecology and Biogeography*, **12**, 403–410.

Kudrna, O., Harpke, A., Lux, K., Pennerstorfer, J., Schweiger, O., Settele, J. & Wiemers, M. (2011) *Distribution Atlas of Butterflies in Europe*. Gesellschaft für Schmetterlingsschutz e.V., Halle, Germany.

Lyons, S.K., Wagner, P.J. & Dzikiewicz, K. (2010) Ecological correlates of range shifts of Late Pleistocene mammals. *Proceedings of the Royal Society of London B-Biological Sciences*, **365**, 3681–3693.

Maclean, I.M.D., Hopkins, J.J., Bennie, J., Lawson, C.R. & Wilson, R.J. (2015) Microclimates buffer the responses of plant communities to climate change. *Global Ecology and Biogeography*, **24**, 1340–1350.

Mair, L., Hill, J.K., Fox, R., Botham, M., Brereton, T. & Thomas, C.D. (2014) Abundance changes and habitat availability drive species' responses to climate change. *Nature Climate Change*, **4**, 127–131.

Mair, L., Thomas, C.D., Anderson, B.J., Fox, R., Botham, M. & Hill, J.K. (2012) Temporal variation in responses of species to four decades of climate warming. *Global Change Biology*, **18**, 2439–2447.

Mason, S.C., Palmer, G., Fox, R., Gillings, S., Hill, J.K., Thomas, C.D. & Oliver, T.H. (2015) Geographical range margins of many taxonomic groups continue to shift polewards. *Biological Journal of the Linnean Society*, **115**, 586–597.

Mattila, N., Kaitala, V., Komonen, A., Päivinen, J. & Kotiaho, J.S. (2011) Ecological correlates of distribution change and range shift in butterflies. *Insect Conservation and Diversity*, **4**, 239–246.

McCain, C.M. & King, S.R.B. (2014) Body size and activity times mediate mammalian responses to climate change. *Global Change Biology*, **20**, 1760–1769.

Meier, R. & Dikow, T. (2004) Significance of specimen databases from taxonomic revisions for estimating and mapping the global species diversity of invertebrates and repatriating reliable specimen data. *Conservation Biology*, **18**, 478–488.

Menéndez, R., González-Megías, A., Lewis, O.T., Shaw, M.R. & Thomas, C.D. (2008) Escape from natural enemies during climate-driven range expansion: a case study. *Ecological Entomology*, **33**, 413–421.

Menéndez, R., González Megías, A., Hill, J.K., Braschler, B., Willis, S.G., Collingham, Y., Fox, R., Roy, D.B. & Thomas, C.D. (2006) Species richness changes lag behind climate change. *Proceedings of the Royal Society B-Biological Sciences*, **273**, 1465–1470.

Oliver, T.H., Marshall, H.H., Morecroft, M.D., Brereton, T., Prudhomme, C. & Huntingford, C. (2015) Interacting effects of climate change and habitat fragmentation on drought-sensitive butterflies. *Nature Climate Change*, **5**, 941–945.

Pacifici, M., Foden, W.B., Visconti, P., Watson, J.E.M., Butchart, S.H.M., Kovacs, K.M., Scheffers, B.R., Hole, D.G., Martin, T.G., Akçakaya, H.R., Corlett, R.T., Huntley, B., Bickford, D., Carr, J.A., Hoffmann, A.A., Midgley, G.F., Pearce-Kelly, P., Pearson, R.G., Williams, S.E., Willis, S.G., Young, B. & Rondinini, C. (2015) Assessing species vulnerability to climate change. *Nature Climate Change*, **5**, 215–224.

Parmesan, C. (2006) Ecological and evolutionary responses to recent climate change. *Annual Review of Ecology, Evolution and Systematics*, **37**, 637–669.

Parmesan, C., Ryrholm, N., Stefanescu, C., Hill, J.K., Thomas, C.D., Descimon, H., Huntley, B., Kaila, L., Kullberg, J., Tammaru, T., Tennent, W.J., Thomas, J.A. & Warren, M. (1999) Poleward shifts in geographical ranges of butterfly species associated with regional warming. *Nature*, **399**, 579–583.

Parmesan, C. & Yohe, G. (2003) A globally coherent fingerprint of climate change impacts across natural systems. *Nature*, **421**, 37–42.

Pateman, R.M., Hill, J.K., Roy, D.B., Fox, R. & Thomas, C.D. (2012) Temperature-dependent alterations in host use drive rapid range expansion in a butterfly. *Science*, **336**, 1028–1030.

Pocock, M.J.O., Roy, H.E., Preston, C.D. & Roy, D.B. (2015) The Biological Records Centre: a pioneer of citizen science. *Biological Journal of the Linnean Society*, **115**, 475–493.

Pollard, E. & Yates, T.J. (1993) *Monitoring butterflies for ecology and conservation*. Chapman & Hall, London.

Poloczanska, E.S., Hoegh-Guldberg, O., Cheung, W., Pörtner, H.-O. & Burrows, M.T. (2014) Cross-chapter box on observed global responses of marine biogeography, abundance, and phenology to climate change. *Climate Change 2014: Impacts, Adaptation, and Vulnerability. Part A: Global and Sectoral Aspects. Contribution of Working Group II to the Fifth Assessment Report of the Intergovernmental Panel of Climate Change* (eds C.B. Field, V.R. Barros, D.J. Dokken, K.J. Mach, M.D. Mastrandrea, T.E. Bilir, M. Chatterjee, K.L. Ebi, Y.O. Estrada, R.C. Genova, B. Girma, E.S. Kissel, A.N. Levy, S. MacCracken, P.R. Mastrandrea & L.L. White), pp. 123–127. Cambridge University Press, Cambridge, United Kingdom and New York, NY, USA.

Powney, G.D. & Isaac, N.J.B. (2015) Beyond maps: a review of the applications of biological records. *Biological Journal of the Linnean Society*, **115**, 532–542.

Pöyry, J., Luoto, M., Heikkinen, R.K., Kuussaari, M. & Saarinen, K. (2009) Species traits explain recent range shifts of Finnish butterflies. *Global Change Biology*, **15**, 732–743.

Rivadeneira, M.M. & Fernández, M. (2005) Shifts in southern endpoints of distribution in rocky intertidal species along the south-eastern Pacific coast. *Journal of Biogeography*, **32**, 203–209.

Rocchini, D., Hortal, J., Lengyel, S., Lobo, J.M., Jiménez-Valverde, A., Ricotta, C., Bacaro, G. & Chiarucci, A. (2011) Accounting for uncertainty when mapping species distributions: The need for maps of ignorance. *Progress in Physical Geography*, **35**, 211–226.

Romo, H., García-Barros, E., Márquez, A.L., Moreno, J.C. & Real, R. (2014) Effects of climate change on the distribution of ecologically interacting species: butterflies and their main food plants in Spain. *Ecography*, **37**, 1063–1072.

Roy, D.B., Rothery, P., Moss, D., Pollard, E. & Thomas, J.A. (2001) Butterfly numbers and weather: predicting historical trends in abundance and the future effects of climate change. *Journal of Animal Ecology*, **70**, 201–217.

Roy, D.B. & Sparks, T.H. (2000) Phenology of British butterflies and climate change. *Global Change Biology*, **6**, 407–416.

Schweiger, O., Heikkinen, R.K., Harpke, A., Hickler, T., Klotz, S., Kudrna, O., Kühn, I., Pöyry, J. & Settele, J. (2012) Increasing range mismatching of interacting species under global change is related to their ecological characteristics. *Global Ecology and Biogeography*, **21**, 88–99.

Settele, J., Kudrna, O., Harpke, A., Kühn, I., van Swaay, C., Verovnik, R., Warren, M., Wiemers, M., Hanspach, J., Hickler, T., Kühn, E., van Halder, I., Veling, K., Vliegenthart, A., Wynhoff, I. & Schweiger, O. (2008) *Climatic Risk Atlas of European Butterflies*. Pensoft Publishers, Sofia, Bulgaria.

Simmons, A.D. & Thomas, C.D. (2004) Changes in dispersal during species' range expansions. *American Naturalist*, **164**, 378–395.

Sorte, C.J.B., Williams, S.L. & Carlton, J.T. (2010) Marine range shifts and species introductions: comparative spread rates and community impacts. *Global Ecology and Biogeography*, **19**, 303–316.

Sunday, J.M., Pecl, G.T., Frusher, S., Hobday, A.J., Hill, N., Holbrook, N.J., Edgar, G.J., Stuart-Smith, R., Barrett, N., Wernberg, T., Watson, R.A., Smale, D.A., Fulton, E.A., Slawinski, D., Feng, M., Radford, B.T., Thompson, P.A. & Bates, A.E. (2015) Species traits and climate velocity explain geographic range shifts in an ocean-warming hotspot. *Ecology Letters*, **18**, 944–953.

Sutherland, W.J., Roy, D.B. & Amano, T. (2015) An agenda for the future of biological recording for ecological monitoring and citizen science. *Biological Journal of the Linnean Society*, **115**, 779–784.

Tang, M., Hardman, C.J., Ji, Y., Meng, G., Liu, S., Tan, M., Yang, S., Moss, E.D., Wang, J., Yang, C., Bruce, C., Nevard, T., Potts, S.G., Zhou, X. & Yu, D.W. (2015) High-throughput monitoring of wild bee diversity and abundance via mitogenomics. *Methods in Ecology and Evolution*, **6**, 1034–1043.

Thomas, C.D. & Lennon, J.J. (1999) Birds extend their ranges northwards. *Nature*, **399**, 213.

Thomas, J.A., Rose, R.J., Clarke, R.T., Thomas, C.D. & Webb, N.R. (1999) Intraspecific variation in habitat availability among ectothermic animals near their climatic limits and their centres of range. *Functional Ecology*, **13** Supplement 1, 55–64.

Turner, W., Rondinini, C., Pettorelli, N., Mora, B., Leidner, A.K., Szantoi, Z., Buchanan, G., Dech, S., Dwyer, J., Herold, M., Koh, L.P., Leimgruber, P., Taubenboeck, H., Wegmann, M., Wikelski, M. & Woodcock, C. (2015) Free and open-access satellite data are key to biodiversity conservation. *Biological Conservation*, **182**, 173–176.

VanDerWal, J., Murphy, H.T., Kutt, A.S., Perkins, G.C., Bateman, B.L., Perry, J.J. & Reside, A.E. (2013) Focus on poleward shifts in species' distribution underestimates the fingerprint of climate change. *Nature Climate Change*, **3**, 239–243.

Walther, G.-R. (2010) Community and ecosystem responses to recent climate change. *Philosophical Transactions of the Royal Society B: Biological Sciences*, **365**, 2019–2024.

Walther, G.-R., Post, E., Convey, P., Menzel, A., Parmesan, C., Beebee, T.J.C., Fromentin, J.-M., Hoegh-Guldberg, O. & Bairlein, F. (2002) Ecological responses to recent climate change. *Nature*, **416**, 389–395.

Warren, M.S., Hill, J.K., Thomas, J.A., Asher, J., Fox, R., Huntley, B., Roy, D.B., Telfer, M.G., Jeffcoate, S., Harding, P., Jeffcoate, G., Willis, S.G., Greatorex-Davies, J.N., Moss, D. & Thomas, C.D. (2001) Rapid responses of British butterflies to opposing forces of climate and habitat change. *Nature*, **414**, 65–69.

Wilson, R.J., Gutiérrez, D., Gutiérrez, J., Martínez, D., Agudo, R. & Monserrat, V.J. (2005) Changes to the elevational limits and extent of species ranges associated with climate change. *Ecology Letters*, **8**, 1138–1146.

Woiwod, I.P. & Harrington, R. (1994) Flying in the face of change: the Rothamsted Insect Survey. *Long-term Experiments in Agricultural and Ecological Sciences* (eds R.A. Leigh & A.E. Johnston). CAB International, Wallingford.

Zuckerberg, B., Woods, A.M. & Porter, W.F. (2009) Poleward shifts in breeding bird distributions in New York State. *Global Change Biology*, **15**, 1866–1883.

3

Experimental Approaches for Assessing Invertebrate Responses to Global Change Factors

Richard L. Lindroth and Kenneth F. Raffa

Department of Entomology, University of Wisconsin-Madison, Madison, WI 53706, U.S.A

Summary

Controlled experiments are an indispensable tool in the arsenal of scientific approaches used to understand the impacts of climate change on invertebrates. Their principal value lies in the capacity to disentangle complex, interactive relationships by quantitatively measuring the effects of one to multiple environmental change factors on endpoints ranging from individual physiology to ecosystem processes. This chapter provides guidance for formulating appropriate experimental designs and devising optimal study systems for such research. We first address issues of scale, espousing the perspective that reductionist, holistic, and integrated approaches are all important, but must be appropriately matched to the questions asked. Small-scale studies afford fine-tuned control, replication and rapid assessment of multiple variables, and thus are particularly useful in elucidating mechanisms. Large-scale studies allow for incorporation of multispecies interactions, time delays, feedbacks, and indirect effects, and thus provide insight into emergent, context-dependent patterns and processes. We then describe commonly encountered problems with experimental design and statistics, emphasizing designs that maximize statistical power and avoid pseudoreplication. Next, we discuss positive and negative aspects of a range of experimental approaches, proceeding from indoor closed systems (e.g. environmental control rooms), to outdoor closed systems (e.g., open-top chambers), to outdoor open systems (e.g., Free Air CO_2 Enrichment [FACE] systems). We conclude with comments about the human dimensions of conducting large, collaborative, and multidisciplinary global change experiments. Effective team assembly, leadership, communication and project management will maximize benefits from the substantial investments required for large global change research projects.

3.1 Introduction

Controlled experimental manipulation has been the bedrock of the scientific method for centuries, affording unparalleled insights into the patterns and processes that define the natural world. More recently, experimentation has been indispensable in attempts to evaluate and anticipate the impacts of global environmental changes on biotic systems. Yet the magnitude, scale, and complexity of global changes pose challenges to the utility of experimental approaches like few other issues faced by humankind.

This chapter describes advantages, disadvantages and constraints in the application of manipulative experimental approaches to understanding the effects of global change on invertebrates, and the impacts invertebrates exert on managed and unmanaged ecosystems. We endeavour to provide guidance for devising appropriate experimental designs and selecting optimal study systems and scales across the range of biological organization, from individual physiology to ecosystem processes. We will focus primarily on the climate change drivers of temperature and precipitation, as well as on atmospheric changes in concentrations of carbon dioxide (CO_2) and tropospheric ozone (O_3). These four factors are not only dominant drivers, but have also been the most thoroughly investigated via experimental means. Research to date has shown that these factors can affect invertebrates both directly, through changes in abiotic conditions (e.g., temperature), or indirectly, via changes in other abiotic (e.g., shading) or biotic (e.g., predation) conditions (Fig. 3.1).

As important as describing what this chapter is, is describing what it is not. We will describe and evaluate a variety of experimental approaches, but will not delineate technical procedures. Details

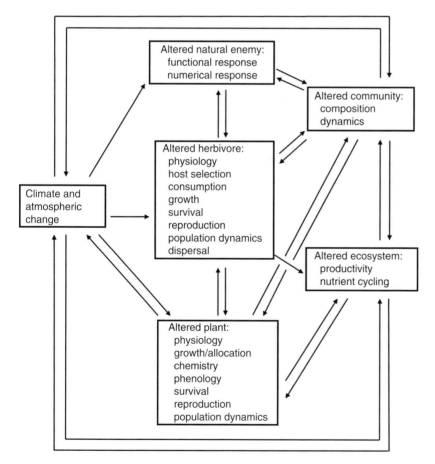

Figure 3.1 The complex of climate and atmospheric change factors that directly and indirectly influence invertebrate biology is best investigated by a spectrum of experimental approaches. The direct effects of global change factors on rapidly responding metrics (e.g., insect physiology, growth) are amenable to study via small-scale, chamber systems. More complex direct and indirect effects on slowly responding metrics (e.g., community composition) are best studied with large-scale, open-air systems. Figure adapted from Lindroth (2010).

of infrastructure design, assembly, and operation can be obtained from the literature cited and references therein. We will likewise not encompass other important scientific methodologies, such as modelling and correlative studies, as these valuable approaches are addressed elsewhere in this volume. We will emphasize terrestrial systems, as these have been most intensively studied, but hope that our recommendations will provide useful guidance for studying freshwater and marine systems. Finally, we will not address other global change factors, such as UV radiation, nitrogen deposition, land use change, and biological invasions. Those concerns, although important, are beyond the focus of this volume on climate and atmospheric change.

3.2   Experimental Scale: Reductionist, Holistic and Integrated Approaches

The appropriate scale at which to investigate ecological phenomena has been a matter of lively debate for decades (Carpenter, 1996; Daehler & Strong, 1996; Drake et al., 1996; Lawton, 1996; Fraser & Keddy, 1997). Central to this debate have been differences in perspectives on the value of precision and accuracy versus realism, and of mechanism versus phenomenon. Small-scale studies afford fine-tuned control, replication and rapid assessment of a diverse array of variables of interest. They have proven particularly useful in testing hypotheses and elucidating mechanisms. Large-scale studies allow for incorporation of multispecies interactions, time delays, feedbacks, and indirect effects, and thus identification of emergent, context-dependent patterns and processes. These approaches are complementary rather than mutually exclusive; the potential impacts of mechanisms identified under tightly controlled conditions can be evaluated in larger-scale, open-air experiments, which might otherwise miss important drivers when assessed against considerable environmental noise. Likewise, putative processes identified from simulation models, post-hoc correlations and meta-analyses can be subjected to controlled tests, which may have initially overlooked these important variables during the experimental design phase.

Global environmental changes are occurring across such vast spatial, temporal, and taxonomic scales that they are sometimes viewed as intractable by manipulative experiments. Indeed, the value of traditional experimentation to anticipate the ecological effects of global change has been criticized: specifically, the relative ease with which mechanisms can be identified under controlled conditions may provide robust biological insights at one scale, yet have little relevance to the functioning of systems at higher scales (Benton et al., 2007). Accurate understanding of complex ecological systems ultimately requires exploration of how the components (e.g., species) interact together, and how they affect, and are affected by, the larger systems in which they are imbedded (Brown et al., 2001). Thus, multi-species interactions, indirect effects and feedbacks are critical to how fine-scale processes may be buffered, amplified or perpetuated across complex systems (Raffa et al., 2008). Conversely, observations drawn from large-scale studies can contribute to untenable conclusions if not supported by an understanding of the underlying mechanisms (Whittaker, 2001). Moreover, global change drivers are not occurring in isolation. Rather, ecological systems are being subjected to multiple, simultaneous drivers, whose interacting effects may be antagonistic, multiplicative or synergistic (Rosenblatt & Schmitz, 2014; Fig. 3.1). Indeed, not only the identity but also the simple number of global change drivers can contribute to variance in responses of ecological interactions (Scherber et al., 2013). Finally, many biological systems involving invertebrates exhibit nonlinear dynamics, in which critical thresholds separate organizational realms with fundamentally distinct dynamics. In such systems, studies that encompass large spatio-temporal scales can fail to detect critical drivers that determine whether or not thresholds are breached, yet are not correlated with ultimate impacts (Raffa et al., 2008).

Our view is that controlled, manipulative experiments, at multiple levels of application (simple microcosms to complex open-air systems), are an invaluable tool in the arsenal of methodologies available to understand the impacts of global change on invertebrates. When integrated with other methodologies, they provide powerful approaches to link pattern with process, and to make mechanistically grounded, realistic predictions. In contrast, when conducted in isolation or using ill-suited or poorly interpreted models, they can fall far short of their stated goals. As John Lawton (1995) concluded about experimental model systems: "Like all tools they do some things well, some things badly, and other things not at all."

When evaluating options for use of experimental approaches, we recommend:

1. *Carefully identify the question.* What, *exactly*, is the subject of interest, and the domain of the hypothesis? Specify what is, and is not, included within the intended field of inference for the study (e.g., taxonomic units, time-frame, community structure).

2. *Match the question asked with an appropriate scale of investigation, and the appropriate model.* Avoid the temptation to misalign methods with questions, relying on what is familiar rather than what is best. As attributed to Abraham Maslow (1966), "When all you have is a hammer, everything looks like a nail." If an appropriate method or model is not available, return to number 1 and refine the question.

3. *Accept that no study can be equally strong from all perspectives and prioritize accordingly.* In studies where process is prioritized over pattern, first reduce, then increase, complexity. Complex systems are sometimes best approached by first isolating critical components and mechanisms. As these come to light, complexity can be increased in a controlled manner, allowing for an increasing range of components and dynamics in a system (Drake et al., 1996).

4. *Imbed small-scale experiments into large-scale projects.* Small-scale, short duration experiments can often be fruitfully incorporated into large-scale, long-term projects to improve understanding of key emerging components and dynamics.

5. *Incorporate experimental results into generalized models.* To expand the relevance of controlled manipulations beyond the experimental system itself, results should be incorporated into generalized, conceptual or mathematical models that provide predictive power for understanding global change impacts on invertebrates. These emergent models should then become the subject of both new manipulative studies and large-scale tests against observed patterns.

3.3   Experimental Design: Statistical Concerns

In our experience, experimental design is optimally conceived by an iterative process. It typically begins with laying out of the ideal experiment, complete with multiple treatments across quantitative gradients, ample replication and sequential sampling, for multiple species. Our initial enthusiasm is then inexorably quelled by the realization that such a plan cannot possibly be executed within the constraints of funding, facilities, personnel and time available. After brief consideration of career changes, we return to the difficult task at hand: how to conduct meaningful, mechanistically grounded yet biologically realistic, research within very real logistical constraints. This again requires a willingness to make tough decisions borne in tradeoffs: Do we sacrifice treatments? Replication? Species? Do we shorten the duration of the study? All are viable options. Throughout the process, our guiding principle has been: *It's better to evaluate a few manipulations well, than many manipulations poorly.*

To that end, we have generally reduced the number of experimental factors in order to maintain a critical threshold of replication.

Most experiments addressing the effects of global change on invertebrates have employed traditional designs (e.g., analysis of variance [ANOVA]), in which variation in measured traits can be partitioned among treatment factors and their interactions. Due to logistical constraints, these designs commonly incorporate only two (control and elevated) or three levels of the global change factor of interest. An alternative, less commonly used approach is a regression (response surface) design, in which a range of treatment levels is employed (e.g., Pelini et al., 2011). Regression designs afford the potential to detect threshold and nonlinear effects of experimental treatments.

Experimental designs should be constructed so as to maximize statistical power, within the constraints of available resources and time. Statistical power is the probability of correctly *rejecting* the null hypothesis (H_0) when it is false (Gotelli & Ellison, 2004). It is defined as 1-β, where β is the probability of a type II statistical error (*accepting* the null hypothesis when it is false). When the value of β is high, statistical power is low, and the null hypothesis (of no treatment effect) is likely to be accepted, even when treatment effects are indeed important (Filion et al., 2000).

Numerous factors, some of which are determined by the experimenter, influence statistical power. Power increases with increases in sample size, number of levels of treatment factors, precision of data collection, and magnitude of treatment effects. Power decreases in relation to the magnitude of error variance. In general, then, researchers should initially select levels of treatments, within ecological realism, that are likely to maximize differences, take full advantage of replication, and employ uniform and precise data collection practices. When cost or availability of controlled environments (e.g., greenhouse rooms, open-top chambers) are limiting, researchers should consider incorporating a second treatment factor (e.g., drought, species) with many levels within each environment. Such split-plot designs afford effective tests of interactions between the treatment factors (Filion et al., 2000).

Given the importance of replication in ascertaining statistical significance, and the limits on replication imposed by logistical constraints, it is not surprising that pseudoreplication (Hurlbert, 1984) remains a problem in global change research. Interestingly, pseudoreplication is an issue with which practitioners of small-scale and large-scale studies share common ground. Identification of "what is the experimental unit?" is not always intuitive. Properly understood, a replicate is that unit to which an experimental treatment is applied. Thus, when treatments such as temperature or CO_2 are applied to controlled chambers, greenhouses, or FACE (Free Air CO_2 Enrichment) rings, the unit of replication for those treatments is the chamber, greenhouse, or ring, respectively. Subunits (e.g., individual plants) within those experimental systems are not independently subjected to the treatment factor, and thus cannot be considered true experimental replicates. Treating such sampling units as experimental units artificially inflates the error degrees of freedom, resulting in a larger, but invalid, F statistic.

If the number of true experimental units (e.g., chambers) is limiting, as is often the case, what is the value of increasing the number of sampling units? Incorporation of multiple sampling units improves the precision of measurement for the experimental unit. Enhanced precision reduces estimates of variance within treatments, and thus increases statistical power. Moreover, some experimental designs (e.g., split plots and split-split plots) incorporate a hierarchy of experimental units (e.g., chamber, soil fertility, plant genotype). As described above, such designs can be powerful tools for identifying multi-level interactions among treatment factors.

In statistical hypothesis testing, type I errors occur when the null hypothesis (H_0) is incorrectly rejected (a "false positive"), whereas type II errors occur when the null hypothesis is incorrectly accepted (a "false negative"). A significance level (α) of 0.05 is the generally recognized convention for rejection of the null hypothesis. Many global change experiments, especially FACE studies, suffer from low statistical power, increasing the likelihood of type II errors. The selection of $\alpha = 0.05$ reflects a widely held agreement within the academic community that type I errors are more serious than type II errors, and hence barriers should be erected against them. This cautionary approach, however, is itself context-dependent. In drug discovery, for example, type II errors can be the more harmful mistake during the early stages of evaluation, because they preclude subsequent analogue synthesis and exploration. Similarly, in global change research, conclusions of "no effect" when indeed such effects exist may have significant societal, let alone scientific, consequences. We therefore support the recommendation of Filion et al. (2000) that a significance level of 0.10 be employed for reporting global change experiments conducted at the mesocosm or larger scale, because these necessarily have inherently low statistical power. Finally, biologists often hold disparate views about when a P value should be considered borderline. Thus, we further recommend that actual P values as well as F statistics and degrees of freedom be reported in global change studies.

3.4   Experimental Endpoints: Match Metrics to Systems

Invertebrates will experience the consequences of global climate and atmospheric change across the entire range of biological organization, from individual physiology to ecosystem function (Fig. 3.1). Manipulative assessments of the effects of global change on invertebrate biology must seek to match the level of inquiry to experimental systems with appropriate control and environmental context. For instance, experiments on how thermal environment directly affects insect respiration can be better conducted in environmental control chambers than in outdoor open-air systems, as temperature can be more precisely controlled in the former. In contrast, studies on how thermal environment affects interspecific competition and trophic interactions may require the level of experimental sophistication and ecological realism afforded only by long-term, multi-species outdoor facilities.

Experimental endpoints of interest to researchers can be categorized loosely as changes in: 1) individual physiology and behavior, 2) population dynamics and microevolution, 3) community structure and biotic interactions, and 4) ecosystem structure and function (Fig. 3.1). Emphases at the individual level include measurement of various performance traits (e.g., feeding efficiency, growth, development, survivorship, reproduction), typically as surrogates for biological fitness. Other metrics of interest include shifts in stress (e.g., thermal) tolerance, phenology, host selection, mate selection, predator avoidance and dispersal. We caution that a narrow focus on one or a few indices of performance can lead to spurious conclusions. For example, a climate-mediated increase in feeding rate may be due to either improved food palatability or reduced food quality and compensatory feeding (Lindroth et al., 1993). Thus fundamentally different processes can yield equivalent patterns. Extrapolations from individual-based data to population and higher levels of organization must be made judiciously, ideally on the basis of multiple performance metrics and with recognition that, due to external processes and internal feedbacks, predicted responses may not actually emerge (Awmack et al., 2004; Diamond et al., 2013). Assessments of the impacts of global change drivers on invertebrate populations focus on demographic rates (e.g., natality, mortality), population growth, long-term population dynamics (e.g., outbreaks, cycles) and shifts in genetic structure (microevolution). Clearly, the temporal and spatial scale of some of these phenomena makes them ill-suited for investigation via controlled experiments. At the community level, studies explore global change impacts on invertebrate community composition, diversity and multi-species interactions (e.g., herbivory, predation, parasitism, mutualism, competition). Finally, at the ecosystem level, experiments address global change impacts on invertebrates as mediators of ecosystem-level processes, such as carbon sequestration and decomposition, in the context of multi-species interactions and direct and indirect effects.

3.5 Experimental Systems: Manipulations From Bottle to Field

Experimental systems comprise one set in a continuum of approaches that range from mathematical models to whole ecosystem studies (Lawton, 1995). Their signature value is the capacity to "disentangle the complexities of nature" (Lawton, 1995) by quantitatively measuring the impact of one to several environmental change factors on biological endpoints of interest. Although the scientific literature is replete with references to "microcosm" and "mesocosm" studies, there is no clear distinction between the two, and either can be employed in indoor or outdoor environments. We will therefore use alternative terminology in our description of the relative advantages and disadvantages of particular methodologies: indoor closed systems, outdoor closed systems, and outdoor open systems.

3.5.1 Indoor Closed Systems

Indoor closed, or chambered, systems range in size and complexity from reach-in incubators, to walk-in environmental control rooms, to greenhouses (Fig. 3.2). They typically provide good control of environmental factors such as temperature and light, while more technically advanced units also control factors such as humidity and trace gas (e.g., CO_2, O_3) levels. Insects are generally contained in cages, or in mesh bags on plants, but are sometimes allowed free movement within the environmental chamber.

Indoor controlled environments have long been a mainstay of climate and atmospheric change research on invertebrates. Historically, most studies have focused on the level of individual organisms, assessing, for example, the effects of temperature on insect physiology (e.g., diapause, thermoregulation; Snodgrass et al., 2012), nutritional ecology (e.g., feeding rates, food processing efficiencies; Lindroth et al., 1997), and individual performance metrics (e.g., growth, development, reproduction; Johns et al., 2003). Fewer studies have been conducted at the population level, and these have typically evaluated impacts on invertebrates with short generation times, small size, and low motility, such as aphids (Bezemer et al., 1998) and collembolans (Meehan et al., 2010). Indoor controlled environments have only rarely been used to assess global change impacts on higher-order trophic interactions (e.g., herbivore–parasitoid interactions; Roth & Lindroth, 1995) or ecosystem processes (e.g., decomposition and nutrient cycling; Del Toro et al., 2015).

Indoor closed environmental systems are powerful, indeed indispensable, tools for experimental analysis of the effects of global change drivers on invertebrate biology. Conceptually, their principal advantages are the capacity to reduce, partition and control the variation resident in natural systems, and to test the mechanisms underlying ecological phenomena of interest. Functionally, indoor closed systems provide relative ease of replication, fine-tuned regulation of environmental variables of interest, complete control of the cast of players, and rapid generation of data. Financial, logistical and technical barriers are generally low.

Significant advantages notwithstanding, indoor closed systems also suffer from numerous conceptual and functional disadvantages. Conceptually, the greatest handicap of chamber experiments is that their small size, reduced complexity and short duration exclude or distort important features of natural communities and ecosystems, such as multi-species interactions, indirect effects, legacy effects, delayed feedbacks, and spatial and temporal heterogeneity (Lawton, 1995; Carpenter, 1996). The organisms evaluated (generally small, with short generation times) may not be representative of

Figure 3.2 Examples of experimental systems used for evaluating the effects of climate and atmospheric change on invertebrates. (A) Controlled environment chamber with wheat, soybean, and fall armyworm (*Spodoptera frugiperda*). Insect cages shown in the rear. (Photo credit: R.L. Lindroth). (B) Greenhouse with hybrid poplar and gypsy moth (*Lymantria dispar*). (Photo credit: R.L. Lindroth). (C) Open-top chamber with aspen, maple and forest tent caterpillar (*Malacosoma disstria*). (Photo credit: R.L. Lindroth). (D) Open-top chamber with ant communities in eastern deciduous forest, USA. (Photo credit: S.L. Pelini). (E) Open-air warming study (B4WarmED) with forest tent caterpillar in southern boreal forest, USA. (Photo credit: M.A. Jamieson). (F) Free Air CO_2 and Ozone Enrichment study (Aspen FACE) with diverse insects in northern deciduous forest, USA. (Photo credit: R. Anderson, Skypixs Aerial Photography, Lake Linden, MI).

those that are not (Lawton, 1995). Chamber studies also have numerous functional disadvantages. Abiotic conditions may be qualitatively and quantitatively different from those in the field. Light is a good example: the full spectrum of sunlight is difficult, if not impossible, to match in indoor environments (including greenhouses), and the photosynthetic photon flux density from chamber lamps is generally well below the level required for photosynthetic light saturation in sun-adapted plants. Closed indoor systems are also not suitable for use with large or highly mobile organisms, such as mature trees, birds and mammals.

Several experimental design problems are commonly encountered in environmental chamber research. First, when controlled chambers or greenhouses are limited in number, proper experimental design presents a challenge. One option to increase replication is to repeat the experiment over time, with experimental treatment randomly applied to chambers and time incorporated as a blocking factor. This approach preserves true experimental replication, but can significantly prolong the duration of a study. In some situations, however, researchers may not have the option to truly replicate treatments over chambers or through time. A common practice is then to switch treatments – and associated sample units (e.g., insect containers, potted plants) – among chambers during the course of a study. This practice is intended to equalize "chamber effects" by distributing them similarly across the sample units. The approach should be used with caution, however, as not only may true chamber differences exist, but those differences may change over time (Potvin & Tardif, 1988). To detect potential chamber effects, sentinel plants with high genetic similarity and environmentally responsive growth traits can be deployed alongside experimental plants (Porter et al., 2015). In instances where pseudoreplication cannot be avoided, we recommend that researchers clearly acknowledge the constraints of the design used, so that readers can draw appropriate inferences from the work.

Another experimental design problem frequently encountered with chamber work occurs when samples are removed from a large outdoor experiment, and subjected to further experimentation in controlled chambers. An example here is the removal of tree leaves from a FACE site for feeding to insects in controlled chambers. The experiment is not started anew with the chamber studies. Rather, whatever experimental design exists in the larger study must carry through to data analysis for the chamber study as well.

3.5.2 Outdoor Closed Systems

Outdoor closed systems incorporate more ecological realism than indoor systems, while retaining some of the advantages of precise control of variables of interest. Prior to the development and deployment of FACE and FACE-like experimental systems in the 1990s, outdoor closed systems provided the leading technology for experimental climate change research, and they remain important today.

A variety of closed and semi-closed systems have been engineered to manipulate specific global change drivers in outdoor environments (Fig. 3.2). These include branch chambers (e.g., Teskey et al., 1991), whole-tree chambers (e.g., Barton et al., 2010), greenhouses (e.g., Marion et al., 1997), rain-manipulation shelters (e.g., Fay et al., 2000) and open-top chambers (e.g., Dickson et al., 2001). All of these systems are suitable for addressing appropriately matched questions about the effects of global change on invertebrates.

Outdoor closed and semi-closed experimental systems afford some important advantages over indoor systems. Foremost among them is the opportunity to conduct experiments in more natural environments. Depending on the type of technology used, these systems can incorporate biotic and abiotic components of natural systems – such as soils, solar radiation, and precipitation – while manipulating specific factors of interest. Because they are generally less costly than open-air systems, closed systems allow for greater replication, more treatments, or longer duration studies (e.g., Shaver & Jonasson, 1999; Mikkelsen et al., 2008; Scherber et al., 2013). The larger systems can accommodate small-stature trees (Roth et al., 1998) or complex plant communities (Owensby et al., 1993). Closed and semi-closed outdoor experimental systems can be used to investigate global change impacts on insect feeding and performance (Roth et al., 1998), population densities (Stiling et al., 2009), predator–prey interactions (Barton & Ives, 2014), and community composition (Pelini et al., 2011).

Despite the utility of outdoor closed and semi-closed experimental systems, several significant disadvantages constrain their use in global change research. First, the relatively small size of the system used can compromise effective application of the treatment of interest, and conclusions drawn therefrom. For example, the branch chamber technique can be problematic because branch treatments may have systemic effects on non-treated branches, and the response of a single branch may differ from that of the entire plant (Liu & Teskey, 1995). Similarly, what appears to be a whole-plant treatment in open-top chambers may not be, as the roots of some plants, especially trees, may extend well beyond the walls of the chamber. Second, chambers can modify abiotic conditions, thereby affecting endpoints of interest. Chambers influence temperature (magnitude, range, variation), moisture (precipitation, humidity), solar radiation (flux and spectral distribution), air movement, and gas composition (trace gas levels) (Kennedy, 1995). Although these changes can be reduced by adjusting the level of chamber closure, or by instrumented microclimate control, they cannot be eliminated. Incorporation of a set of unchambered control plots into experimental designs can help to identify chamber effects. Third, chambers affect the movements of mobile organisms, both those "in, trying to get out" and those "out, trying to get in". We have found, for example, that populations of spider mites and aphids can increase to outbreak levels inside open-top chambers, while they are not detectable outside the chambers. Such 'greenhouse pests' are widely known to commercial glasshouse industries. Conversely, Richardson et al. (2000) reported that densities of hemipteran insects were lower inside than outside of open-top chambers. Treated chambers, and the vegetation they contain, may function as "islands" that attract or deter insects, a topic addressed more fully in the next section.

3.5.3 Outdoor Open Systems

Outdoor open systems are the premier experimental approach for investigating the impacts of global change drivers on invertebrates in ecologically realistic communities and ecosystems (Fig. 3.2). What differentiates these systems from outdoor closed systems is, simply, the absence of chamber walls. The engineering and technology required for open systems, however, are generally much more sophisticated.

Outdoor open systems have been used primarily to study the impacts of drought, warming, and trace gases (CO_2, O_3) on biotic communities. Open-air manipulation of precipitation can be achieved with the use of retractable rain-out shelters (Hatfield et al., 1990). Temperature control is typically accomplished via aboveground infrared lamp arrays and belowground heating cables (Rich et al., 2015). Impacts of elevated levels of trace gases – usually CO_2, and in a few sites, O_3 – are investigated through the use of Free Air CO_2 Enrichment (FACE) systems (Dickson et al., 2000; Ort et al., 2006). FACE experimental systems typically fumigate entire stands of plants, ranging from pasture grasses to forest trees, although techniques for fumigating individual trees have also been developed (e.g., web-FACE [Hättenschwiler & Schafellner, 2004]). Outdoor open-air systems are generally constructed for the primary purpose of evaluating climate and atmospheric change effects on plants, not invertebrates. Yet invertebrates are ubiquitous in these systems, and not only respond to treatments, but can mediate the effects of those treatments on the experimental plant communities of interest (Couture et al., 2015).

The open nature of these outdoor systems, coupled with their spatial and temporal scale, provide the best option for manipulating climate and atmospheric change in "natural" settings. FACE systems, in particular, incorporate multiple treatment factors and many of the multi-species interactions, direct and indirect effects, and positive, negative and delayed feedbacks, that mediate invertebrate community- and ecosystem-level dynamics (Tylianakis et al., 2008; Robinson et al., 2012; Rosenblatt & Schmitt, 2014). FACE systems provide a platform for genetic screening of crop varieties at a scale relevant for yield estimates (Ainsworth et al., 2008). And they allow for assessment of ontogenetic variation in responses of long-lived organisms, such as trees, to global change factors (Couture et al., 2014).

The principal disadvantage of most open-air systems, especially FACE systems, is their large construction and operational costs. Experimental designs consequently tend to have minimal replication, reducing experimental power. Further, many habitat types, and even biomes (e.g., arctic tundra, tropical forest), are poorly represented among the world's major open-air experiments. Alternative technologies and less expensive sources of CO_2 may ultimately reduce the costs of FACE experiments (Ainsworth et al., 2008), but they will likely remain among the most expensive of manipulative ecological experiments to execute long term.

Several other problems confront investigators seeking to use large, open-air systems (especially FACE systems) for invertebrate research. First, invertebrate work is nearly always subordinate to the larger objectives (e.g., climate change impacts on forest carbon sequestration) of the experimental programme. This reality can sometimes complicate invertebrate studies – such as when plots at the Aspen FACE site (Rhinelander, Wisconsin, USA) were sprayed with insecticide to reduce impacts of a regional outbreak of forest tent caterpillars (*Malacosoma disstria*) (R.L. Lindroth, personal experience). Second, it can be difficult to differentiate direct from indirect effects of global change factors on invertebrate biology (Fig. 3.1). Are changes in insect performance metrics a direct response to experimental treatments of interest? Or are they responses to altered microclimate, or host quality, which in turn are due to experimental treatments? Coupling small-scale, microcosm studies with large-scale open-air experiments can help to control unwanted variation and thereby differentiate direct from indirect effects (Couture & Lindroth, 2012; Jamieson et al., 2015). Third, researchers working on population- and higher-level studies in open-air systems face the significant and largely irresolvable problem of the "island effect". In the future, global change drivers will influence ecosystems at a regional scale, not at the scale of isolated experimental plots. Thus, mobile organisms in the vicinity of current open-air experiments encounter the unnatural situation of choosing or avoiding habitats altered by the global change factors under manipulation. The experimental plots become islands that may preferentially attract and retain, or repel, mobile invertebrates. The issue is particularly problematic for studies of invertebrate populations and communities, and was addressed head-on in a commentary by Moise and Henry (2010). Researchers can reduce the problems posed by potential island effects by focusing their work on species with limited motility, imbedding cage studies within treatment plots, and by clearly linking plot-level treatments to the mechanisms underlying changes in invertebrate performance. Such choices, however, may force compromises with the selection of experimental systems, so the breadth of interpretation should be modified accordingly.

3.6 Team Science: the Human Dimension

The multidisciplinary nature of complex, large-scale, global change experiments requires diverse teams of scientists with complementary expertise. For example, Aspen-FACE hosted over

60 investigators, not including numerous postdoctoral associates and graduate students (http://aspenface.mtu.edu/investigators.htm). Although the researchers who lead such efforts are undoubtedly scientifically astute, few have formal training in project management and organizational leadership. In our experience, some of the more significant challenges to the successful execution of large-scale science arise not from the vicissitudes of the science per se, but from ineffective organizational management.

3.6.1 Personnel

Most research teams are initially assembled in relation to a specific project proposal. Team members should have complementary scientific expertise, but also demonstrably strong interpersonal skills and a commitment to teamwork. The primary leader or director of the group should be clearly identified. As the design and construction of a new experimental research site get underway, a steering (governance) committee should be assembled. Responsibilities of committee members, terms of service, and the process for securing membership on the committee should be delineated. Once a project becomes operational, additional project team members may be added because of identified areas of research need or upon request from individuals in the broader scientific community. In both cases, addition of project members should be vetted through the steering committee. Particularly large projects should employ a science coordinator, whose job is to manage, with direction from the steering committee, use of the experimental site by diverse research groups. Large projects should also employ an operational manager, responsible for maintenance of the physical infrastructure at the site.

3.6.2 Guiding Principles

Early in the development of a large, collaborative research project, the steering committee should develop a set of guiding principles to which all participants in the project commit to adhere. Foremost among these principles should be a set of primary and secondary research priorities. Clearly articulated priorities will aid in assessments of which projects are subordinate to others when the research goals of individual participants conflict. Other policies that the steering committee should develop include guidelines for access to shared data, authorship of collaborative publications, and approval of press releases to the news media.

3.6.3 Operation and Communication

Policies that direct how individual research studies are approved and executed will facilitate scientific interactions and overall success of complex global change projects. The steering committee should develop a policy on site use, addressing participant fee structures, safety training, and sample collection and storage protocols. Similar policies should be directed toward standards for data stewardship and archiving. The need for regular and effective communication, via means such as newsletters, blog posts, and investigator meetings, cannot be over-emphasized.

3.7 Conclusions

The environmental and societal consequences of rapid and pervasive global environmental change are daunting, yet the prospects for experimental science to help predict biological responses, and inform adaptation strategies, are exciting and energizing. Although technical advances are proceeding rapidly, significant conceptual challenges remain. Thoughtful framing of scientific questions, appropriate matching to a spectrum of complementary scientific methodologies, and wise stewardship of human resources promise to markedly advance our understanding of global change impacts on invertebrate systems.

Acknowledgements

We are indebted to the many students and colleagues with whom we have had the privilege to conduct global change research over several decades, as well as our experiences – successful and less so – that shaped the perspectives presented here. Our work has been funded principally by the U.S. Department of Energy, the U.S. Department of Agriculture, the National Science Foundation and the University of Wisconsin College of Agricultural and Life Sciences. Our most recent collaborative venture was funded by the National Institute of Food and Agriculture, U.S. Department of Agriculture (AFRI project 2011-67013-30147). We also thank Philip Smith for proofreading this chapter.

References

Ainsworth, E.A., Beier, C., Calfapietra, C., Ceulemans, R., Durand-Tardif, M., Farquhar, G.D., Godbold, D.L., Hendrey, G.R., Hickler, T., Kaduk, J., Karnosky, D.F., Kimball, B.A., Koerner, C., Koornneef, M., Lafarge, T., Leakey, A.D.B., Lewin, K.F., Long, S.P., Manderscheid, R., McNeil, D.L., Mies, T.A., Miglietta, F., Morgan, J.A., Nagy, J., Norby, R.J., Norton, R.M., Percy, K.E., Rogers, A., Soussana, J.F., Stitt, M., Weigel, H.J. & White, J.W. (2008) Next generation of elevated [CO_2] experiments with crops: a critical investment for feeding the future world. *Plant Cell and Environment*, **31**, 1317–1324.

Awmack, C.S., Harrington, R. & Lindroth, R.L. (2004) Aphid individual performance may not predict population responses to elevated CO_2 or O_3. *Global Change Biology*, **10**, 1414–1423.

Barton, B.T. & Ives, A.R. (2014) Species interactions and a chain of indirect effects driven by reduced precipitation. *Ecology*, **95**, 486–494.

Barton, C.V.M., Ellsworth, D.S., Medlyn, B.E., Duursma, R.A., Tissue, D.T., Adams, M.A., Eamus, D., Conroy, J.P., McMurtrie, R.E., Parsby, J. & Linder, S. (2010) Whole tree chambers for elevated atmospheric CO_2 experimentation and tree scale flux measurements in southeastern Australia: the Hawkesbury Forest Experiment. *Agricultural and Forest Meteorology*, **150**, 941–951.

Benton, T.G., Solan, M., Travis, J.M.J. & Sait, S.M. (2007) Microcosm experiments can inform global ecological problems. *Trends in Ecology & Evolution*, **22**, 516–521.

Bezemer, T.M., Jones, T.H. & Knight, K.J. (1998) Long-term effects of elevated CO_2 and temperature on populations of the peach potato aphid *Myzus persicae* and its parasitoid *Aphidius matricariae*. *Oecologia*, **116**, 128–135.

Brown, J.H., Whitham, T.G., Ernest, S.K.M. & Gehring, C.A. (2001) Complex species interactions and the dynamics of ecological systems: long-term experiments. *Science*, **293**, 643–650.

Carpenter, S.R. (1996) Microcosm experiments have limited relevance for community and ecosystem ecology. *Ecology*, **77**, 677–680.

Couture, J.J., Holeski, L.M., Lindroth, R.L. (2014) Long-term exposure to elevated CO_2 and O_3 alters aspen foliar chemistry across developmental stages. *Plant, Cell and Environment*, **37**, 758–765.

Couture, J.J. & Lindroth, R.L. (2012) Atmospheric change alters performance of an invasive forest insect. *Global Change Biology*, **18**, 3543–3557.

Couture, J.J., Meehan, T.D., Kruger, E.L. & Lindroth, R.L. (2015) Insect herbivory alters impact of atmospheric change on northern temperate forests. *Nature Plants*, 10.1038/nplants.2015.16

Daehler, C.C. & Strong, D.R. (1996) Can you bottle nature? The roles of microcosms in ecological research. *Ecology*, **77**, 663–664.

Del Toro, I., Ribbons, R.R. & Ellison, A.M. (2015) Ant-mediated ecosystem functions on a warmer planet: effects on soil movement, decomposition and nutrient cycling. *Journal of Animal Ecology*, **84**, 1233–1241.

Diamond, S.E., Penick, C.A., Pelini, S.L., Ellison, A.M., Gotelli, N.J., Sanders, N.J. & Dunn, R.R. (2013) Using physiology to predict the responses of ants to climatic warming. *Integrative and Comparative Biology*, **53**, 965–974.

Dickson, R.E., Coleman, M.D., Pechter, P. & Karnosky, D. (2001) Growth and crown architecture of two aspen genotypes exposed to interacting ozone and carbon dioxide. *Environmental Pollution*, **115**, 319–334.

Dickson, R.E., Lewin, K.F., Isebrands, J.G., Coleman, M.D., Heilman, W.E., Riemenschneider, D.E., Sober, J., Host, G.E., Zak, D.R., Hendrey, G.R., Pregitzer, K.S. & Karnosky, D.F. (2000) Forest atmosphere carbon transfer storage (FACTS II) the aspen free-air CO_2 and O_3 enrichment (FACE) project: an overview. *General Technical Report NC-214*. USDA Forest Service, North Central Research Station, Rhinelander, WI.

Drake, J.A., Huxel, G.R. & Hewitt, C.L. (1996) Microcosms as models for generating and testing community theory. *Ecology*, **77**, 670–677.

Fay, P.A., Carlisle, J.D., Knapp, A.K., Blair, J.M. & Collins, S.L. (2000) Altering rainfall timing and quantity in a mesic grassland ecosystem: design and performance of rainfall manipulation shelters. *Ecosystems*, **3**, 308–319.

Filion, M., Dutilleul, P. & Potvin, C. (2000) Optimum experimental design for free-air carbon dioxide enrichment (FACE) studies. *Global Change Biology*, **6**, 843–854.

Fraser, L.H. & Keddy, P. (1997) The role of experimental microcosms in ecological research. *Trends in Ecology & Evolution*, **12**, 478–481.

Gotelli, N.J. & Ellison, A.M. (2004) *A Primer of Ecological Statistics*. Sinauer Associates, Inc., Sunderland, MA.

Hatfield, P.M., Wright, G.C. & Tapsall, W.R. (1990) A large, retractable, low cost and re-locatable rain out shelter design. *Experimental Agriculture*, **26**, 57–62.

Hurlbert, S.H. (1984) Pseudoreplication and the design of ecological field experiments. *Ecological Monographs*, **54**, 187–211.

Hättenschwiler, S. & Schafellner, C. (2004) Gypsy moth feeding in the canopy of a CO_2 enriched mature forest. *Global Change Biology*, **10**, 1899–1908.

Jamieson, M.A., Schwartzberg, E.G., Raffa, K.F., Reich, P.B. & Lindroth, R.L. (2015) Experimental climate warming alters aspen and birch phytochemistry and performance traits for an outbreak insect herbivore. *Global Change Biology*, **21**, 2698–2710.

Johns, C.V., Beaumont, L.J. & Hughes, L. (2003) Effects of elevated CO_2 and temperature on development and consumption rates of *Octotoma championi* and *O. scabripennis* feeding on *Lantana camara*. *Entomologia Experimentalis et Applicata*, **108**, 169–178.

Kennedy, A.D. (1995) Simulated climate change: are passive greenhouses a valid microcosm for testing the biological effects of environmental perturbations? *Global Change Biology*, **1**, 29–42.

Lawton, J.H. (1995) Ecological experiments with model systems. *Science*, **269**, 328–331.

Lawton, J.H. (1996) The Ecotron facility at Silwood Park: the value of "big bottle" experiments. *Ecology*, **77**, 665–669.

Lindroth, R.L. (2010) Impacts of elevated atmospheric CO_2 and O_3 on forests: phytochemistry, trophic interactions, and ecosystem dynamics. *Journal of Chemical Ecology*, **36**, 2–21.

Lindroth, R.L., Kinney, K.K. & Platz, C.L. (1993) Responses of deciduous trees to elevated atmospheric CO_2: productivity, phytochemistry, and insect performance. *Ecology*, **74**, 763–777.

Lindroth, R.L., Klein, K.A., Hemming, J.D.C. & Feuker, A.M. (1997) Variation in temperature and dietary nitrogen affect performance of the gypsy moth (*Lymantria dispar* L.). *Physiological Entomology*, **22**, 55–64.

Liu, S. & Teskey, R.O. (1995) Response of foliar gas exchange to long-term elevated CO_2 concentrations in mature loblolly pine trees. *Tree Physiology*, **15**, 351–359.

Marion, G.M., Henry, G.H.R., Freckman, D.W., Johnstone, J., Jones, G., Jones, M.H., Lévesque, E., Molau, U., Mølgaard, P., Parsons, A.N., Svoboda, J. & Virginia, R.A. (1997) Open-top designs for manipulating field temperature in high-latitude ecosystems. *Global Change Biology*, **3**, 20–32.

Maslow, A.H. (1966) *The Psychology of Science*. Harper & Row, New York.

Meehan, T.D., Crossley, M.S. & Lindroth, R.L. (2010) Impacts of elevated CO_2 and O_3 on aspen leaf litter chemistry and earthworm and springtail productivity. *Soil Biology & Biochemistry*, **42**, 1132–1137.

Mikkelsen, T.N., Beier, C., Jonasson, S., Holmstrup, M., Schmidt, I.K., Ambus, P., Pilegaard, K., Michelsen, A., Albert, K., Andresen, L.C., Arndal, M.F., Bruun, N., Christensen, S., Danbæk, S., Gundersen, P., Jørgensen, P., Linden, L.G., Kongstad, J., Maraldo, K., Priemé, A., Riis-Nielsen, T., Ro-Poulsen, H., Stevnbak, K., Selsted, M.B., Sørensen, P., Larsen, K.S., Carter, M.S., Ibrom, A., Martinussen, T., Miglietta, F. & Sverdrup, H. (2008) Experimental design of multifactor climate change experiments with elevated CO_2, warming and drought: the CLIMAITE project. *Functional Ecology*, **22**, 185–195.

Moise, E.R.D. & Henry, H.A.L. (2010) Like moths to a street lamp: exaggerated animal densities in plot-level global change field experiments. *Oikos*, **119**, 791–795.

Ort, D.R., Ainsworth, E.A., Aldea, M., Allen, D.J., Bernacchi, C.J., Berenbaum, M.R., Bollero, G.A., Cornic, G., Davey, P.A., Dermody, O., Dohleman, F.G., Hamilton, J.G., Heaton, E.A., Leakey, A.D.B., Mahoney, J., Mies, T.A., Morgan, P.B., Nelson, R.L., O'Neil, B., Rogers, A., Zangerl, A.R., Zhu, X.-G., DeLucia, E.H. & Long, S.P. (2006) SoyFACE: the effects and interactions of elevated [CO_2] and [O_3] on soybean. In: Nösberger J, Long SP, Norby RJ, Stitt M, Hendry GR, Blum H, editors. Managed ecosystems and CO_2: case studies, processes and perspectives. Ecological Studies, **187**. New York: Springer Verlag; p. 71–86.

Owensby, C.E., Coyne, P.I., Ham, J.M., Auen, L.M. & Knapp, A.K. (1993) Biomass production in a tallgrass prairie ecosystem exposed to ambient and elevated levels of CO_2. *Ecological Applications*, **3**, 644–653.

Pelini, S.L., Bowles, F.P., Ellison, A.M., Gotelli, N.J., Sanders, N.J. & Dunn, R.R. (2011) Heating up the forest: open-top chamber warming manipulation of arthropod communities at Harvard and Duke forests. *Methods in Ecology and Evolution*, **2**, 534–540.

Porter, A.S., Evans-FitzGerald, C., McElwain, J.C., Yiotis, C. & Elliott-Kingston, C. (2015). How well do you know your growth chambers? Testing for chamber effect using plant traits. *Plant Methods*, **11**, 44.

Potvin, C. & Tardif, S. (1988) Sources of variability and experimental designs in growth chambers. *Functional Ecology*, **2**, 123–130.

Raffa, K.F., Aukema, B.H., Bentz, B.J., Carroll, A.L., Hicke, J.A., Turner, M.G. & Romme, W.H. (2008) Cross-scale drivers of natural disturbances prone to anthropogenic amplification: the dynamics of bark beetle eruptions. *Bioscience*, **58**, 501–517.

Rich, R.L., Stefanski, A., Montgomery, R.A., Hobbie, S.E., Kimball, B.A. & Reich, P.B. (2015) Design and performance of combined infrared canopy and belowground warming in the B4WarmED (Boreal Forest Warming at an Ecotone in Danger) experiment. *Global Change Biology*, **21**, 2334–2348.

Richardson, S.J., Hartley, S.E. & Press, M.C. (2000) Climate warming experiments: are tents a potential barrier to interpretation? *Ecological Entomology*, **25**, 367–370.

Robinson, E.A., Ryan, G.D. & Newman, J.A. (2012) A meta-analytical review of the effects of elevated CO_2 on plant–arthropod interactions highlights the importance of interacting environmental and biological variables. *New Phytologist*, **194**, 321–336.

Rosenblatt, A.E. & Schmitz, O.J. (2014) Interactive effects of multiple climate change variables on trophic interactions: a meta-analysis. *Climate Change Responses*, **1**, 8.

Roth, S.K. & Lindroth, R.L. (1995) Elevated atmospheric CO_2: effects on phytochemistry, insect performance and insect-parasitoid interactions. *Global Change Biology*, **1**, 173–182.

Roth, S., Lindroth, R.L., Volin, J.C. & Kruger, E.L. (1998) Enriched atmospheric CO_2 and defoliation: effects on tree chemistry and insect performance. *Global Change Biology*, **4**, 419–430.

Scherber, C., Gladbach, D.J., Stevnbak, K., Karsten, R.J., Schmidt, I.K., Michelsen, A., Albert, K.R., Larsen, K.S., Mikkelsen, T.N., Beier, C. & Christensen, S. (2013) Multi-factor climate change effects on insect herbivore performance. *Ecology and Evolution*, **3**, 1449–1460.

Shaver, G.R. & Jonasson, S. (1999) Response of arctic ecosystems to climate change: results of long-term field experiments in Sweden and Alaska. *Polar Research* **18**, 245–252.

Snodgrass, G.L., Jackson, R.E., Perera, O.P., Allen, K.C. & Luttrell, R.G. (2012) Effect of food and temperature on emergence from diapause in the tarnished plant bug (Hemiptera: Miridae). *Environmental Entomology*, **41**, 1302–1310.

Stiling, P., Moon, D., Rossi, A., Hungate, B.A. & Drake, B. (2009) Seeing the forest for the trees: long-term exposure to elevated CO_2 increases some herbivore densities. *Global Change Biology*, **15**, 1895–1902.

Teskey, R.O., Dougherty, P.M. & Wiselogel, A.E. (1991) Design and performance of branch chambers suitable for long-term ozone fumigation of foliage in large trees. *Journal of Environmental Quality*, **20**, 591–595.

Tylianakis, J.M., Didham, R.K., Bascompte, J. & Wardle, D.A. (2008) Global change and species interactions in terrestrial ecosystems. *Ecology Letters*, **11**, 1351–1363.

Whittaker, J.B. (2001) Insects and plants in a changing atmosphere. *Journal of Ecology*, **89**, 507–518.

4

Transplant Experiments – a Powerful Method to Study Climate Change Impacts

Sabine S. Nooten[1] and Nigel R. Andrew[2]

[1] *Hawkesbury Institute for the Environment, Western Sydney University, NSW 2751, Australia*
[2] *Centre for Behavioural and Physiological Ecology, Zoology, University of New England, Armidale, NSW 2351, Australia*

Summary

Transplant experiments are a direct test of what might happen in the future as species or entire communities are moved out of their current climate into a location with novel climate conditions. Here we assess our current understanding of climate change adaptation responses using transplant experiments. Firstly, we assess the current knowledge on species and community responses to climate change. We then identify the way climate change responses have been carried out to date with an emphasis on transplant experiments including: adaptation to a warmer climate; potential of range shifts; changes in phenology; shifts in species interactions; disentangling genotypic and phenotypic responses; and shifts in communities. Further, we assess transplant experiments that specifically assess invertebrate responses using network analyses and conclude with an assessment of what is missing in the current approaches and the way forward with future transplant experiments.

4.1 Global Climate Change

Human-induced global climate change is already affecting species and ecosystems, mainly by increasing temperatures, altering precipitation patterns, and increasing the frequency of severe weather events (IPCC, 2012, 2013). Over the last century, mean global surface temperatures have increased by ~0.85°C (IPCC, 2013), with a rate of 0.2°C per decade in the last 25 years (Allison et al., 2009; Burrows et al., 2011). This rate of warming can be expressed as velocity of climate change, which averages over the last half century at 27.3 km per decade across global land surfaces (Burrows et al., 2011). The velocity of climate change is heterogeneous across the Earth, highest values being estimated in flat land areas such as low lying grasslands and forests, mangroves and deserts, and lowest velocity in mountainous areas (Loarie et al., 2009; Burrows et al., 2011). Climate change is now considered a major threat to global biodiversity, next to habitat destruction and invasive species (Thomas et al., 2004; Maclean & Wilson, 2011).

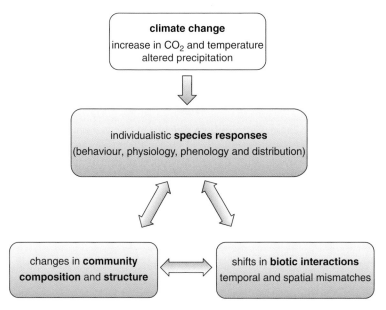

Figure 4.1 Flowchart showing known species' response to climate change impacts.

4.2 Climate Change Impacts on Species

Profound impacts are expected on the distribution, abundance and ecology of virtually all species (Fig. 4.1) (Hughes, 2000; Walther et al., 2002; Parmesan & Yohe, 2003; Root et al., 2003; Pármesan, 2006; Rosenzweig et al., 2008) and many species are predicted to be at risk of extinction (Thomas et al., 2004; Hannah, 2012). Species can respond via adaptation or extinction; adaptation consists of (i) adjusting to the new climate in situ, or (ii) moving to remain in the suitable climate, that is, shifting their range. Adaptive responses include genetic change, and changes in physiology, behaviour, distribution and/or phenology (Hughes, 2000; Root et al., 2003; Pármesan, 2006; Rosenzweig et al., 2008; Visser, 2008; Hoffmann & Sgrò, 2011; Parmesan et al., 2011). Assuming that macro climate is a major factor that determines species distribution (MacArthur, 1972), species will have to move latitudinally and/or altitudinally to track climate change and stay within their climatically preferred niche (see Palmer & Hill, this volume). Or they will have to move to another micro-climate within their current niche. In other words, this 'climate envelope' represents conditions under which populations can persevere in the face of competitors and natural enemies (Thomas et al., 2004). Indeed many species have already responded to macro-climate changes by shifting their geographic ranges either polewards, with an average of 6.1 km per decade, or to high elevations – by 6.1 m per decade (Parmesan & Yohe, 2003). Shifts in phenology have already occurred with an average spring advancement in life cycle events of 2.8 days per decade (Root et al., 2003; Parmesan, 2007; Rosenzweig et al., 2008). A meta-analysis conducted across multiple taxa showed that the magnitude of spring advancements is variable and that amphibians, birds and butterflies showed the greatest spring advancement, whereas herbs, grasses and shrubs showed the least (Parmesan, 2007).

Recent studies have identified insects as particularly vulnerable to warming temperatures (Deutsch et al., 2008; Netherer & Schopf, 2010; Wilson & Maclean, 2011), because many stages within insect life cycles are cued and inhibited by environmental factors such as temperature (Bale et al., 2002; Andrew & Terblanche, 2013; Boardman & Terblanche, 2015). High sensitivity to climatic changes also makes insects potentially useful indicators of a changing climate (Hodkinson & Bird, 1998). The impacts of climate change on insects may be direct, via effects on their physiology and behaviour, or indirect, via climate-induced changes in species with which they interact, such as host plants (Bale et al., 2002; Andrew et al., 2013b). Some insects have already responded to climate change with changes in phenology, especially with the advancement of spring events such as first spring flights (e.g., Forister & Shapiro, 2003; Kearney et al., 2010). Many insects have also responded with changes in distribution, either latitudinal (e.g., Parmesan et al., 1999; Musolin, 2007), or altitudinal (e.g., Wilson et al., 2007; Merrill et al., 2008). Increased numbers of generations per year have been observed in some species, leading to increased population sizes (Altermatt, 2010). These types of responses may result in a higher frequency of insect outbreaks, with potential threats to agriculture and forestry (Cornelissen, 2011).

4.3 Climate Change Impacts on Communities

Idiosyncratic species responses to climate change will affect biotic interactions (e.g. competition, predator–prey and host–parasite) with some existing relationships becoming increasingly decoupled (Tylianakis et al., 2008; Thackeray et al., 2010; Hughes, 2012; Hentley & Wade, this volume). Present day plant–insect interactions may be disrupted because the greater mobility and climate sensitivity of many insects means they have greater capacity to adapt to climate change than their hosts, either through range shifts or changes in phenology. These differences in responses may lead to pronounced mismatches in trophic interactions, for example, differential changes in phenology leading to temporal mismatches in interactions between plants and herbivores (Visser & Both, 2005; Parmesan, 2007), or between plants and insect pollinators (Memmott et al., 2007; Hegland et al., 2009). Differential shifts in distribution and phenology of interacting species within communities have already resulted in pronounced changes in community composition and structure (reviewed in Walther, 2010). For example, the disassembly of communities due to climate driven range shifts is already apparent across various ectotherm and endotherm species in temperate and tropical communities (Sheldon et al., 2011). These same range shifts can also lead to the assembly of new communities (González-Megías et al., 2008; Thomas, 2010). Temperature increases in particular, have been found to have profound impacts on community composition. Studies have reported compositional changes in detrital arthropod communities (Lessard et al., 2011) and herbivore insect communities (Musolin, 2007; Villalpando et al., 2009) associated with warming. In some cases, entire herbivore communities have responded to climate change with uphill range shifts of the component species, while retaining a similar composition (Wilson et al., 2007). These differential responses to climate change make generalisations difficult across multiple species.

4.4 Common Approaches to Study Climate Change Impacts

There are several approaches to study impacts of climate change and infer generalisations and predictions for future scenarios (see Andrew et al., 2013b). Many predictions on individual species are derived from species distribution models (SDMs), which are based on the statistical relationship between species' occurrences and environmental variables (Pearson et al., 2002; Beaumont et al., 2007; Roubicek et al., 2010). These models are useful tools for understanding relative species' vulnerabilities to climate change but they have well known limitations, including the fact that they typically

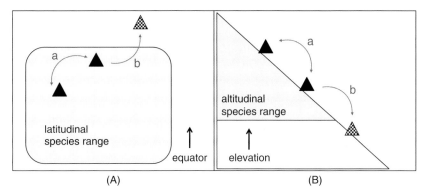

Figure 4.2 Schematic illustrating latitudinal (A) and altitudinal (B) transplant experiments. Grey areas indicate current species range; filled triangles show species transplanted within their range (a), hatched triangles show species moved outside of their range (b) into a warmer climate to simulate future conditions.

do not incorporate the effects of species interactions into their predictions (see review by Guisan & Thuiller, 2005; Van der Putten, Macel, & Visser, 2010). Manipulative experiments in controlled environments (e.g., glasshouses) have also been used to assess the potential impacts of factors such as increased temperature and elevated atmospheric CO_2 (Johns et al., 2003; Coll & Hughes, 2008) but they tend to be on a short term base and limited in replication and the numbers of factors tested. Other studies have used thermal tolerance curves of insects to predict impacts of warming on insect populations (Deutsch et al., 2008; Andrew, 2013a).

The challenges of predicting future impacts of climate change for individual species are significant. To study multiple species at the community level, however, these challenges are orders of magnitude greater. Potential changes in communities have been assessed using field surveys along latitudinal (e.g., Andrew & Hughes, 2004; Andrew & Hughes, 2005a; Andrew & Hughes, 2005b), and altitudinal gradients (e.g., Colwell et al., 2008; Chen et al., 2009; Garibaldi et al., 2011). A few studies have also related long term changes in communities to environmental changes (e.g., Voigt et al., 2003), but such data sets are relatively rare. Other studies have used artificial manipulation of the microclimate, that is, local warming experiments, such as open top chambers in the field to assess impacts of increased temperature on communities in situ (Barton et al., 2009; Pelini et al., 2011). While these approaches have proved useful in many respects, they suffer from several shortcomings. In particular, generally species are investigated in isolation (e.g., modelling approach), or a small subset of parameters is investigated in a relatively short time frame and under controlled conditions (e.g., glasshouse approach). These shortcomings can often be overcome by studying future climate change impacts using transplant experiments. In these, species or groups of species are moved to new climatic habitats and their response measured in the field (Fig. 4.2). This approach has proven very powerful, but despite their potential as predictive tools, relatively few such studies have been performed in relation to climate change.

4.5 Transplant Experiments – a Powerful Tool to Study Climate Change

Transplant experiments are a direct test of what might happen in the future as species or entire communities are moved out of their current climate into a location with novel climate conditions. Such experiments allow the monitoring of responses of single species, changes in multi trophic interactions and impacts on community composition and structure. In this way, a variety of questions can be addressed: (i) the potential of species to adapt to a warmer climate; (ii) the potential for pole- or upward range shifts; (iii) shifts in phenology, especially the likelihood of temporal mismatches and (iv) alterations of species interactions; (v) disentangling genotype vs phenotype responses; and (vi) shifts in communities. The following paragraphs describe the versatile use of transplant experiments.

4.5.1 Can Species Adapt to a Warmer Climate?

To assess whether species are able to adapt to a warmer climate, organisms can be transplanted outside of their range – either latitudinal or altitudinal (Fig. 4.2) – and responses, in relation to survival, general fitness, recruitment and reproduction can be monitored in the field. This approach is principally similar to a glasshouse experiment with the benefit of testing responses under more natural conditions. Ideally both approaches should be used in conjunction, however are rarely applied (but see Pelini et al., 2009). Many stages within insect life cycles are cued by temperature, thus higher temperature can benefit certain life cycle stages, such as larval development time and pupae mass, especially in temperate climates (Bale et al., 2002). However, responsiveness to temperature cues can differ among species and populations. To test for the assumption that warming enhances peripheral poleward populations, two butterfly species with contrasting degrees of host specialization and similar geographical distribution were reciprocally transplanted between their range core and edge area at the Pacific Northwest USA (Pelini et al., 2009). The survival of the specialist species, the Propertius duskywing (*Erynnis propertius*), was increased by warming, whereas summer temperatures decreased the survival of the generalist species, the Anise swallowtail (*Papilio zelicaon*). These results indicate that warming may not facilitate polewards range shifts by enhancing peripheral populations.

4.5.2 The Potential of Range Shifts

To address range shift questions, species are generally transplanted outside of their range (latitudinal or altitudinal) into a geographical location, which is predicted to be suitable for this species to occupy in the future. Transplant experiments can be useful tools to experimentally investigate the potential of pole- or upwards range expansion, equator wards range contraction or contractions on both ends. Once species are moved from potential source populations to new locations (e.g., to test for polewards range expansion one would use species from the polewards edge), their potential to survive and establish viable populations can then be monitored over time. In this way, species responses to the new climate conditions can be measured under natural field conditions. This approach has the advantage that multiple factors influencing range shifts, including effects of abiotic factors and biotic interactions, can be assessed simultaneously. Thus transplant experiments provide a more direct test of how species might fare under future climate conditions than the more widely used modelling techniques. Ideally, transplant experiments investigating range shifts could be used to test/verify predictions of species distribution models, but to our knowledge, so far no such attempt has been made.

Both, abiotic and biotic factors can potentially limit pole- or upwards range shifts of species into new locations. For example, the polewards range expansion of the Sachem skipper (*Atalopedes campestris*) at the Pacific Northwest, USA appeared to be limited by cooler minimum temperatures, which lengthened butterfly development time during summer months and minimised survival of larval overwintering stages (Crozier, 2004). The availability of food sources, such as appropriate host plants for herbivorous insects limited their expansion beyond the current range latitudinally for the Propertius duskywing (*Erynnis propertius*) in the Pacific Northwest USA (Hellmann et al., 2008) and altitudinally for the black-veined white butterfly (*Aporia crataegi*) in Spain (Merrill et al., 2008). A comparative test between contrasting species inhabiting different ecological niches in terms of host plant use breadth, showed idiosyncratic abilities for potential poleward range shifts, as the

Propertius duskywing (*Erynnis propertius*), a specialist butterfly, was more strongly limited by host plant availability than the Anise swallowtail (*Papilio zelicaon*), the generalist species (Hellmann et al., 2008). On the other hand, current biotic interactions can change: UK brown argus butterfly (*Aricia agestis*) adapted to new biotic conditions in the expanding range by switching host plants (Buckley & Bridle, 2014). These different trends in individual species responses can make generalisations very difficult.

The capacity to adapt to new environmental conditions during range shifts generally varies to some degree – not only among species but also within species across their distribution. Therefore, species can be transplanted from various source populations within their current range to suitable new locations to investigate variations in their ability to adapt. Generally, a high adaptation capacity is thought to enable species to shift their ranges at the polewards edge so they thus might better cope with a changing climate. This is exemplified by the poleward range expansion of the wasp spider (*Argiope bruennichi*) over the last century across eastern Europe: a shift in temperature tolerance curves enabled the overwintering of individuals in the cooler climates of the expansion range made possible by the species' recent ability to adapt to cooler temperatures (Krehenwinkel & Tautz, 2013). On the other hand, UK brown argus butterfly (*Aricia agestis*) showed a loss of adaptive variation in host plant use within its recently expanded range across the UK; the butterfly's ability to change its host plant use in the expanding range, was then itself associated with the inability to switch back to use the ancestral host plant (Buckley & Bridle, 2014). Such a narrowing of the adaptation ability in range expanding populations might have consequences for further polewards range shifts.

4.5.3 Changes in the Timing of Events

To address questions about phenology and potential temporal mismatches, which can arise based on individualistic responses of interacting species to climate change, organisms can for example be transplanted between cool and warm sites within their current range. Potential shifts in the timing of spring events, for example, flowering of plants and emergence of insects, or foliage appearance and herbivore emergence, can concurrently be monitored under field conditions. More importantly, multiple environmental factors, which may act as cues to initiate spring events can be assessed. So far, most studies used theoretical approaches by combining empirical data, such as observational data on the timing of spring events collected over a range of species, years and geographical locations, and climate models to estimate shifts in phenology under future climate scenarios (e.g., Memmott et al., 2007; Parmesan, 2007). An elegant way to supplement such findings would be to conduct transplant experiments to verify these predictions. Alternatively, one can use transplant experiments to deliver the baseline data for building models to more reliably predict potential changes in the timing of spring events of interacting partners (e.g., Forrest & Thomson, 2011). The potential risk of decoupling the seasonal timing of flowering onset in subalpine forbs and the emergence of native bees was investigated using an altitudinal transplant experiment in the Rocky Mountains (Forrest & Thomson, 2011). Eight bee species were transplanted reciprocally between high and low elevation sites in their overwintering stages inside 'bee hotels' or nest boxes. The emergence of the transplanted bees was earlier at low elevation sites and matched the flowering onset of the local alpine plants. From a range of potential environmental factors, temperature was found to be the main cue for both interacting partners. Besides detecting temperature as main cue, they also found that there was a strong effect of local environmental conditions on bee emergence and no site of origin effect (Forrest & Thomson, 2011). This shows that well designed transplant experiments can be used to answer several questions at the same time, which is probably highly desirable given the relatively high costs and long time frame required for this type of field experiments.

4.5.4 Shifts in Species Interactions

To address questions relating to species interactions and spatial mismatches that can arise based on idiosyncratic responses of interacting partners to climate change, species can be transplanted in a range of different ways, depending on the question: (1) to investigate changes in species interactions under warmer climate scenarios, organisms can be transplanted outside of their range into a warmer climate; (2) to investigate the role of species interactions, for example, plant–insect interactions, in driving range dynamics, species can be transplanted into the expanding range, and (3) to disentangle effects of local environments and genotype on species interactions, species can be transplanted reciprocally between contrasting locations within their current distribution (Fig. 4.2).

Within the multitude of species interactions, including mutualism, competition, pollination and predation, herbivory is one of the most fundamental processes on earth, as interacting species – plants and their herbivorous insects – encompass ~50% of the described species on earth (Strong et al., 1984; Chapman, 2009). Their interactions are essential elements affecting community structure and food webs worldwide (Coley, 1998). Understanding how climate change will affect these interactions is crucial, and to this end a range of transplant experiments has been carried out. To assess whether herbivory pressure will increase in a warmer climate, a common and widespread understorey shrub *Acacia falcata* was transplanted in Australia beyond its equatorial range limit into a warmer climate (Andrew & Hughes, 2007), and herbivory rates and type of damage compared to plants within their current range; plants at current and warmer sites experienced similar herbivory pressure, in terms of total amount and dominant damage types, such as foliage chewing and sap sucking (Fig. 4.3a). As this study investigated one plant species, the question arose how generalizable this result was. Therefore, a follow up study was carried out using multiple species: four locally common understorey shrub species from major plant families were transplanted into the centre of their current range and ~600 km beyond their current equatorial range limit to simulate the climate of ~30 years in the future (Nooten & Hughes, 2014). Total leaf herbivory, accumulated after one year in the field, ranged from ~3–10% across all plant species, showing there was no consistent pattern between control and warmer sites. The proportion of herbivore damage types, however, was relatively consistent between warm and control sites for three species (Fig. 4.3b,c,d,e). Hence, the results of these experiments indicate that making general predictions about how future climates will affect herbivory may be challenging, and that transplant experiments provide a useful complement to laboratory experiments and gradient studies in the search for understanding of future impacts on species interactions.

To investigate the role of plant–insect interactions in driving range dynamics, plant species can be transplanted within the current range and beyond into the expanding range; impacts of the main herbivores can then be investigated over time and under natural field conditions. In New Zealand, the common understorey Kawakawa tree (*Macropiper excelsum*), was transplanted in this fashion, and the plants survived beyond their current range limit and even showed enhanced growth, which was attributed to the lack of its main herbivore, the caterpillar of the Kawakawa looper (*Cleora scriptaria*) (Lakeman-Fraser & Ewers, 2013). Thus, this study showed that 'enemy release' may facilitate the establishment of plant species in novel areas, if they arrive there before their herbivorous insects. Conversely, holly (*Ilex aquifolium*) a common evergreen understorey shrub in Europe, transplanted in a similar fashion as described above, showed no release from its main herbivore, holly leaf-miner (*Phytomyza ilicis*), which was present throughout current and expanding range (Skou et al., 2011); the presence of the leaf miner throughout the range was at least partly attributed to the dispersal abilities and the use of garden cultivars of *Ilex* in the expanding range; the study also showed that the rate of successful emergence of adults from mines, was higher at the range limit. However both

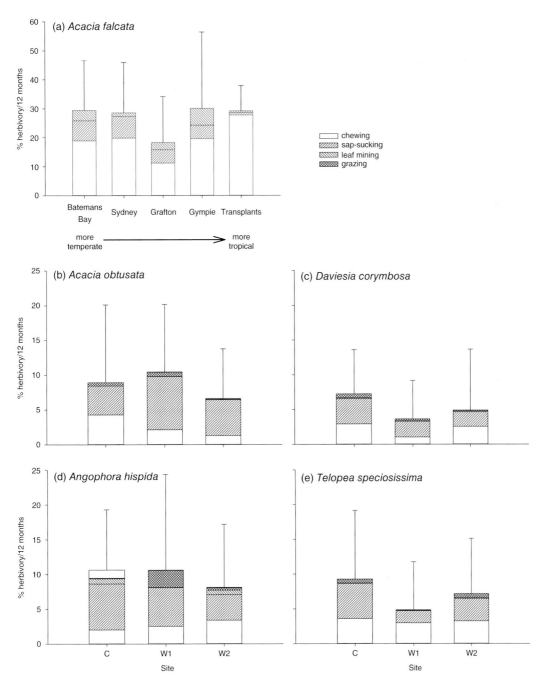

Figure 4.3 Redrawn percentage herbivory data/12 months from the latitudinal extent of five species from Eastern Australia. (a) *Acacia falcata* at four locations along its coastal range (Batemans Bay, Sydney, Grafton, Gympie) and transplant sites 200 km towards the tropics from Andrew and Hughes (2007); (b → e) four locally common understorey shrub species from major plant families transplanted into the centre of their current range and two sites ~600 km beyond their current equatorial range limit (denoted C, W1 and W2 respectively) from Nooten and Hughes (2014).

studies used the transplant experiment approach to not only look at range dynamics at large spatial scales (i.e., range shift/expansion) but also investigated factors, influencing the plants–insect interactions at a smaller local spatial scale. For example, transplants at forest interiors experienced more herbivore pressure than those at forest edge positions, a trend that disappeared in the expanding range (Lakeman-Fraser & Ewers, 2013). Such a shift in species interactions can potentially have far reaching consequences for communities.

4.5.5   Disentangling Genotypic and Phenotypic Responses

To disentangle the effects of 'site of origin' and environmental factors on species interaction strength, plant species can be transplanted reciprocally between contrasting regions of their distribution, and changes in species interaction strength can be monitored over time under field conditions. In Argentina, seedlings of the southern beech (*Nothofagus pumillo*) were transplanted reciprocally at high and low elevation sites and herbivory was assessed (Garibaldi et al., 2011); transplants at low elevation sites experienced more herbivory pressure, which was more pronounced on plants originating from low altitudes. Thus species interaction strength, measured as herbivory pressure, changed in a simulated warmer climate. Both 'environmental' factors and 'site of origin' treatments exhibited an effect, with 'environmental' factors being more pronounced. Conversely, the indirect effect of spiders (*Pisaurina mira*) on plants, via predation on a generalist grasshopper (*Melanoplus femurrubrum*) showed a stronger 'site of origin effect' and no influence of environmental factors (Barton, 2011). Spiders were translocated reciprocally in the northeastern United States at three sites spanning a temperature gradient of ~5°C to assess their indirect effect on plant biomass via top-down control. Spiders originating from warmer sites showed higher thermal response curves than those from cooler sites. Spiders originating from cooler sites caught less prey and thus exerted less top-down control, which was partly attributed to their lower thermal tolerance, indicating that spiders regulate their physiology and predation behaviour to maintain food web function across a range of temperatures similar to that predicted by global change models.

4.5.6   Shifts in Communities

To address questions relating to community effects in the light of climate change, arising through the individualistic responses of community members, organisms can be transplanted either into a warmer climate or reciprocally among sites within their current climate (Fig. 4.2). Impacts on communities can generally be assessed by choosing a focal taxon and then investigating it in two ways: firstly, in relation to species composition and secondly in relation to community structure by assigning functional guilds, that is, feeding types (Root, 1973; Simberloff & Dayan, 1991). In this way multiple aspects of a community can be investigated in regards to a changing climate. So far, several types of ecological communities have been investigated using transplant experiments, including herbivore insect communities (Andrew & Hughes, 2007; Heimonen et al., 2014; Nooten, Andrew, & Hughes, 2014), soil invertebrate communities (Briones et al., 1997; Todd et al., 1999; Budge et al., 2011) and rarely insect communities comprised of herbivores and non-herbivores (Nooten et al., 2014). The challenges of investigating climate change effects on communities are considerable and multiple benefits are associated with choosing the transplant experiment approach over glasshouse, growth chamber experiments or gradient studies. For example, impacts on communities can be investigated over a longer time frame to disentangle effects of rapid adaptation and long-term evolutionary responses. Furthermore, effects of multiple factors – abiotic and biotic – can be assessed simultaneously and most importantly factors driving community assembly can be investigated in situ. Transplant experiments can offer a useful tool to investigate the roles of host plant characteristics (Strong et al., 1984) and climatic factors (MacArthur, 1972) as potential drivers of plant–herbivore community assembly under warmer climate conditions. To this end, *Acacia falcata* was transplanted in eastern Australia beyond its equatorial range limit into a warmer climate and the Coleoptera and Hemiptera herbivore community composition and structure was investigated (Andrew & Hughes, 2007); community composition changed profoundly between plants in a warmer climate and those within the current range. Feeding guild structure however, remained surprisingly consistent among plants at warm and current sites (Fig. 4.4a), supporting the idea that host plant identity plays a major role in herbivore guild structure. A follow-up study tested the generalizability of this remarkable result, by conducting a multispecies transplant experiment in southeast Australia. Eight plant species from three major Australian plant families (Fabaceae, Myrtaceae and Proteaceae) were transplanted (i) into a warmer climate, ~600 km beyond their current equatorial range limit, to simulate future climate conditions, and (ii) back into the range centre as a control (Nooten et al., 2014) The herbivore community composition on all eight transplanted species showed an almost complete turnover between plants at warm and control sites, further corroborating similar results found earlier by Andrew and Hughes (2007). The herbivore guild structure was found to be similar between warm and control sites for six out of eight plant species (Fig. 4.4b,c,d,e,f,g,h,i) (Nooten et al., 2014). This study also revealed that other factors play an important role in driving the community assembly in a warmer climate, such as the nature of the habitat, including plant community composition and structural vegetation complexity. Additionally, the composition of the newly assembled insect communities also differed significantly from those of congeneric native plant species at the warm transplant sites (Andrew & Hughes, 2007; Nooten et al., 2014). Findings from both studies suggest that phytophagous insect species, which migrate polewards to track their climatic niche, may colonise new host plants by replacing species within the same functional guild (Andrew & Hughes, 2007; Nooten et al., 2014).

In northern Europe, the widespread birch (*Betula pendula*), originating from a range of different source populations was transplanted in Finland to assess effects of 'site of origin' and environmental factors in relation to their susceptibility to insects (Heimonen et al., 2014); herbivore community composition differed between sites but also showed variations between sites and study years. Another important factor in driving herbivore community assembly was the geographical distance of source population to transplant site: the similarity of the communities among the populations decreased with increasing latitudinal distance between source populations and transplant sites, indicating the influence of the genotype, which is thought to be more similar in birch populations that grow geographically closer to each other.

Whilst most studies looked at insect communities associated with common plant species, to date only one transplant experiment is known to have investigated the herbivore community on rare and threatened plants: In Western Australia, Moir et al. (2012) transplanted three rare *Banksia* species into warmer and drier locations. This longer-term experiment revealed that the herbivore community on plants at the new location changed to closely resemble the communities associated with nearby growing closely related *Banksia* species.

These findings suggest that with the current rate of warming a new suite of herbivore insect species might colonise plants within their current range over the coming decades and that polewards moving herbivores will also be able to colonise novel host plant genotypes. These transplant experiments, conducted in the northern and southern hemispheres, suggest that profound changes in the composition of herbivorous insect communities on host plants can be expected in the future (Andrew & Hughes, 2007; Heimonen et al., 2014; Nooten et al., 2014), but that structural elements of the community – the distribution of feeding guilds – might retain some consistency (Andrew & Hughes, 2007; Nooten et al., 2014). These results suggest that insect community structure and composition can be driven by both climate and host plant identity.

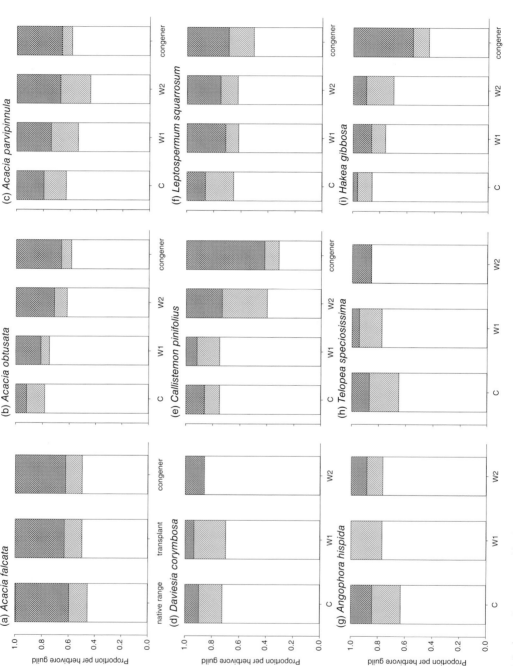

Figure 4.4 Redrawn Proportion of herbivore guilds from the latitudinal extent of nine host plant species on the East Coast of Australia. (a) *Acacia falcata* combined (native range) and transplanted *A. falcata* (transplant) and a congener at sites 200 km towards the tropics from Andrew & Hughes (2007); (b → i) eight locally common species transplanted into the centre of their current range (c) and at two sites ~600 km beyond their current equatorial range limit (W1 and W2) and an associated congener (if present) from Nooten and Hughes (2014). Key: phloem feeders (open bar); leaf chewers (diagonal bars): mesophyll

To investigate effects of a warmer climate on belowground invertebrate communities, soil cores can be transplanted reciprocally between sites of high and low elevation (Briones et al., 1997) and latitudes (Sohlenius & Bostrom, 1999; Todd et al., 1999). To study the effects of drought, soil cores can be transplanted longitudinal between dry and wet sites (O'Lear & Blair, 1999). In the UK, soil cores were transplanted reciprocally between high and low elevation sites, under the addition of a precipitation manipulation experiment to study the effects of temperature and soil moisture on the vertical distribution of the invertebrate community (Briones et al., 1997); Soil invertebrate taxa, including Enchytraeids, Diptera larvae and Tardigrades, responded idiosyncratically to changes in temperature and soil moisture, with some species retreating into deeper soil layers in response to reduced soil moisture and other species increasing in numbers in response to warming, resulting in shifts in community composition throughout the soil layers.

In tall prairie ecosystems in the United States, soil cores were transplanted in a reciprocal fashion longitudinally between mesic and xeric sites to assess effects of altered soil water availability on the vertical distribution of the soil micro-arthropod community in terms of composition (O'Lear & Blair, 1999) and on a functional-level in terms of feeding guild structure (Todd et al., 1999). Micro-arthropod taxa responded idiosyncratically to higher soil water availability as abundance of specific groups decreased, leading to a change in micro-arthropod composition (O'Lear & Blair, 1999). Similarly, functional guild level responses were complex and depended on the guild type, but herbivore taxa responded the strongest and positively, whereas microbial feeding taxa responded variably and slightly negatively to wetter conditions. In northern Sweden, peat blocks were transplanted to a warmer climate, and once more nematode taxa responded idiosyncratically, as some taxa increased in abundance at the warmer site (Sohlenius & Bostrom, 1999). Other factors, including soil structure and vegetation had a stronger influence on community composition than the moderate level of warming simulated by the experiment. These experiments overall show similar species-specific responses to those of aboveground taxa, with more tolerant species increasing in numbers (Briones et al., 1997; O'Lear & Blair, 1999), and at functional guild level responses depend on the feeding type, with herbivore guilds showing the strongest and most consistent positive responses to increased soil-moisture (Todd et al., 1999).

4.6 Transplant Experiment Trends Using Network Analysis

When reviewing the transplant experiment literature, a total of 69 studies were found, using the search terms 'climate change', 'transplant experiment' and 'climate change', 'translocation' in Web of Science and Scopus in December 2014. Of the 69 transplant studies directly related to climate change research across all taxa, we found 22 studies assessed invertebrates (Nooten & Andrew, 2015), many of which we have presented above. To assess the interactions between study taxon, climatic drivers, and location of studies we analysed our literature database using a network analysis. Interaction networks are a useful tool to assess the current state of knowledge within a discipline as they visually integrate key structures and information provided in manuscripts. We used *Cytoscape* 3.2.1 (Shannon et al., 2003) to develop the networks by combining two nodes with an edge connection. The types of research that these 22 studies represent are shown in the network analyses (Figs. 4.5 and 4.6).

Continent and Climatic driver were characterised as nodes and Transplant type was identified as the edge between nodes for the network produced in Fig. 4.5. Number of publications is labelled on each edge. Nine of these studies were based in North America, seven in Europe, five in Australia and one in South America. Temperature was the sole key climatic driver tested for 13 of these studies; six of the studies assessed both Temperature and Precipitation; two assessed only precipitation and

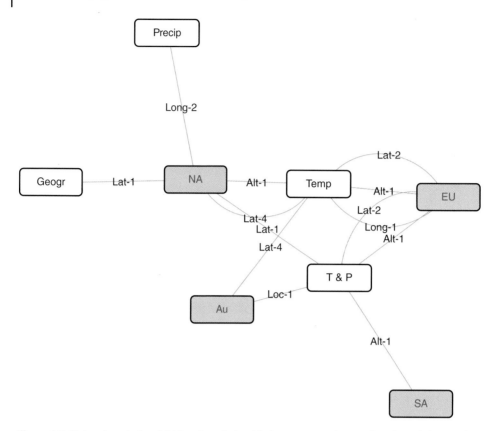

Figure 4.5 Network analysis exhibiting the relationship between continents of study and climatic drivers tested (nodes) and their interactions via transplant study type (edges). Continent (shaded: North America –NA; Europe – EU; Australia – AU; South America – SA) and Climatic driver (Temperature – Temp; Temperature and Precipitation - T & P; Precipitation – Precip; and Geographic – Geogr) were characterised as nodes and Transplant type (Latitude – Lat; Altitude – Alt; Longitude – long; Local – Loc) was identified as the edge between nodes. Number on each edge indicates number of papers (out of 22 papers assessed) showing that connection between Nodes.

one was a geographical study (multiple continents). Of the types of transplants carried out 14 were latitudinal, four were altitudinal, three were longitudinal, and one was locational. Latitudinal studies assessing temperature change were the most common in North America and Australia (four studies each) and in Europe (two studies) (Fig. 4.5).

Transplant experiments assessing temperature changes along latitudinal gradients are the most active manipulations currently carried out on invertebrates (Fig. 4.6). These include five studies where insects were transplanted, four studies where plants were transplanted and insect assessed, and one treatment assessing invertebrates in soil cores. Temperature and precipitation changes were assessed along latitudinal gradients in two studies where insects were transplanted, and one study with plants transplanted; along altitudinal gradients, temperature and precipitation was assessed with one of either: a soil core and plant; a plant; or a locational transplant. Two insect transplants were carried out along an altitudinal gradient assessing temperature change; and two soil core transplants were carried out along a longitudinal gradient assessing changes in precipitation.

These large-scale macroclimatic assessments are critical for assessing how invertebrates will respond to large-scale changes in climate, and they will also enable insects to interact with a wide

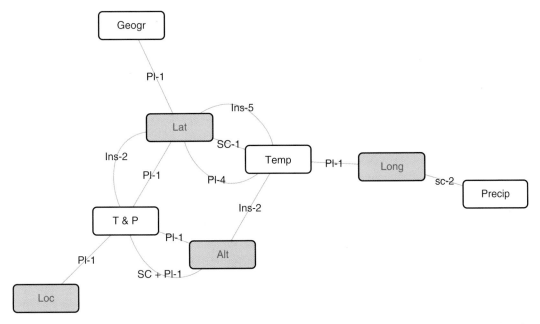

Figure 4.6 Network analysis exhibiting the relationship between transplant type and climatic drivers tested (nodes) and their interactions via the taxa transplanted (edges). Transplant type (shaded: Latitude – Lat; Altitude – Alt; Longitude – long; Local – Loc) and Climatic driver (Temperature – Temp; Temperature and Precipitation - T & P; Precipitation – Precip; and Geographic – Geogr) were characterised as nodes and taxa transplanted (Plant – Pl; Insect – Ins; Soil cores – sc; Soil cores and plants – Sc + Pl) was identified as the edge between nodes. Number on each edge indicates number of papers (out of 22 papers assessed) showing that connection between Nodes.

range of microclimatic variables. Microclimatic variables are less well understood in terms of generalised invertebrate responses to climate change. The role that ecology, behaviour, and physiology of individuals within populations and the intricate interactions at local scales (at which organisms live), makes for very complex reactions and reduces the strength of predictions to be made across taxa.

In summary, transplants experiments are a versatile tool to investigate a multitude of questions about climate change impacts, ranging from individualistic responses of single species, spatial and temporal shifts in species interactions and shifts in community composition and functional structure. While field transplant experiments are very time- and labour-intensive and relatively rarely used, they offer a valuable complement to other commonly used approaches to study climate change, including species distribution modelling, observations along gradients and glasshouse experiments. In a few studies, transplant experiments were used in conjunction with a gradient study (e.g., O'Lear & Blair, 1999; Andrew & Hughes, 2007; Garibaldi et al., 2011; Merrill et al., 2008), a growth chamber experiment (Pelini et al., 2009), or used as baseline data to develop models (e.g., Forrest & Thompson, 2011) for understanding and predicting future climate-change impacts. The frequency of finding idiosyncratic species responses to climate change further illustrates the importance of field-based experiments using multiple species at the same time. A multispecies approach is relatively easier to carry out in investigating soil fauna, as soil cores containing the entire community can be translocated conveniently (Briones et al., 1997; O'Lear & Blair, 1999; Todd et al., 1999), but rarely used to investigate aboveground insect communities, but see Nooten et al. (2014) and Moir et al. (2012) who translocated multiple plant species at the same time.

4.7 What's Missing in Our Current Approaches? Next Steps for Implementing Transplant Experiments

To date, there is a range of experiments and approaches that could be incorporated into transplant experiments to enhance our understanding of how species and communities might respond to a changing climate. This includes incorporating microclimate variation, including behavioural and physiological adaptations, climatic variability, and species interactions.

Woods et al. (2015) defined three axes of microclimatic variation: abiotic vs biotic; amplification versus buffering; and long versus short temporal and spatial scales. Abiotic microclimates are influenced by different structures in the environment such as rocks, soils, topography, and plant canopies. Biotic microclimates are influenced by nearby organisms, such as social insect nests, insect herbivores influenced by leaf surface temperature and humidity via stomatal opening, and leaf miners. Both abiotic and biotic environments can be manipulated to some extent by organisms to find their most favourable microclimate, making responses to macroclimate warming more difficult to assess and predict.

Microclimate may buffer or amplify macroclimates, but this may not happen as a linear relationship. Andrew et al. (2013a) found that air temperatures stayed consistent over a three hour period (around 22°C) from a local weather station, however ant temperatures (as measured by a thermocouple attached directly to a meat ants' (*Iridomyrmex purpureus*) exoskeleton) were at least 10°C higher and showed variation between 52 and 31°C. In addition, ibuttons on the surface (but not in direct sunlight) varied between 9°C and 56°C during summer and -5.5°C and 26.5°C during winter. Below the surface (15 cm) temperatures ranged from 11°C and 46°C in summer and 0.5°C and 22.5°C during winter; average air temperatures in Armidale range between 12.7°C and 26.8°C during summer and -0.4°C and 12.9°C during winter.

Spatial and temporal extents of microclimates are critical especially in relation to the organisms being assessed. For example, a wingless aphid will not move further than a few leaves on a plant within a few weeks, whereas a locust may travel hundreds of kilometres within a few weeks. Invertebrates accessing areas with complex topography and sparse vegetation canopies are more likely to expose themselves to a wider range of microclimates. Such changes cannot be predicted along large geographical gradients. Heterogeneity and spatial structure within an organism's micro-environment is critical when assessing thermoregulation, movement and energetics of invertebrates (Sears & Angilletta, 2015).

Behavioural thermal regulation is also critical for invertebrates. Moving to an environment where animals are not exposed to extreme temperatures is critical, and this would include localities in their current distribution in which they are not exposed to temperatures above their thermal safety margin for long periods of time. If animals are exposed, they may move quicker to return to safety and their nests. Meat ants (*I. purpureus*) forage for short periods of time on ground surfaces (up to 56°C) well above their critical thermal maximums (45.3°C); the hotter the ground temperatures, the faster they run, particularly back to the nest. The western horse lubber grasshopper (*Taeniopoda eques*) moves between vegetation and soil during the day to attain its optimum temperature of 35.2°C: roosting on the plants overnight, moving to the ground in mornings (to warm up), then returning to the vegetation during the middle of the day (to stay cool), back to open ground in the afternoon (for warmth), and then returning to vegetation at dusk for protection (Whitman, 1987).

A critical assumption in the transplant experiment literature is that the biggest impacts on invertebrates due to the average temperature increase will be that they need to move into higher elevations or latitudes to stay in their climatic envelope. However, not all species move at the same rate, and indeed some species may not expand their ranges (Gilman et al., 2010).

To take the experimentation further, an understanding of where the target transplanted species, or species of interest that will interact with the transplanted species, lie within food web and trophic levels is critical. For example ecosystem engineers in complex habitats may play maintenance roles with other species, but in a novel environment they may develop roles to re-design the landscape. Or indeed they may change their biology: an example of this is an invasive red-legged earth mite (RLEM) introduced from South Africa into Australia. In the Australian environment the RLEM increased its upper thermal threshold and was able to recover from cold stress more rapidly than the population in South Africa, effectively going through a complete niche shift (Hill et al., 2013). Even though this study was cross continental, it does suggest that physiological and behavioural responses of invertebrates need to be considered when assessing abilities to adapt to a new environment. In an animal's current distribution, its realised niche (Begon et al., 2006) may be constrained by competition, predation, parasitism, and/or unique resources; when moved to a new environment it is necessary to know if an invertebrate is still within its fundamental niche, or if it has been moved outside this, or indeed if niche constraints have shifted. The interactions with temperature and biotic interactions are critical.

As temperatures increase, biotic interactions may also change, and competition and dispersal differences may result in very different communities to those that are currently seen (Gilman et al., 2010; Urban et al., 2012). In order to assess species responses to climate change, Gilman et al. (2010) identified that interactions among species must be taken into account. They came up with a framework of species distributions in six open community interactions along a climatic gradient. For an interaction between a specialised enemy and victim, the enemy's range is limited as it cannot disperse farther than the victim; however, the victim can disperse further than its enemy, and with enemy release it can increase in density. For an interaction between two mutualists, they both limit each other's ability to track a changing climate envelope. For two species (one warm-adapted and the other cold-adapted) under exploitative competition (there is only a narrow range of coexistence), the expansion of the cold-adapted species moves it outside the region of interaction, however, the range of the cold-adapted species is reduced with the expansion of the warm-adapted species. In a predator–prey food chain the food chain predators cannot track changing climates beyond the dispersal-limits of the prey. Under apparent competition with two prey species and a predator, prey species 2 has a positive effect on the predator to reduce the abundance of prey species 1. Once the prey species expand beyond the range of the predator, both prey species increase in abundance. For keystone predation between a predator and two asymmetric competitive prey species, once both move beyond their keystone predator, the competitive species will exclude the subordinate.

Urban et al. (2012) modelled competition, niche breadth and dispersal ability among species along a climatic gradient. They found that climate change impacts were most disruptive when species had narrow niches, low dispersal rates of species within the community, and when there was high variation in dispersal rates among species. In particular those species that track climatic changes were the best dispersers which would, in turn, out-compete the slower dispersers and cause their local extinction. Thus in transplant experiments dispersal capability and competition would ideally be incorporated before, during and after transplantation.

Transplant experiments are one of the most powerful experimental methods that can be used to assess if populations will be able to adapt to a changing climate in the field. We believe that future assessments of climate change effects on invertebrates by moving taxa or habitats into novel environments using transplant experiments are critical. These experimental manipulations allow an assessment of their capabilities to adapt and respond to climatic variation incorporating microclimate variation, including behavioural and physiological adaptations, climatic variability and species interactions. These responses are critical, and deepen our understanding of biotic responses to climate change.

Acknowledgements

We are grateful to Philip Smith for proofreading this chapter.

References

Allison, I., Bindoff, N.L., Bindschadler, R.A., Cox, P.M., de Noblet, N., England, M.H., Francis, J.E., Gruber, N., Haywood, A.M., Karoly, D.J., Kaser, G., Le Quéré, C., Lenton, T.M., Mann, M.E., McNeil, B.I., Pitman, A.J., Rahmstorf, S., Rignot, E., Schellnhuber, H.J., Schneider, S.H., Sherwood, S.C., Somerville, R.C.J., Steffen, K., Steig, E.J., Visbeck, M. & Weaver, A.J. (2009) *The Copenhagen Diagnosis, 2009: Updating the World on the Latest Climate Science.* The University of New South Wales Climate Change Research Centre (CCRC), Sydney, Australia.

Altermatt, F. (2010) Climatic warming increases voltinism in European butterflies and moths. *Proceedings of the Royal Society B-Biological Sciences*, **277**, 1281–1287.

Andrew, N.R., Hart, R.A., Jung, M.-P., Hemmings, Z. & Terblanche, J.S. (2013a) Can temperate insects take the heat? A case study of the physiological and behavioural responses in a common ant, *Iridomyrmex purpureus* (Formicidae), with potential climate change. *Journal of Insect Physiology*, **59**, 870–880.

Andrew, N.R., Hill, S.J., Binns, M., Bahar, M.H., Ridley, E.V., Jung, M.-P., Fyfe, C., Yates, M. & Khusro, M. (2013b) Assessing insect responses to climate change: What are we testing for? Where should we be heading? *PeerJ*, **1**, e11.

Andrew, N.R. & Hughes, L. (2004) Species diversity and structure of phytophagous beetle assemblages along a latitudinal gradient: predicting the potential impacts of climate change. *Ecological Entomology*, **29**, 527–542.

Andrew, N.R. & Hughes, L. (2005a) Arthropod community structure along a latitudinal gradient: implications for future impacts of climate change. *Austral Ecology*, **30**, 281–297.

Andrew, N.R. & Hughes, L. (2005b) Diversity and assemblage structure of phytophagous Hemiptera along a latitudinal gradient: predicting the potential impacts of climate change. *Global Ecology and Biogeography*, **14**, 249–262.

Andrew, N.R. & Hughes, L. (2007) Potential host colonization by insect herbivores in a warmer climate: a transplant experiment. *Global Change Biology*, **13**, 1539–1549.

Andrew, N.R. & Terblanche, J.S. (2013) The response of insects to climate change. *Climate of Change: Living in a Warmer World* (ed J. Salinger), pp. 38–50. David Bateman Ltd., Auckland.

Bale, J.S., Masters, G.J., Hodkinson, I.D., Awmack, C., Bezemer, T.M., Brown, V.K., Butterfield, J., Buse, A., Coulson, J.C., Farrar, J., Good, J.E.G., Harrington, R., Hartley, S., Jones, T.H., Lindroth, R.L., Press, M.C., Symrnioudis, I., Watt, A.D. & Whittaker, J.B. (2002) Herbivory in global climate change research: direct effects of rising temperature on insect herbivores. *Global Change Biology*, **8**, 1–16.

Barton, B.T. (2011) Local adaptation to temperature conserves top-down control in a grassland food web. *Proceedings of the Royal Society B-Biological Sciences*, **278**, 3102–3107.

Barton, B.T., Beckerman, A.P. & Schmitz, O.J. (2009) Climate warming strengthens indirect interactions in an old-field food web. *Ecology*, **90**, 2346–2351.

Beaumont, L.J., Pitman, A.J., Poulsen, M. & Hughes, L. (2007) Where will species go? Incorporating new advances in climate modelling into projections of species distributions. *Global Change Biology*, **13**, 1368–1385.

Begon, M., Townsend, C.R. & Harper, J.L. (2006) *Ecology: From Individuals to Ecosystems.* Blackwell, Melbourne.

Boardman, L. & Terblanche, J.S. (2015) Oxygen safety margins set thermal limits in an insect model system. *Journal of Experimental Biology*, **218**, 1677–1685.

Briones, M.J.I., Ineson, P. & Piearce, T.G. (1997) Effects of climate change on soil fauna; responses of enchytraeids, diptera larvae and tardigrades in a transplant experiment. *Applied Soil Ecology*, **6**, 117–134.

Buckley, J. & Bridle, J.R. (2014) Loss of adaptive variation during evolutionary responses to climate change. *Ecology Letters*, **17**, 1316–1325.

Budge, K., Leifeld, J., Egli, M. & Fuhrer, J. (2011) Soil microbial communities in (sub)alpine grasslands indicate a moderate shift towards new environmental conditions 11 years after soil translocation. *Soil Biology and Biochemistry*, **43**, 1148–1154.

Burrows, M.T., Schoeman, D.S., Buckley, L.B., Moore, P., Poloczanska, E.S., Brander, K.M., Brown, C., Bruno, J.F., Duarte, C.M., Halpern, B.S., Holding, J., Kappel, C.V., Kiessling, W., O'Connor, M.I., Pandolfi, J.M., Parmesan, C., Schwing, F.B., Sydeman, W.J. & Richardson, A.J. (2011) The pace of shifting climate in marine and terrestrial ecosystems. *Science*, **334**, 652–655.

Chapman, A.D. (2009) *Numbers of Living Species in Australia and the World*. Australian Biological Resources Study, Canberra.

Chen, I.-C., Shiu, H.-J., Benedick, S., Holloway, J.D., Chey, V.K., Barlow, H.S., Hill, J.K. & Thomas, C.D. (2009) Elevation increases in moth assemblages over 42 years on a tropical mountain. *Proceedings of the National Academy of Sciences of the United States of America*, **106**, 1479–1483.

Coley, P.D. (1998) Possible effects of climate change on plant/herbivore interactions in moist tropical forests. *Climatic Change*, **39**, 455–472.

Coll, M. & Hughes, L. (2008) Effects of elevated CO_2 on an insect omnivore: a test for nutritional effects mediated by host plants and prey. *Agriculture, Ecosystems and Environment*, **123**, 271–279.

Colwell, R.K., Brehm, G., Cardelús, C.L., Gilman, A.C. & Longino, J.T. (2008) Global warming, elevational range shifts, and lowland biotic attrition in the wet tropics. *Science*, **322**, 258–261.

Cornelissen, T. (2011) Climate change and its effects on terrestrial insects and herbivory patterns. *Neotropical Entomology*, **40**, 155–163.

Crozier, L. (2004) Warmer winters drive butterfly range expansion by increasing survivorship. *Ecology*, **85**, 231–241.

Deutsch, C.A., Tewksbury, J.J., Huey, R.B., Sheldon, K.S., Ghalambor, C.K., Haak, D.C. & Martin, P.R. (2008) Impacts of climate warming on terrestrial ectotherms across latitude. *Proceedings of the National Academy of Sciences of the United States of America*, **105**, 6668–6672.

Forister, M.L. & Shapiro, A.M. (2003) Climatic trends and advancing spring flight of butterflies in lowland California. *Global Change Biology*, **9**, 1130–1135.

Forrest, J.R.K. & Thomson, J.D. (2011) An examination of synchrony between insect emergence and flowering in the Rocky Mountains. *Ecological Monographs*, **81**, 469–491.

Garibaldi, L.A., Kitzberger, T. & Chaneton, E.J. (2011) Environmental and genetic control of insect abundance and herbivory along a forest elevational gradient. *Oecologia*, **167**, 117–129.

Gilman, S.E., Urban, M.C., Tewksbury, J., Gilchrist, G.W. & Holt, R.D. (2010) A framework for community interactions under climate change. *Trends in Ecology & Evolution*, **25**, 325–331.

González-Megías, A., Menéndez, R., Roy, D., Brereton, T. & Thomas, C.D. (2008) Changes in the composition of British butterfly assemblages over two decades. *Global Change Biology*, **14**, 1464–1474.

Guisan, A. & Thuiller, W. (2005) Predicting species distribution: offering more than simple habitat models. *Ecology Letters*, **8**, 993–1009.

Hannah, L. (ed.) (2012) *Saving a Million Species: Extinction Risk from Climate Change*. Island Press, Washington DC.

Hegland, S.J., Nielsen, A., Lázaro, A., Bjerknes, A.-L. & Totland, Ø. (2009) How does climate warming affect plant-pollinator interactions? *Ecology Letters*, **12**, 184–195.

Heimonen, K., Valtonen, A., Kontunen-Soppela, S., Keski-Saari, S., Rousi, M., Oksanen, E. & Roininen, H. (2014) Colonization of a host tree by herbivorous insects under a changing climate. *Oikos*, **124**, 1013–1022.

Hellmann, J.J., Pelini, S.L., Prior, K.M. & Dzurisin, J.D.K. (2008) The response of two butterfly species to climatic variation at the edge of their range and the implications for poleward range shifts. *Oecologia*, **157**, 583–592.

Hill, M.P., Chown, S.L. & Hoffmann, A.A. (2013) A predicted niche shift corresponds with increased thermal resistance in an invasive mite, *Halotydeus destructor*. *Global Ecology and Biogeography*, **22**, 942–951.

Hodkinson, I.D. & Bird, J. (1998) Host-specific insect herbivores as sensors of climate change in arctic and Alpine environments. *Arctic and Alpine Research*, **30**, 78–83.

Hoffmann, A.A. & Sgrò, C.M. (2011) Climate change and evolutionary adaptation. *Nature*, **470**, 479–485.

Hughes, L. (2000) Biological consequences of global warming: Is the signal already apparent? *Trends in Ecology and Evolution*, **15**, 56–61.

Hughes, L. (2012) Climate change impacts on species interactions: assessing the threat of cascading extinctions. *Saving a Million Species: Extinction Risk from Climate Change* (ed. L. Hannah), pp. 337–359. Island Press, Washington DC.

IPCC (2012) Summary for Policymakers. *Managing the Risks of Extreme Events and Disasters to Advance Climate Change Adaptation* (eds C.B. Field, V. Barros, T.F. Stocker, D. Qin, D.J. Dokken, K.L. Ebi, M.D. Mastrandrea, K.J. Mach, G.-K. Plattner, S.K. Allen, M. Tignor, and P.M. Midgley). A Special Report of Working Groups I and II of the Intergovernmental Panel on Climate Change. 1–21. Cambridge University Press, Cambridge and New York.

IPCC (2013) Summary for Policymakers. In: *Climate Change 2013: The Physical Science Basis. Contribution of Working Group I to the Fifth Assessment Report of the Intergovernmental Panel on Climate Change* (eds T.F. Stocker, D. Qin, G.-K. Plattner, M. Tignor, S. K. Allen, J. Boschung, A. Nauels, Y. Xia, V. Bex and P.M. Midgley). Cambridge University Press, Cambridge and New York.

Johns, C.V., Beaumont, L.J. & Hughes, L. (2003) Effects of elevated CO_2 and temperature on development and consumption rates of *Octotoma championi* and *O. scabripennis* feeding on *Lantana camara*. *Entomologia Experimentalis et Applicata*, **108**, 169–178.

Kearney, M.R., Briscoe, N.J., Karoly, D.J., Porter, W.P., Norgate, M. & Sunnucks, P. (2010) Early emergence in a butterfly causally linked to anthropogenic warming. *Biology Letters*, **6**, 674–677.

Krehenwinkel, H. & Tautz, D. (2013) Northern range expansion of European populations of the wasp spider *Argiope bruennichi* is associated with global warming-correlated genetic admixture and population-specific temperature adaptations. *Molecular Ecology*, **22**, 2232–2248.

Lakeman-Fraser, P. & Ewers, R.M. (2013) Enemy release promotes range expansion in a host plant. *Oecologia*, **172**, 1203–1212.

Lessard, J.P., Sackett, T.E., Reynolds, W.N., Fowler, D.A. & Sanders, N.J. (2011) Determinants of the detrital arthropod community structure: the effects of temperature and resources along an environmental gradient. *Oikos*, **120**, 333–343.

Loarie, S.R., Duffy, P.B., Hamilton, H., Asner, G.P., Field, C.B. & Ackerly, D.D. (2009) The velocity of climate change. *Nature*, **462**, 1052–1055.

MacArthur, R.H. (1972) *Geographical Ecology: Patterns in the Distribution of Species*. Harper & Row, New York.

Maclean, I.M.D. & Wilson, R.J. (2011) Recent ecological responses to climate change support predictions of high extinction risk. *Proceedings of the National Academy of Sciences of the United States of America*, **108**, 12337–12342.

Memmott, J., Craze, P.G., Waser, N.M. & Price, M.V. (2007) Global warming and the disruption of plant–pollinator interactions. *Ecology Letters*, **10**, 710–717.

Merrill, R.M., Gutiérrez, D., Lewis, O.T., Gutiérrez, J., Diez, S.B. & Wilson, R.J. (2008) Combined effects of climate and biotic interactions on the elevational range of a phytophagous insect. *Journal of Animal Ecology*, **77**, 145–155.

Moir, M.L., Vesk, P.A., Brennan, K.E.C., Poulin, R., Hughes, L., Keith, D.A., McCarthy, M.A. & Coates, D.J. (2012) Considering Extinction of Dependent Species during Translocation, Ex Situ Conservation, and Assisted Migration of Threatened Hosts. *Conservation Biology*, **26**, 199–207.

Musolin, D.L. (2007) Insects in a warmer world: ecological: physiological and life-history responses of true bugs (Heteroptera) to climate change. *Global Change Biology*, **13**, 1565–1585.

Netherer, S. & Schopf, A. (2010) Potential effects of climate change on insect herbivores in European forests – general aspects and the pine processionary moth as specific example. *Forest Ecology and Management*, **259**, 831–838.

Nooten, S.S. & Andrew, N.R. (2015) Transplant Experiments: Cytoscape data. *Figshare*. https://dx.doi.org/10.6084/m9.figshare.2010225.v2.

Nooten, S.S., Andrew, N.R. & Hughes, L. (2014) Potential impacts of climate change on insect communities: A transplant experiment. *PLOS ONE*, **9**, e85987.

Nooten, S.S. & Hughes, L. (2014) Potential impacts of climate change on patterns of insect herbivory on understorey plant species: A transplant experiment. *Austral Ecology*, **39**, 668–676.

O'Lear, H.A. & Blair, J.M. (1999) Responses of soil microarthropods to changes in soil water availability in tallgrass prairie. *Biology and Fertility of Soils*, **29**, 207–217.

Pármesan, C. (2006) Ecological and evolutionary responses to recent climate change. *Annual Review of Ecology, Evolution and Systematics*, **37**, 637–669.

Parmesan, C. (2007) Influences of species, latitudes and methodologies on estimates of phenological response to global warming. *Global Change Biology*, **13**, 1860–1872.

Parmesan, C., Duarte, C., Poloczanska, E., Richardson, A.J. & Singer, M.C. (2011) Overstretching attribution. *Nature Climate Change*, **1**, 2–4.

Parmesan, C., Ryrholm, N., Stefanescu, C., Hill, J.K., Thomas, C.D., Descimon, H., Huntley, B., Kaila, L., Kullberg, J., Tammaru, T., Tennent, W.J., Thomas, J.A. & Warren, M. (1999) Poleward shifts in geographical ranges of butterfly species associated with regional warming. *Nature*, **399**, 579–583.

Parmesan, C. & Yohe, G. (2003) A globally coherent fingerprint of climate change impacts across natural systems. *Nature*, **421**, 37–42.

Pearson, R.G., Dawson, T.P., Berry, P.M. & Harrison, P.A. (2002) SPECIES: A Spatial Evaluation of Climate Impact on the Envelope of Species. *Ecological Modelling*, **154**, 289–300.

Pelini, S.L., Bowles, F.P., Ellison, A.M., Gotelli, N.J., Sanders, N.J. & Dunn, R.R. (2011) Heating up the forest: open-top chamber warming manipulation of arthropod communities at Harvard and Duke Forests. *Methods in Ecology and Evolution*, **2**, 534–540.

Pelini, S.L., Dzurisin, J.D.K., Prior, K.M., Williams, C.M., Marsico, T.D., Sinclair, B.J. & Hellmann, J.J. (2009) Translocation experiments with butterflies reveal limits to enhancement of poleward populations under climate change. *Proceedings of the National Academy of Sciences of the United States of America*, **106**, 11160–11165.

Root, R.B. (1973) Organization of a plant arthropod association in simple and diverse habitats: the fauna of collards (*Brassica oleracea*). *Ecological Monographs*, **43**, 95–124.

Root, T.L., Price, J.T., Hall, K.R., Schneider, S.H., Rosenzweig, C. & Pounds, J.A. (2003) Fingerprints of global warming on wild animals and plants. *Nature*, **421**, 57–60.

Rosenzweig, C., Karoly, D., Vicarelli, M., Neofotis, P., Wu, Q.G., Casassa, G., Menzel, A., Root, T.L., Estrella, N., Seguin, B., Tryjanowski, P., Liu, C.Z., Rawlins, S. & Imeson, A. (2008) Attributing physical and biological impacts to anthropogenic climate change. *Nature*, **453**, 353–357.

Roubicek, A.J., VanDerWal, J., Beaumont, L.J., Pitman, A.J., Wilson, P. & Hughes, L. (2010) Does the choice of climate baseline matter in ecological niche modelling? *Ecological Modelling*, **221**, 2280–2286.

Sears, M.W. & Angilletta, M.J. Jr. (2015) Costs and benefits of thermoregulation revisited: Both the heterogeneity and spatial structure of temperature drive energetic costs. *The American Naturalist*, **185**, E94–E102.

Shannon, P., Markiel, A., Ozier, O., Baliga, N.S., Wang, J.T., Ramage, D., Amin, N., Schwikowski, B. & Ideker, T. (2003) Cytoscape: A software environment for integrated models of biomolecular interaction networks. *Genome Research*, **13**, 2498–2504.

Sheldon, K.S., Yang, S. & Tewksbury, J.J. (2011) Climate change and community disassembly: impacts of warming on tropical and temperate montane community structure. *Ecology Letters*, **14**, 1191–1200.

Simberloff, D. & Dayan, T. (1991) The guild concept and the structure of ecological communities. *Annual Review of Ecology and Systematics*, **22**, 115–143.

Skou, A.-M.T., Markussen, B., Sigsgaard, L. & Kollmann, J. (2011) No evidence for enemy release during range expansion of an evergreen tree in northern Europe. *Environmental Entomology*, **40**, 1183–1191.

Sohlenius, B. & Bostrom, S. (1999) Effects of global warming on nematode diversity in a Swedish tundra soil — a soil transplantation experiment. *Nematology*, **1**, 695–709.

Strong, D.R., Lawton, J.H. & Southwood, T.R.E. (1984) *Insects on Plants. Community Patterns and Mechanisms*. Blackwell Scientific, Oxford.

Thackeray, S.J., Sparks, T.H., Frederiksen, M., Burthe, S., Bacon, P.J., Bell, J.R., Botham, M.S., Brereton, T.M., Bright, P.W., Carvalho, L., Clutton-Brock, T., Dawson, A., Edwards, M., Elliott, J.M., Harrington, R., Johns, D., Jones, I.D., Jones, J.T., Leech, D.I., Roy, D.B., Scott, W.A., Smith, M., Smithers, R.J., Winfield, I.J. & Wanless, S. (2010) Trophic level asynchrony in rates of phenological change for marine, freshwater and terrestrial environments. *Global Change Biology*, **16**, 3304–3313.

Thomas, C.D. (2010) Climate, climate change and range boundaries. *Diversity and Distributions*, **16**, 488–495.

Thomas, C.D., Cameron, A., Green, R.E., Bakkenes, M., Beaumont, L.J., Collingham, Y.C., Erasmus, B.F.N., de Siqueira, M.F., Grainger, A., Hannah, L., Hughes, L., Huntley, B., van Jaarsveld, A.S., Midgley, G.F., Miles, L., Ortega-Huerta, M.A., Peterson, A.T., Phillips, O.L. & Williams, S.E. (2004) Extinction risk from climate change. *Nature*, **427**, 145–148.

Todd, T.C., Blair, J.M. & Milliken, G.A. (1999) Effects of altered soil-water availability on a tallgrass prairie nematode community. *Applied Soil Ecology*, **13**, 45–55.

Tylianakis, J.M., Didham, R.K., Bascompte, J. & Wardle, D.A. (2008) Global change and species interactions in terrestrial ecosystems. *Ecology Letters*, **11**, 1351–1363.

Urban, M.C., Tewksbury, J.J. & Sheldon, K.S. (2012) On a collision course: competition and dispersal differences create no-analogue communities and cause extinctions during climate change. *Proceedings of the Royal Society B-Biological Sciences*, **279**, 2072–2080.

Van der Putten, W.H., Macel, M. & Visser, M.E. (2010) Predicting species distribution and abundance responses to climate change: why it is essential to include biotic interactions across trophic levels. *Philosophical Transactions of the Royal Society of London. Series B, Biological Sciences*, **365**, 2025–2034.

Villalpando, S.N., Williams, R.S. & Norby, R.J. (2009) Elevated air temperature alters an old-field insect community in a multifactor climate change experiment. *Global Change Biology*, **15**, 930–942.

Visser, M.E. (2008) Keeping up with a warming world; assessing the rate of adaptation to climate change. *Proceedings of the Royal Society B-Biological Sciences*, **275**, 649–659.

Visser, M.E. & Both, C. (2005) Shifts in phenology due to global climate change: the need for a yardstick. *Proceedings of the Royal Society B-Biological Sciences*, **272**, 2561–2569.

Voigt, W., Perner, J., Davis, A.J., Eggers, T., Schumacher, J., Bährmann, R., Fabian, B., Heinrich, W., Köhler, G., Lichter, D., Marstaller, R. & Sander, F.W. (2003) Trophic levels are differentially sensitive to climate. *Ecology*, **84**, 2444–2453.

Walther, G.-R. (2010) Community and ecosystem responses to recent climate change. *Philosophical Transactions of the Royal Society of London. Series B, Biological Sciences*, **365**, 2019–2024.

Walther, G.-R., Post, E., Convey, P., Menzel, A., Parmesan, C., Beebee, T.J.C., Fromentin, J.-M., Hoegh-Guldberg, O. & Bairlein, F. (2002) Ecological responses to recent climate change. *Nature*, **416**, 389–395.

Whitman, D.W. (1987) Thermoregulation and daily activity patterns in a black desert grasshopper, *Taeniopoda eques. Animal Behaviour*, **35**, 1814–1826.

Wilson, R.J., Gutiérrez, D., Gutiérrez, J. & Monserrat, V.J. (2007) An elevational shift in butterfly species richness and composition accompanying recent climate change. *Global Change Biology*, **13**, 1873–1887.

Wilson, R.J. & Maclean, I.M.D. (2011) Recent evidence for the climate change threat to Lepidoptera and other insects. *Journal of Insect Conservation*, **15**, 259–268.

Woods, H.A., Dillon, M.E. & Pincebourde, S. (2015) The roles of microclimatic diversity and of behavior in mediating the responses of ectotherms to climate change. *Journal of Thermal Biology*, **54**, 86–97.

Part II

Friends and Foes: Ecosystem Service Providers and Vectors of Disease

5

Insect Pollinators and Climate Change

Jessica R. K. Forrest

Department of Biology, University of Ottawa, Ottawa, Canada

Summary

Many of the world's plants depend on insects for pollination. Climate change has led to shifts in these insects' phenologies, as warming temperatures have caused bees and butterflies to emerge earlier in the spring and (in some cases) extend their flight seasons. A few recent studies have pointed to climate change as a cause of pollinator declines, but other environmental changes (habitat loss, pesticides, pathogens) are generally suspected to have more important impacts on pollinator populations and have been the subject of more research. The mechanisms by which climate change might be causing observed population declines are so far undemonstrated, but insect pollinators can be affected both directly and indirectly (i.e., via their interactions with other species) by climate change. Insect pollinators, like other ectotherms, increase their activity and development rates with temperature, up to species-specific maxima beyond which performance declines abruptly. Large-bodied pollinators, in particular, can overheat and be unable to fly at high temperatures. Other known costs of warm temperatures for pollinators include depletion of energetic reserves, reduced adult body size, and (sometimes) increased mortality rates. Warm temperatures could also affect pollinators by altering plants' abilities to produce floral resources (nectar and pollen), by disrupting temporal synchronization or spatial overlap of plants and pollinators, or by changing the frequency of interactions with natural enemies. So far, however, there are few examples of these indirect mechanisms affecting pollinator populations, and no examples involving bees. There is clearly room for further research on responses of bee populations to the changing climate, including their capacity for adaptive evolutionary change.

5.1   Introduction

Close to 90% of the world's flowering plant species and 75% of the major crops on which humans depend are reliant on animals for pollination (Klein et al., 2007; Ollerton et al., 2011). Insects make up the vast majority of these animal pollinators. Although any insect may incidentally transfer pollen among flowers, the primary pollinators are many types of flies, butterflies and moths; some wasps, beetles, and thrips; and, of course, bees (Willmer, 2011). Adults of these groups visit flowers to feed on nectar or pollen, and in the process can effect cross-pollination. Most bees, in addition to requiring nectar to fuel adult flight, also collect pollen to feed their offspring; for this reason, they are the

Global Climate Change and Terrestrial Invertebrates, First Edition. Edited by Scott N. Johnson and T. Hefin Jones.

most frequent flower-visitors in many ecosystems. A few bee species are managed (e.g., a few species of bumble bee, *Bombus* spp.; the alfalfa leafcutter bee, *Megachile rotundata*; the alkali bee, *Nomia melanderi*; mason bees in the genus *Osmia*) for the express purpose of crop pollination. The critical role of the European honey bee, *Apis mellifera*, in crop pollination is well known, but recent studies have made clear that wild pollinators can play at least as important a role in agriculture as the honey bee (Garibaldi et al., 2011; Garibaldi et al., 2013).

How is climate change affecting pollinating insects? Given the multidimensional nature of climate change and the many taxa and ecosystems affected, there can be no simple answer. In addition, because pollinating insects are almost universally holometabolous (having distinct egg, larva, pupa, and adult stages), different life stages may experience and respond to climate in very different ways (see Kingsolver et al., 2011). In this chapter, I first review the existing evidence of climate-change effects on pollinator populations. I then discuss the mechanisms that may be driving these patterns, which I divide into *direct* and *indirect* effects of climate change. *Direct* effects are those driven by the changing abiotic conditions themselves – warmer temperatures, droughts, more extreme precipitation events, and so on. I will focus on the warmer temperatures, which are the best-studied aspect of climate change and a phenomenon that applies broadly, though to varying degrees, across most of the globe. Changes in precipitation, in contrast, are difficult to forecast and region-specific (IPCC, 2014), and documented responses of pollinators to long-term changes in precipitation patterns are rare. *Indirect* effects of climate change are those driven by changing interactions with other species – food plants, predators, or parasites. Likely because pollinating insects are a guild defined by their ecological interaction with plants, the indirect threats of climate change – specifically, the threat of altered interactions with the plants they pollinate – have received a great deal of attention (see Scaven & Rafferty, 2013; Straka & Starzomski, 2014). However, the direct effects are likely to be important as well.

This chapter will focus on bees as much as the available studies allow. However, I also rely heavily on studies of butterflies and other Lepidoptera, which – although generally less important as pollinators – have been better studied with respect to climate change. While many flowers are visited by butterflies and moths, and a sizeable minority of flowering plants are seemingly adapted for butterfly or moth pollination (Willmer, 2011), these insects are non-pollinating herbivores as larvae, and their ecosystem-level impacts as herbivores may often exceed their impacts as pollinators – from a plant's perspective, they may do more harm than good! Furthermore, many lepidopterans are not flower visitors, even as adults: several moth species are non-feeding, and some butterflies feed mainly on fruit or honeydew (Krenn, 2010). A similar caveat applies to studies of flies, some of which are important pollinators and many of which are not. Bees, in contrast, are almost invariably nectar-feeders as adults and reliant on floral resources throughout their lives (although the taxon includes numerous kleptoparasitic species – brood parasites that do not collect their own pollen but instead invade nests provisioned by other bees). Unfortunately, there are comparatively few long-term datasets on bees, and so far there have been few ecophysiological studies on bees conducted in a climate-change context.

5.2 The Pattern: Pollinator Populations and Climate Change

5.2.1 Phenology

Shifts in seasonal activity patterns (phenology) in response to warming have provided some of the earliest and best-documented evidence of climate-change impacts on populations (Table 5.1). Butterflies have long attracted the attention of professional and amateur entomologists; for this reason, they

Table 5.1 Known or suggested effects of global warming on insect pollinators. See text for discussion.

Type of climate-change impact	Response variable	Demonstrated relationship between rising temperature and response variable	Change observed through time?[a]	Pollinator taxa studied	Studies
Direct	Phenology (emergence or first appearance)	usually negative (earlier)	Y	bees, butterflies	Forister & Shapiro (2003), Gordo & Sanz (2005, 2006), Sparks et al. (2010), Bartomeus et al. (2011), Ovaskainen et al. (2013)
	Phenology (flight-season duration)	positive	Y	butterflies	Roy & Sparks (2000)
	Voltinism	positive (more generations)	Y	butterflies	Roy & Sparks (2000), Altermatt (2010)
	Activity level (feeding or foraging rate)	positive and negative	N[b]	bees, butterflies	Herrera (1995), Cameron et al. (1996), Higgins et al. (2014)
	Body size	negative	N	bees, hawkmoths	Davidowitz et al. (2004), Radmacher & Strohm (2010)
	Development rate	positive and negative	N	bees, butterflies	Stephen (1965), Kingsolver et al. (2011), O'Neill et al. (2011), Radmacher & Strohm (2011)
	Energetic expenditures or mass loss	usually positive	N	bees, butterflies	Sgolastra et al. (2010), Williams et al. (2012), Fründ et al. (2013), Stuhldreher et al. (2014)
	Mortality rate	positive and negative	N	bees, butterflies, fig wasps	Whitfield & Richards (1992), Bosch & Kemp (2003, 2004), Radmacher & Strohm (2011), Sgolastra et al. (2011), Jevanandam et al. (2013), Stuhldreher et al. (2014), Bennett et al. (2015a)
Indirect	Flower production	positive and negative	N	NA	Mu et al. (2015); earlier studies reviewed by Scaven & Rafferty (2013)
	Nectar production (per flower)	positive and negative	N	NA	Mu et al. (2015), Takkis et al. (2015); earlier studies reviewed by Scaven & Rafferty (2013)
	Pollen production (per flower)	negative	N	NA	Prasad et al. (2002), Pressman et al. (2002), Koti et al. (2005)
	Temporal overlap with flowers	negative	Y	flies (Muscidae, Chironomidae)	Høye et al. (2013)
	Spatial overlap with host plants	negative (simulated)	N	butterflies	Schweiger et al. (2008, 2012)
	Activity or prevalence of natural enemies	positive and negative	N	bees, butterflies	Stone & Willmer (1989), van Nouhuys & Lei (2004), James (2005)

a) i.e., has there been a documented long-term change in the response variable?

b) Change in activity levels over time not observed, but Higgins et al. (2014) documented evolutionary change in the thermal performance curve.

have provided some of the best long-term data for assessing the effects of climate change on insect phenology. These data have shown that, since the 1970s, several butterfly species in Europe and North America are appearing earlier in the year, by as much as 16 days earlier per decade (range: 16 days earlier to 4 days later per decade), apparently in response to rising temperatures (Forister & Shapiro, 2003; Gordo & Sanz, 2005; Gordo & Sanz, 2006; see Parmesan, 2007, for review and synthesis of earlier studies). Long-term data from Japan have not shown advances in phenology, but this appears to be an artefact of declining populations of the butterfly under study rather than a true failure to shift seasonal activity in response to climate change (Ellwood et al., 2012).

Advances in phenology have also been observed in other pollinating taxa in temperate regions. Since the 1970s or 1980s, honey bees (*Apis mellifera*) in Spain and Poland (but not Austria; Scheifinger et al., 2005), have been making their first appearances progressively earlier (by up to 13.5 d/decade) in the spring (Gordo & Sanz, 2006; Sparks et al., 2010). Among the 10 spring-active North American bee species studied by Bartomeus et al. (2011), phenology had advanced by approximately 10 days since 1880, at a maximum rate of 1.4 d/decade (for *Colletes inequalis*); most of the advance (approximately 7 days) had occurred since 1970. First appearances of bumble bees (*Bombus* spp.) at a Russian locality have also been occurring earlier, at a rate of 3.1 d/decade, since the 1970s (Ovaskainen et al., 2013). Bee phenologies were strongly correlated with springtime temperatures in all of these studies, suggesting that warmer springs were causing earlier activity, earlier development, or earlier termination of diapause in these bees – which included species that overwinter as diapausing adults (*Osmia, Bombus* spp.), as non-diapausing adults (*Apis*), or as larvae (*Andrena, Colletes* spp.). In contrast, Iler et al. (2013) reported no detectable change in phenology of syrphid flies at their study site in the Rocky Mountains of Colorado, USA, between 1992 and 2011. However, this lack of trend reflected the lack of directional change in climate parameters over the studied time period rather than insensitivity of phenology to climate. In fact, syrphid phenology was strongly correlated with date of snowmelt, which is showing a long-term trend toward earlier dates in this region (Iler et al., 2013).

In addition to beginning their flight seasons earlier in spring, many pollinators have increased the duration of their activity periods. Several UK butterfly species have lengthened their flight seasons as temperatures have risen (Roy & Sparks, 2000). While this pattern has been observed even in some univoltine species (those with only a single generation per year), the most dramatic increases (in some cases by >10 d/decade) have been in those species that have been able to fit an additional generation into the lengthening growing season (Roy & Sparks, 2000). Among European butterflies and moths, the incidence of second (or subsequent) generations has increased significantly in the period since 1980 compared with preceding decades (Altermatt, 2010). Bumble bees (*Bombus* spp.) in northern Europe and North America typically have an annual colony life cycle; that is, they have a single queen-to-queen generation per year; however, wintertime bumble bee (*Bombus terrestris*) activity has recently been reported in UK cities, apparently in response to warm temperatures and flowering of cultivated plants in urban gardens (Stelzer et al., 2010). These winter-active bees may represent an attempted second generation by the normally univoltine British subspecies of *B. terrestris*. At present, it is unclear if the second generation is able to successfully produce new reproductives (males and queens) (Stelzer et al., 2010).

Long-term data on phenologies of tropical pollinators are scarce, and in habitats without distinct growing seasons, simply characterizing phenology – let alone documenting long-term changes – can be challenging (see Valtonen et al., 2013b). Furthermore, because phenology is less likely to be a function of temperature at lower latitudes (Wolda, 1988), climate warming is unlikely to produce a consistent (and therefore readily detectable) effect on pollinator phenology. As a result, we know little about how pollinators in tropical forests and deserts are shifting their seasonal activity patterns in response to climate change.

5.2.2 Range Shifts

Some pollinating insects have experienced shifts in geographic distribution toward higher latitudes and altitudes. Again, the best data are available for butterflies of the northern hemisphere, many of which have exhibited clear northward and upward range shifts (Parmesan, 1996; Parmesan et al., 1999; Hill et al., 2002; Konvicka et al., 2003; White & Kerr, 2006; Pöyry et al., 2009; Forister et al., 2010; Mair et al., 2012). These shifts imply that conditions at many species' poleward and higher-elevational range limits have improved while, at least in some cases, equatorward and lower-elevational limits have deteriorated (equatorward limits are less often investigated than poleward limits). However, some butterfly species have failed to track warming climates by expanding their ranges northward, apparently because suitable habitat for colonization is unavailable within newly climatically appropriate areas (Warren et al., 2001). Bumble bees (*Bombus* spp.) across North America and Europe have shown northward shifts in their southern range limits, consistent with climate change; however, these species have not, as a group, expanded their northward range (Kerr et al., 2015). Bumble bees are characteristically cool-climate pollinators, thought to have diversified during a period of global cooling ca. 34 Mya (Hines, 2008). They are large, hairy bees, capable of raising their body temperatures well above ambient and thus able to forage at lower temperatures than many other pollinators (Heinrich, 1979). These characteristics make bumble bees important pollinators in cooler climates (they are most diverse in mountainous and high-latitude areas of the northern hemisphere); however, these same traits make bumble bees a group likely to suffer from global warming – as the Kerr et al. (2015) study suggests.

5.2.3 Declining Populations

Recent, broad-scale declines in pollinator populations (particularly bees) have been attributed predominantly to habitat loss, pesticide use, and the spread of pathogens (Cameron et al., 2011; Szabo et al., 2012; Vanbergen et al., 2013; Goulson et al., 2015; Rundlöf et al., 2015; Senapathi et al., 2015). Climate change has not generally been considered a dominant threat to pollinators, although several authors have suggested that climate-change impacts are likely to worsen in the future and may interact with other stressors (Le Conte & Navajas, 2008; Batley & Hogendoorn, 2009; Potts et al., 2010; Schweiger et al., 2010; González-Varo et al., 2013; Vanbergen et al., 2013; Goulson et al., 2015). Regardless, population declines or extinctions in warmer parts of species' ranges are a component of some of the range shifts mentioned above. So far, climate-change-linked extirpations of pollinators have been best documented in butterflies. For example, three of four alpine-specialist butterfly species monitored at a site in California's Sierra Nevada mountains have become less frequent since the 1970s, perhaps because of changing climate (Forister et al., 2010). A pair of local extinctions (of the Bay checkerspot, *Euphydryas editha bayensis*) at a site near the California coast was convincingly attributed to increasing variability in precipitation, an expected consequence of global climate change, occurring on a background of regional habitat loss (McLaughlin et al., 2002).

Among bees, there is little evidence so far that local population declines and extinctions are the result of climate warming. This may simply reflect a lack of investigation, as researchers' efforts have been primarily directed at other (and perhaps more important) threats to these insects. Bee species losses from a site in Illinois, USA, may owe something to climate change and disruptions of plant and pollinator phenologies (Burkle et al., 2013), but other local habitat changes cannot be ruled out as the primary cause. However, the widespread retraction of southern range limits among northern-hemisphere bumble bees, noted above, suggests that local declines are occurring in warmer parts of these species' ranges (Kerr et al., 2015).

Declines of some fly populations have also been attributed to climate change. In Greenland, where summer temperatures have risen by approximately 2.5°C since the 1990s, there has been a concomitant >60% decline in abundance of muscid and chironomid Diptera, many of which are common flower-visitors in the Arctic (Høye et al., 2013). The Diptera decline has occurred over the same period as a ~5 d shortening of the already short flowering season (~26 d for the plant species observed), and fly abundance in a given year is correlated with the extent of temporal overlap between fly and flower phenology in the previous year. This correlation suggests a causal relationship – flies may be declining because of declines in their adult food source (nectar) or declines in their favoured basking sites (flowers) – but other interpretations are possible, since some Diptera are non-feeding as adults, and their larval ecologies are diverse. Regardless, it is likely that some correlate of climate change (perhaps affecting larval habitat) is responsible.

5.3   The Process: Direct Effects of Climate Change

5.3.1   Warmer Growing-Season Temperatures

Because of the temperature-dependence of metabolic rates (Gillooly et al., 2001), activity levels and development rates of insects increase with increasing temperature, above some lower threshold temperature, up to a species- or population-specific optimum. This speeding-up of metabolic processes explains most of the phenological advancements observed as climates have warmed over the last few decades (see section 5.2.1). Warmer temperatures would allow faster development and emergence by pollinators that have overwintered in situ, provided diapause has been terminated (Stephen, 1965; White et al., 2009; Forrest & Thomson, 2011; Valtonen et al., 2013a), as well as faster migration to summer habitat by migratory species (some butterflies).

Faster development at higher temperatures is a likely cause of increased voltinism. Longer growing seasons can make it *possible* for some insect populations to achieve an additional generation at a given locality, but insects typically must make a "decision" about whether to attempt an additional generation, or whether instead to enter winter diapause, well before the end of the season (Tauber et al., 1986). If an insect fails to receive a diapause cue, such as a particular ("critical") photoperiod, by a given developmental stage, it may develop directly to the next generation rather than enter diapause (e.g., Stephen, 1965; Tobin et al., 2008). If the insect does receive the diapause signal, the subsequent generation will not appear until the following growing season. The precise cues controlling diapause induction, and thus voltinism, are unknown for most pollinators, but it is clear that, as the climate warms, many pollinators are now experiencing those cues at a different point in their life cycle. Importantly, the early-season conditions that allow direct development must be accompanied by a prolongation of the growing season and a continued availability of food plants in order for the second generation to successfully reproduce. Longer growing seasons are occurring throughout much of the northern hemisphere (Linderholm, 2006), but there can be local exceptions. For example, Van Dyck et al. (2015) have suggested that local declines in the European butterfly *Lasiommata megera* are the result of an attempted (failed) third generation in regions where late-summer conditions have become warm enough to prevent diapause – but where season length is still insufficient to support a complete third generation.

Many pollinators are unlikely to benefit from an increase in summer temperatures and growing season length by shortening their generation times. Pollen-specialist (oligolectic) bees, or other pollinators with specialized larval diets, would not normally encounter suitable food plants if they attempted a second generation outside their usual activity season – that is, unless the host plants themselves had greatly lengthened their flowering or leaf-out period. Thus, increases in voltinism seem most likely for pollinators with generalist larval diets – although evolutionary transitions in diet might occur in conjunction with altered voltinism. An exception might be the high-altitude bees that frequently require two years to complete development to adulthood (Torchio & Tepedino, 1982; Forrest & Thomson, 2011). These species might more often achieve a generation in a single year if temperatures during development were warmer; however, the exact cues triggering a one-year life cycle in these bees are unknown.

The overall temperature-dependence of fitness-related rates can be described using thermal performance curves, obtained under laboratory conditions where temperatures can be held constant (see Fig. 5.1). "Performance" can be any fitness-related measure, such as activity level, development rate, or reproductive output. If the performance measure is the population's intrinsic rate of increase, thermal performance curves can be used to forecast population growth rates under a given temperature regime (e.g., Deutsch et al., 2008). However, such forecasts assume that features of the environment other than temperature remain constant, that organisms in fact experience the prescribed temperature regime (rather than avoiding temperature change by behavioural thermoregulation, for example), and that there is no adaptation or acclimation of the thermal performance profile. I will revisit these assumptions below.

There seem to be no whole-life-cycle thermal performance curves for pollinators. Nevertheless, thermal performance curves do exist for components of fitness in a few butterflies (*Colias* spp.), the hawkmoth *Manduca sexta* (Kingsolver et al., 2011; Higgins et al., 2014), and some managed bee species such as the alkali bee *Nomia melanderi* and the alfalfa leafcutter bee *Megachile rotundata* (Stephen, 1965; O'Neill et al., 2011; Fig. 5.1). These show the typical asymmetric hump-shaped relationship with temperature, with performance (here, measured as feeding rate or rate of development

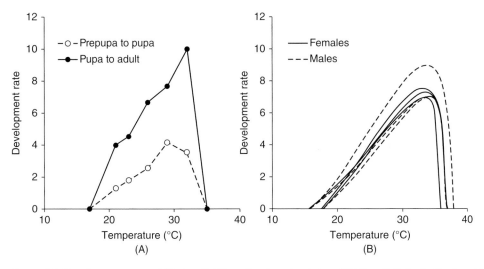

Figure 5.1 Development rates of (A) the alkali bee, *Nomia melanderi*, and (B) the alfalfa leafcutter bee, *Megachile rotundata*, as a function of rearing temperature. Values are 100/t, where t is the mean development time in days. In (A), development times are from the end of overwintering (as a prepupa) until pupation (dashed line, open circles), or from the start of the pupal stage until adult emergence from the pupal skin (solid line, filled circles). In (B), each curve was fitted to data on development from overwintering (as a prepupa) until adult emergence for males (dashed lines) and females (solid lines) in each of three study years. Data in (A) from Stephen (1965); in (B) from O'Neill et al. (2011).

to adulthood) typically peaking at temperatures in the realm of what the insects would experience in nature. Field studies, which investigate responses to naturally occurring temperature variation, are less likely than laboratory experiments to capture the upper, descending part of the thermal performance relationship. For example, in a study of two butterfly species near their northern range limit, Hellmann et al. (2008) showed increasing larval biomass production with increasing temperature throughout the range of naturally experienced temperatures.

Several studies have investigated temperature effects on aspects of bee performance and development, either using naturally observed variation in the field (typically investigating effects of temperature on activity levels) or using a smaller number of manipulated temperatures in the laboratory (typically investigating effects of temperature on survival and reproduction). For example, Cameron et al. (1996) found that activity levels of female *Melissodes rustica*, a ground-nesting bee, increased with increasing soil surface temperatures under field conditions (temperature mean $\pm$ s.e. = 31.7 $\pm$ 1.4°C). Similarly, Herrera (1995) observed that as ambient temperature increased from 14° to 24°C, *Andrena bicolor* bees increased their percentage of time spent on flowers, and thus potentially pollinating, from ~25% to ~90%. In general, the high thoracic temperatures required for insect flight mean that ambient temperatures must be high, or that a pollinator be able to behaviourally thermoregulate (e.g., by basking), or, for those bees (e.g., *Anthophora*, *Bombus*, *Osmia* spp.) and hawkmoths (Sphingidae) capable of facultative endothermy, that energetic stores be sufficient to allow shivering thermogenesis (Willmer & Stone, 2004). Warmer ambient temperatures reduce the amount of energy required to raise body temperature to the level necessary for flight. During development, warm temperatures are also beneficial – up to a point. Cold temperatures during development can not only slow a bee's progress to adulthood but also lower survival, harm subsequent adult performance, or limit reproductive output (Vogt, 1986; Whitfield & Richards, 1992; Becher et al., 2009; Bennett et al., 2015a).

Like all organisms, pollinators have upper critical temperatures above which they cannot function. These upper limits are often in the range of 45–50°C (Undurraga & Stephen, 1980; Kingsolver & Watt, 1983; Willmer & Stone, 2004). Even if air temperatures do not reach these levels, radiant heat from the sun or hot earth, combined with the metabolic heat generated by flight, can threaten to put pollinators' internal temperatures into this lethal zone (e.g., Chappell, 1982; 1984). Thus, some pollinators – particularly larger species, which are generally less able to dissipate heat – tend to avoid activity during the warmest parts of the day (Willmer, 1983; Muniz et al., 2013; but see Chappell, 1982). These prohibitively hot periods are likely to lengthen as the Earth warms and restrict foraging opportunities for some pollinators. By extrapolating from current diel activity patterns of various pollinators, Rader et al. (2013) have projected that watermelon crops will receive fewer visits from honey bees and squash bees (*Peponapis pruinosa*) as mean temperatures rise, because, in their northeastern U.S. study area, most visits by these pollinators to watermelon flowers occurred below air temperatures of 30°C. On the other hand, their data predict no loss of pollination services (and even some gains) by several other bee taxa, including bumble bees (*Bombus impatiens*). Some pollinators will be able to escape the heat by shifting their activity earlier or later in the day, but there are limits. While some pollinators are adapted for foraging at night (e.g., many hawkmoths [Sphingidae]) or in dim light (e.g., several lineages of bees, particularly from deserts and the tropics; Wcislo & Tierney, 2009), most bees and other pollinating insects are strictly diurnal. Extreme heat experienced at the larval stage – when pollinators have limited capacity to escape temperature extremes – can also be lethal, although heat sensitivity again varies by species (e.g., Barthell et al., 2002).

Even if we leave aside extreme heat stress, high temperatures can increase mortality rates in adult pollinators, just as they increase other biological rates. For example, fig wasp (Agaonidae) longevity was reduced with increasing temperature in a laboratory study (Jevanandam et al., 2013), although the implications for lifetime fitness and pollination in the field are unclear. In mason bees (*Osmia* spp.), warm temperatures during larval development or during adult dormancy (in the season before adult emergence) have been associated with adult mortality, possibly owing to consumption of energetic reserves (Radmacher & Strohm, 2011; Sgolastra et al., 2011).

Most insects reach smaller adult sizes when reared at warmer temperatures (Kingsolver & Huey, 2008). The pattern holds for those pollinator species that have been studied (Davidowitz et al., 2004; Radmacher & Strohm, 2010). Red mason bees (*Osmia bicornis*) reared at 30°C, for example, weighed less than half as much at the cocoon stage as conspecifics reared at 20°C (Radmacher & Strohm, 2010). This plastic response to warmer developmental temperatures may therefore reinforce the expected community-level shift in favour of smaller species at warmer temperatures (because of the superior heat-dissipation abilities of smaller bees, noted above). Scaven and Rafferty (2013) have discussed the possible consequences for pollination of a community of smaller pollinators, suggesting that foraging distances, and hence pollen-transport distances, might be reduced. However, pollinating insects as a group span orders of magnitude in size; even within a species, body size can vary substantially (between sexes and castes, and in response to larval food supply). In this context, the mass changes caused by within-species responses to temperature may be ecologically insignificant.

Climate change will mean that air temperatures increase on average. But organisms do not experience average air temperatures; rather, they experience highly local, instantaneously changing temperatures, which vary seasonally and which can be modulated by solar radiation, convective heat loss by wind, and (in some species) endothermy or active heat dissipation (see Willmer & Stone, 2004). In addition, by taking advantage of warm or cool microhabitats, pollinators can exert considerable control over their internal temperature. Advances in phenology in response to warm temperatures during development should also moderate the temperatures experienced during later developmental stages (e.g., see Bennett et al., 2015b). Pollinator eggs, and the larvae of many pollinators (e.g., bees), cannot actively disperse to avoid heat, but mothers have options in where they construct nests (bees) or oviposit (moths, butterflies, flies). For example, ground-nesting bees can construct nests in deeper or shallower soil (Cane & Neff, 2011), cavity-nesting bees can nest in sun or shade (e.g., Hranitz et al., 2009), and moths and butterflies can oviposit on shaded or sunlit leaf surfaces (e.g., Bonebrake et al., 2010). Variation in height of oviposition allows eggs of *Euphydryas editha* to experience temperatures from >5°C cooler to >20°C warmer than ambient (Bennett et al., 2015b). There are of course limits to how well behaviour can mitigate the effects of global warming, and many behaviours may be energetically costly or entail other risks. Nevertheless, they add to the complexity of forecasting climate-change impacts on pollinators.

5.3.2 Warmer Winters and Reductions in Snowpack

Winter warming has been a conspicuous aspect of climate change in many parts of the world, but its ecological consequences are relatively unstudied (Williams et al., 2014). For some pollinators, termination of winter diapause requires – or is hastened by – a period of cold temperatures. Shorter chilling periods can delay subsequent emergence from dormancy (Johansen & Eves, 1973; Bosch & Kemp, 2003; Bosch & Kemp, 2004; Forrest & Thomson, 2011). Thus, early onset of spring temperatures can partially counteract the phenological effects of warmer springs and summers, resulting in less of an advance in pollinator phenology than one would expect based on when winter ends. However, *warmer* winters (as opposed to *shorter* ones) can hasten springtime emergence, at least in some bees and butterflies (Bosch & Kemp, 2004; Williams et al., 2012; Fründ et al., 2013;

Stuhldreher et al., 2014). The effects of winter warming on pollinator phenology are therefore likely to depend on the details of when the warming occurs, as well as the species involved.

Warmer temperatures in winter can also affect pollinator survival or body condition on emergence. All else being equal, shorter and warmer winters are likely to benefit honey bees (*Apis mellifera*), which actively thermoregulate through the winter (instead of entering dormancy as most other insects do) and must consume large quantities of honey to maintain colony warmth (Seeley & Visscher, 1985). Shorter winters also benefit diapausing queens of the bumble bee *Bombus terrestris* (Beekman et al., 1998). In contrast, very short, or very long but warm winters, reduce survival in *Osmia* bees (Bosch & Kemp, 2003; Bosch & Kemp, 2004). The latter response is likely due to depletion of energy reserves or desiccation, both effects that have been observed in the skipper *Erynnis propertius* overwintered under warmer-than-normal conditions (Williams et al., 2012). Greater mass loss at warmer overwintering temperatures has also been observed in several megachilid bees and the butterfly *Erebia medusa*, in which larval mortality also increased with temperature (Sgolastra et al., 2010; Fründ et al., 2013; Stuhldreher et al., 2014). Although the adult fitness consequences of these mass losses were not measured, body mass is strongly associated with fecundity in ectotherms (Roff, 1992), so we expect negative effects of overwinter mass loss. However, not all pollinators are harmed by warm winters. Williams et al. (2012) and Fründ et al. (2013) have suggested that species with deeper winter diapause (i.e., a more depressed metabolic rate during winter dormancy) are less affected by warm temperatures. Thus, pollinators that overwinter as relatively active stages (e.g., adults, as in *Osmia* and *Bombus* spp.) may be more susceptible to mass loss than those that overwinter as pupae or eggs. Oddly, Fründ et al. (2013) observed that two bee species (of the nine they studied) lost less mass at higher-than-normal winter temperatures, suggesting that some pollinators could benefit from warmer winters.

Warmer winters in normally snowy regions can mean reduced winter and springtime snowpack (although this depends on the net effects of rising temperatures and altered precipitation regimes; e.g., Mote et al., 2005). Loss of insulating snow cover can expose overwintering insects to potentially harmful temperature variation, including – ironically – extreme cold (Pauli et al., 2013). At present, it is unknown if loss of snow poses a threat to pollinators. Bees that overwinter aboveground in cold climates, such as some cavity-nesting Megachilidae, can withstand temperatures well below −20°C during diapause (Krunić & Stanisavljević, 2006). Cold-hardiness is gradually lost in the spring, potentially leaving bees vulnerable to late-winter temperature swings; but even post-diapausing bees can remain remarkably tolerant of cold temperatures (e.g., Sheffield, 2008). These findings suggest that aboveground-nesting bees are quite well adapted to exposure to cold-temperature extremes. We might expect pollinators that overwinter at ground level in normally snowy parts of the world to be more sensitive to loss of snowpack, and some correlative data support this idea: population growth rate in the subalpine butterfly *Speyeria mormonia* is positively associated with duration of spring snow cover, perhaps because snow protects overwintering larvae from early-spring frosts (Boggs & Inouye, 2012). Similarly, population growth rate of the alpine butterfly *Parnassius smintheus* is negatively associated with "extreme" winter weather (both warm and cold), the frequency of which has increased over the last century (Roland & Matter, 2013). One suggested mechanism for this association is mortality of overwintering eggs, which are normally snow-covered and thereby protected from freezing and desiccation, but which may more often be exposed to harmful conditions in extreme winters. Unfortunately, I am unaware of any data on the cold-tolerance of ground-nesting bees, which we might also expect to be harmed by loss of snow cover.

5.4   The Process: Indirect Effects of Climate Change

5.4.1   Interactions with Food Plants

Flowering times of many north-temperate plant populations have shifted conspicuously earlier over the last several decades (e.g., Fitter & Fitter, 2002; Calinger et al., 2013; Ovaskainen et al., 2013). This widespread trend has led to concerns that phenologies of pollinators might not shift in parallel with the phenologies of the plants on which they depend, leading to temporal mismatch between flowering and activity periods of pollinators (Memmott et al., 2007; Hegland et al., 2009). Mismatches of this kind might result in pollinators being active at times of the season when their food plants are scarce, and consequently starving or failing to reproduce.

So far, there is little evidence that mismatch of this type has occurred – although it must be noted that research effort is heavily biased toward north-temperate and Arctic ecosystems. Pollinator phenologies have largely shifted earlier in parallel with flowering phenology (Bartomeus et al., 2011; Ovaskainen et al., 2013), despite some differences between plant and pollinator phenology in their responses to temperature variation (Forrest & Thomson, 2011; Kharouba & Vellend, 2015). Some plants do seem to be at risk of growing temporal mismatch with their pollinators – particularly plants that are pollinated by only a small number of flower-visiting species (Thomson, 2010; Kudo & Ida, 2013; Robbirt et al., 2014). While the converse – pollinators suffering fitness consequences of mismatch with flowering of their food plants – seems likely, it has not yet been conclusively demonstrated, perhaps simply because studies of pollinator fitness in the field are challenging (Forrest, 2015). The Greenland study of Høye et al. (2013) and the Illinois study of Burkle et al. (2013) show long-term evidence consistent with an effect of temporal mismatch with flowers on pollinator populations. As noted above, however, the mechanisms behind the declining pollinator populations are uncertain in both studies. In contrast, the local extinctions of Bay checkerspots (*Euphydryas editha bayensis*) studied by McLaughlin et al. (2002) were convincingly linked to temporal mismatch between the butterfly and the food plant – but in that case, the larval food plant, not the nectar resource.

Whether or not phenology of pollinators is becoming less synchronized with that of their food plants, pollinators might increasingly experience resource shortages if their food plants are negatively affected by climate change – becoming scarce, or producing less nectar – or if shifts in the geographic ranges of food plants do not mirror those of the pollinators. The potential for some of these indirect effects has been reviewed by Scaven and Rafferty (2013), who suggested that changes in floral abundance and nectar production (both increases and decreases with warmer temperatures have been observed; Table 5.1) might affect pollinator energetics, and, ultimately, fitness. Indeed, Boggs and Inouye (2012) observed higher population growth in the fritillary *Speyeria mormonia* in years when its preferred nectar plant, *Erigeron speciosus*, was most available. These authors concluded that *S. mormonia* benefited from late snowmelt in part because *E. speciosus* flower buds are frost-sensitive, and because bud-killing frosts occur more frequently in years of early snowmelt (Boggs & Inouye, 2012). The trend toward earlier snowmelt at their study site thus may pose a threat to the butterfly.

In several crop plants, including some that are pollinated by insects, pollen production is strongly depressed by high temperatures (Table 5.1). For example, Pressman et al. (2002) observed a 47% reduction in pollen per flower in tomato plants subjected to a high-temperature regime (32/28°C versus 26/22°C). If plants do not adapt to warmer temperatures, the consequent reductions in pollen production could harm bee populations. In reality, there is no indication that pollen availability is declining in field settings, or that it is likely to decline at the scale of whole plant communities (which are shifting in phenology and in species composition at the same time) – the scale that would matter for pollinators. However, it must be noted that pollen production of animal-pollinated plants is not often quantified, and any long-term declines might easily be missed.

As the planet warms, there is a risk that the climatic niches of some pollinators will no longer coincide geographically with those of their food plants. Projections of the future geographic distributions of several European butterflies, based on ecological niche modelling, suggest that most species will continue to have their range limits determined by climatic constraints and land use rather than by availability of larval host plants (Schweiger et al., 2012). There are exceptions, however – particularly for butterflies specializing on plants with small geographic distributions. Food shortages are predicted for *Boloria titania*, for example, because the range of its sole larval host plant (*Polygonum bistorta*) is expected to contract under future climate scenarios and may overlap only a small portion of the projected future range of the butterfly (Schweiger et al., 2008). The gravity of the future spatial mismatch depends on the extent to which the two taxa are able to disperse to newly available climatic niche space. It is so far unknown if any bee pollinators are at risk of spatial mismatch with floral host plants, since the necessary climate modelling has not been conducted; furthermore, the pollen-host relationships of many bees are undocumented.

5.4.2 Interactions with Natural Enemies

Numerous pathogenic microbes, parasitic mites, nest parasites, parasitoids, and predators attack pollinating insects, and some of these have become infamous for their suspected role in bee declines – particularly declines of honey bees and bumble bees (Cameron et al., 2011; Goulson et al., 2015). Should we expect natural enemies to have stronger impacts on pollinator populations as a result of climate change? Le Conte and Navajas (2008) have discussed the possible ways in which climate change might exacerbate honey bee disease and parasite load, including infection by the brood-parasitic mite *Tropilaelaps*. However, as with pollinators themselves, generalizations about how their parasites will fare under climate change are difficult. For example, chalkbrood (*Ascosphaera* spp.), a parasitic fungus that attacks many bee species, can be limited by both cold and hot temperatures, and different phases of its life cycle have different temperature sensitivities (James, 2005). To the extent that parasites have been selected to use the same thermal niche as their hosts, we might predict that parasites and their hosts should respond to climate warming in a similar way. However, this is not always the case.

Brood parasites (kleptoparasites), which take over bee brood chambers and typically kill the host bee's offspring, can be important sources of mortality in bee populations. For example, Torchio (1979) reported mortality rates as high as 74% in a managed population of leafcutter bees (*Megachile rotundata*) owing to kleptoparasite attack. The kleptoparasites that have been studied appear to benefit more from warmer temperatures than their hosts do. Stone and Willmer (1989) noted that the four kleptoparasitic bee species in their study had lower warm-up rates for their size than the non-parasitic species they investigated, and others have reported kleptoparasites foraging only during the warmest parts of the day – while their hosts were less selective (Straka & Bogusch, 2007; Rozen et al., 2009). It is unclear why kleptoparasites should prefer, or require, warmer temperatures for activity, but one could infer from these observations that kleptoparasites will benefit more than their hosts from warmer daytime temperatures. However, we know little about the upper thermal limits for kleptoparasite activity. Kleptoparasites may tolerate higher temperatures overall than their hosts, or they may simply have a narrower thermal window for activity.

Phenological synchronization between hosts and parasites could also be affected by warming. van Nouhuys and Lei (2004) studied a metapopulation of the fritillary *Melitaea cinxia* attacked by a braconid wasp parasitoid, *Cotesia melitaearum*, and observed that the butterfly caterpillars were better able to escape parasitism when spring weather was cool but sunny. This occurred because the dark-coloured caterpillars were able to raise their body temperatures (and develop rapidly to the "safe" pupal stage) using behavioural thermoregulation, while the white wasp pupae were not. Parasitoids were therefore more likely to colonize new host populations following warm springs, when their adult stage was better synchronized with their larval hosts (van Nouhuys & Lei, 2004). Although there was no evidence in this case that parasitoids regulated host butterfly populations, one can imagine that such top-down control could materialize if a string of warm springs allowed parasitoid populations to build up. However, in another, similar fritillary–braconid system, Klapwijk et al. (2010) were unable to relate population dynamics of the host or the parasitoid to interannual climate variation, despite differential effects of warmth and shade on development rates of the butterfly and the wasp. Overall, there seems to be no reason to expect natural enemies of pollinators to become better synchronized with their hosts in general, although phenologies may converge in some systems.

5.5   Synthesis, and the View Ahead

When considering the impacts of climate change (or other environmental changes) on pollinator populations, it can be useful to think about which factors currently limit population sizes (see Roulston & Goodell, 2011), and whether these factors will become more or less limiting as the Earth warms. All else being equal, ectotherms whose population growth rates are currently limited by cold temperatures should benefit from warmer temperatures. This is the scenario one could forecast for many insects of temperate, polar, and alpine habitats, based solely on thermal performance curves and changing mean temperatures (Deutsch et al., 2008). Some pollinator populations may indeed conform to this optimistic expectation.

However, many pollinators are likely limited by factors other than cool temperatures – or will become progressively more limited by other factors as temperatures rise. This includes bumble bees, charismatic members of the cool-temperate pollinator fauna that function well at low ambient temperatures and whose populations are likely limited in much of their range by natural enemies (see Cameron et al., 2011), availability of food plants and nesting sites (Goulson et al., 2005; Bommarco et al., 2012; Dupont et al., 2011), and, perhaps, pesticides (Rundlöf et al., 2015). Spatial patterns of bumble bee decline and persistence in the UK suggest that climatic suitability and food availability jointly limit some bumble bee populations (Williams et al., 2007). The failure of several British butterfly species to take advantage of warmer temperatures via range expansion suggests that these species, too, are more limited by habitat requirements than by cool temperatures (Warren et al., 2001). Pollinators in warm climates, where high temperatures already constrain foraging opportunities, are of course unlikely to benefit from further warming. Furthermore, because thermal performance curves are non-linear and typically asymmetric (e.g., Fig. 5.1), increases in temperature variability could produce worse-than-expected effects on insect fitness (Vasseur et al., 2014).

Persistence of pollinator populations under climate change, as with other organisms, will depend heavily on their capacity for adaptation. While the enormous existing variety in pollinator life histories, nesting or oviposition strategies, and diets is evidence of past evolution in response to changing conditions, it is not certain that pollinator populations already shrunken by other threats will keep pace with current rates of change. One optimistic note is provided by the well-studied *Colias* butterflies, which have shown rapid evolution of thermal tolerances in response to recent warming (Higgins et al., 2014). However, some pollinators are likely to face constraints on, and limits to, adaptation. The skipper *Erynnis propertius* appears to be constrained in its ability to shift its range in response to warming, because locally adapted populations lack suitable host plants (oaks; *Quercus* spp.) beyond their current range limits (Hellmann et al., 2008; Pelini et al., 2010). In contrast, populations of the brown argus butterfly (*Aricia agestis*) at the expanding (northward) range margin in the UK have shifted to a new larval host plant, but this has apparently occurred at the expense of phenotypic variation – which may limit the potential for ongoing range and host shifts (Buckley & Bridle, 2014).

In summary, climate change may not be the primary global threat to pollinator populations at present. Many pollinators in temperate regions could even benefit from moderate warming, via increased development rates and increased opportunities for foraging and oviposition. However, these benefits may be counteracted by increased mortality rates and changes in interactions with other species. Temperature extremes – hot and cold – also pose a threat, both through their direct effects on pollinators and their indirect effects on food plants; and these extremes are likely to become more frequent as climate continues to change. There is a clear lack of research on tropical pollinators' responses to climate change, and this is one area where further work would be useful. In addition, some of the most interesting and hopeful research so far has addressed pollinator "coping mechanisms" – whether evolutionary change (Higgins et al., 2014) or plastic behavioural variation (Bennett et al., 2015b). Although butterflies have proven to be excellent model systems for this work, more research is still needed on whether and how the world's pre-eminent pollinators – the bees – are adapting to the changing climate.

Acknowledgements

I thank three anonymous reviewers for their helpful comments on the manuscript and Philip Smith for proofreading this chapter.

References

Altermatt, F. (2010) Climatic warming increases voltinism in European butterflies and moths. *Proceedings of the Royal Society B*, **277**, 1281–1287.

Barthell, J.F., Hranitz, J.M., Thorp, R.W. & Shue, M.K. (2002) High temperature responses in two exotic leafcutting bee species: *Megachile apicalis* and *M. rotundata* (Hymenoptera: Megachilidae). *Pan-Pacific Entomologist*, **78**, 235–246.

Bartomeus, I., Ascher, J.S., Wagner, D., Danforth, B.N., Colla, S., Kornbluth, S. & Winfree, R. (2011) Climate-associated phenological advances in bee pollinators and bee-pollinated plants. *Proceedings of the National Academy of Sciences of the United States of America*, **108**, 20654–20659.

Batley, M. & Hogendoorn, K. (2009) Diversity and conservation status of native Australian bees. *Apidologie*, **40**, 347–354.

Becher, M.A., Scharpenberg, H. & Moritz, R.F.A. (2009) Pupal developmental temperature and behavioral specialization of honeybee workers (*Apis mellifera* L.). *Journal of Comparative Physiology A*, **195**, 673–679.

Beekman, M., van Stratum, P. & Lingeman, R. (1998) Diapause survival and post-diapause performance in bumblebee queens (*Bombus terrestris*). *Entomologia Experimentalis et Applicata*, **89**, 207–214.

Bennett, M.M., Cook, K.M., Rinehart, J.P., Yocum, G.D., Kemp, W.P. & Greenlee, K.J. (2015a) Exposure to suboptimal temperatures during metamorphosis reveals a critical developmental window in the solitary bee, *Megachile rotundata*. *Physiological and Biochemical Zoology: Ecological and Evolutionary Approaches*, **88**, 508–520.

Bennett, N.L., Severns, P.M., Parmesan, C. & Singer, M.C. (2015b) Geographic mosaics of phenology, host preference, adult size and microhabitat choice predict butterfly resilience to climate warming. *Oikos*, **124**, 41–53.

Boggs, C.L. & Inouye, D.W. (2012) A single climate driver has direct and indirect effects on insect population dynamics. *Ecology Letters*, **15**, 502–508.

Bommarco, R., Lundin, O., Smith, H.G. & Rundlöf, M. (2012) Drastic historic shifts in bumble-bee community composition in Sweden. *Proceedings of the Royal Society B*, **279**, 309–315.

Bonebrake, T.C., Boggs, C.L., McNally, J.M., Ranganathan, J. & Ehrlich, P.R. (2010) Oviposition behavior and offspring performance in herbivorous insects: consequences of climatic and habitat heterogeneity. *Oikos*, **119**, 927–934.

Bosch, J. & Kemp, W.P. (2003) Effect of wintering duration and temperature on survival and emergence time in males of the orchard pollinator *Osmia lignaria* (Hymenoptera: Megachilidae). *Environmental Entomology*, **32**, 711–716.

Bosch, J. & Kemp, W.P. (2004) Effect of pre-wintering and wintering temperature regimes on weight loss, survival, and emergence time in the mason bee *Osmia cornuta* (Hymenoptera: Megachilidae). *Apidologie*, **35**, 469–479.

Buckley, J. & Bridle, J.R. (2014) Loss of adaptive variation during evolutionary responses to climate change. *Ecology Letters*, **17**, 1316–1325.

Burkle, L.A., Marlin, J.C. & Knight, T.M. (2013) Plant-pollinator interactions over 120 years: loss of species, co-occurrence and function. *Science*, **339**, 1611–1615.

Calinger, K.M., Queenborough, S. & Curtis, P.S. (2013) Herbarium specimens reveal the footprint of climate change on flowering trends across north-central North America. *Ecology Letters*, **16**, 1037–1044.

Cameron, S.A., Lozier, J.D., Strange, J.P., Koch, J.B., Cordes, N., Solter, L.F. & Griswold, T.L. (2011) Patterns of widespread decline in North American bumble bees. *Proceedings of the National Academy of Sciences of the United States of America*, **108**, 662–667.

Cameron, S.A., Whitfield, J.B., Hulslander, S.L., Cresko, W.A., Isenberg, S.B. & King, R.W. (1996) Nesting biology and foraging patterns of the solitary bee *Melissodes rustica* (Hymenoptera: Apidae) in northwest Arkansas. *Journal of the Kansas Entomological Society*, **69**, S260–S273.

Cane, J.H. & Neff, J.L. (2011) Predicted fates of ground-nesting bees in soil heated by wildfire: Thermal tolerances of life stages and a survey of nesting depths. *Biological Conservation*, **144**, 2631–2636.

Chappell, M.A. (1982) Temperature regulation of carpenter bees (*Xylocopa californica*) foraging in the Colorado Desert of southern California. *Physiological Zoology*, **55**, 267–280.

Chappell, M.A. (1984) Temperature regulation and energetics of the solitary bee *Centris pallida* during foraging and intermale mate competition. *Physiological Zoology*, **57**, 215–225.

Davidowitz, G., D'Amico, L.J. & Nijhout, H.F. (2004) The effects of environmental variation on a mechanism that controls insect body size. *Evolutionary Ecology Research*, **6**, 49–62.

Deutsch, C.A., Tewksbury, J.J., Huey, R.B., Sheldon, K.S., Ghalambor, C.K., Haak, D.C. & Martin, P.R. (2008) Impacts of climate warming on terrestrial ectotherms across latitude. *Proceedings of the National Academy of Sciences of the United States of America*, **105**, 6668–6672.

Dupont, Y.L., Damgaard, C. & Simonsen, V. (2011) Quantitative historical change in bumblebee (*Bombus* spp.) assemblages of red clover fields. *PLOS ONE*, **6**, e25172.

Ellwood, E.R., Diez, J.M., Ibáñez, I., Primack, R.B., Kobori, H., Higuchi, H. & Silander, J.A. (2012) Disentangling the paradox of insect phenology: are temporal trends reflecting the response to warming? *Oecologia*, **168**, 1161–1171.

Fitter, A.H. & Fitter, R.S.R. (2002) Rapid changes in flowering time in British plants. *Science*, **296**, 1689–1691.

Forister, M.L., McCall, A.C., Sanders, N.J., Fordyce, J.A., Thorne, J.H., O'Brien, J., Waetjen, D.P. & Shapiro, A.M. (2010) Compounded effects of climate change and habitat alteration shift patterns of butterfly diversity. *Proceedings of the National Academy of Sciences of the United States of America*, **107**, 2088–2092.

Forister, M.L. & Shapiro, A.M. (2003) Climatic trends and advancing spring flight of butterflies in lowland California. *Global Change Biology*, **9**, 1130–1135.

Forrest, J.R.K. (2015) Plant–pollinator interactions and phenological change: what can we learn about climate impacts from experiments and observations? *Oikos*, **124**, 4–13.

Forrest, J.R.K. & Thomson, J.D. (2011) An examination of synchrony between insect emergence and flowering in Rocky Mountain meadows. *Ecological Monographs*, **81**, 469–491.

Fründ, J., Zieger, S.L. & Tscharntke, T. (2013) Response diversity of wild bees to overwintering temperatures. *Oecologia*, **173**, 1639–1648.

Garibaldi, L.A., Steffan-Dewenter, I., Kremen, C., Morales, J.M., Bommarco, R., Cunningham, S.A., Carvalheiro, L.G., Chacoff, N.P., Dudenhöffer, J.H., Greenleaf, S.S., Holzschuh, A., Isaacs, R., Krewenka, K., Mandelik, Y., Mayfield, M.M., Morandin, L.A., Potts, S.G., Ricketts, T.H., Szentgyörgyi, H., Viana, B.F., Westphal, C., Winfree, R. & Klein, A.-M. (2011) Stability of pollination services decreases with isolation from natural areas despite honey bee visits. *Ecology Letters*, **14**, 1062–1072.

Garibaldi, L.A., Steffan-Dewenter, I., Winfree, R., Aizen, M.A., Bommarco, R., Cunningham, S.A., Kremen, C., Carvalheiro, L.G., Harder, L.D., Afik, O., Bartomeus, I., Benjamin, F., Boreux, V., Cariveau, D., Chacoff, N.P., Dudenhöffer, J.H., Freitas, B.M., Ghazoul, J., Greenleaf, S., Hipólito, J., Holzschuh, A., Howlett, B., Isaacs, R., Javorek, S.K., Kennedy, C.M., Krewenka, K.M., Krishnan, S., Mandelik, Y., Mayfield, M.M., Motzke, I., Munyuli, T., Nault, B.A., Otieno, M., Petersen, J., Pisanty, G., Potts, S.G., Rader, R., Ricketts, T.H., Rundlöf, M., Seymour, C.L., Schüepp, C., Szentgyörgyi, H., Taki, H., Tscharntke, T., Vergara, C.H., Viana, B.F., Wanger, T.C., Westphal, C., Williams, N. & Klein, A.M. (2013) Wild pollinators enhance fruit set of crops regardless of honey bee abundance. *Science*, **339**, 1608–1611.

Gillooly, J.F., Brown, J.H., West, G.B., Savage, V.M. & Charnov, E.L. (2001) Effects of size and temperature on metabolic rate. *Science*, **293**, 2248–2251.

González-Varo, J.P., Biesmeijer, J.C., Bommarco, R., Potts, S.G., Schweiger, O., Smith, H.G., Steffan-Dewenter, I., Szentgyörgyi, H., Woyciechowski, M. & Vilà, M. (2013) Combined effects of global change pressures on animal-mediated pollination. *Trends in Ecology & Evolution*, **28**, 524–530.

Gordo, O. & Sanz, J.J. (2005) Phenology and climate change: a long-term study in a Mediterranean locality. *Oecologia*, **146**, 484–495.

Gordo, O. & Sanz, J.J. (2006) Temporal trends in phenology of the honey bee *Apis mellifera* (L.) and the small white *Pieris rapae* (L.) in the Iberian Peninsula (1952–2004). *Ecological Entomology*, **31**, 261–268.

Goulson, D., Hanley, M.E., Darvill, B., Ellis, J.S. & Knight, M.E. (2005) Causes of rarity in bumblebees. *Biological Conservation*, **122**, 1–8.

Goulson, D., Nicholls, E., Botías, C. & Rotheray, E.L. (2015) Bee declines driven by combined stress from parasites, pesticides, and lack of flowers. *Science*, **347**, 1255957.

Hegland, S.J., Nielsen, A., Lázaro, A., Bjerknes, A.-L. & Totland, Ø. (2009) How does climate warming affect plant-pollinator interactions? *Ecology Letters*, **12**, 184–195.

Heinrich, B. (1979) *Bumblebee Economics*. Harvard University Press, Cambridge, MA, USA.

Hellmann, J.J., Pelini, S.L., Prior, K.M. & Dzurisin, J.D.K. (2008) The response of two butterfly species to climatic variation at the edge of their range and the implications for poleward range shifts. *Oecologia*, **157**, 583–592.

Herrera, C.M. (1995) Floral biology, microclimate, and pollination by ectothermic bees in an early-blooming herb. *Ecology*, **76**, 218–228.

Higgins, J.K., MacLean, H.J., Buckley, L.B. & Kingsolver, J.G. (2014) Geographic differences and microevolutionary changes in thermal sensitivity of butterfly larvae in response to climate. *Functional Ecology*, **28**, 982–989.

Hill, J.K., Thomas, C.D., Fox, R., Telfer, M.G., Willis, S.G., Asher, J. & Huntley, B. (2002) Responses of butterflies to twentieth century climate warming: implications for future ranges. *Proceedings of the Royal Society B*, **269**, 2163–2171.

Hines, H.M. (2008) Historical biogeography, divergence times, and diversification patterns of bumble bees (Hymenoptera: Apidae: *Bombus*). *Systematic Biology*, **57**, 58–75.

Høye, T.T., Post, E., Schmidt, N.M., Trøjelsgaard, K. & Forchhammer, M.C. (2013) Shorter flowering seasons and declining abundance of flower visitors in a warmer Arctic. *Nature Climate Change*, **3**, 759–763.

Hranitz, J.M., Barthell, J.F., Thorp, R.W., Overall, L.M. & Griffith, J.L. (2009) Nest site selection influences mortality and stress responses in developmental stages of *Megachile apicalis* Spinola (Hymenoptera: Megachilidae). *Environmental Entomology*, **38**, 484–492.

Iler, A.M., Inouye, D.W., Høye, T.T., Miller-Rushing, A.J., Burkle, L.A. & Johnston, E.B. (2013) Maintenance of temporal synchrony between syrphid flies and floral resources despite differential phenological responses to climate. *Global Change Biology*, **19**, 2348–2359.

IPCC (2014) *Climate Change 2014: Synthesis Report. Contribution of Working Groups I, II and III to the Fifth Assessment Report of the Intergovernmental Panel on Climate Change*. IPCC, Geneva, Switzerland.

James, R.R. (2005) Temperature and chalkbrood development in the alfalfa leafcutting bee, *Megachile rotundata*. *Apidologie*, **36**, 15–23.

Jevanandam, N., Goh, A.G.R. & Corlett, R.T. (2013) Climate warming and the potential extinction of fig wasps, the obligate pollinators of figs. *Biology Letters*, **9**, 20130041.

Johansen, C.A. & Eves, J.D. (1973) Effects of chilling, humidity and seasonal conditions on emergence of the alfalfa leafcutting bee. *Environmental Entomology*, **2**, 23–26.

Kerr, J.T., Pindar, A., Galpern, P., Packer, L., Potts, S.G., Roberts, S.M., Rasmont, P., Schweiger, O., Colla, S.R., Richardson, L.L., Wagner, D.L., Gall, L.F., Sikes, D.S. & Pantoja, A. (2015) Climate change impacts on bumblebees converge across continents. *Science*, **349**, 177–180.

Kharouba, H.M. & Vellend, M. (2015) Flowering time of butterfly nectar food plants is more sensitive to temperature than the timing of butterfly adult flight. *Journal of Animal Ecology*, **84**, 1311–1321.

Kingsolver, J.G. & Huey, R.B. (2008) Size, temperature, and fitness: three rules. *Evolutionary Ecology Research*, **10**, 251–268.

Kingsolver, J.G. & Watt, W.B. (1983) Thermoregulatory strategies in *Colias* butterflies: thermal stress and the limits to adaptation in temporally varying environments. *American Naturalist*, **121**, 32–55.

Kingsolver, J.G., Woods, H.A., Buckley, L.B., Potter, K.A., MacLean, H.J. & Higgins, J.K. (2011) Complex life cycles and the responses of insects to climate change. *Integrative and Comparative Biology*, **51**, 719–732.

Klapwijk, M.J., Gröbler, B.C., Ward, K., Wheeler, D. & Lewis, O.T. (2010) Influence of experimental warming and shading on host–parasitoid synchrony. *Global Change Biology*, **16**, 102–112.

Klein, A.-M., Vaissière, B.E., Cane, J.H., Steffan-Dewenter, I., Cunningham, S.A., Kremen, C. & Tscharntke, T. (2007) Importance of pollinators in changing landscapes for world crops. *Proceedings of the Royal Society B*, **274**, 303–313.

Konvicka, M., Maradova, M., Benes, J., Fric, Z. & Kepka, P. (2003) Uphill shifts in distribution of butterflies in the Czech Republic: effects of changing climate detected on a regional scale. *Global Ecology and Biogeography*, **12**, 403–410.

Koti, S., Reddy, K.R., Reddy, V.R., Kakani, V.G. & Zhao, D.L. (2005) Interactive effects of carbon dioxide, temperature, and ultraviolet-B radiation on soybean (*Glycine max* L.) flower and pollen morphology, pollen production, germination, and tube lengths. *Journal of Experimental Botany*, **56**, 725–736.

Krenn, H.W. (2010) Feeding mechanisms of adult Lepidoptera: structure, function, and evolution of the mouthparts. *Annual Review of Entomology*, **55**, 307–327.

Krunić, M.D. & Stanisavljević, L.Ž. (2006) Supercooling points and diapause termination in overwintering adults of orchard bees *Osmia cornuta* and *O. rufa* (Hymenoptera: Megachilidae). *Bulletin of Entomological Research*, **96**, 323–326.

Kudo, G. & Ida, T.Y. (2013) Early onset of spring increases the phenological mismatch between plants and pollinators. *Ecology*, **94**, 2311–2320.

Le Conte, Y. & Navajas, M. (2008) Climate change: impact on honey bee populations and diseases. *Revue scientifique et technique - Office internationale des épizooties*, **27**, 499–510.

Linderholm, H.W. (2006) Growing season changes in the last century. *Agricultural and Forest Meteorology*, **137**, 1–14.

Mair, L., Thomas, C.D., Anderson, B.J., Fox, R., Botham, M. & Hill, J.K. (2012) Temporal variation in responses of species to four decades of climate warming. *Global Change Biology*, **18**, 2439–2447.

McLaughlin, J.F., Hellmann, J.J., Boggs, C.L. & Ehrlich, P.R. (2002) Climate change hastens population extinctions. *Proceedings of the National Academy of Sciences of the United States of America*, **99**, 6070–6074.

Memmott, J., Craze, P.G., Waser, N.M. & Price, M.V. (2007) Global warming and the disruption of plant–pollinator interactions. *Ecology Letters*, **10**, 710–717.

Mote, P.W., Hamlet, A.F., Clark, M.P. & Lettenmaier, D.P. (2005) Declining mountain snowpack in western North America. *Bulletin of the American Meteorological Society*, **86**, 39–49.

Mu, J., Peng, Y., Xi, X., Wu, X., Li, G., Niklas, K.J. & Sun, S. (2015) Artificial asymmetric warming reduces nectar yield in a Tibetan alpine species of Asteraceae. *Annals of Botany*, **116**, 899–906.

Muniz, J.M., Pereira, A.L.C., Valim, J.O.S. & Campos, W.G. (2013) Patterns and mechanisms of temporal resource partitioning among bee species visiting basil (*Ocimum basilicum*) flowers. *Arthropod-Plant Interactions*, **7**, 491–502.

O'Neill, K.M., O'Neill, R.P., Kemp, W.P. & Delphia, C.M. (2011) Effect of temperature on post-wintering development and total lipid content of alfalfa leafcutting bees. *Environmental Entomology*, **40**, 917–930.

Ollerton, J., Winfree, R. & Tarrant, S. (2011) How many flowering plants are pollinated by animals? *Oikos*, **120**, 321–326.

Ovaskainen, O., Skorokhodova, S., Yakovleva, M., Sukhov, A., Kutenkov, A., Kutenkova, N., Shcherbakov, A., Meyke, E. & del Mar Delgado, M. (2013) Community-level phenological response to climate change. *Proceedings of the National Academy of Sciences of the United States of America*, **110**, 13434–13439.

Parmesan, C. (1996) Climate and species' range. *Nature*, **382**, 765–766.

Parmesan, C. (2007) Influences of species, latitudes and methodologies on estimates of phenological response to global warming. *Global Change Biology*, **13**, 1860–1872.

Parmesan, C., Ryrholm, N., Stefanescu, C., Hill, J.K., Thomas, C.D., Descimon, H., Huntley, B., Kaila, L., Kullberg, J., Tammaru, T., Tennent, W.J., Thomas, J.A. & Warren, M. (1999) Poleward shifts in geographical ranges of butterfly species associated with regional warming. *Nature*, **399**, 579–583.

Pauli, J.N., Zuckerberg, B., Whiteman, J.P. & Porter, W. (2013) The subnivium: a deteriorating seasonal refugium. *Frontiers in Ecology and the Environment*, **11**, 260–267.

Pelini, S.L., Keppel, J.A., Kelley, A.E. & Hellmann, J.J. (2010) Adaptation to host plants may prevent rapid insect responses to climate change. *Global Change Biology*, **16**, 2923–2929.

Potts, S.G., Biesmeijer, J.C., Kremen, C., Neumann, P., Schweiger, O. & Kunin, W.E. (2010) Global pollinator declines: trends, impacts and drivers. *Trends in Ecology & Evolution*, **25**, 345–353.

Pöyry, J., Luoto, M., Heikkinen, R.K., Kuussaari, M. & Saarinen, K. (2009) Species traits explain recent range shifts of Finnish butterflies. *Global Change Biology*, **15**, 732–743.

Prasad, P.V.V., Boote, K.J., Allen, L.H. Jr., & Thomas, J.M.G. (2002) Effects of elevated temperature and carbon dioxide on seed-set and yield of kidney bean (*Phaseolus vulgaris* L.). *Global Change Biology*, **8**, 710–721.

Pressman, E., Peet, M.M. & Pharr, D.M. (2002) The effect of heat stress on tomato pollen characteristics is associated with changes in carbohydrate concentration in the developing anthers. *Annals of Botany*, **90**, 631–636.

Rader, R., Reilly, J., Bartomeus, I. & Winfree, R. (2013) Native bees buffer the negative impact of climate warming on honey bee pollination of watermelon crops. *Global Change Biology*, **19**, 3103–3110.

Radmacher, S. & Strohm, E. (2010) Factors affecting offspring body size in the solitary bee *Osmia bicornis* (Hymenoptera, Megachilidae). *Apidologie*, **41**, 169–177.

Radmacher, S. & Strohm, E. (2011) Effects of constant and fluctuating temperatures on the development of the solitary bee *Osmia bicornis* (Hymenoptera: Megachilidae). *Apidologie*, **42**, 711–720.

Robbirt, K.M., Roberts, D.L., Hutchings, M.J. & Davy, A.J. (2014) Potential disruption of pollination in a sexually deceptive orchid by climatic change. *Current Biology*, **24**, 2845–2849.

Roff, D.A. (1992) *The Evolution of Life Histories: Theory and Analysis*. Chapman & Hall, New York.

Roland, J. & Matter, S.F. (2013) Variability in winter climate and winter extremes reduces population growth of an alpine butterfly. *Ecology*, **94**, 190–199.

Roulston, T.H. & Goodell, K. (2011) The role of resources and risks in regulating wild bee populations. *Annual Review of Entomology*, **56**, 293–312.

Roy, D.B. & Sparks, T.H. (2000) Phenology of British butterflies and climate change. *Global Change Biology*, **6**, 407–416.

Rozen, J.G. Jr, Straka, J. & Rezkova, K. (2009) Oocytes, larvae, and cleptoparasitic behavior of *Biastes emarginatus* (Hymenoptera: Apidae: Nomadinae: Biastini). *American Museum Novitates*, **3667**, 1–15.

Rundlöf, M., Andersson, G.K.S., Bommarco, R., Fries, I., Hederström, V., Herbertsson, L., Jonsson, O., Klatt, B.K., Pedersen, T.R., Yourstone, J. & Smith, H.G. (2015) Seed coating with a neonicotinoid insecticide negatively affects wild bees. *Nature*, **521**, 77–80.

Scaven, V.L. & Rafferty, N.E. (2013) Physiological effects of climate warming on flowering plants and insect pollinators and potential consequences for their interactions. *Current Zoology*, **59**, 418–426.

Scheifinger, H., Koch, E. & Winkler, H. (2005) Results of a first look into the Austrian animal phenological records. *Meteorologiche Zeitschrift*, **14**, 203–209.

Schweiger, O., Biesmeijer, J.C., Bommarco, R., Hickler, T., Hulme, P.E., Klotz, S., Kühn, I., Moora, M., Nielsen, A., Ohlemüller, R., Petanidou, T., Potts, S.G., Pyšek, P., Stout, J.C., Sykes, M.T., Tscheulin, T., Vilà, M., Walther, G.-R., Westphal, C., Winter, M., Zobel, M. & Settele, J. (2010) Multiple stressors on biotic interactions: how climate change and alien species interact to affect pollination. *Biological Reviews*, **85**, 777–795.

Schweiger, O., Heikkinen, R.K., Harpke, A., Hickler, T., Klotz, S., Kudrna, O., Kühn, I., Pöyry, J. & Settele, J. (2012) Increasing range mismatching of interacting species under global change is related to their ecological characteristics. *Global Ecology and Biogeography*, **21**, 88–99.

Schweiger, O., Settele, J., Kudrna, O., Klotz, S. & Kühn, I. (2008) Climate change can cause spatial mismatch of trophically interacting species. *Ecology*, **89**, 3472–3479.

Seeley, T.D. & Visscher, P.K. (1985) Survival of honeybees in cold climates: the critical timing of colony growth and reproduction. *Ecological Entomology*, **10**, 81–88.

Senapathi, D., Carvalheiro, L.G., Biesmeijer, J.C., Dodson, C.-A., Evans, R.L., McKerchar, M., Morton, R.D., Moss, E.D., Roberts, S.P.M., Kunin, W.E. & Potts, S.G. (2015) The impact of over 80 years of land cover changes on bee and wasp pollinator communities in England. *Proceedings of the Royal Society B*, **282**, 20150294.

Sgolastra, F., Bosch, J., Molowny-Horas, R., Maini, S. & Kemp, W.P. (2010) Effect of temperature regime on diapause intensity in an adult-wintering Hymenopteran with obligate diapause. *Journal of Insect Physiology*, **56**, 185–194.

Sgolastra, F., Kemp, W.P., Buckner, J.S., Pitts-Singer, T.L., Maini, S. & Bosch, J. (2011) The long summer: Pre-wintering temperatures affect metabolic expenditure and winter survival in a solitary bee. *Journal of Insect Physiology*, **57**, 1651–1659.

Sheffield, C.S. (2008) Summer bees for spring crops? Potential problems with *Megachile rotundata* (Fab.) (Hymenoptera : Megachilidae) as a pollinator of lowbush blueberry (Ericaceae). *Journal of the Kansas Entomological Society*, **81**, 276–287.

Sparks, T.H., Langowska, A., Głazaczow, A., Wilkaniec, Z., Bieńkowska, M. & Tryjanowski, P. (2010) Advances in the timing of spring cleaning by the honeybee *Apis mellifera* in Poland. *Ecological Entomology*, **35**, 788–791.

Stelzer, R.J., Chittka, L., Carlton, M. & Ings, T.C. (2010) Winter active bumblebees (*Bombus terrestris*) achieve high foraging rates in urban Britain. *PLOS ONE*, **5**, e9959.

Stephen, W.P. (1965) Temperature effects on the development and multiple generations in the alkali bee, *Nomia melanderi* Cockerell. *Entomologia Experimentalis et Applicata*, **8**, 228–240.

Stone, G.N. & Willmer, P.G. (1989) Warm-up rates and body temperatures in bees: the importance of body size, thermal regime and phylogeny. *Journal of Experimental Biology*, **147**, 303–328.

Straka, J. & Bogusch, P. (2007) Description of immature stages of cleptoparasitic bees *Epeoloides coecutiens* and *Leiopodus trochantericus* (Hymenoptera: Apidae: Osirini, Protepeolini) with remarks to their unusual biology. *Entomologica Fennica*, **18**, 242–254.

Straka, J. & Starzomski, B.M. (2014) Humming along or buzzing off? The elusive consequences of plant-pollinator mismatches. *Journal of Pollination Ecology*, **13**, 129–145.

Stuhldreher, G., Hermann, G. & Fartmann, T. (2014) Cold-adapted species in a warming world - an explorative study on the impact of high winter temperatures on a continental butterfly. *Entomologia Experimentalis et Applicata*, **151**, 270–279.

Szabo, N.D., Colla, S.R., Wagner, D.L., Gall, L.F. & Kerr, J.T. (2012) Do pathogen spillover, pesticide use, or habitat loss explain recent North American bumblebee declines? *Conservation Letters*, **5**, 232–239.

Takkis, K., Tscheulin, T., Tsalkatis, P. & Petanidou, T. (2015) Climate change reduces nectar secretion in two common Mediterranean plants. *AoB PLANTS*, **7**, plv111.

Tauber, M.J., Tauber, C.A. & Masaki, S. (1986) *Seasonal Adaptations of Insects*. Oxford University Press, New York.

Thomson, J.D. (2010) Flowering phenology, fruiting success, and progressive deterioration of pollination in an early-flowering geophyte. *Philosophical Transactions of the Royal Society B*, **365**, 3187–3199.

Tobin, P.C., Nagarkatti, S., Loeb, G. & Saunders, M.C. (2008) Historical and projected interactions between climate change and insect voltinism in a multivoltine species. *Global Change Biology*, **14**, 951–957.

Torchio, P.F. (1979) An eight-year field study involving control of *Sapyga pumila* Cresson (Hymenoptera: Sapygidae), a wasp parasite of the alfalfa leafcutter bee, *Megachile pacifica* Panzer. *Journal of the Kansas Entomological Society*, **52**, 412–419.

Torchio, P.F. & Tepedino, V.J. (1982) Parsivoltinism in three species of *Osmia* bees. *Psyche*, **89**, 221–238.

Undurraga, J.M. & Stephen, W.P. (1980) Effect of temperature on development and survival in post-diapausing alfalfa leafcutting bee prepupae and pupae (*Megachile rotundata* (F.): Hymenoptera: Megachilidae). I. High temperatures. *Journal of the Kansas Entomological Society*, **53**, 669–676.

Valtonen, A., Leinonen, R., Pöyry, J., Roininen, H., Tuomela, J. & Ayres, M.P. (2013a) Is climate warming more consequential towards poles? The phenology of Lepidoptera in Finland. *Global Change Biology*, **20**, 16–27.

Valtonen, A., Molleman, F., Chapman, C.A., Carey, J.R., Ayres, M.P. & Roininen, H. (2013b) Tropical phenology: bi-annual rhythms and interannual variation in an Afrotropical butterfly assemblage. *Ecosphere*, **4**, 36.

Van Dyck, H., Bonte, D., Puls, R., Gotthard, K. & Maes, D. (2015) The lost generation hypothesis: could climate change drive ectotherms into a developmental trap? *Oikos*, **124**, 54–61.

van Nouhuys, S. & Lei, G.C. (2004) Parasitoid–host metapopulation dynamics: the causes and consequences of phenological asynchrony. *Journal of Animal Ecology*, **73**, 526–535.

Vanbergen, A.J., Baude, M., Biesmeijer, J.C., Britton, N.F., Brown, M.J.F., Brown, M., Bryden, J., Budge, G.E., Bull, J.C., Carvell, C., Challinor, A.J., Connolly, C.N., Evans, D.J., Feil, E.J., Garratt, M.P., Greco, M.K., Heard, M.S., Jansen, V.A.A., Keeling, M.J., Kunin, W.E., Marris, G.C., Memmott, J., Murray, J.T., Nicolson, S.W., Osborne, J.L., Paxton, R.J., Pirk, C.W.W., Polce, C., Potts, S.G., Priest, N.K., Raine, N.E., Roberts, S., Ryabov, E.V., Shafir, S., Shirley, M.D.F., Simpson, S.J., Stevenson, P.C., Stone, G.N., Termansen, M. & Wright, G.A. (2013) Threats to an ecosystem service: pressures on pollinators. *Frontiers in Ecology and the Environment*, **11**, 251–259.

Vasseur, D.A., DeLong, J.P., Gilbert, B., Greig, H.S., Harley, C.D.G., McCann, K.S., Savage, V., Tunney, T.D. & O'Connor, M.I. (2014) Increased temperature variation poses a greater risk to species than climate warming. *Proceedings of the Royal Society B*, **281**, 20132612.

Vogt, F.D. (1986) Thermoregulation in bumblebee colonies. II. Behavioral and demographic variation throughout the colony cycle. *Physiological Zoology*, **59**, 60–68.

Warren, M.S., Hill, J.K., Thomas, J.A., Asher, J., Fox, R., Huntley, B., Roy, D.B., Telfer, M.G., Jeffcoate, S., Harding, P., Jeffcoate, G., Willis, S.G., Greatorex-Davies, J.N., Moss, D. & Thomas, C.D. (2001) Rapid responses of British butterflies to opposing forces of climate and habitat change. *Nature*, **414**, 65–69.

Wcislo, W.T. & Tierney, S.M. (2009) Behavioural environments and niche construction: the evolution of dim-light foraging in bees. *Biological Reviews*, **84**, 19–37.

White, J., Son, Y. & Park, Y.-L. (2009) Temperature-dependent emergence of *Osmia cornifrons* (Hymenoptera: Megachilidae) adults. *Journal of Economic Entomology*, **102**, 2026–2032.

White, P. & Kerr, J.T. (2006) Contrasting spatial and temporal global change impacts on butterfly species richness during the 20th century. *Ecography*, **29**, 908–918.

Whitfield, G.H. & Richards, K.W. (1992) Temperature-dependent development and survival of immature stages of the alfalfa leafcutter bee, *Megachile rotundata* (Hymenoptera: Megachilidae). *Apidologie*, **23**, 11–23.

Williams, C.M., Hellmann, J. & Sinclair, B.J. (2012) Lepidopteran species differ in susceptibility to winter warming. *Climate Research*, **53**, 119–130.

Williams, C.M., Henry, H.A.L. & Sinclair, B.J. (2014) Cold truths: how winter drives responses of terrestrial organisms to climate change. *Biological Reviews*, **90**, 214–235.

Williams, P.H., Araújo, M.B. & Rasmont, P. (2007) Can vulnerability among British bumblebee (*Bombus*) species be explained by niche position and breadth? *Biological Conservation*, **138**, 493–505.

Willmer, P. (2011) *Pollination and Floral Ecology*. Princeton University Press, Princeton, NJ, USA.

Willmer, P.G. (1983) Thermal constraints on activity patterns in nectar-feeding insects. *Ecological Entomology*, **8**, 455–469.

Willmer, P.G. & Stone, G.N. (2004) Behavioral, ecological, and physiological determinants of the activity patterns of bees. *Advances in the Study of Behavior*, **34**, 347–466.

Wolda, H. (1988) Insect seasonality: why? *Annual Review of Ecology and Systematics*, **19**, 1–18.

6

Climate Change Effects on Biological Control in Grasslands

Philippa J. Gerard and Alison J. Popay

AgResearch, Ruakura Research Centre, Private Bag, Hamilton 3123, New Zealand

Summary

The impacts of climate change on grassland communities are likely to be diverse, complex and difficult to predict. The direct effects of increased CO_2, higher temperatures and variable rainfall on invertebrate herbivores and their biocontrol agents together with a myriad of indirect effects mediated via their plant hosts need to be considered. Plant diversity is a key determinant of invertebrate species richness and fertile modified grasslands are more vulnerable to species loss than natural grasslands. In addition, introduced and endemic predators and parasitoids, as well as plant defences and symbionts, play vital roles in determining weed and pest abundance in these grassland ecosystems. Thus as a consequence of climate variability, increased incidences of pest outbreaks are predicted through disruption of predator–prey relationships. Climate change attributes can also change the concentrations of plant secondary metabolites produced by plants and symbiotic fungi, both in the short term within the plant and over time through plant adaptation, that may alter the strength of the protection they provide against insect pests. The networks of interacting species in grasslands form complex food webs, all parts of which may be impacted by climate change, with sensitivities increasing with trophic level. Within these food webs, climate change may have both bottom-up and top-down influences but can also affect invertebrate diversity, behaviour and distribution independently of the plant community. The evidence suggests so far that the multiple effects induced by climate change can vary at the species level. The relatively rapid response of grasslands to changing conditions, however, facilitates the further exploration and quantification of these effects that is required to better predict future changes.

6.1 Introduction

Natural grasslands are ecological ecosystems dominated by grasses and maintained by a combination of seasonal drought, occasional fires, and grazing by large mammals. They include the tropical savannahs, the temperate prairies of North America and montane grasslands and tundra. Grasslands are a very large carbon sink at world level and are as important as forests in reducing the CO_2 concentration in the atmosphere ('t Mannetje, 2007b).

Global Climate Change and Terrestrial Invertebrates, First Edition. Edited by Scott N. Johnson and T. Hefin Jones.
© 2017 John Wiley & Sons, Ltd. Published 2017 by John Wiley & Sons, Ltd.

Throughout geological time, the amounts of grassland on the planet surface has expanded and contracted with climate change. During the Jurassic and Cretaceous periods (206–265 million years ago (Ma)) the climate was more uniformly warm and moist than at present and tropical rainforests were widespread. During the Miocene period (23.8–5.3 Ma) grasslands expanded into huge areas (e.g., prairies in the USA, steppes in Eurasia and pampas and llanos in South America). The Quaternary period (1.8 Ma till now) has been more variable with some twenty-two different ice ages with periodicities of about 100,000 years. Eighteen thousand years ago, both the Sahara and Amazonian regions were covered in extensive areas of grasslands and only in the last 10,000 years has rainforest covered the Amazonian region and the desert expanded into the Sahara ('t Mannetje, 2007a).

Since Neolithic times, mankind has manipulated grasslands, in particular in temperate regions where forests have been cleared (Kuneš et al., 2015), crops and sown pastures established in the higher fertility areas, and extensive grazing undertaken in areas not suitable for other forms of agriculture. These grasslands are vital to food security, providing feed for ruminants used for meat and milk production. In 2000, 22% of the Earth's ice-free land surface (28.0 million km^2) was estimated to be in grasslands grazed by domesticated animals (Ramankutty et al., 2008). From 2001 to 2013, 97 million hectares of new pastureland replaced forests throughout Latin America, mainly in central Brazil, western Paraguay, and northern Guatemala (Graesser et al., 2015). In contrast, the area of permanent grasslands and pastures in Europe is declining. For example in the UK, the areas of permanent and temporary grassland were 45 and 40%, respectively, lower in 2000 than in 1900.

Climate change and agricultural management are highly interconnected in determining the future of these grassland ecosystems. While intensive agricultural systems have the potential to adapt to these changing conditions (e.g., irrigation, cultivar selection), low-input pastoral systems may be affected more seriously (Fuhrer, 2003). Climate change is likely to affect managed grasslands in many ways, from more variable pasture productivity and quality and greater risk of pest and weed outbreaks to more frequent and longer droughts, more intense rainfall events, and greater risks of soil erosion (Stokes et al., 2008). The ecosystem impacts may be the result of direct effects at the organism level, or indirect effects at the system level, for instance, through shifts in nutrient cycling and disturbance of predator–prey relationships.

Introduced and endemic predators and parasitoids, as well as plant resistance and endophytic fungal symbionts, play vital roles in determining weed and pest abundance in managed grassland and forage systems. In turn, grasslands are a source of natural enemies for nearby annual crops (Parry et al., 2015). With pastoralists reliant on non-insecticidal controls to maintain forage production and quality and the trend towards lower and more targeted use of insecticides in crops, the need for effective biological control systems in grasslands is unlikely to diminish.

Biological control in managed grasslands varies in complexity. The species-rich semi-natural grasslands of Central Europe and North America have rich networks of indigenous predator, parasitoid, pathogens and plant defences (Joern & Laws, 2013). In contrast, the predominantly perennial ryegrass (*Lolium perenne*)–white clover (*Trifolium repens*) pastures in New Zealand rely on introduced classical biological agents and grass endophytes for protection (Gerard et al., 2012; Thom et al., 2012). Reviews by Cock et al. (2013), Jeffs and Lewis (2013) and Thomson et al. (2010) have identified the many ways in which changes in temperature, precipitation and CO_2 will affect the fitness and efficacy of biological control agents, including via changes in plant biomass and quality, species distributions, synchrony with hosts and management strategies adopted by farmers. In this chapter we focus on the likely responses of grassland biocontrol systems to climate change, particularly on the indirect changes in these systems brought about by alterations in the underlying plant community.

6.2 Changes in Plant Biodiversity

Plant species richness is a significant predictor of invertebrate species richness. Reductions in grassland plant diversity have been shown to result in cascading effects on soil food webs (density and diversity of soil organisms) and functions (Eisenhauer et al., 2013), increased vulnerability to invasions by weeds, enhanced spread of plant fungal diseases, and changes in the richness and structure of insect communities (Joern & Laws, 2013; Knops et al., 1999). Changing climate patterns cause directional loss of plant species diversity with fertile, productive disturbed grasslands more vulnerable to such loss than low fertility, natural grasslands (Grime et al., 2000). Individually, warming has been observed to have a lesser effect on plant biodiversity than elevated CO_2 (eCO_2) and rainfall (Zavaleta et al., 2003a), but the largest change occurred when all three attributes were combined (e.g. 50% increase in forb abundance in California grasslands (Zavaleta et al., 2003b). Different trophic levels appear to have different sensitivities to climate, with higher trophic levels (e.g. parasitoids) being more susceptible to change (Voigt et al., 2003).

While transitory weather events can initiate sudden changes in plant diversity, when changes in precipitation patterns are sustained across years, feedbacks and species interactions can reverse initial community trajectories.

Many studies have documented the negative effects of drought on Lepidoptera food plant and nectar sources and flow-on impacts on diversity (Robinson et al., 2012 and references therein). However, managed grasslands can be manipulated to promote food web robustness. For example, the addition of drought-tolerant plant species prevented beetle extinctions in a lowland pasture in southern England (Woodcock et al., 2012).

Increases in both temperature and CO_2 are generally predicted to favour C_4 grasses over C_3 grasses, especially when soil moisture is limited. Models of the impact of climate change on current temperate grasslands within North and South America predict that the relative abundance of C_4 grasses will increase more than 10% at the expense of C_3 grasses (Morgan et al., 2011). There were essentially no areas where C_4 grasses decreased in abundance, and C_3 grasses declined throughout with the exception of the north-western Great Plains of the United States and Canada, and north central Argentina (Epstein et al., 2002). The shift towards C_4 grasses and warming is likely to cause increases in associated subtropical pests in temperate pastures, especially in regions where the pests have arrived without their suite of natural enemies. An example is African black beetle (*Heteronychus arator* Fabricius) that can cause severe damage to the forage grasses, ryegrass (*Lolium* spp.) and tall fescue (*Schedonorus phoenix* (Schreb.) Dumort.,) in Australia and New Zealand but reaches higher populations in pastures containing C_4 grasses such paspalum (*Paspalum dilatatum* Poir.) and kikuyu (*Pennisetum clandestinum* Hochst.) (King et al., 1981). In New Zealand, African black beetle first established in the northernmost regions and has expanded its range south over the last 30–40 years, as has kikuyu (Bell et al. 2011). The combination of the current prevalence of C_4 grasses in Waikato pastures (Tozer et al. 2012), warmer climate and more active El Niño and La Niña weather events has shifted the status of this pest from sporadic to persistent in the northern North Island (Gerard et al., 2013).

6.3 Multitrophic Interactions and Food Webs

Grassland biological control systems are part of complex food webs made up of networks of interacting species operating at different trophic levels above- and belowground, all of which may be impacted by climate change (Fig. 6.1). While most work on climate change affecting biological control

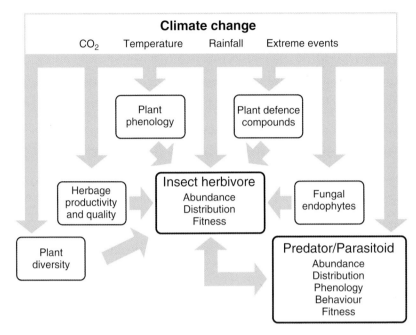

Figure 6.1 Major pathways through which climate change may impact on grassland biological control systems.

systems are for pairs of species interacting in isolation to the wider community, the relative simplicity of grassland ecosystems means they are relatively well represented in empirical studies attempting to quantify how different trophic levels are likely to respond to climate change (Barton & Schmitz, 2009; de Sassi et al., 2012; Laws & Joern, 2015; Tylianakis & Binzer, 2014).

Generally, grassland systems support the enemies hypothesis (Root, 1973) by which increased diversity in plant and insect communities results in more complex networks and more efficient and constant checks and balances on insect herbivore populations (Zhang & Adams, 2011). This complexity should buffer and mitigate the effect of climatic variations. However, agricultural intensification tends to produce food webs with low complexity and uneven interaction strengths (Tylianakis & Binzer, 2014). As a consequence, trophic interactions may be disrupted more easily by climate changes in sown or disturbed pasture communities than in undisturbed natural communities where greater complexity may be buffering these effects. Voigt et al. (2007) showed that climatic sensitivity increases with trophic level and that a disturbed grassland community is not only characterized by a higher climatic sensitivity but also by fewer and weaker interactions than an undisturbed grassland community.

Climate change can have both bottom-up (Case study 6.1) and top-down influences (Case study 6.2) on foodwebs. For example, Barton et al. (2009) showed that plant production in an old-field was not directly affected by temperature or precipitation, but the strength of top-down indirect effects of spider predation on grasshopper foraging. Even though warming increased grasshopper feeding, the presence of spiders increased grasshopper forb consumption relative to grasses. The magnitude of this effect on grasses and forbs increased with temperature at a rate of approximately 30% and 40% per 1°C, respectively.

Case study 6.1 Impact of drought on aboveground–belowground interactions operating in a multi-species plant community (Johnson et al., 2011)

Reduced summer rainfall is a predicted outcome of climate change in certain areas. In a large-scale field study the interactive effects of earthworms (*Aporrectodea caliginosa* Savigny) and summer drought on a plant community containing a grass and two forbs and their influence on populations of the aphid *Rhopalosiphum padi* L. and its parasitoid, *Aphidius ervi* Haliday were determined. Drought increased shoot nitrogen concentrations but decreased plant biomass, the latter being partially mitigated by the presence of earthworms. Drought reduced aphid abundance by over 50% and these effects were exacerbated when earthworms were present, especially in the monocultures. The density of *R. padi* was positively correlated with changes in the abundance of its parasitoid, *Aphidius ervi*. In addition, drought had a negative impact on *A. ervi* abundance beyond its impacts on aphid density, suggesting reduced prey quality as well as quantity. This study demonstrated the effect of predicted climate change on plant-mediated interactions between earthworms and aboveground multitrophic groups. These effects were seen to differ between monocultures and multi-species plant communities, suggesting that changes in aboveground–belowground linkages in response to drought may influence plant communities in the future.

Case study 6.2 Impact of warming and nitrogen deposition on a New Zealand native grassland (de Sassi et al., 2012; de Sassi & Tylianakis, 2012).

Several experiments were carried out in a native grassland system to test how biomass at the plant, insect herbivore, and natural enemy levels respond to the interactive effects of two key global change drivers: warming and nitrogen deposition. Herbivore biomass on average doubled in response to temperature and to a lesser amount with nitrogen. In contrast, parasitoids showed no significant response to the treatments resulting in reduced parasitism rates and a community dominated by herbivores. It was found that the parasitoids interacted less evenly within their host range and increasingly focused on abundant and high-quality (i.e., larger) hosts. These results provide evidence that climate change attributes can significantly alter food-web structure leading to reductions in top-down regulation. This in turn is likely to cascade to other fundamental ecosystem processes and supports the general concern of increasing herbivore pest outbreaks in a warmer world.

While gradual increases in temperature, eCO_2 and soil aridity may cause shifts in food webs, climatic fluctuations may create greater disruptions to host–parasitoid webs, as they may be affected both directly and indirectly by changes in host availability. Stireman et al. (2005) found a negative association between parasitism frequency in caterpillars and climatic (especially precipitation) variability. They interpreted this to be the result of increased lags and disconnections between herbivores and their enemy populations that occur as climatic variability increases.

Invertebrate biodiversity may also respond to climate attributes independently from the plant community. For example, only temperature, not eCO_2 and rainfall, affected insect community composition in an old-field plant community in eastern Tennessee, with corresponding reductions of diversity measures at higher trophic levels (i.e., parasitoids) (Villalpando et al., 2009). Climate, not vertical changes in vegetation structure due to the effects of an invasive grass (*Andropogon gayanus* Kunth.), determined invertebrate species richness and abundance in undisturbed savannas in northern Australia (Parr et al., 2010).

6.3.1 Warming and Predator Behaviour

The open nature of grasslands means invertebrates are exposed to greater thermal extremes than those under tree and shrub habitats. For example, the monthly temperature range is approximately 10°C greater for grasslands than under woodland canopy in September (autumn), and 5°C greater in January (winter) in North Yorkshire, UK (Suggitt et al., 2011). Therefore, phenotypic plasticity and behavioural responses of predatory species in seeking thermal refuges may determine the resilience of species interactions, grassland food web structure and function and subsequent efficacy of grasslands biological control systems (Schmitz & Barton, 2014). For example, the effect of a single climate change attribute (elevated temperature) on predator behaviour can alter biological control, sometimes in unexpected ways. For instance, in a field study on corn leaf aphid (*Rhopalosiphum maidis* (Fitch)) warming broke down the ant–aphid mutualism, exposing them to attack by ladybird larvae and counterintuitively reducing the abundance of this agricultural aphid pest (Barton & Ives, 2014).

The concept of habitat domain (the microhabitat a predator or prey preferentially occupies in space and time) is a useful framework to understand how climate change may influence predator–prey interactions (Schmitz & Barton, 2014). Sit and wait predators (e.g., web building spiders) have relatively narrow domains while active hunters (e.g., ladybirds) have broader domains. Prey also may have narrow or broad habitat domains depending on their mobility and feeding preferences. The effect of warming on a predator–prey interaction would not change if both had similar phenotypic plasticity and retained the same overlap in domains when seeking thermal refuge. However, if one species was less heat tolerant than another, it would be forced relatively lower in the canopy, which would change the existing domain overlap and may create new interactions by encroaching on the habitat domains of other species (Case study 6.3).

Climate warming may increase the relative abundance and diversity of species with active, rather than passive, hunting styles in grasslands. This is supported by an altitudinal and longitudinal study by Gibb et al. (2015) of foliage-living spider assemblages associated with relatively undisturbed kangaroo grass *Themeda triandra* grasslands (>70% cover) along a 900 km climatic gradient in south-eastern

---

Case study 6.3 Temperature and the plant–grasshopper–spider food chain (Barton, 2010, 2011; Barton & Schmitz, 2009)

The interactions between thermal refugia, spiders and grasshoppers have been explored in grassland ecosystems. Warming strengthened the top-down pressure by affecting spiders and grasshoppers differently. Experimental warming had no direct effect on grasshoppers and they maintained their position in the canopy. However, spiders moved their habitat domain lower in the canopy to seek thermal refuge. Therefore while grasshopper and spider habitat domains had a high degree of overlap in cooler, ambient conditions, warming decreased the overlap between predator and prey, allowing nymphal grasshoppers to increase daily feeding time. This had three impacts:

- Enhanced grasshopper growth, reproductive fitness, and survival
- Increased grass biomass and decreased forb biomass
- The shift of spiders lower in the vegetation canopy also increased intra-guild predation between spider species with consequent species extinction.

These results highlight the importance of understanding how key biotic and abiotic factors combine to influence species interactions.

Australia. The mean annual temperature varied from 9.3°C to 19.4°C and mean annual precipitation ranged from 385 to 1659 mm. They showed that larger spider species and species that were active hunters were more common in warmer climates. Invertebrate behavioural traits and their association with the environment are relatively unstudied and further research may reveal generalities that provide greater predictive power across a range of assemblages.

6.3.2 Herbage Productivity and Quality

Climate change models are predicting lower soil moisture over southwest North America, Central America, the Mediterranean, Australia, and Southern Africa in all seasons (Wang, 2005) which will have direct effect on grassland herbage production in these regions. However, these losses may be partially compensated for by other climate change attributes such as warmer temperature (Cullen et al., 2012) and increasing CO_2 (Hanley et al., 2004; Morgan et al., 2011).

Concurrent warming and eCO_2 conditions are expected to increase plant production and lead to a change in herbage nutritive value due to increased non-structural soluble carbohydrate content (AbdElgawad et al., 2014; Lilley et al., 2001). While nitrogen (N) fixation in legumes is increased under eCO_2, N concentration in foliage is decreased due to dilution from increased carbohydrates and other metabolic responses with possible flow-on implications for species fitness and competition in the grassland community (Lilley et al., 2001). However, responses vary with plant species, temperature and time. Elevated CO_2 favours growth of legumes (via feedbacks on nitrogen acquisition) over grasses in low fertility systems (Hanley et al., 2004; Lazzarotto et al., 2010) and C_4 grasses over C_3 grasses in native grasslands (Morgan et al., 2011). The importance of long-term studies was demonstrated by Cantarel et al. (2013) who showed that continuous, multi-year exposure to projected climate conditions eventually had a negative impact on aboveground biomass in their grassland study system. They concluded that this may be driven by changes in the relative abundance of plant functional groups and soil nutrient availability.

In general, because of the changes in the nutritional quality of their host plants, the abundance and fitness of insect herbivores decreases in eCO_2, and their consumption rate and development time increases (Stiling & Cornelissen, 2007; Walter et al., 2012). While this may lead to more damage by pests, it does not mean that most weed biological control agents will become more effective. For example, survival, development time and adult size of the chrysomelid *Gratiana boliviana* Spaeth, the biological control agent for tropical soda apple *Solanum viarum* Dunal, is negatively affected by eCO_2 (Diaz et al., 2012). In contrast, neither eCO_2 nor temperature affected the fitness or consumption rate of the beetles *Octotoma championi* Baly and *O. scabripennis* Guérin-Méneville, biological control agents of *Lantana camara* L. (Johns et al., 2003).

The impact on insect predators and parasitoids preying on herbivores feeding on plants impacted by eCO_2 is also highly variable. Elevated CO_2 may have no impact, a negative effect due to small host size, or be beneficial by decreasing the immune response of insect herbivores against endoparasitoids with no detrimental effects on the biocontrol agent (Yin et al., 2014). Thus responses to climate change at this trophic level is dependent on the individual species involved.

6.3.3 Plant Defence Compounds

Plant secondary metabolites stored in the plant tissue can protect the plant by reducing herbivory, while herbivore-induced plant volatiles can attract natural enemies. On the other hand, some insect herbivores sequester secondary metabolites to protect themselves against predation and parasitism; for example, larvae of the SW USA grassland moth *Grammia incorrupta* Hy. Edwards, deters predators by eating a mixture of plants containing iridoid glycosides and pyrrolizidine alkaloids (Mason et al., 2014). Climate change attributes can change the concentrations of these plant secondary metabolites both in the short term within the plant and over time through plant adaptation, with repercussions for grassland foodwebs (Fig. 6.2). For instance, the combination of drought and warming increased the levels of terpenes, catechin and indole acetic acid accumulated in shoots of the perennial grasses, *Holcus lanatus* L. and *Alopecurus pratensis* L. (Gargallo-Garriga et al., 2015). In general, eCO_2 enhances the production of phenolic compounds, and warming increases the production of volatile organic compounds (Bidart-Bouzat & Imeh-Nathaniel, 2008).

Of species used in extensive forage systems, the development of pest resistance based on plant secondary metabolites is probably most advanced in *Medicago* species. In this genus, saponins are the main defence compounds, but flavonoids and amino acids are also linked to herbivore resistance (Ryalls et al., 2013). Saponins have been shown to increase in response to both eCO_2 (Agrell et al., 2004) and warming (Dyer et al., 2013). Therefore climate change is likely to impact positively on the susceptibility of cultivars bred for resistance based on these compounds. However, Johnson

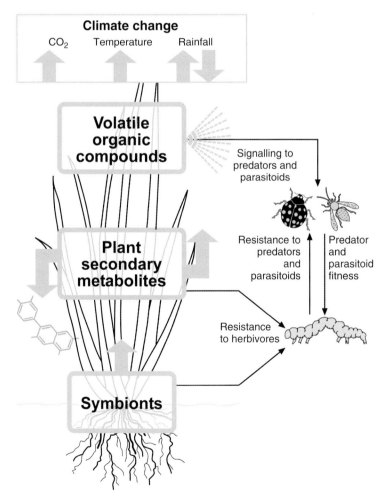

Figure 6.2 Main pathways through which climate change may impact on defence compounds that protect grassland plants from invertebrate herbivory.

et al. (2014) found contrasting differences between cultivars where aphid resistance was based on low levels of essential amino acids. Under eCO_2, levels of lysine, phenylalanine and tyrosine increased in a resistant cultivar and aphid colonisation rose from 22% at ambient CO_2 to 78%, and reproduction rates from 1.1 to 4.3 nymphs week^{-1}. In contrast, they found phenylalanine and histidine levels fell in a moderately resistant cultivar when held at eCO_2, which then became more resistant compared with plants at aCO_2.

How changing levels of plant secondary metabolites will impact on grassland biological control systems will vary with each foodweb. For example, reduced host size in response to increased secondary metabolites following exposure to eCO_2 resulted in reduced fitness of the endoparasitoid *Diaeretiella rapae* (McIntosh), which specialises on aphids attacking brassicas (Klaiber et al., 2013). In contrast, eCO_2 had no effect on the whitefly *Bemisia tabaci* (Gennadius) infesting Bt cotton and its parasitoid *Encarsia formosa* (Gahan) (Wang et al., 2014).

6.3.4 Fungal Endophytes

Two major forage grasses, tall fescue and ryegrass, are often infected with species of *Epichloë*, symbiotic fungi that reside entirely within their hosts with no external stage and are maternally transmitted by seed. The relationship is often regarded as mutualistic with the fungi gaining nutrients and protection while conferring biotic and abiotic stress tolerance on its host. The endophytes protect their host grass from herbivory via the production of alkaloids, which vary according to the species and strain of fungus. While all known alkaloids have some effect on insects, two also cause disorders in livestock. Agriculture has sought to exploit the natural diversity of endophytes to utilise those species and strains which provide resistance to insect pests, while minimising the toxicity to livestock.

Like other relationships in the grassland ecosystem, grass–endophyte symbioses are sensitive to climate change factors, such as eCO_2, warming, and changing precipitation. Alkaloid concentrations in plants are modified by host genotype, but more importantly in a climate change context, also by environmental conditions such as drought and temperature. While any increases in alkaloid concentration may provide plants with more robust protection against insect pests, they may also cause more problems to grazing livestock where toxic alkaloids are present. Conversely, if alkaloids are decreased under climate change scenarios, insect resistant properties resulting from their presence will be diminished. Evidence for responses, however, tend to vary depending on individual effects of temperature, precipitation, heat and their interaction as well as interactions with N fertilisation. In perennial ryegrass (*Lolium perenne*) infected with *E. festucae* var. *lolii* (formerly *Neotyphodium lolii*), concentrations of peramine – and to a lesser extent ergovaline – interacted with N fertilisation to reduce alkaloid concentrations but such effects were not apparent under eCO_2 (Hunt et al., 2005). These effects suggest that endophytes will retain, and under some scenarios perhaps increase, efficacy against pest species in low-precipitation grasslands, but may be less effective in protecting these agronomically-important grasses in higher rainfall areas.

The advantages that the endophyte provides can increase the productivity and persistence of the host resulting in a more competitive plant. In managed pasture systems in New Zealand, where endophytes have been studied extensively, the pressure of insect attack can influence plant composition in favour of endophyte-infected plants, increasing infection rates in the field (Thom et al., 2012). While the increase in endophyte levels is thought to be driven mainly by herbivores feeding on endophyte-free plants, there is also evidence that eCO_2 itself may increase rates of infection in the field (Brosi et al., 2011). The competitive influence of endophyte-infected grasses can also reduce the presence of other species in pasture such as clover and weeds (Thom et al., 2013). With eCO_2 favouring legume growth, clover content may increase in endophyte-infected pastures as a result of climate change, although this effect may be countered by the drought protection conferred by endophyte. *Epichloë coenophiala* increased tiller and biomass production in tall fescue under high temperatures and enhanced recovery from drought, although this was dependent on plant genotype (Bourguignon et al., 2015). Increased drought tolerance can also alter the natural geographic range of plants as has been shown for endophyte-infection of wild *Bromus laevipes* Shear that has enabled populations of this plant to extend its range by thousands of square kilometres into drier habitats (Afkhami et al., 2014). In ryegrass, changes in plant physiology that result in better plant performance of endophyte-infected plants under drought conditions have not been conclusively demonstrated. However, reduced root feeding damage by African black beetle larvae in endophyte-infected plots improved plant survival and recovery from moisture stress.

The effects of endophytes in reducing insect populations have implications for diversity of insects in pastures at different trophic levels. Populations of insect herbivores that are sensitive to endophytes are generally lower in pastures with a high content of endophyte-infected grasses (e.g., Argentine stem weevil (*Listronotus bonariensis* Kuschel)) but less is known about the performance of their natural enemies.

As with plant secondary metabolites, the alkaloids endophytes produce may be sequestered by certain herbivores for use as part of their own defensive repertoire against predators and parasitoids (de Sassi et al., 2006; Faeth, 2002). In these cases, natural enemies may decrease and herbivore abundance and biodiversity increase (Jani et al., 2010). Therefore, endophytes play an important role in determining grassland biodiversity and the relative abundance of herbivore, natural enemy and plant species.

The plant damage caused by pests may also be accentuated by other responses of the grass–endophyte symbiosis to climate change. Endophyte-infected grasses are sensitive to temperature increases which may reduce their tiller numbers but not those of endophyte-free grasses (Brosi et al., 2011). Similarly differential changes in nutritive value of foliage may alter insect response to endophyte infection. For example, endophyte-free ryegrass had 40% lower concentrations of soluble protein under eCO_2 than under ambient CO_2, but this effect was largely absent in endophyte-infected plants (Hunt et al., 2005). While climate change will induce adaptation processes in grass–endophyte symbioses over time, the more rapid development of plant cultivar–endophyte combinations adapted to the projected climates is warranted for managed agricultural systems.

6.3.5 Changes in Plant Phenology

Climate change can advance the reproductive phenology of grassland plant species which will in turn affect herbage quality and when nectar, pollen and seeds are available for invertebrates. For example, dates of first flowering of the same varieties of white clover (*T. repens* L.) recorded at Aberystwyth, UK for 40 years were found to have advanced by approximately 7.5 days per decade since 1978 (Williams & Abberton, 2004).

Synchrony with target plant phenology can make or break the effectiveness of a weed biological control agent. *Centaurea diffusa* Lam., a problematic invader in much of the western United States, flowered earlier and had faster seed head development under eCO_2 and increased temperature. This resulted in a better phenological match and an increased impact of the biological control agent *Larinus minutus* (Coleoptera: Curculionidae) (Reeves et al., 2015). In contrast, precipitation determines the efficacy of the moth *Epiblema strenuana* Walker against *Parthenium hysterophorus* L. (Asteraceae), a major weed in the prime grazing country in Queensland, Australia. Insufficient rainfall caused asynchrony between weed germination and *E. strenuana* emergence in three years out of four (Dhileepan, 2003). With climate change scenarios predicting decreased rainfall in agricultural regions of Queensland, we could therefore see reductions in the efficiency of *P. hysterophorus* biocontrol by *E. strenuana*.

6.4   Greater Exposure to Extreme Events

Global climate change suggests an increase in extreme weather events and in the frequency and amplitude of El Niño/Southern Oscillation phenomena that are related to the occurrence of drought, fire and changes in vegetation dynamics in grasslands in many regions. Drought decreases species richness, whereas fire increases richness by allowing the establishment of fugitive species from seed in the seed bank (Ghermandi & Gonzalez, 2009). Grassland weed and pest outbreaks have been related to major El Niño events, for example, repeated invasion of the serpentine grassland by the non-native grass *Bromus hordeaceus* L. in Californian grasslands (Hobbs et al., 2007).

Droughts and floods are followed frequently by pest outbreaks through the disruption of natural enemies. Field data indicate that parasitoids are generally more sensitive than their hosts, and lag behind in population recovery (Thomson et al., 2010 and references therein). In NZ, outbreaks of a native scarab *Costelytra zealandica* (White) often occur in pasture two to four years after a severe drought (East & Willoughby, 1980). Pathogens that limit *C. zealandica* larval populations have limited survival in dry soils (O'Callaghan et al., 1989) and severe droughts that reduce larval populations to low densities result in low infection rates in the next couple of generations. Both effects reduce the amount of pathogen inoculum available for infecting the host grass grub.

With the relative lack of canopy, temperature variation – in particular exposure to extreme temperatures – may have unexpected consequences on predator–prey relationships. Harmon et al. (2009) found contrasting effects when two predatory ladybird species were exposed to an increased frequency of episodic heat shocks: predation by *Coccinella septempunctata* L. added to the detrimental effect of the heat treatments on pea aphid (*Acyrthosiphon pisum* Harris) population growth, whereas *Harmonia axyridis* (Pallas) lessened the effect. A model describing the effect of warming on grasshopper–spider daily activity periods (overlaps of habitat domain over time) indicated that increasing temperatures led to increasing grasshopper populations, whereas increased variation in daily temperature tended to lower densities and increase predation (Logan et al., 2006; see also Case Study 6.3).

6.4.1   Changes in Precipitation

Rainfall can be a key driver of ecosystem functioning in semi-arid and arid grasslands. It impacts on plant community dynamics, which in turn influences the community composition and structure of higher trophic levels, especially insects, in grasslands.

To understand the impacts of rainfall on grassland ecosystems, it is essential to undertake long-term experiments that extend beyond the immediate responses. While certain species might react very quickly to climate change in terms of activity densities, they in turn will cause other shifts in abundance patterns (Buchholz et al., 2013).

Zhu et al. (2014) manipulated rainfall in a meadow steppe over 3 years and measured the effects on insect diversity, abundance, and trophic structure. As expected, rainfall enhanced aboveground biomass, particularly grasses, whereas the decreased water availability significantly reduced biomass. Both increased and decreased rainfall caused declines in insect species richness and abundance owing to vegetation-mediated effects. The trophic levels responded differently, with lower herbivore abundance but unchanged abundance of predators and parasitoids. Thus changes in precipitation in this system appeared to generate an insect community that is increasingly dominated by secondary consumers (Zhu et al., 2014).

In Southern England, plots exposed to predicted future rainfall patterns experienced a decline in the abundance of spiders, suggesting that bottom-up effects of reduced prey availability and/or lower plant biomass (via effects on hunting/shelter availability) may flow through the food chain to affect higher trophic levels. There were also a greater number of Hymenoptera which may have been driven by changes to flowering times, the availability of nectar and pollen or simply a transient behavioural effect of mobile species (Lee et al., 2014). While there are Hymenoptera that attack spiders, and vice versa, the study went to Order only, so changes in these interactions have yet to be explored in this system.

6.4.2 Drought Effects

Changing rainfall patterns associated with climate change are expected to result in greater frequency and longer droughts in many grassland-dominated regions. While the effects of drought stress on plants are relatively well studied, the indirect effects on predators and parasitoids are less well understood.

Aphid parasitism and predation rates have been found to be lower on drought-stressed plants (Aslam et al., 2013; Barton & Ives., 2014; Tariq et al., 2013). Generally, this is related to changes in host availability. In the Gramineae pest, *Rhopalosiphum padi* L., the lower parasitism rate by *Aphidius ervi* under drought was attributed to a lower incidence of nymphs and more adults, the latter being more difficult to parasitize (Aslam et al., 2013). Case study 6.4 shows how indirect effects can echo through a community and affect many species, including some that may not have been directly affected by the perturbation.

6.5   Range Changes

The ranges of plants and animals are moving in response to recent changes in climate. Owing to topographic effects, the velocity of range change is lowest in mountainous biomes (0.08 km yr^{-1}) and highest in flooded grasslands (1.26 km yr^{-1}), mangroves and deserts. Temperate grasslands at 0.59 km yr^{-1} have a higher rate than the global mean of 0.42 km yr^{-1} (A1B emission scenario) (Loarie et al., 2009).

Case study 6.4 A chain of indirect effects driven by reduced precipitation (Barton & Ives, 2014)

A simulated drought was imposed on alfalfa plants infested with spotted aphid *Therioaphis maculata* (Buckton) and pea aphid *Acyrthosiphon pisum* Harris. While water stress in alfalfa had no direct effect on spotted aphids, it lowered the population growth rate of pea aphids. Because ladybeetle predators were attracted to high pea aphid densities, predator densities were lower in drought treatments. Consequently, spotted aphid densities were released from top-down control (apparent competition) in drought treatments and reached densities three times higher than spotted aphids in ambient treatments with high pea aphid densities. Thus, drought affected spotted aphids in the interaction chain: drought → alfalfa → pea aphids → predators → spotted aphids. This result illustrates the lengthy path that indirect effects of climate change may take through a community, as well as the importance of community-level experiments in determining the net effect of climate change.

Warming will allow natural enemies to expand their range, potentially benefiting biological control in regions that are currently too cold. An example is the control of alligator weed, *Alternanthera philoxeroides* by *Agasicles hygrophila* in China. Elevated temperatures enabled the beetle to dramatically decrease *A. philoxeroides* growth (Lu et al., 2013). However, the target species may also expand its range towards the poles in response to warming. In such cases where weeds tolerate cold better than their natural enemies, the geographical gap between plant and herbivorous insect ranges may not disappear but will shift to higher latitudes.

During range expansion, herbivores on the invasion front may outstrip their natural enemies, which in turn, enhances their rate of range expansion. This has been observed in the brown argus butterfly, *Aricia agestis* (Denis and Schiffermüller), a species that has expanded northward in Britain during the last 30 years in association with climate warming (Menéndez et al., 2008). Although the same six parasitoid species were found in both new and established parts of its range, *A. agestis* larvae suffered lower mortality from parasitoids in newly colonised areas. However, parasitism rates by the specialist parasitoid *Cotesia astrarches* (Marshall) was highest in newly colonised areas, suggesting that the enemy release hypothesis should apply to all trophic levels.

6.6   Greater Exposure to Pest Outbreaks

Increased incidences of pest outbreaks are predicted in grasslands through disruption of predator–prey relationships due to increased climate variability (Stireman et al., 2005), and warmer temperatures enabling more generations and better winter survival. In addition, phenotypic plasticity is a key trait of successful pest species, and they may be relatively more able to cope with higher, more variable temperatures under climate change than their natural enemies. Further, eCO$_2$ can alter above–belowground interactions. While eCO$_2$ reduces oviposition by adults of the weevil *Sitona obsoletus* Gmelin it is extremely beneficial for *S. obsoletus* larvae living belowground, due to the enhanced nodulation by white clover, its favoured host plant. Climate change may, therefore, enhance biological nitrogen fixation by clover, but potential benefits may be undermined by larger populations of the pest belowground (Johnson & McNicol, 2010).

Grassland management may amplify or dampen the likelihood of an outbreak. The fitness of the grassland caterpillar *Gynaephora menyuanensis* Yan & Chou and subsequent damage in an alpine meadow on the Tibetan Plateau were exacerbated by the combination of grazing and warming (Cao et al., 2015). In comparing pest outbreak reports from Western Australia, where detailed records are available from the mid-1990s, the relative incidence of armyworms, aphids and vegetable weevils has decreased, while the incidence of pasture cockchafers, mite pests, the springtail *Sminthurus viridis* L. and snails has increased. Possible drivers were drier conditions, exacerbated by climate change, that reduced the build-up of migratory species from inland Australia, as well as the increased adoption of minimum and no-tillage systems in order to retain soil moisture (Hoffmann et al., 2008).

6.7   Non-Target Impacts

Classical biological control has uncertainties; irrespective of prior host range testing, inserting an exotic species into communities can produce multiple outcomes due to context-specific factors. These will be further confounded by climate change drivers.

The Moroccan ecotype of the braconid parasitoid *Microctonus aethiopoides* Loan, introduced into New Zealand for biological control of the lucerne pest *Sitona discoideus* Gyllenhal, is known to attack several non-target native weevil species. Currently, the levels of parasitism in these species in native grasslands appeared to be constrained by low temperatures and by the frequency of sub-zero temperatures. However, projected future temperatures may reduce this constraint (Ferguson et al. 2016).

A shift in plant life history from perennial to annual has resulted in non-target damage on a native plant *A. sessilis* (Linn.) DC. (Amaranthaceae), in China by *A. hygrophila* Selman & Vogt, (Coleoptera: Chrysomelidae), a biological control agent introduced to control alligator weed *Alternanthera philoxeroides* (Mart.) Griseb. (Lu et al., 2015).

6.8 Conclusion

The effects of climate change on grassland ecosystems are complex. However, compared to many other terrestrial ecosystems, they are easily manipulated, and respond relatively rapidly to changing conditions. Therefore, they offer researchers the opportunity to explore and quantify the multiple effects climate change may have on terrestrial invertebrate communities and food webs in field locations. Such understanding will ultimately enable informed decisions in ecosystem management in a changing world, from conservation of biodiversity to economic and environmentally sustainable agricultural practices. With grasslands grazed by domesticated animals covering over a fifth of the Earth's ice-free land surface (Ramankutty et al., 2008) and playing a key role in global food security, carbon storage and carbon sequestration (O'Mara, 2012), such research is vital to our planet's future.

Acknowledgements

The authors thank Pauline Hunt for the artwork and Philip Smith for proofreading this chapter.

References

AbdElgawad, H., Peshev, D., Zinta, G., Van Den Ende, W., Janssens, I.A. & Asard, H. (2014) Climate extreme effects on the chemical composition of temperate grassland species under ambient and elevated CO_2: A comparison of fructan and non-fructan accumulators. *PLOS ONE*, **9**, e92044.

Afkhami, M.E., McIntyre, P.J. & Strauss, S.Y. (2014) Mutualist-mediated effects on species' range limits across large geographic scales. *Ecology Letters*, **17**, 1265–1273.

Agrell, J., Anderson, P., Oleszek, W., Stochmal, A. & Agrell, C. (2004) Combined Effects of Elevated CO_2 and Herbivore Damage on Alfalfa and Cotton. *Journal of Chemical Ecology*, **30**, 2309–2324.

Aslam, T.J., Johnson, S.N. & Karley, A.J. (2013) Plant-mediated effects of drought on aphid population structure and parasitoid attack. *Journal of Applied Entomology*, **137**, 136–145.

Barton, B.T. (2010) Climate warming and predation risk during herbivore ontogeny. *Ecology*, **91**, 2811–2818.

Barton, B.T. (2011) Local adaptation to temperature conserves top-down control in a grassland food web. *Proceedings of the Royal Society B: Biological Sciences*, **278**, 3102–3107.

Barton, B.T., Beckerman, A.P. & Schmitz, O.J. (2009) Climate warming strengthens indirect interactions in an old-field food web. *Ecology*, **90**, 2346–2351.

Barton, B.T. & Ives, A.R. (2014) Species interactions and a chain of indirect effects driven by reduced precipitation. *Ecology*, **95**, 486–494.

Barton, B.T. & Schmitz, O.J. (2009) Experimental warming transforms multiple predator effects in a grassland food web. *Ecology Letters*, **12**, 1317–1325.

Bell, N.L., Townsend, R.J., Popay, A.J., Mercer, C.F. & Jackson, T.A. (2011). Black beetle: lessons from the past and options for the future. *Pasture Persistence Symposium, Grassland Research and Practice Series No. 15* (ed. C.F. Mercer), pp. 119–124. New Zealand Grassland Association, Dunedin.

Bidart-Bouzat, M.G. & Imeh-Nathaniel, A. (2008) Global change effects on plant chemical defenses against insect herbivores. *Journal of Integrative Plant Biology*, **50**, 1339–1354.

Buchholz, S., Rolfsmeyer, D. & Schirmel, J. (2013) Simulating small-scale climate change effects – lessons from a short-term field manipulation experiment on grassland arthropods. *Insect Science*, **20**, 662–670.

Cantarel, A.A.M., Bloor, J.M.G. & Soussana, J.-F. (2013) Four years of simulated climate change reduces above-ground productivity and alters functional diversity in a grassland ecosystem. *Journal of Vegetation Science*, **24**, 113–126.

Cao, H., Zhao, X., Wang, S., Zhao, L., Duan, J., Zhang, Z., Ge, S. & Zhu, X. (2015) Grazing intensifies degradation of a Tibetan Plateau alpine meadow through plant–pest interaction. *Ecology and Evolution*, **5**, 2478–2486.

Cock, M.J.W., Biesmeijer, J.C., Cannon, R.J.C., Gerard, P.J., Gillespie, D., Jiménez, J.J., Lavelle, P.M. & Raina, S.K. (2013) The implications of climate change for positive contributions of invertebrates to world agriculture. *CAB Reviews: Perspectives in Agriculture, Veterinary Science, Nutrition and Natural Resources*, **8**, 028.

Cullen, B.R., Eckard, R.J. & Rawnsley, R.P. (2012) Resistance of pasture production to projected climate changes in south-eastern Australia. *Crop and Pasture Science*, **63**, 77–86.

de Sassi, C., Müller, C.B. & Krauss, J. (2006) Fungal plant endosymbionts alter life history and reproductive success of aphid predators. *Proceedings of the Royal Society B: Biological Sciences*, **273**, 1301–1306.

de Sassi, C., Staniczenko, P.P.A. & Tylianakis, J.M. (2012) Warming and nitrogen affect size structuring and density dependence in a host–parasitoid food web. *Philosophical Transactions of the Royal Society B: Biological Sciences*, **367**, 3033–3041.

de Sassi, C. & Tylianakis, J.M. (2012) Climate change disproportionately increases herbivore over plant or parasitoid biomass. *PLOS ONE*, **7**, e40557.

Dhileepan, K. (2003) Seasonal variation in the effectiveness of the leaf-feeding beetle *Zygogramma bicolorata* (Coleoptera: Chrysomelidae) and stem-galling moth *Epiblema strenuana* (Lepidoptera: Tortricidae) as biocontrol agents on the weed *Parthenium hysterophorus* (Asteraceae). *Bulletin of Entomological Research*, **93**, 393–401.

Diaz, R., Manrique, V., He, Z. & Overholt, W.A. (2012) Effect of elevated CO_2 on tropical soda apple and its biological control agent *Gratiana boliviana* (Coleoptera: Chrysomelidae). *Biocontrol Science and Technology*, **22**, 763–776.

Dyer, L.A., Richards, L.A., Short, S.A. & Dodson, C.D. (2013) Effects of CO_2 and Temperature on Tritrophic Interactions. *PLOS ONE*, **8**, e62528.

East, R. & Willoughby, B.E. (1980) The effects of pasture defoliation in summer on grass grub (*Costelytra zealandica*) populations. *New Zealand Journal of Agricultural Research*, **23**, 547–562.

Eisenhauer, N., Dobies, T., Cesarz, S., Hobbie, S.E., Meyer, R.J., Worm, K. & Reich, P.B. (2013) Plant diversity effects on soil food webs are stronger than those of elevated CO_2 and N deposition in a long-term grassland experiment. *Proceedings of the National Academy of Sciences of the United States of America*, **110**, 6889–6894.

Epstein, H.E., Gill, R.A., Paruelo, J.M., Lauenroth, W.K., Jia, G.J. & Burke, I.C. (2002) The relative abundance of three plant functional types in temperate grasslands and shrublands of North and South America: Effects of projected climate change. *Journal of Biogeography*, **29**, 875–888.

Ferguson, C.M., Kean, J.M., Barton, D.M. & Barratt, B.I.P. (2016) Ecological mechanisms for non-target parasitism by the Moroccan ecotype of *Microctonus aethiopoides* Loan (Hymenoptera: Braconidae) in native grassland. *Biological Control*, **96**, 28–38.

Faeth, S.H. (2002) Are endophytic fungi defensive plant mutualists? *Oikos*, **98**, 25–36.

Fuhrer, J. (2003) Agroecosystem responses to combinations of elevated CO_2, ozone, and global climate change. *Agriculture, Ecosystems and Environment*, **97**, 1–20.

Gargallo-Garriga, A., Sardans, J., Pérez-Trujillo, M., Oravec, M., Urban, O., Jentsch, A., Kreyling, J., Beierkuhnlein, C., Parella, T. & Peñuelas, J. (2015) Warming differentially influences the effects of drought on stoichiometry and metabolomics in shoots and roots. *New Phytologist*, **207**, 591–603.

Gerard, P.J., Bell, N.L., Eden, T.M., King, W.M., Mapp, N.R., Pirie, M.R., & Rennie, G.M. (2013) Influence of pasture renewal, soil factors and climate on black beetle abundance in Waikato and Bay of Plenty. *Proceedings of the New Zealand Grassland Association*, **75**, 235–240.

Gerard, P.J., Vasse, M. & Wilson, D.J. (2012) Abundance and parasitism of clover root weevil (*Sitona lepidus*) and Argentine stem weevil (*Listronotus bonariensis*) in pastures. *New Zealand Plant Protection*, **65**, 180–185.

Ghermandi, L. & Gonzalez, S. (2009) Diversity and functional groups dynamics affected by drought and fire in Patagonian grasslands. *Ecoscience*, **16**, 408–417.

Gibb, H., Muscat, D., Binns, M.R., Silvey, C.J., Peters, R.A., Warton, D.I. & Andrew, N.R. (2015) Responses of foliage-living spider assemblage composition and traits to a climatic gradient in *Themeda* grasslands. *Austral Ecology*, **40**, 225–237.

Graesser, J., Aide, T.M., Grau, H.R. & Ramankutty, N. (2015) Cropland/pastureland dynamics and the slowdown of deforestation in Latin America. *Environmental Research Letters*, **10**, 034017.

Grime, J.P., Brown, V.K., Thompson, K., Masters, G.J., Hillier, S.H., Clarke, I.P., Askew, A.P., Corker, D. & Kielty, J.P. (2000) The response of two contrasting limestone grasslands to simulated climate change. *Science*, **289**, 762–765.

Hanley, M.E., Trofimov, S. & Taylor, G. (2004) Species-level effects more important than functional group-level responses to elevated CO_2: Evidence from simulated turves. *Functional Ecology*, **18**, 304–313.

Harmon, J.P., Moran, N.A. & Ives, A.R. (2009) Species Response to Environmental Change: Impacts of Food Web Interactions and Evolution. *Science*, **323**, 1347–1350.

Hobbs, R.J., Yates, S. & Mooney, H.A. (2007) Long-term data reveal complex dynamics in grassland in relation to climate and disturbance. *Ecological Monographs*, **77**, 545–568.

Hoffmann, A.A., Weeks, A.R., Nash, M.A., Mangano, G.P. & Umina, P.A. (2008) The changing status of invertebrate pests and the future of pest management in the Australian grains industry. *Australian Journal of Experimental Agriculture*, **48**, 1481–1493.

Hunt, M.G., Rasmussen, S., Newton, P.C.D., Parsons, A.J. & Newman, J.A. (2005) Near-term impacts of elevated CO_2, nitrogen and fungal endophyte-infection on *Lolium perenne* L. growth, chemical composition and alkaloid production. *Plant, Cell and Environment*, **28**, 1345–1354.

Jani, A.J., Faeth, S.H. & Gardner, D. (2010) Asexual endophytes and associated alkaloids alter arthropod community structure and increase herbivore abundances on a native grass. *Ecology Letters*, **13**, 106–117.

Jeffs, C.T. & Lewis, O.T. (2013) Effects of climate warming on host–parasitoid interactions. *Ecological Entomology*, **38**, 209–218.

Joern, A. & Laws, A.N. (2013) Ecological mechanisms underlying arthropod species diversity in grasslands. *Annual Review of Entomology*, **58**, 19–36.

Johns, C.V., Beaumont, L.J. & Hughes, L. (2003) Effects of elevated CO_2 and temperature on development and consumption rates of *Octotoma championi* and *O. scabripennis* feeding on *Lantana camara*. *Entomologia Experimentalis et Applicata*, **108**, 169–178.

Johnson, S.N. & McNicol, J.W. (2010) Elevated CO_2 and aboveground-belowground herbivory by the clover root weevil. *Oecologia*, **162**, 209–216.

Johnson, S.N., Ryalls, J.M.W. & Karley, A.J. (2014) Global climate change and crop resistance to aphids: Contrasting responses of lucerne genotypes to elevated atmospheric carbon dioxide. *Annals of Applied Biology*, **165**, 62–72.

Johnson, S.N., Staley, J.T., McLeod, F.A.L. & Hartley, S.E. (2011) Plant-mediated effects of soil invertebrates and summer drought on above-ground multitrophic interactions. *Journal of Ecology*, **99**, 57–65.

King, P.D., Mercer, C.F. & Meekings, J.S. (1981) Ecology of black beetle, *Heteronychus arator* (Coleoptera: Scarabaeidae) – population studies. *New Zealand Journal of Agricultural Research*, **24**, 87–97.

Klaiber, J., Najar-Rodriguez, A.J., Dialer, E. & Dorn, S. (2013) Elevated carbon dioxide impairs the performance of a specialized parasitoid of an aphid host feeding on *Brassica* plants. *Biological Control*, **66**, 49–55.

Knops, J.M.H., Tilman, D., Haddad, N.M., Naeem, S., Mitchell, C.E., Haarstad, J., Ritchie, M.E., Howe, K.M., Reich, P.B., Siemann, E. & Groth, J. (1999) Effects of plant species richness on invasion dynamics, disease outbreaks, insect abundances and diversity. *Ecology Letters*, **2**, 286–293.

Kuneš, P., Svobodová-Svitavská, H., Kolář, J., Hajnalová, M., Abraham, V., Macek, M., Tkáč, P. & Szabó, P. (2015) The origin of grasslands in the temperate forest zone of east-central Europe: Long-term legacy of climate and human impact. *Quaternary Science Reviews*, **116**, 15–27.

Laws, A.N. & Joern, A. (2015) Predator–prey interactions are context dependent in a grassland plant–grasshopper–wolf spider food chain. *Environmental Entomology*, **44**, 519–528.

Lazzarotto, P., Calanca, P., Semenov, M. & Fuhrer, J. (2010) Transient responses to increasing CO_2 and climate change in an unfertilized grass–clover sward. *Climate Research*, **41**, 221–232.

Lee, M.A., Manning, P., Walker, C.S. & Power, S.A. (2014) Plant and arthropod community sensitivity to rainfall manipulation but not nitrogen enrichment in a successional grassland ecosystem. *Oecologia*, **176**, 1173–1185.

Lilley, J.M., Bolger, T.P., Peoples, M.B. & Gifford, R.M. (2001) Nutritive value and the nitrogen dynamics of *Trifolium subterraneum* and *Phalaris aquatica* under warmer, high CO_2 conditions. *New Phytologist*, **150**, 385–395.

Loarie, S.R., Duffy, P.B., Hamilton, H., Asner, G.P., Field, C.B. & Ackerly, D.D. (2009) The velocity of climate change. *Nature*, **462**, 1052–1055.

Logan, J.D., Wolesensky, W. & Joern, A. (2006) Temperature-dependent phenology and predation in arthropod systems. *Ecological Modelling*, **196**, 471–482.

Lu, X., Siemann, E., He, M., Wei, H., Shao, X. & Ding, J. (2015) Climate warming increases biological control agent impact on a non-target species. *Ecology Letters*, **18**, 48–56.

Lu, X., Siemann, E., Shao, X., Wei, H. & Ding, J. (2013) Climate warming affects biological invasions by shifting interactions of plants and herbivores. *Global Change Biology*, **19**, 2339–2347.

Mason, P.A., Bernardo, M.A. & Singer, M.S. (2014) A mixed diet of toxic plants enables increased feeding and anti-predator defense by an insect herbivore. *Oecologia*, **176**, 477–486.

Menéndez, R., González-Megías, A., Lewis, O.T., Shaw, M.R. & Thomas, C.D. (2008) Escape from natural enemies during climate-driven range expansion: A case study. *Ecological Entomology*, **33**, 413–421.

Morgan, J.A., LeCain, D.R., Pendall, E., Blumenthal, D.M., Kimball, B.A., Carrillo, Y., Williams, D.G., Heisler-White, J., Dijkstra, F.A. & West, M. (2011) C_4 grasses prosper as carbon dioxide eliminates desiccation in warmed semi-arid grassland. *Nature*, **476**, 202–205.

O'Callaghan, M., Jackson, T.A. & Noonan, M.J. (1989) Ecology of *Serratia entomophila* in soil. *Proceedings of the 5th Australaisian Grassland Invertebrate Conference* (ed. P.P. Stahle), pp. 69–75. D.D. Printing, Victoria.

O'Mara, F.P. (2012) The role of grasslands in food security and climate change. *Annals of Botany*, **110**, 1263–1270.

Parr, C.L., Ryan, B.J. & Setterfield, S.A. (2010) Habitat complexity and invasive species: The impacts of gamba grass (*Andropogon gayanus*) on invertebrates in an Australian tropical savanna. *Biotropica*, **42**, 688–696.

Parry, H.R., Macfadyen, S., Hopkinson, J.E., Bianchi, F.J.J.A., Zalucki, M.P., Bourne, A. & Schellhorn, N.A. (2015) Plant composition modulates arthropod pest and predator abundance: Evidence for culling exotics and planting natives. *Basic and Applied Ecology*, **16**, 531–543.

Ramankutty, N., Evan, A.T., Monfreda, C. & Foley, J.A. (2008) Farming the planet: 1. Geographic distribution of global agricultural lands in the year 2000. *Global Biogeochemical Cycles*, **22**, GB1003.

Reeves, J.L., Blumenthal, D.M., Kray, J.A. & Derner, J.D. (2015) Increased seed consumption by biological control weevil tempers positive CO_2 effect on invasive plant (*Centaurea diffusa*) fitness. *Biological Control*, **84**, 36–43.

Robinson, N., Armstead, S. & Bowers, M.D. (2012) Butterfly community ecology: The influences of habitat type, weather patterns, and dominant species in a temperate ecosystem. *Entomologia Experimentalis et Applicata*, **145**, 50–61.

Root, R.B. (1973) Organization of a plant-arthropod association in simple and diverse habitats: The fauna of collards (*Brassica oleracea*). *Ecological Monographs*, **43**, 95–124.

Ryalls, J.M.W., Riegler, M., Moore, B.D. & Johnson, S.N. (2013) Biology and trophic interactions of lucerne aphids. *Agricultural and Forest Entomology*, **15**, 335–350.

Schmitz, O.J. & Barton, B.T. (2014) Climate change effects on behavioral and physiological ecology of predator–prey interactions: Implications for conservation biological control. *Biological Control*, **75**, 87–96.

Stiling, P. & Cornelissen, T. (2007) How does elevated carbon dioxide (CO_2) affect plant–herbivore interactions? A field experiment and meta-analysis of CO_2-mediated changes on plant chemistry and herbivore performance. *Global Change Biology*, **13**, 1823–1842.

Stireman, J.O. III, Dyer, L.A., Janzen, D.H., Singer, M.S., Lill, J.T., Marquis, R.J., Ricklefs, R.E., Gentry, G.L., Hallwachs, W., Coley, P.D., Barone, J.A., Greeney, H.F., Connahs, H., Barbosa, P., Morais, H.C. & Diniz, I.R. (2005) Climatic unpredictability and parasitism of caterpillars: Implications of global warming. *Proceedings of the National Academy of Sciences of the United States of America*, **102**, 17384–17387.

Stokes, C.J., Ash, A. & Howden, S.M. (2008) Climate change impacts on Australian rangelands. *Rangelands*, **30**, 40–45.

Suggitt, A.J., Gillingham, P.K., Hill, J.K., Huntley, B., Kunin, W.E., Roy, D.B. & Thomas, C.D. (2011) Habitat microclimates drive fine-scale variation in extreme temperatures. *Oikos*, **120**, 1–8.

't Mannetje, L. (2007a) Climate change and grasslands through the ages: An overview. *Grass and Forage Science*, **62**, 113–117.

't Mannetje, L. (2007b) The role of grasslands and forests as carbon stores. *Tropical Grasslands*, **41**, 50–54.

Tariq, M., Wright, D.J., Bruce, T.J.A. & Staley, J.T. (2013) Drought and Root Herbivory Interact to Alter the Response of Above-Ground Parasitoids to Aphid Infested Plants and Associated Plant Volatile Signals. *PLOS ONE*, **8**, e69013.

Thom, E.R., Popay, A.J., Hume, D.E. & Fletcher, L.R. (2012) Evaluating the performance of endophytes in farm systems to improve farmer outcomes – A review. *Crop and Pasture Science*, **63**, 927–943.

Thomson, L.J., Macfadyen, S. & Hoffmann, A.A. (2010) Predicting the effects of climate change on natural enemies of agricultural pests. *Biological Control*, **52**, 296–306.

Tylianakis, J.M. & Binzer, A. (2014) Effects of global environmental changes on parasitoid–host food webs and biological control. *Biological Control*, **75**, 77–86.

Tozer, K.N., Minneé, E. & Cameron, C.A. (2012) Resistance of New Zealand dairy pastures to ingress of summer-active annual grass weeds. *Crop and Pasture Science*, **63**, 1026–1033.

Villalpando, S.N., Williams, R.S. & Norby, R.J. (2009) Elevated air temperature alters an old-field insect community in a multifactor climate change experiment. *Global Change Biology*, **15**, 930–942.

Voigt, W., Perner, J., Davis, A.J., Eggers, T., Schumacher, J., Bährmann, R., Fabian, B., Heinrich, W., Köhler, G., Lichter, D., Marstaller, R. & Sander, F.W. (2003) Trophic levels are differentially sensitive to climate. *Ecology*, **84**, 2444–2453.

Voigt, W., Perner, J. & Jones, T.H. (2007) Using functional groups to investigate community response to environmental changes: Two grassland case studies. *Global Change Biology*, **13**, 1710–1721.

Walter, J., Hein, R., Auge, H., Beierkuhnlein, C., Löffler, S., Reifenrath, K., Schädler, M., Weber, M. & Jentsch, A. (2012) How do extreme drought and plant community composition affect host plant metabolites and herbivore performance? *Arthropod-Plant Interactions*, **6**, 15–25.

Wang, G. (2005) Agricultural drought in a future climate: results from 15 global climate models participating in the IPCC 4th assessment. *Climate Dynamics*, **25**, 739–753.

Wang, G.-H., Wang, X.-X., Sun, Y.-C. & Ge, F. (2014) Impacts of elevated CO_2 on *Bemisia tabaci* infesting Bt cotton and its parasitoid *Encarsia formosa*. *Entomologia Experimentalis et Applicata*, **152**, 228–237.

Williams, T.A. & Abberton, M.T. (2004) Earlier flowering between 1962 and 2002 in agricultural varieties of white clover. *Oecologia*, **138**, 122–126.

Woodcock, B.A., Bullock, J.M., Nowakowski, M., Orr, R., Tallowin, J.R.B. & Pywell, R.F. (2012) Enhancing floral diversity to increase the robustness of grassland beetle assemblages to environmental change. *Conservation Letters*, **5**, 459–469.

Yin, J., Sun, Y. & Ge, F. (2014) Reduced plant nutrition under elevated CO_2 depresses the immunocompetence of cotton bollworm against its endoparasite. *Scientific Reports*, **4**, 4538.

Zavaleta, E.S., Shaw, M.R., Chiariello, N.R., Mooney, H.A. & Field, C.B. (2003a) Additive effects of simulated climate changes, elevated CO_2, and nitrogen deposition on grassland diversity. *Proceedings of the National Academy of Sciences of the United States of America*, **100**, 7650–7654.

Zavaleta, E.S., Shaw, M.R., Chiariello, N.R., Thomas, B.D., Cleland, E.E., Field, C.B. & Mooney, H.A. (2003b) Grassland responses to three years of elevated temperature, CO_2, precipitation, and N deposition. *Ecological Monographs*, **73**, 585–604.

Zhang, Y. & Adams, J. (2011) Top-down control of herbivores varies with ecosystem types. *Journal of Ecology*, **99**, 370–372.

Zhu, H., Wang, D., Wang, L., Fang, J., Sun, W. & Ren, B. (2014) Effects of altered precipitation on insect community composition and structure in a meadow steppe. *Ecological Entomology*, **39**, 453–461.

7

Climate Change and Arthropod Ectoparasites and Vectors of Veterinary Importance

Hannah Rose Vineer[1], Lauren Ellse[2] and Richard Wall[2]

[1] *School of Veterinary Sciences, Life Sciences Building, University of Bristol, UK*
[2] *School of Biological Sciences, Life Sciences Building, University of Bristol, UK*

Summary

Arthropod ectoparasites are important worldwide in terms of both economic and welfare impacts. Many ectoparasites with off-host life-cycle stages are vulnerable to climate change as invertebrate life-history processes such as development are temperature- and moisture-dependent. Changes in the phenology and distribution of tick species have been observed in recent years, and significant changes in the seasonal patterns of risk of blowfly strike are predicted under scenarios of climate change. Other ectoparasites such as mites spend their entire life-cycle on the host and are therefore relatively protected from the direct impacts of climate change. The impact of climate change on invertebrates is often considered in terms of increases in development rates and subsequent increases in disease risk. However, complex interactions between development, mortality, and farmer behaviour can result in non-linear effects of increasing temperatures on parasite abundance and disease risk. Predictive models have demonstrated the importance of husbandry as a driver of ectoparasite epidemiology and the impact of future changes in management strategies must be considered in climate impact assessments. Host–parasite and host–vector–pathogen interactions play a major role in determining disease dynamics and evidence from endoparasite species suggests that the nature and timing of these interactions may change with altered climate. Although there is potential for these changes to result in an increase in ectoparasite abundance and disease incidence, strategic changes to husbandry practices also provide opportunities to mitigate impacts of climate change on parasite risk. Predictive models provide a valuable tool to evaluate alternative management strategies.

7.1   Introduction

Ectoparasitic arthropods live on, puncture, or burrow into the surface of their host's epidermis, to feed or shelter, which results in direct damage to skin and other subcutaneous tissues. The presence of ectoparasites and their salivary and faecal antigens can stimulate immune responses, in some individuals, leading to hypersensitivity. Feeding may also result in significant blood loss, secondary infestation, pruritis, excoriation, alopecia and, in some cases, ultimately death. The behaviour of ectoparasites may also cause harm indirectly, particularly when present at high intensities, causing disturbance, increasing levels of behaviour such as rubbing, and leading to reduced time spent

Global Climate Change and Terrestrial Invertebrates, First Edition. Edited by Scott N. Johnson and T. Hefin Jones.
© 2017 John Wiley & Sons, Ltd. Published 2017 by John Wiley & Sons, Ltd.

grazing or ruminating and, in some cases, self-wounding (Berriatua et al., 2001). As a result, ectoparasitic infections are of major welfare and economic concern and are a constraint to efficient animal production, affecting growth rates, fertility, birth weights and incurring direct costs in their control (Wall, 2007). Importantly, some ectoparasites also act as vectors of protozoa, bacteria, viruses, cestodes (tapeworms) and nematodes (roundworms), many of which may be zoonotic. Although relatively few in number, through their direct and indirect effects on their hosts, the various species of arthropod ectoparasite have a major effect on the husbandry, productivity and welfare of livestock (Rehbein et al., 2003; Nieuwhof & Bishop, 2005; Wright et al., 2014).

In the last 50 years, there have been marked technical, social, economic and political changes which have altered the nature of animal husbandry and parasite control in many parts of the world. Perhaps the most notable of these changes has been the development of powerful new synthetic insecticides and endectocides such as the organophosphates, macrocyclic lactones and insect growth regulators that have allowed farmers to depend less on expensive, labour intensive, physical control measures, for example crutching sheep (removal of soiled wool around the tail and anus) to reduce the risk of infestation of living animals by the blowflies *Lucilia sericata* or *Lucilia cuprina* (known as myiasis). However, although these insecticides were and in many cases remain major contributors to improved animal welfare and cost-effective animal husbandry, the almost inevitable development of resistance to these compounds is driving the need for more strategic thinking about their use. For example, resistance to organophosphates and pyrethroids has been reported in a wide range of species of flies, lice and ticks (Ellse et al., 2012; Heath & Levot, 2015; Sands et al., 2015) and resistance is now a major barrier to effective ectoparasite control in some areas. The incidence and prevalence of ectoparasites may also have been altered by changes in husbandry associated with increased productivity such as high protein diets, reduced genetic diversity, large-scale movement of animals and increased stocking rates. The impact of these factors and outbreaks of endemic disease, the threats of exotic disease introduction and changes in arthropod abundance and distribution have been further exacerbated by concern associated with changes in climate (Olesen & Bindi, 2002; Morgan & Wall, 2009).

Climate change predictions indicate that global temperatures may increase by up to 4.8°C in the next hundred years (IPCC, 2013). Generally, arthropods have short generation times and high reproductive capacities. Therefore, their response to changes in climate can result in large and rapid population fluctuations. In addition, many arthropod species are capable of high levels of movement which may lead to changes in distribution and range (Stange & Ayres, 2010). While the potential impact of climate change on the distribution and prevalence of arthropod ectoparasites and vectors of disease incidence is of very real and immediate concern, the differential effects of climate change on mortality, development and transmission are likely to vary widely depending on the species in question (Rose et al., 2015) and within species in different geographical regions (Baker & Oormazdi, 1978). As a result, unambiguous links between climate change and disease incidence are difficult to predict or demonstrate.

Climate warming may lead to increased numbers of arthropod parasites and vectors, and prolonged periods during which conditions are favourable for arthropod survival and pathogen transmission. In addition, the vector potential of each individual arthropod may be enhanced by biological changes stimulated by temperature. Similarly, warming may make it more likely that tropical and subtropical species, for example ticks, will establish in more temperate regions. However, the effects of changing climate may also have negative effects on arthropod abundance. For example, higher summer temperatures are likely to decrease the survival of off-host stages of obligate parasites, thereby reducing environmentally mediated transmission. Similarly, warmer winters may increase the metabolic rate during diapause which may increase mortality of seasonal parasites. In addition, since most arthropod parasites are at least as limited by humidity as by temperature, effects of climate change will depend critically on the interaction of both these climatic factors. Hence, while general predictions concerning the effects of climate change and in particular global increases in temperature on parasite epidemiology, which are based on the impacts of accelerated development on increases in abundance or distribution, might in themselves be valid, actual effects in terms of the incidence of disease and production loss will be complex and are modulated by many factors. These include biological processes such as host immunity and even behaviour by farmers and their responses to perceived changing threat or loss of income. Factors affecting the rate of detection of parasites and associated disease, thresholds for treatment, and changes in husbandry, are likely to strongly affect the nature of farmer responses to altered parasite epidemiology, and their effectiveness. Interactions between these factors and parasite biology are very system specific. Hence, the net effect of climate change may be complex and far from easily predicted (Morgan & Wall, 2009).

This chapter will introduce some of the key concepts associated with climate change impacts in the context of ectoparasites and vectors of veterinary importance, particularly those associated with animal husbandry, using specific examples that have been the focus of significant research efforts – primarily ticks, mange mites and the blowfly *L. sericata.*

7.2 Parasite–Host Interactions

The nature of the problems caused by any arthropod ectoparasite and whether infestation results in disease will be highly dependent on the nature of its interaction with its host. For ectoparasites that live permanently on the host, with no free-living element of the life cycle, such as lice and mites, superficially it might be expected that impacts of changing climate will be limited, because these arthropods live in a highly stable buffered environment within the host's pelage. Any effect on such parasites may occur through changes to the length of the grazing and housing periods during the year. For example, climate change is likely to have only limited direct impact on the sheep scab mite, *Psoroptes ovis,* which causes psoroptic mange in sheep, since this parasite completes its entire life cycle on the host, protected by the relatively stable microclimate in the sheep's fleece. However, indirect effects of changes to housing periods (increasing contact rate), altered shearing regimes (changing the microclimate within the fleece), and altered applications of endectocides in response to changes in the epidemiology of endoparasites, may affect the seasonal incidence and prevalence of psoroptic mange in sheep flocks.

For the majority of arthropod parasites and vectors that spend a large proportion of their lives within the environment and are not closely associated with the microclimate of the host skin, climate change is likely to have more profound direct effects on their populations. In the first instance, the effect of climate warming on development rates is likely to lead to an increased number of generations, increasing abundance and prolonged periods during which conditions are favourable for survival and transmission (Wall & Ellse, 2011). However, this may be offset by increased mortality rates as a result of drought and increased temperatures (Rose & Wall, 2011).

Climate change, and in particular climate warming, may make it more likely that tropical and subtropical species will establish in more temperate areas. For example, it has been suggested that an increase in summer temperature of 2–3°C, which is predicted in the next 50 years by many of the Intergovernmental Panel on Climate Change (IPCC) scenarios, could result in the permanent establishment of *Rhipicephalus sanguineus* in many locations where it is currently a temporary alien and found only in heated or sheltered buildings such as kennels (Gray et al., 2009). In some instances, the overall effects of climate on ectoparasites may be further complicated by differential effects of climate on host and parasite and their potential for adaptation.

There is currently no evidence for this for ectoparasites, but endoparasitic nematodes offer a useful illustration of potential impacts on host and parasite. The eggs of the nematode *Nematodirus battus* hatch on pasture *en masse* in spring, coinciding with the peak availability of naïve hosts (young lambs). Mathematical modelling suggests that in some temperate regions increasing temperatures and earlier temperature-driven hatching of *N. battus* may have led to fewer outbreaks due to a mismatch between host (lambing date) and parasite phenology (Gethings et al., 2015).

The interaction between climatic events, phenology and seasonality can also affect the transmission of vector-borne pathogens, population dynamics of host, vector and pathogen, and disease dynamics in the host. In the transmission of tick-borne encephalitis (TBE) by the tick, *Ixodes ricinus*, moisture stress levels may affect the feeding patterns. If moisture stress (e.g., drought or high potential evaporation rates) occurs early in the season when ticks are searching for food (known as questing) in April–June, the nymphal ticks may feed on small rodents nearer the ground and increase the rodent reservoir of TBE (Randolph & Storey, 1999), whereas if high levels of moisture stress occur later in the season (June–August) then the ticks may starve for want of sufficient hosts prior to the winter when questing generally ceases (Randolph et al., 2002; Randolph, 2004). Hence, an increase in temperature and extreme weather events predicted under most climate change scenarios, are likely to affect the incidence and severity of vector-borne disease in many locations.

Global warming is also likely to result in an unpredictable climate with the incidence of unusual or unseasonal weather likely to be more frequent. Extreme weather events can lead to establishment of vector-borne disease in new locations. Since a European heatwave in 2010, West Nile Virus (WNV) has become endemic in southeast Europe with summer outbreaks of disease in equines and humans in subsequent years despite the temperatures being lower than in 2010 (Paz & Semenza, 2013). Establishment of vector-borne disease in a new area is the result of complex interplay between hosts, vectors and pathogens. For WNV, birds are the reservoir host. Therefore, climate mediated deviations in the flight path of infected migratory birds may initially introduce the pathogen to a new location. Native bird species may then be infected by mosquito vectors and become amplifiers for the virus if the temperature is sufficient. Finally, a bridge vector such as *Culex pipiens,* which feeds on both birds and mammals, must be present for mammalian transmission (Paz & Semenza, 2013). Each stage of this transmission cycle can be affected by numerous biotic, environmental and seasonal factors.

7.3   Evidence of the Impacts of Climate on Ectoparasites and Vectors

Empirical studies of the effects of climate change on ectoparasites are hard to demonstrate, partly due to the long time-scales involved. Understandably, many studies have relied on historical surveillance datasets. However, these data can have drawbacks in that the frequency and quality of sampling effort may not be consistent throughout the decades. Therefore, the robustness of conclusions drawn from these studies should be considered. Nevertheless, existing data strongly suggest that the distribution, abundance and phenology of some ectoparasites has changed markedly in recent years and this appears to be particularly evident in some of the intermittent ectoparasite species, such as the tick *I. ricinus*, which is highly sensitive to climate variation (Kirby et al., 2004). In northern latitudes, *I. ricinus* is typically active in spring and early summer, quiescent in mid-summer and usually has a secondary peak of activity in later summer or autumn. However, studies in Germany have shown that unfed *I. ricinus,* which had previously been kept in the laboratory, were able to survive and quest throughout the winter where previously the winter temperatures would have been too cold to support tick activity. Moreover, in this study, both nymphs and adult *I. ricinus* were collected in January of the same year (Dautel et al., 2008). This suggests that the risk period for tick bites and the associated disease transmission may be becoming a year round issue in areas where temperature induced seasonal diapause was previously induced in the tick population. There is also considerable evidence for the altitudinal and latitudinal expansion of *I. ricinus* in central and northern Europe where it is considered to have moved northwards (Lindgren et al., 2000; Daniel et al., 2003; Kirby et al., 2004; Pietzsch et al., 2005). Data collected by questionnaire from 20 districts in Sweden showed a large increase in tick abundance in the west-central regions where previously ticks had been rare. This was validated by blanket dragging in 54 regions along the perceived latitudinal boundary for ticks (Tälleklint & Jaenson, 1998; Lindgren et al., 2000). These studies found that in northern areas the winters were milder in the 1990s compared to the 1980s, with fewer days having a minimum temperature under $-12°C$ and that this temperature change was significantly correlated with an increase in tick abundance, whereas, in southern areas the rise in temperature had produced longer periods conducive for tick development and activity during the spring and autumn (Lindgren et al., 2000).

In the Šumava National Park in the Czech Republic the expansion of the threshold altitude for ticks has been well documented. Previously the maximum altitude in which ticks were found on small mammals was shown to be 800 m above sea level (a.s.l.) in studies carried out in 1957 and 1979–1980 (Daniel et al., 2003). In 1981–1983 experimental field studies demonstrated that *I. ricinus* could not survive at low temperatures above 700 m a.s.l. as it resulted in depletion of fat supplies and therefore, death before completion of the life-cycle (Daniel, 1993). However, in the same region in 2001, 400 ticks were collected from 55 working dogs which had remained permanently within an altitude range of 1000–1100 m a.s.l. In addition, there has been an increase in recorded cases of tick-borne diseases such borreliosis (Lyme disease), caused by the spirochaete *Borrelia burgdorferi* s.l., and TBE in this region of the Czech Republic at altitudes where infection risk was previously thought to be low. These reports coincide with meteorological events which favour *I. ricinus* survival and questing activity (Daniel et al., 2003, 2004; Danielová et al., 2006).

In four habitats in Eastern Russia the prevalence of ticks was measured in 1 km transects over a 35 year period. The number of ticks was considered to have increased significantly in the more recent 25 years with some study sites seeing an increase from <1 tick per 1 km transect to approximately 18 per 1 km transect. The authors postulate that the probable cause of this increase is a combination of climatic and anthropogenic factors (Korotkov et al., 2015).

As alluded to above, these changes in tick distribution and abundance have been associated with changes in disease incidence, for example, the development of new foci of TBE in Scandinavia. The incidence of TBE in the Stockholm county region of Sweden has increased between 1960 and 1998; high disease incidence was associated with two consecutive mild winters in the years previous to the outbreak as well as extended spring and autumn tick activity periods in the year prior to high disease incidence (Lindgren & Gustafson, 2001). The emergence of tick-borne pathogens has also been reported in Norway and Denmark, which the authors correlate with climatic factors such as daily temperature (Skarpaas et al., 2004; Skarphédinsson et al., 2005).

The geographic distribution of ticks in Great Britain is estimated to have expanded by 17% and the abundance of ticks to have increased at 73% of locations surveyed over recent decades (Scharlemann et al., 2008). Associated with this, human cases of Lyme disease apparently increased by 30-fold between 1999 and 2008 in Scotland (Health Protection Scotland, 2009). Lyme borreliosis has also seen an increase in incidence in the United States and many European countries, with some rates increasing two or three-fold. Similarly the detection of pathogenic tick-borne *Rickettsia* is becoming more common, implicating a range of tick species in a number of new geographic areas. Similarly, the distribution of the European meadow tick, *Dermacentor reticulatus,* an important vector of canine babesiosis in Europe, is also believed to have extended northwards and populations have become established in Poland (Zygner et al., 2009), Belgium (Beugnet & Marie, 2009), Germany (Dautel et al., 2006), the Netherlands (Matjila et al., 2005) and in southern England (Jameson & Medlock, 2011; Smith et al., 2011). In addition, a significant increase in the abundance of *D. reticulatus* has occurred since the 1950s, as has the incidence of *Babesia canis* infections in Hungary (Sréter et al., 2005).

However, it is difficult to differentiate climate-mediated changes in vector distributions from underlying changes in host abundance. For example, over the last 20–30 years populations of roe deer, *Capreolus capreolus,* have also expanded in areas of increased tick abundance (Tälleklint & Jaenson, 1998). Nevertheless, studies carried out in Eastern Russia have provided good evidence that *I. ricinus* populations are increasing independently of changes in wildlife abundance (Korotkov et al., 2015). Furthermore, it is difficult to disentangle range expansions and shifts due to changes in the distribution of suitable habitat from colonisation of habitat that was previously suitable for parasite persistence but remained free of disease for reasons such as isolation or national biosecurity (e.g., the Pet Travel Scheme in the UK). However, the spread of some vector-borne diseases does not necessarily rely on range expansion of specific vectors. In some cases, resident parasites species may become competent disease vectors if the climatic conditions are suitable and they have sufficient exposure to the pathogen.

The change in distribution of some pathogens transmitted by dipteran vectors has been purported to be due to an increase in temperature. In Europe, the northern shift of various serotypes of bluetongue virus (BTV) and the establishment of WNV was attributed to an increase in temperature in these regions (Wittmann et al., 2002; Purse et al., 2005; Paz, 2006; Paz & Semenza, 2013). When pathogens have an obligatory development phase within the vector (extrinsic incubation period), the rate of pathogen development is also temperature-dependent. Increases in temperature above a certain threshold allow the pathogen to complete its development much faster and this increases the potential for onward transmission. As mosquito vectors for WNV have relatively short lifespans, summer heatwaves increase the likelihood of infected mosquitos carrying pathogenic quantities of virus. Indeed, outbreaks of the disease in Romania and New York City have both occurred after extremely hot summer weather (Savage et al., 1999; Epstein & Defilippo, 2001). In addition, temperature can affect the competence of the vector by altering physical characteristics of the arthropod. It has been demonstrated that other species of *Culicoides*, not usually known as bluetongue vectors, can show an increase in competency as vectors for BTV at higher temperatures due to physical changes to their gut permitting the leakage of virus into the haemocoel, allowing onward transmission. In combination with potential range expansion of existing vectors this allows further expansion of the disease loci (Wittmann et al., 2002; Purse et al., 2005).

7.4 Impact of Human Behaviour and Husbandry on Ectoparasitism

In addition to interactions between animal hosts, their parasites and climate, human behaviour plays an important role in mediating parasite challenge and moderating disease dynamics. In some cases this directly affects parasitism and associated vector-borne disease in humans. For example, in the Czech Republic in 2006 there was an extremely high incidence of human TBE: 1029 cases were reported compared with an average of 619 cases per year which had been observed for the previous 12 years (Daniel et al., 2008). The probable cause of this increase was considered to be a combination of vectoral and human factors mediated by a cool, wet summer and a warm, wet spring. The unseasonably high humidity allowed the activity of the tick vector, *I. ricinus*, to continue throughout the summer and also resulted in an increase in mushroom picking by local communities, as the weather produced good crops of wild mushrooms and was not favourable for other, low risk summer activities, such as swimming (Daniel et al., 2008). Human behaviour also affects ectoparasitism in domestic animals through animal husbandry. For example louse infestations in donkeys are associated with winter housing (Ellse et al., 2014).

Climate change may therefore indirectly affect veterinary ectoparasites through changes in farmer or pet owner behaviours. This might include altered management practices such as the timing of reproduction, housing, nutrition, breed selection, and management interventions such as shearing or chemical treatment (Morgan & Wall, 2009; discussed in more detail in the following section).

Changes in husbandry can also be optimised to mitigate negative impacts of climate change on parasite risk and the incidence of disease. For example, fleece length and condition (presence of faecal material, fleece rot and high moisture content) are significant risk factors for the infestation of sheep by the larvae of *L. sericata* or *L. cuprina*. The date ewes are sheared has a significant impact on the cumulative incidence of blowfly strike due to the reduced risk of infestation in shorn ewes. In a northern European scenario, earlier shearing is likely to significantly reduce the cumulative incidence of blowfly strike in ewes (Wall & Ellse, 2011). Although the flexibility to initiate earlier shearing is often constrained by lambing dates and the impact of early shearing on the ewe's body condition (as more energy is diverted to thermoregulation), in future, earlier and seasonal lambing, and warmer springs may be conducive to (and necessitate) early shearing, providing benefits in terms of reduced risk of blowfly strike.

Changing patterns of grass growth might provoke altered grazing patterns, with effects on the seasonality of parasite infection (Phelan et al., 2016). Such changes may lead to concurrent changes in husbandry as farmers adapt to altered seasonal availability of grass. Some parasites (e.g., lice) are transmitted more efficiently in housed or constrained livestock, whereas for others longer grazing seasons might be expected to promote infestation. Similarly, practices associated with sustainable intensification such as large-scale rearing units and indoor confinement may also generate the potential for an increased incidence of ectoparasitic disease.

Temporary host absence through housing has a strong seasonal forcing effect on the dynamics of infection in cattle, so changes in livestock management might be expected to have dominant effects on the epidemiology of this system, even without climatic effects on free-living stages. These are difficult to predict because management assumptions made in relevant mathematical models are simplistic. As well as increased temperature, altered rainfall patterns could affect both husbandry and parasite biology. Turn-out of cattle onto pasture in temperate areas, for example, requires reasonably dry pasture as well as grass growth, potentially affecting exposure to parasites at the start and end of the grazing season. The timing of transmission is important not only to the overall abundance of parasite populations, but also potentially to the patho-physiological consequences of infection, which depend on host age, nutritional and immune status, and seasonality in the state of the free-living stages.

A change in the perception of disease risk may also lead to changes in approach to intervention, with perhaps a greater willingness to treat prophylactically or to intervene with treatment earlier. However, this is also dependent on policy, surveillance efforts and education and is therefore likely to further complicate the identification of a link between climate and disease risk.

Finally, it must also be borne in mind when considering the impact of any changes in climate in a veterinary context, that livestock or pets face a range of parasitic and other diseases and interactions between them are common. Thus, climate-driven changes in the epidemiology of one parasitic disease may depend on several other parasitic diseases and interdependent husbandry factors. For instance, in some habitats, increases in burdens of gastrointestinal nematodes in sheep early in the grazing season (Rose et al., 2015) are likely to increase susceptibility to blowfly myiasis through increased fleece soiling, irrespective of effects on blowfly populations. In contrast, a predicted decrease in summer transmission of gastrointestinal nematodes and management interventions to reduce parasitic gastroenteritis (Rose et al., 2015) may decrease the risk of blowfly strike at this time of year.

Hence, the factors affecting the farmer/pet-owner–vector interactions are often as complex as those affecting the biology of the vectors or disease agents, and need to be taken into account when attempting to predict the effects of climate change. Attempts to predict the effects of climate change must include the behaviour and motivation of animal owners and managers.

7.5 Farmer Intervention as a Density-Dependent Process

In the absence of antagonistic effects as a result of altered animal husbandry, or limitation by host immunity, increased parasite abundance as a result of climate change might theoretically lead to unconstrained increases in disease incidence and production loss. However, this is unrealistic, since we would expect farmers to intervene before effects of parasites on their livestock become overwhelming. This process is essentially density-dependent, since the strength of the effect on parasite populations through control would increase with parasite abundance (Morgan & Wall, 2009). As such, farmer intervention might share some dynamic properties with other density-dependent processes in parasite population dynamics. Farmer intervention has its own properties, however, including a very pronounced trigger point based on observed effects, and frequent application of treatment to a group of animals irrespective of individual parasite burdens. It would be of considerable value to model this process, to generate insights into expected farmer behaviour when faced with altered parasite epidemiology and how effective this will be in limiting the impact of disease.

Advice on parasite control will only be taken up if it makes sense in the context of farm-level economics. Although the role of economics in the selection of parasite control strategies is a core part of farming and veterinary practice and has long been appreciated by parasitologists, scant scientific attention has been paid to this subject in recent decades as control has relied on routine chemical treatment. Resistance to chemicals is increasing in many parasite taxa, and it is likely that in dealing with the challenges of climate change, fewer parasite control options will be available to farmers. Including farm economics as well as the dynamics of diagnosis and intervention in models of parasite epidemiology will increase their relevance to control in the field.

7.6 Predicting Future Impacts of Climate Change on Ectoparasites and Vectors

Quantitatively predicting the future impact of climate change on parasites and vectors of veterinary importance can be valuable, helping to guide the direction of future research, targeting surveillance efforts where they are needed most, and evaluating potential mitigation strategies in the face of change. But, given the range of issues outlined above, this is a complex undertaking. In recent decades considerable effort has been devoted to this aim, driven by access to detailed ecological, climatic and environmental datasets, improved computational power and data sharing.

Changes in the spatial distribution of parasites and vectors are likely to have the greatest perceived impact where species establish in new regions, particularly if the parasite's hosts are ill-equipped to adapt to the new challenge. Stochastic events such as inadvertent introductions or novel species, are of course hard to predict. Climate change impacts on invertebrates are likely to be greatest (and

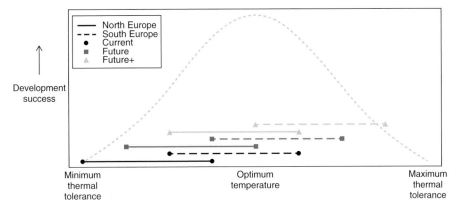

Figure 7.1 Ectoparasite developmental success at a range of temperatures (grey dotted line), varying between the species' minimum thermal tolerance and maximum thermal tolerance. Development success is highest around the species-specific optimum temperature. Horizontal lines correspond to the range of temperatures represented in Figures 7.2 and 7.4 for North (solid lines) and South (dashed lines) Europe under current (black circles) and future climatic conditions (grey squares and triangles). Future (dark grey squares) and future+ (light grey triangles) scenarios correspond to low and high climate change scenarios respectively (e.g., IPCC's RCP 2.6 and RCP 8.5 scenario projections for the 2080s).

more predictable) at the edge of the range of environmental tolerances for particular species (Lafferty, 2009). The scenario of range expansion is widely cited as the most tractable and observable impact of climate change, as invertebrate development typically ceases below a species-specific temperature threshold (Fig. 7.1; minimum thermal tolerance) and therefore species may move and establish in new regions once temperatures rise above this threshold. On the other hand, however, species existing in environments that are at the upper end of their thermal and moisture tolerances (Fig. 7.1; maximum thermal tolerance) may experience an increase in mortality rates with climate change and therefore range contractions and shifts are also potential consequences.

Seasonal variation in ectoparasite and vector abundance, and the incidence of disease is common, due to the impact of temperature on key life-history processes such as development, mortality, and host-seeking behaviour (e.g., tick questing behaviour) and the dependence of many species on seasonally available water for breeding. Ectoparasites have developed adaptations to survive seasonal variation, even where temperatures fall below their metabolic requirements. For example, the blowfly, *L. sericata,* enters diapause during the winter period in northern Europe, when temperatures fall below the development threshold (Wall et al., 1992). It is possible that current predictions of climate warming will increase the period during the year when conditions are suitable for development and survival for many species, increasing the period of transmission and number of generations per year (Wall & Ellse, 2011) and increasing the incidence of disease (Rose & Wall, 2011).

However, as temperature increases from the minimum thermal tolerance of the species, development rate increases and mortality decreases up to an optimal point (which varies depending on a species' evolutionary history and ecological adaptation). Thereafter, development rates continue to increase, but so do mortality rates (Rose et al., 2015). This results in a characteristic bell-shaped curve of development success (Fig. 7.1). This can also be interpreted as transmission potential, as an increase in development success equates to an increase in the number of infective/transmissible stages. The shape and position of this curve is highly species-specific (and may vary for locally adapted populations within species). The climate-dependent seasonal dynamics of a parasite population and relative risk of disease will therefore depend on their thermal (and moisture) tolerance profile (Fig. 7.1) and

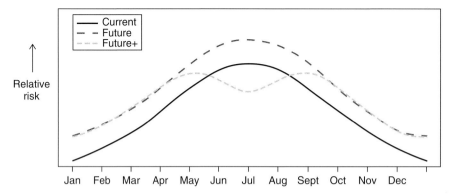

Figure 7.2 Potential change in the seasonal pattern of relative risk of hypothetical parasitic or vector-borne disease transmission in northern Europe under current and future climatic conditions. Future (dark grey dashed line) and future+ (light grey dotted line) scenarios show the potential change due to low and high climate change scenarios (e.g., IPCC's RCP2.6 and RCP8.5 scenario projections for the 2080s) respectively. Under current climatic conditions (black solid line) temperatures are sub-optimal (see Fig. 7.1) and therefore a low to moderate increase in temperature (dark grey dashed line) results in a year-round increase in relative risk of disease. A greater increase in temperature (light grey dotted line) results in an increase in relative risk for the majority of the year but takes the parasite above its optimal temperature for development success during the summer months, resulting in a "summer dip" in risk. The overall impact of this is that increasing temperatures do not necessarily translate to increases in disease risk.

geographic location, which determines the range of climatic conditions experienced. Thus for a hypothetical parasite, conditions experienced, for example in Northern Europe, may lie between its minimum thermal tolerance and optimal temperature, which, under current climatic conditions results in very little development and survival at the height of winter and a peak in abundance and relative risk of transmission during the summer (Fig. 7.1, black solid line; Fig. 7.2, black solid line). Under low to moderate climate change scenarios such as the Intergovernmental Panel on Climate Change's (IPCC, 2013) RCP 2.6 (Representative Concentration Pathway 2.6), the parasite may experience conditions closer to its optimum temperature resulting in a year-round increase in relative risk of transmission (Fig. 7.1, dark grey squares; Fig. 7.2, dark grey dashed line). However, under more extreme climate change scenarios such as the IPCC's RCP 8.5, the parasite experiences temperatures above its optimum and a "summer dip" is seen at the height of summer (Fig. 7.1, light grey triangles; Fig. 7.2, light grey dotted line). This pattern is emergent in many climate change impact evaluations of parasitic invertebrates in temperate regions (Rose & Wall, 2011; Rose et al., 2015, 2016) as a consequence of a trade-off between development and mortality at higher temperatures, which are often accompanied by limited moisture availability.

Where parasites currently experience temperatures spanning their optimal range the bimodal pattern of seasonal abundance is already observed. Based on simulations on other invertebrate parasites (gastrointestinal nematodes; Rose et al., 2016), further temperature increases which take the parasite beyond the optimum range are expected to accentuate this bimodal pattern.

The consequences of these complex patterns of response to climate change are that in many regions the predicted impact is not simply an increase in year-round risk or abundance, but may be a net increase or decrease in risk composed of seasonal peaks and troughs. Since husbandry is an important driving factor in the epidemiology of many ectoparasitic diseases, husbandry and targeted interventions will play an important role in mitigating the impacts of climate change (Morgan & Wall, 2009). Where future seasonal patterns of risk in one region are predicted to align with current patterns of risk in another, e.g. future risk in North Europe may become similar to current risk in South

Europe, husbandry practices employed in these areas may be adopted as a starting point for mitigation measures.

Clearly, predictive models present an opportunity to evaluate alternative management scenarios and develop optimum management strategies. However, their development and implementation in a veterinary context has been limited to date due to the lack of the detailed knowledge of parasite

Figure 7.3 Predicted probability of blowfly strike in Great Britain in 2003/2004 (first column), based on reported cases of blowfly strike, mean monthly air temperature and mean monthly rainfall. The model was projected onto future mean air temperature and rainfall data to predict the future probability of blowfly strike in the 2080s under the low (IPCC B1) and high (IPCC A1fi) emissions scenarios of climate change. Output shown is redrawn from data underlying predictions reported by Rose & Wall (2011). Black indicates a probability of blowfly strike of 0, and white indicates a probability of blowfly strike of 1.

ecology, host–parasite interactions and the husbandry data that is necessary to generate meaningful predictions. Sufficient detail does exist, however, to develop models to simulate incidence of myiasis in sheep by the blowfly *L. sericata* in the UK in the context of changing climate and management interventions based on the known relationships between risk factors for blowfly strike, temperature and rainfall. Originally developed as a decision support tool, one model was used to evaluate strategic control measures under scenarios of climate change and predicted that under a scenario of up to a 3°C increase in temperature, a significant increase in the cumulative incidence of blowfly strike was likely and that combinations of strategic insecticide treatment of ewes and lambs, trap deployment and strategic shearing of ewes could mitigate these climate-driven changes and reduce the magnitude of the peak blowfly incidence to below current levels (Wall & Ellse, 2011). A further empirical model considered the spatial pattern of disease dynamics (Rose & Wall, 2011). Under a scenario of low greenhouse gas emissions, by the 2080s, relatively little change in the length of the season of blowfly transmission was predicted for Great Britain overall, but the probability of infestation increased significantly. By reducing the spatial scale of predictions it was apparent that this increase was due to a lengthening of the period of transmission in northern regions, a pattern of change similar to the dark grey dashed line in Fig. 7.2. Interpretation became more complex under a scenario of high emissions, whereby a smaller increase in the probability of infestation was predicted, despite an overall extended transmission season. Under the high emissions scenario significantly more spatial variation was seen with a year-round increase in the probability of risk in the North (similar to the dark grey dashed line in Fig. 7.2) and a "summer-dip" in probability of strike in the South East (similar to the light grey dotted line in Fig. 7.2), accounting for the modest overall increase in probability of strike (Rose & Wall, 2011; Fig. 7.3). This example also demonstrates the importance of considering the scale, or resolution, in the design and interpretation of predictive models and their output as higher resolution spatial models were needed in the case of *L. sericata* to identify the spatial variation underlying the parasite's overall response to climate change.

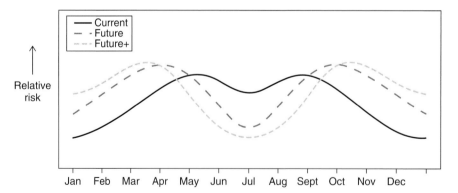

Figure 7.4 Potential change in the seasonal pattern of relative risk of parasitic or vector-borne disease transmission in Southern Europe under current and future climatic conditions. Future (dark grey dashed line) and future+ (light grey dotted line) scenarios show the potential change due to low and high climate change scenarios (e.g., IPCC's RCP2.6 and RCP8.5 scenarios) respectively. Figure 7.4 adapted from Rose et al. (2016). Under current climatic conditions (black solid line) temperatures span the optimal temperature for parasite development success (see Fig. 7.1) with development possible over the cooler winter months but increases in mortality rates in the summer resulting in a "summer dip" in disease risk. An increase in temperature (dark grey and light grey) takes the parasite further beyond the optimal temperature for development success (Fig. 7.1) resulting in an increase in winter risk but a corresponding increase in magnitude of the "summer dip".

Acknowledgements

We are grateful to Philip Smith for proofreading this chapter.

References

Baker, K.P. & Oormazdi, H. (1978) Geographical variation of the optimum temperature for *in vitro* rearing of *Bovicola bovis*. *Veterinary Parasitology*, **4**, 91–93.

Berriatua, E., French, N.P., Broster, C.E., Morgan, K.L. & Wall, R. (2001) Effect of infestation with *Psoroptes ovis* on the nocturnal rubbing and lying behaviour of housed sheep. *Applied Animal Behaviour Science*, **71**, 43–55.

Beugnet, F. & Marié, J.L. (2009) Emerging arthropod-borne diseases of companion animals in Europe. *Veterinary Parasitology*, **163**, 298–305.

Daniel, M. (1993) Influence of the microclimate on the vertical distribution of the tick *Ixodes ricinus* (L.) in Central Europe. *Acarologia*, **34**, 105–113.

Daniel, M., Danielová, V., Kříž, B. & Kott, I. (2004) An attempt to elucidate the increased incidence of tick-borne encephalitis and its spread to higher altitudes in the Czech Republic. *International Journal of Medical Microbiology*, **293**, 55–62.

Daniel, M., Danielová, V., Kříž, B., Jirsa, A. & Nožička, J. (2003) Shift of the tick *Ixodes ricinus* and tick-borne encephalitis to higher altitudes in central Europe. *European Journal of Clinical Microbiology & Infectious Diseases*, **22**, 327–328.

Daniel, M., Kříž, B., Danielová, V. & Beneš, Č. (2008) Sudden increase in tick-borne encephalitis cases in the Czech Republic, 2006. *International Journal of Medical Microbiology*, **298**, 81–87.

Danielová, V., Rudenko, N., Daniel, M., Holubová, J., Materna, J., Golovchenko, M. & Schwarzová, L. (2006) Extension of *Ixodes ricinus* ticks and agents of tick-borne diseases to mountain areas in the Czech Republic. *International Journal of Medical Microbiology*, **296**, 48–53.

Dautel, H. , Dippel, C. Oehme, R., Hartelt, K. & Schettler, E. (2006) Evidence for an increased geographical distribution of *Dermacentor reticulatus* in Germany and detection of *Rickettsia* sp. RpA4. *International Journal of Medical Microbiology*, **296**, 149–156.

Dautel, H., Dippel, C., Kämmer, D., Werkhausen, A. & Kahl, O. (2008) Winter activity of *Ixodes ricinus* in a Berlin forest. *International Journal of Medical Microbiology*, **298**, 50–54.

Ellse, L., Burden, F. & Wall, R. (2012) Pyrethroid tolerance in the chewing louse *Boviola (Werneckiella) ocellatus*. *Veterinary Parasitology*, **188**, 134–139.

Ellse, L., Burden, F.A. & Wall, R. (2014) Seasonal infestation of donkeys by lice: Phenology, risk factors and management. *Veterinary Parasitology*, **203**, 303–309.

Epstein, P.R. & Defilippo, C. (2001) West Nile Virus and drought. *Global Change and Human Health*, **2**, 105–107.

Gethings, O.J., Rose, H., Mitchell, S., Van Dijk, J. & Morgan, E.R. (2015). Asynchrony in host and parasite phenology may decrease disease risk in livestock under climate warming: *Nematodirus battus* in lambs as a case study. *Parasitology*, **142**, 1306–1317.

Gray, J.S., Dautel, H., Estrada-Peña, A., Kahl, O. & Lindgren, E. (2009) Effects of climate change on ticks and tick-borne diseases in Europe. *Interdisciplinary Perspectives on Infectious Diseases*, **2009**, 593232.

Health Protection Scotland (2009) Lyme Disease, Scotland, Annual Totals. Retrieved from http://www.documents.hps.scot.nhs.uk/giz/10-year-tables/lyme.pdf

Heath, A.C.G. & Levot, G.W. (2015) Parasiticide resistance in flies, lice and ticks in New Zealand and Australia: mechanisms, prevalence and prevention. *New Zealand Veterinary Journal*, **63**, 199–210.

IPCC (2013) Annex I: Atlas of Global and Regional Climate Projections (eds G.J. van Oldenborgh, M. Collins, J. Arblaster, J.H. Christensen, J. Marotzke, S.B. Power, M. Rummukainen & T. Zhou). *Climate Change 2013: The Physical Science Basis. Contribution of Working Group I to the Fifth Assessment Report of the Intergovernmental Panel on Climate Change* (eds T.F. Stocker, D. Qin, G-K. Plattner, M. Tignor, S.K. Allen, J. Boschung, A. Nauels, Y. Xia, V. Bex & P.M. Midgley). Cambridge University Press, Cambridge, UK.

Jameson, L.J. & Medlock, J.M. (2011) Tick surveillance in Great Britain. *Vector-borne and Zoonotic Diseases*, **11**, 403–412.

Kirby, A.D., Smith, A.A., Benton, T.G. & Hudson, P.J. (2004) Rising burden of immature sheep ticks (*Ixodes ricinus*) on red grouse (*Lagopus lagopus scoticus*) chicks in the Scottish uplands. *Medical and Veterinary Entomology*, **18**, 67–70.

Korotkov, Y., Kozlova, T. & Kozlovskaya, L. (2015) Observations on changes in abundance of questing *Ixodes ricinus*, castor bean tick, over a 35-year period in the eastern part of its range (Russia, Tula region). *Medical and Veterinary Entomology*, **29**, 129–136.

Lafferty, K.D. (2009) The ecology of climate change and infectious diseases. *Ecology*, **90**, 888–900.

Lindgren, E. & Gustafson, R. (2001) Tick-borne encephalitis in Sweden and climate change. *Lancet*, **358**, 16–18.

Lindgren, E., Tälleklint, L. & Polfeldt, T. (2000) Impact of climatic change on the northern latitude limit and population density of the disease-transmitting European tick *Ixodes ricinus*. *Environmental Health Perspectives*, **108**, 119–123.

Matjila, T.P., Nijhof, A.M., Taoufik, A, Houwers, D., Teske, E., Penzhorn, B.L., de Lange, T. & Jongejan, F. (2005) Autochthonous canine babesiosis in The Netherlands. *Veterinary Parasitology*, **131**, 23–29.

Morgan, E.R. & Wall, R. (2009) Climate change and parasitic disease: farmer mitigation? *Trends in Parasitology*, **25**, 308–313.

Nieuwhof, G.J. & Bishop, S.C. (2005) Costs of the major endemic diseases of sheep in Great Britain and the potential benefits of reduction in disease impact. *Animal Science*, **81**, 23–29.

Olesen, J.E. & Bindi, M. (2002) Consequences of climate change for European agricultural productivity, land use and policy. *European Journal of Agronomy*, **16**, 239–262.

Paz, S. (2006) The West Nile Virus outbreak in Israel (2000) from a new perspective: the regional impact of climate change. *International Journal of Environmental Health Research*, **16**, 1–13.

Paz, S. & Semenza, J.C. (2013) Environmental drivers of West Nile Fever epidemiology in Europe and Western Asia – a review. *International Journal of Environmental Research and Public Health*, **10**, 3543–3562.

Phelan, P., Morgan, E.R., Rose, H., Grant, J. & O'Kiely, P. (2016) Predictions of future grazing season length for European dairy, beef and sheep farms based on regression with bioclimatic variables. *Journal of Agricultural Science*, **154**, 765–781

Pietzsch, M.E., Medlock, J.M., Jones, L., Avenell, D., Abbott, J., Harding, P. & Leach, S. (2005) Distribution of *Ixodes ricinus* in the British Isles: investigation of historical records. *Medical and Veterinary Entomology*, **19**, 306–314.

Purse, B.V, Mellor, P.S., Rogers, D.J., Samuel, A.R., Mertens, P.P.C. & Baylis, M. (2005) Climate change and the recent emergence of bluetongue in Europe. *Nature Reviews Microbiology*, **3**, 171–181.

Randolph, S.E. & Storey, K. (1999) Impact of microclimate on immature tick-rodent host interactions (Acari: Ixodidae): Implications for parasite transmission. *Journal of Medical Entomology*, **36**, 741–748.

Randolph, S.E., Green, R., Hoodless, A. & Peacey, M. (2002) An empirical quantitative framework for the seasonal population dynamics of the tick *Ixodes ricinus*. *International Journal for Parasitology*, **32**, 979–989.

Randolph, S.E. (2004) Evidence that climate change has caused "emergence" of tick-borne diseases in Europe? *International Journal of Medical Microbiology Supplements*, **293**, 5–15.

Rehbein, S., Visser, M., Winter, R., Trommer, B., Matthes, H.-F., Maciel, A.E. & Marley, S.E. (2003) Productivity effects of bovine mange and control with ivermectin. *Veterinary Parasitology*, **114**, 267–284.

Rose, H. & Wall, R. (2011) Modelling the impact of climate change on spatial patterns of disease risk: sheep blowfly strike by *Lucilia sericata* in Great Britain. *International Journal for Parasitology*, **41**, 739–746.

Rose, H. Caminade, C., Bolajoko, M.B., Phelan, P., van Dijk, J., Baylis, M., Williams, D. & Morgan, E.R. (2016) Climate-driven changes to the spatio-temporal distribution of the parasitic nematode, *Haemonchus contortus*, in sheep in Europe. *Global Change Biology*, **22**, 1271–1285.

Rose, H., Wang, T., van Dijk, J. & Morgan, E.R. (2015) GLOWORM-FL: A simulation model of the effects of climate and climate change on the free-living stages of gastrointestinal nematode parasites of ruminants. *Ecological Modelling*, **297**, 232–245.

Sands, B., Ellse, L., Mitchell, S., Sargison, N.D. & Wall, R. (2015) First report of deltamethrin tolerance in the cattle chewing louse *Bovicola bovis* in the UK. *Veterinary Record*, **176**, 231.

Savage, H.M., Ceianu, C., Nicolescu, G., Karabatsos, N., Lanciotti, R., Vladimirescu, A., Laiv, L., Ungureanu, A., Romanca, C. & Tsai, T.F. (1999) Entomologic and avian investigations of an epidemic of West Nile fever in Romania in 1996, with serologic and molecular characterization of a virus isolate from mosquitoes. *The American Journal of Tropical Medicine and Hygiene*, **61**, 600–611.

Scharlemann, J.P.W., Johnson, P.J., Smith, A.A., Macdonald, D.W. & Randolph, S.E. (2008). Trends in ixodid tick abundance and distribution in Great Britain. *Medical and Veterinary Entomology*, **22**, 238–247.

Skarpaas, T., Ljøstad, U. & Sundøy, A. (2004) First human cases of tickborne encephalitis, Norway. *Emerging Infectious Diseases*, **10**, 2241–2243.

Skarphédinsson, S., Jensen, P.M. & Kristiansen, K. (2005) Survey of tickborne infections in Denmark. *Emerging Infectious Diseases*, **11**, 1055–1061.

Smith, F.D., Ballantyne, R., Morgan, E.R., & Wall, R. (2011) Prevalence, distribution and risk associated with tick infestation of dogs in Great Britain. *Medical and Veterinary Entomology*, **25**, 377–384.

Sréter, T., Széll, Z. & Varga, I. (2005) Spatial distribution of *Dermacentor reticulatus* and *Ixodes ricinus* in Hungary: evidence for change? *Veterinary Parasitology*, **128**, 347–351.

Stange, E.E. & Ayres, M.P. (2010) *Climate change impacts: insects. Encyclopedia of Life Sciences.* John Wiley & Sons, Ltd, Chichester.

Tälleklint, L. & Jaenson, T.G.T. (1998) Increasing geographical distribution and density of *Ixodes ricinus* (Acari: Ixodidae) in Central and Northern Sweden. *Journal of Medical Entomology*, **35**, 521–526.

Wall, R. (2007) Ectoparasites: Future challenges in a changing world. *Veterinary Parasitology* **148**, 62–74.

Wall, R. & Ellse, L.S. (2011) Climate change and livestock parasites: integrated management of sheep blowfly strike in a warmer environment. *Global Change Biology*, **17**, 1770–1777.

Wall, R., French, N. & Morgan, K.L. (1992) Effects of temperature on the development and abundance of the sheep blowfly *Lucilia sericata* (Diptera: Calliphoridae). *Bulletin of Entomological Research*, **82**, 125–131.

Wittmann, E.J., Mellor, P.S. & Baylis, M. (2002) Effect of temperature on the transmission of orbiviruses by the biting midge, *Culicoides sonorensis. Medical and Veterinary Entomology*, **16**, 147–156.

Wright, N., Phythian, C., Phillips, K. & Morgan, M. (2014) Sheep health, welfare and production planning 3. Using financial indicators. *In Practice*, **36**, 191–198.

Zygner, W., Górski, P. & Wedrychowicz, H. (2009) Detection of the DNA of *Borrelia afzelii, Anaplasma phagocytophilum* and *Babesia canis* in blood samples from dogs in Warsaw. *Veterinary Record*, **164**, 465–467.

8

Climate Change and the Biology of Insect Vectors of Human Pathogens

Luis Fernando Chaves[1,2]

[1] *Nagasaki University Institute of Tropical Medicine (NEKKEN), Sakamoto 1-12-4, Nagasaki, Japan*
[2] *Programa de Investigación en Enfermedades Tropicales (PIET), Escuela de Medicina Veterinaria, Universidad Nacional, Apartado Postal 304-3000, Heredia, Costa Rica*

Summary

Changes in insect vectors of human pathogens under climate change span their ecology, evolution and interaction with the pathogens they transmit, all critical factors for disease transmission. Given that vector-borne diseases account for almost one fifth of the total burden of infectious diseases, understanding the impacts of climate change on vector biology is fundamental to reduce current levels of transmission. This knowledge is also important to predict any potential exacerbation in vector-borne disease transmission that could emerge as the planet keeps warming, and that could also interfere with ongoing efforts to eliminate major vector-borne diseases. This chapter reviews different aspects of vector biology that are sensitive to meteorological conditions, looking at patterns that emerge under temperature gradients but also as a result of changes in the variability of weather conditions. Main topics include the interaction of vectors with pathogens, vector phenology and development, as well as the ecological dynamics of medically important insects. Reflecting the bias about insect vector knowledge, most of the examples come from studies on mosquitoes, but, where available, relevant information from other medically important insect taxa is included. Useful concepts to study impacts of climate change on insect vectors of disease are illustrated with a case study on changes in the distribution of two major globally invasive vectors: *Aedes albopictus* and *Aedes japonicus*, facing the emergence of *Aedes flavopictus*, formerly a rare species along an altitudinal gradient in Nagasaki, Japan. Finally, Schmalhausen's law, the principle stating that organism resilience to "normal" changes in the environment is decreased when forced towards the limits of any of its dimensions of existence, is presented as a tool that can be useful to understand ecological and evolutionary changes in insect vectors, and other invertebrates, following climate change.

8.1 Introduction

Vector-borne diseases can be due to the infection by many kinds of pathogens, including viruses, parasitic protozoa and worms (Patz & Olson, 2006a). The linking feature of these infectious diseases is their transmission by insects, or other arthropods, that bite vertebrate animals to consume blood as a major energetic resource for their reproduction (Edman, 1988; Lehane, 2005). Pathogen transmission, in general, occurs when infected arthropods bloodfeed[1] on the vertebrate host, and also when arthropods acquire pathogens from infected vertebrate hosts via bloodfeeding (Chaves et al., 2010; Rabinovich et al., 2011). These features set apart vectors of disease from other invertebrates that can also cause health problems through their defensive bites (e.g., ants), invasive tissue development (e.g., blowflies) or venomous stings (e.g., scorpions and jellyfish) all of which occur independently of bloodfeeding (Cupo, 2015).

The importance of vectors for society is major, given that vector-borne diseases account for 17% of the burden from all infectious diseases affecting humans (WHO, 2016), and that ongoing changes in vector-borne disease transmission are likely to have been influenced by climate change (Altizer et al., 2013). Here, however, it is essential to clarify that vector-borne disease transmission, especially when affecting humans, is an eco-social phenomenon (Levins, 1995; Levins & Lopez, 1999), where many factors, beyond the biology of vectors and parasites under changing environments, play key roles on emerging patterns of disease transmission that have occurred along with global warming (Levins et al., 1994). This is important since oversimplification of "other" factors is potentially misleading to understand the impact of climate change on the transmission of vector-borne diseases (Reiter, 2001; Chaves & Koenraadt, 2010; Hurtado et al., 2014). Nevertheless, given the nature of this volume, this chapter will be focused on the impacts of changing climate on arthropod vectors of disease, their relationship with the pathogens they transmit, and the implications of these phenomena for disease transmission, *ceteris paribus*, that is, assuming everything else stays the same.

Insects and other arthropod vectors of pathogens are among the many invertebrates that have changed and are expected to keep changing their biology in response to climate change (Parmesan, 2006; Colinet et al., 2015; Parham et al., 2015). In the case of human disease vectors, new and emerging climatic patterns have been shown to impact their distribution (Campbell et al., 2015; Proestos et al., 2015), their physiology and development (Uvarov, 1931; Huffaker, 1944; Bar-Zeev, 1958a), phenology (Taylor, 1981; Reisen et al., 2010; Chaves et al., 2013b), life history (Janisch, 1932; Reisen et al., 1984; Chaves et al., 2011b; Carrington et al., 2013a), population dynamics (Pascual et al., 2006; Chaves et al., 2012b; Chaves et al., 2014b), interactions with other species (Juliano, 2009; Chaves & Koenraadt, 2010), interaction with the pathogens they transmit (Chaves et al., 2011a; Lambrechts et al., 2011; Zouache et al., 2014), genetics and evolution (Armbruster & Conn, 2006; Kaufman & Fonseca, 2014; Egizi et al., 2015), and the invasion and expansion into new habitats (Lounibos, 2002; Juliano & Lounibos, 2005; Mogi et al., 2015). Moreover, field observations have shown, for example, that infection of *Culex* spp. mosquitoes with West Nile virus (WNV) increases with temperature (Ruiz et al., 2010) and rainfall variability in a season (Chaves et al., 2011a). It has also been noted, for example, that epidemics of WNV in the United States (Reisen et al., 2006; Hartley et al., 2012) and malaria in the East African Highlands (Paaijmans et al., 2009) have been associated with unusually high temperatures and El Niño events or a high Indian Ocean Dipole Mode (Chaves et al., 2012a; Chaves et al., 2012c). To some extent, all of these changes are related to the ectothermic nature of arthropod vectors of disease, that is, the basic fact that arthropods regulate their temperature following that of the environment (Uvarov, 1931; Colinet et al., 2015). But, before starting a more detailed description of the impacts of climate change on medically important taxa, it is imperative to clarify that there are two aspects of climate change as a force generating a changing environment that can alter the biology of disease vectors. There has been an increase in the global mean temperature, which

---

1 But there are exceptions, for example, *Trypanosoma cruzi*, the etiologic agent of Chagas disease, is not transmitted by biting during bloodfeeding, but by the post bloodmeal "defecation" of kissing bugs (Lehane, 2005).

Box 8.1 Warming, Weather Variability and Climate Patterns at Nagasaki City, Japan

Nagasaki City has a temperate seasonal climate, with four clear seasons (Isida 1969). The city has had a working meteorological station since July 1878 (WMO Station ID 47817, http://www.data.jma.go.jp/obd/stats/data/en/smp/index.html). During this period, monthly rainfall (Fig. 8.1A) and temperature (mean, Fig. 8.1B, maximum Fig. 8.1C and minimum Fig. 8.1D) have been recorded. The mean temperature has increased 2.1°C (Fig. 8.1B), the maximum temperature 1.4°C (Fig. 8.1C) and the minimum temperature 2.7°C (Fig. 8.1D), rendering this place a suitable location to study impacts of climate change on insects. These trends also reflect what is defined as "global warming" in the introduction. Rainfall seasonally peaks during June (Fig. 8.1A), with a smaller peak in September, and temperature seasonally peaks during the months of July and August (Fig. 8.1B-D). In Figures 8.1B to 8.1D, it also can be seen how the winter months (especially January and February) are more platykurtic than all other months, illustrating the concept of "weather variability" from the introduction. Temperature rarely drops below 0°C during the winter, which is relatively dry and where snowfall is extremely rare (Isida, 1969). During 1989, the year when a study on mosquitoes along the altitudinal gradient of Mt Konpira in Nagasaki was performed (Zea Iriarte et al., 1991), it is important to notice that weather was drier than usual (Fig. 8.1A) while temperature was slightly hotter than usual (Fig. 8.1B-D). By contrast 2014, the year when the data of the case study presented in this chapter were collected, was a wet and relatively cool year. This pattern highlights one of the problems of comparing sparse records on fauna presence to assess the impacts of climate change on the distribution and abundance of insects, where little can be said about impacts of warming trends when samples might have come from years with unusual weather patterns.

---

is usually referred as "global warming" (McMichael et al., 2006). In addition to the rising temperatures, there has also been an increase in the variability of meteorological conditions, which will be referred through this chapter as "weather variability" (Chaves & Koenraadt, 2010), which can also impact the biology of vectors and the diseases they transmit. Box 8.1 illustrates both components of climate change by looking at climatic records of Nagasaki, Japan.

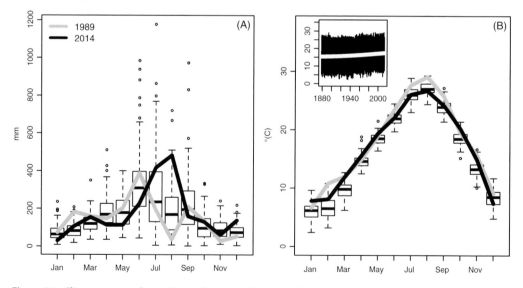

Figure 8.1 Climate seasonality at Nagasaki, Japan (A) rainfall (B) mean temperature (C) maximum temperature (D) minimum temperature. Lines indicate the seasonal trajectories of 1989 and 2014, see inset legend of panel (A) for guidance. inset panels in (B), (C) and (D) show temperature trends from January 1879 to December 2014.

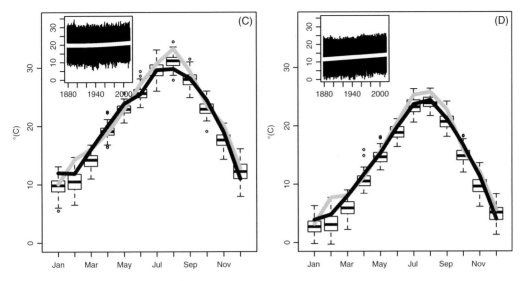

Figure 8.1 (*Continued*)

This chapter will review different aspects of the biology of vectors of human pathogens where studies have shown impacts of climate change. Topics will include the interaction of vectors with pathogens, phenology and population dynamics. Major trends for each topic will be illustrated with examples from several medically important taxa, but mainly focusing on mosquitoes, since they are the most widely studied taxa among invertebrates with medical importance given their role as vectors of malaria (Lindsay & Birley, 1996), dengue (Gubler, 2001) and other emerging vector-borne diseases (Weaver & Reisen, 2010; Reiner et al., 2013). The chapter will then introduce a case study that illustrates some of the complexities present when untangling the potential role of climate change on the distribution of disease vectors. Finally, the chapter will end highlighting Schmalhausen's Law, the principle that the resilience of organisms is decreased when forced towards the limits of any of its dimensions of existence (Chaves & Koenraadt, 2010), to understand the impacts of climate change on arthropod vectors of disease.

8.2 Interaction with Pathogens

The basic pattern that has been observed regarding the impact of temperature on pathogens developing inside vectors of disease is that the extrinsic incubation period (EIP), that is, the time it takes for a vector to be able to infect vertebrate hosts once it has been infected with a pathogen, shortens as temperature rises. The pattern has been widely observed in mosquitoes (Diptera: Culicidae), where *Plasmodium* spp., the parasitic protozoa causative of malaria, decrease their EIP in an accelerated (i.e., nonlinear) manner as temperature increases in the environment where they infect *Anopheles* spp. mosquitoes (Patz & Olson, 2006b). But these observations have not been only limited to protozoa, similar patterns have also been observed for several viruses, for example, in Yellow Fever virus infecting *Haemagogus* spp. where it was recorded that minimum EIP length decreases with increasing temperatures (Bates & Roca-García, 1946). Similar patterns, have also been reported for Eastern Equine Encephalitis in *Aedes triseriatus* (Chamberlain & Sudia, 1955), Japanese Encephalitis virus (JEV) in *Culex (Cx) tritaeniorhynchus* (Takahashi, 1976), WNV in *Cx pipiens* (Dohm et al.,

2002) and *Cx tarsalis* (Reisen et al., 2006), Western Equine Encephalomyelitis virus and St. Louis Encephalitis virus (SLEV) in *Cx tarsalis* (Reisen et al., 1993), SLEV in *Cx quinquefasciatus* (Hurlbut, 1973), Murray Valley Encephalitis virus in *Cx annulirostris* (Kay et al., 1989) and dengue virus in *Aedes aegypti* (Watts et al., 1986; Salazar et al., 2007). For virus infections in mosquitoes, however, it also has been observed that infection titres diminish if mosquitoes are suddenly changed to significantly lower temperatures (Chamberlain & Sudia, 1955; Kay et al., 1989; Kay & Jennings, 2002) or when the temperatures are too high over several days, for example, over a week (Kay & Jennings, 2002; Reisen et al., 2006). The duration of the EIP as function of temperature also seems to have some autonomy of the pathogen density used to infect disease vectors. For example, *Aedes vigilax* can be infected by Ross River virus (RRV) at similar rates independently of the initial titre of RRV used for infection, where the EIP diminish with increasing temperature, yet by the seventh day post infection all mosquitoes had similar titres and infection rates. Nevertheless, after 14 days it was observed that for mosquitoes infected with low virus doses the infection was highest at the lowest temperature of 18°C, and lowest at the high temperature treatment of 32°C (Kay & Jennings, 2002).

Weather variability can also play a role on the interaction between pathogens and vectors. For dengue virus serotypes I and II it has been observed that daily temperature ranges at which *Aedes aegypti* mosquitoes are held under experimental conditions can decrease transmission if it has a large variance when temperatures are high, as mosquito mortality can be increased (Lambrechts et al., 2011). On the other hand it might increase transmission at lower temperatures, by easing virus development (Carrington et al., 2013b). Similar patterns have been also reported for *Plasmodium chabaudi*, a rodent malaria parasite, on *Anopheles stephensi* mosquitoes under experimental conditions (Paaijmans et al., 2010).

Other factors that can influence the duration of the EIP are related to the pathogens themselves. For example, it is well known that *Plasmodium vivax* has a more accelerated EIP than *P. falciparum* across the same range of temperatures (Patz & Olson, 2006b). Likewise, it has also been observed that genotype WN02 of WNV, detected for the first time in 2001 spread across the United States and was more efficient than the first described genotype, NY99, at infecting and being transmitted by *Cx pipiens* mosquitoes, with the fitness advantage of genotype WN02 increasing as function of the product of environmental temperature and time post infection (Kilpatrick et al., 2008). Similarly, different vector species can have different responses to the same pathogen. Recently, it was observed that infection of *Ae aegypti* and *Ae albopictus* with serotypes I, II, III and IV of dengue virus was similar, but dengue virus RNA concentration was higher in the abdomen of *Ae albopictus*, yet infectious saliva was more common in *Ae aegypti* (Whitehorn et al., 2015). Also, raising the temperature and nutrition of pre-adult stages can have an impact on the vectorial competence, that is, the ability of a vector to develop a pathogen. In a classical study Takahashi (1976) found that *Cx tritaeniorhynchus* mosquitoes raised on a nutrient rich diet were less likely to be infected with JEV. Recently, Alto and Bettinardi (2013) found that *Ae albopictus* mosquitoes raised at lower temperatures were less likely to be infected with dengue virus serotype I. Both the genetic background of vectors and pathogens can interact with environmental temperature in shaping vectorial competence patterns. A study found differential patterns of transmission efficiency when mosquito colonies of *Ae albopictus* from six different geographical origins were exposed to the infection with two strains of Chikungunya virus and kept at two temperatures. Basically, no general pattern was observed, as infectious saliva six days after infection not always increased with temperature and patterns were often opposite depending on the Chikungunya virus strain or the origin of the mosquito colony (Zouache et al., 2014).

Studies on vectors other than mosquitoes are relatively few, yet results follow a similar pattern where EIP decreases with increasing temperature. Examples include Bluetongue virus and haemorrhagic disease of deer virus in the biting midge *Culicoides sonorensis* (Diptera: Ceratopogonidae) (Wittmann et al., 2002); *Onchocerca volvulus*, the worm causing river blindness, in the black fly *Simulium damnosum* s.l. (Diptera: Simuliidae) (Cheke et al., 2015); *Trypanosoma cruzi*, the pathogen causing Chagas disease, in the kissing bug *Triatoma infestans* (Heteroptera: Reduviidae) (Asin & Catala, 1995). In general, the shortening of the EIP with hotter temperatures has also been observed for *Leishmania* spp. parasites in sand flies (Diptera: Psychodidae). Nevertheless, *Leishmania peruviana*, a parasite transmitted at high altitudes in South America, more frequently infected, and with higher parasitic loads, the sand fly *Lutzomyia longipalpis* at 20°C when compared with 26°C, supporting the idea of local adaption for transmission at the low temperature of high altitudes (Hlavacova et al., 2013). This latter observation also supports that, more generally, responses to climate change might reflect genotype by environment interactions of vectors and pathogens.

8.3 Physiology, Development and Phenology

Insects and other arthropods develop as function of the temperature in their surrounding environment (Uvarov, 1931; Colinet et al., 2015). This lack of independence from the environment restricts their phenology, that is, seasonal activity, by constraining their physiology, development and life cycle, which also limits their potential to transmit pathogens in seasonal environments (Ruiz et al., 2010). In general, the development of invertebrates is a non-linear function of temperature (Taylor, 1981; Huey & Berrigan, 2001). The development for any given insect species increases with temperature above a minimum threshold, because extreme low temperatures are lethal for insects (Bar-Zeev, 1958a), then development reaches a maximum and then decreases until reaching a second threshold when high temperatures become lethal for insects (Huey & Berrigan, 2001). Along these lines, several studies have been done for insect vectors of disease, where observations can fit the asymmetric "n" trajectory described before (Huey & Berrigan, 2001), and which can be described by many mathematical models, whose description is beyond the goals of this chapter. In these studies, cohorts of insects are studied at constant temperature, as has been done for several mosquitoes (Diptera: Culicidae) like *Culex restuans* (Reisen et al., 1984), *Aedes aegypti* and *Cx quinquefasciatus* (Rueda et al., 1990), *Anopheles arabiensis* (Oliva et al., 2012) and *An gambiae* (Lyimo et al., 1992). In general, similar studies have been conducted for many mosquito species, and the conclusions are similar (Huffaker, 1944; Bar-Zeev, 1958a,b; Afrane et al., 2007) and also for black flies (Cheke et al., 2015). Nevertheless, little effort has been done at synthesizing patterns. One exception is the case of *Ae aegypti*, where a meta-analysis showed temperature to be the major factor governing its developmental rate across the globe (Couret & Benedict, 2014). Another species where patterns have been systematically studied is the *Culex pipiens* complex, the house mosquito which is globally distributed, where a comprehensive meta-analysis showed that the decrease in the developmental rate with temperature was always present, but the rearing density of insects was a major factor explaining differences in survival at different temperatures; survival non-linearly decreased with density and in environments where temperature was variable (Couret, 2013). The reason why fluctuating temperatures increased the mortality of *Culex pipiens* complex could be related to the different autonomy from the environment that insects develop with age (Janisch, 1932). For example, it has been reported that high temperatures in the development of early stage larval cohorts of *Cx quinquefasciatus* led to a higher mortality than if similar temperatures are experienced at later stages (Chaves et al., 2011b). Similar patterns have also been observed for *Ae aegypti* where increasing variability around a mean temperature can reduce different fitness components for this dengue vector (Carrington et al., 2013a; Carrington et al., 2013c). These patterns are probably general, since similar observations have been made for the kissing bug *Rhodnius prolixus*, where development increased with temperature until reaching a maximum, and where cyclical fluctuations in environmental temperature increased the mortality of early nymphal stages, with higher mortality observed as the variance of the fluctuations increased (Luz et al., 1999).

Given the temperature dependence of arthropod vectors of disease, unsurprisingly, an earlier phenology has been observed towards the tropics and shorter season length towards the poles (Taylor, 1981). This clear pattern has been observed in the adult phase of mosquitoes with extended latitudinal distributions, like *Cx pipiens* (Barker et al., 2010; Chaves et al., 2013b), *Cx tarsalis* (Barker et al., 2010) and *Armigeres subalbatus* (Chaves et al., 2015) to name a few species where the phenology has been studied/compared across latitudinal gradients.

Finally, with global warming it has been observed that mosquitoes might be changing their overwintering strategies. For example, *Culex* spp. mosquitoes showed earlier gonotrophic activity in warm winters, where the gonotrophic activity was demonstrated by the more frequent collection of gravid, parous and host seeking female mosquitoes (Reisen et al., 2010). Similarly, across altitudinal and deforestation gradients, hotter environments shorten the duration of the gonotrophic cycle of *Anopheles gambiae* (Afrane et al., 2005). An interesting and related phenomenon has also been reported for *Ae albopictus*, where it has been observed that diapausing eggs laid by females reared under short day length (10 h light:14 h dark) conditions are more autonomous from day length to terminate diapause and more likely to follow temperature signals in populations from locations close to the tropics as opposed to populations from high latitudes (Tsunoda et al., 2015). These phenomena raise questions about whether overwintering strategies of mosquitoes and other vectors might be changing in response to global warming, for example, replacing reproductive/egg diapause with quiescence (Reisen et al., 2010; Tsunoda et al., 2015).

8.4 Population Dynamics, Life History and Interactions with Other Vector Species

In the previous section it was already mentioned that environmental temperature is a major factor shaping the time of the year, that is, seasonality, when bloodfeeding insects are active and transmit pathogens. But, temperature and other weather components could also modify vector abundance patterns, an important parameter for vector-borne disease transmission (Fig. 8.2A), as derived from studies looking at the positive association between mosquito abundance and malaria transmission (Smith et al., 2010) and sand fly abundance and cutaneous leishmaniasis transmission (Chaves et al., 2014a). For example, mosquito population dynamics across different places in the planet are, in general, associated with rainfall and/or temperature changes. Changes in weather factors lead (or force), often with lags, abundance changes in mosquitoes (Reisen et al., 1992; Suwonkerd et al., 1996; Scott et al., 2000; Minakawa et al., 2002). Similar patterns have also been observed for sand flies (Chaniotis et al., 1971; Morrison et al., 1995; Chaves et al., 2013a) and biting midges (Chuang et al., 2000). Nevertheless, independently of vector abundance changes via the forcing from weather changes, mosquito population dynamics are density-dependent, that is, changes in vector abundance are self-regulated (Yang et al., 2008a; Yang et al., 2008b; Chaves et al., 2012b; Hoshi et al., 2014a; Chaves et al., 2015). Thus, impacts of warming and weather variability could be leading to the emergence of new patterns in the population dynamics of vectors because of the non-linear disruption that weather changes impose on the density-dependent regulation of a population (Chaves et al., 2012b). For example, a mathematical model suggested that rainfall frequencies similar to those of the reproductive cycle of *Culex* spp. mosquitoes could resonate and produce mosquito outbreaks, that is, sudden increases

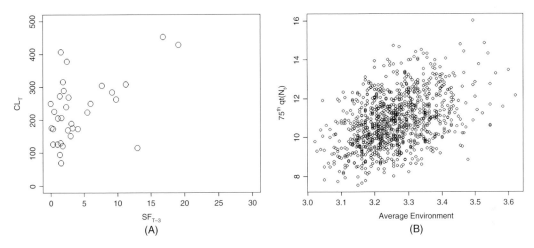

Figure 8.2 Climate change and population dynamics of vectors of human diseases (A) association between monthly cutaneous leishmaniasis cases (CL$_T$) and *Lutzomyia trapidoi* (SF$_{T-3}$) in Panamá. There is a three month lag between *Lu. trapidoi* (time – 3 months) and the cutaneous leishmaniasis cases (time). (B) 75th quantile of the distribution of simulated weekly population size of *Aedes aegypti* (N$_t$) as function of the average environment. In the X axis a larger value means a more frequent oscillation between enviroments that change the life history parameters of *Aedes aegypti*. Panel (A) is redrawn from Chaves et al. (2014a), Panel (B) is redrawn from Chaves et al. (2014b).

in mosquito abundance (Shaman & Day, 2007). The resonance leading to population outbreaks is an intuitive ecological phenomenon, which is named that way because of an analogy with dynamical systems in physics (Nisbet & Gurney, 1976). For example, if a mosquito has a gonotrophic cycle of 5 d, which occurs in a landscape where large rainfall, which ultimately translates in the generation of new habitats for oviposition and larval development, occurs every 5 d, the time concordance between optimal conditions for mosquito larval development and recruitment of new individuals can lead to an unusual large abundance of adult mosquitoes. Those outbreaks would be less likely to occur if habitat generation (rainfall) and oviposition (gonotrophic cycle) had different frequencies. Similarly, a stage-structured temperature dependent life cycle model for *An gambiae* showed the size of mosquito population oscillation could be amplified by small differences in temperature (Pascual et al., 2006). It has been suggested that more platykurtic environmental variables, that is, those where the variance around the mean is proportionally bigger (Box 8.1), can be predictive of their importance at forcing population dynamics across different locations (Chaves et al., 2012b). This is illustrated by *Ae aegypti*, whose abundance in Thailand was associated with temperature, while in Puerto Rico forcing was associated with rainfall, both variables being the more platykurtic at each location (Chaves et al., 2012b). A stage structured model (Chaves et al., 2014b) also suggested that more variable environments, defined as an increased occurrence of environmental conditions increasing the fitness of emerging mosquitoes, are more likely to produce *Ae aegypti* outbreaks (Fig. 8.2B).

Mechanisms for the emergence of new population dynamics patterns could arise from changes in the life history of vectors. For example, a classical study on the life history of *Cx restuans* showed that both high temperatures and densities could decrease the fitness of mosquitoes by both increasing mortality, but also by a carry-over effect, that is, one where the impact of a change is observed in the next generation, by decreasing the size of emerging adult mosquitoes, which is related to a decrease in fecundity for females (Reisen et al., 1984). Similar patterns have also been recorded for *An gambiae* (Lyimo et al., 1992). And these patterns likely reflect changes that occur in nature, as seasonal body size changes have been recorded for *Cx nigripalpus* in the southern United States (Day et al.,

1990b) and *Ae albopictus* in Japan (Suzuki et al., 1993) where average mosquito body size is reduced during the hottest months in the season. Regarding "weather variability" impacts on vector life history, it has been observed that increasing the variance of environmental temperature can negatively impact life history parameters of *Ae aegypti*, decreasing the overall fitness of this species (Carrington et al., 2013c). The stage structured mathematical model by Chaves et al. (2014b) indeed showed that sequential changes in life history parameters could explain the occurrence of *Ae aegypti* outbreaks following hot weather spells.

Changes in temperature and rainfall as part of the "weather variability" component of climate change might also play a role in the metapopulation dynamics of vectors at the landscape level. For *Cx pipiens* it has been observed that temperature could synchronize population dynamics across the latitudinal gradient of Jeju-do in South Korea, whose more localized population dynamics are nevertheless sensitive to rainfall (Chaves et al., 2013b). Rainfall is an important factor synchronizing adult abundance of *Anopheles walkeri* in swampy environments of New Jersey, United States (Shaman et al., 2006) and the abundance and bloodfeeding of *Cx nigripalpus* in Florida, United States (Day & Curtis, 1989). Similarly, both temperature and rainfall affect oviposition and habitat selection by *Cx quinquefasciatus* (Chaves & Kitron, 2011; Nguyen et al., 2012) and *Cx nigripalpus* (Day et al., 1990a).

Finally, temperature has also been observed as an important factor explaining the co-occurrence of vectors (Juliano, 2009). An elegant study by Lounibos et al. (2010) showed that both *Ae albopictus* and *Ae aegypti* co-exist in cemeteries of Florida, southern United States, despite the superior competitive ability of *Ae albopictus* over *Ae aegypti* in artificial container conditions (Juliano et al., 2004). This is because *Ae aegypti* eggs are better able to deal with hotter environments than *Ae albopictus*, which otherwise has excluded *Ae aegypti* from cemeteries which experience cooler temperatures.

8.5 Case Study of Forecasts for Vector Distribution Under Climate Change: The Altitudinal Range of *Aedes albopictus* and *Aedes japonicus* in Nagasaki, Japan

It has been observed that *Ae albopictus*, a major invasive vector species worldwide (Futami et al., 2015), has been moving towards higher latitudes in Japan as temperatures have increased with global warming (Kobayashi et al., 2002; Mogi & Tuno, 2014). The ability of *Ae albopictus* to develop well in hot and humid environments has been suggested as a major factor explaining its invasive success across the planet (Mogi et al., 2015). Observations like these have propelled research on the future distribution, as global temperature warms further. Forecasts about future distribution have been common among most vector taxa, including mosquitoes (Proestos et al., 2015), kissing bugs (Medone et al., 2015), sand flies (Moo-Llanes et al., 2013) among others. These forecasts tend to make predictions based on the association between the occurrence of a vector at, generally, coarsely grained spatial scales, around 5 km², in relation to the dispersal of vectors, which, in general, is at least one order of magnitude below (Saldaña et al., 2015) and temperature estimates for the same area. These predictions use linear and nonlinear methods to predict future distributions given records of occurrence/absence for a vector across and expected temperature changes. Nevertheless, these methods tend to oversimplify other factors explaining the distribution of vectors, as it will be illustrated with finely grained spatial data about mosquito presence in Mt Konpira at Nagasaki city, Japan.

Nagasaki city, Japan is an ideal place to study climate change impacts on insect vectors of disease that have become global invasive species. As shown in Box 8.1 and Figure 8.1 the city has had a significant warming over the last century. There has been an increase in air temperature of around

2°C, since 1878 when weather patterns started to be recorded, and around 0.6°C in the last quarter of a century, that is, from 1989 to 2014. Both *Ae albopictus*, a major vector of dengue and other emerging arboviruses affecting humans, and *Ae japonicus*, a competent vector for JEV and WNV (Kaufman & Fonseca, 2014), are common species in Nagasaki city. Previous studies done along Mt Konpira (Zea Iriarte et al., 1991; Tsuda et al., 1994) recorded the presence of both *Ae albopictus* and *Ae japonicus* across an altitudinal gradient, observations with historical value for comparison with current distribution patterns. The baseline records of *Ae albopictus* and *Ae japonicus* in Mt Konpira (Zea Iriarte et al., 1991; Tsuda et al., 1994) offer an unique opportunity to study changes in the distribution and abundance of these two invasive vectors in their native range, by rendering plausible comparisons both on a historical scale and across the temperature gradient generated by the altitudinal range of the mountain (Chaves & Koenraadt, 2010). Thus, historical records were evaluated to test the hypothesis that temperature can drive the expansion of *Ae albopictus* (Mogi et al., 2015), since a rise in temperature would be expected to be accompanied by an expansion of the altitudinal range of *Ae albopictus*, which was restricted up to 250 m in 1989 (Zea Iriarte et al., 1991) and to also test whether *Ae japonicus*, by contrast, has contracted its altitudinal range (Kaufman & Fonseca, 2014) by increasing its lower altitudinal limit, which was set at around 110 m in 1989 (Zea Iriarte et al., 1991).

To compare the distribution of *Aedes* spp. mosquitoes along the altitudinal gradient of Mt Konpira, data collected in 1989 was extracted from Zea Iriarte et al. (1991). Fresh mosquito data was collected following a sampling methodology where 4th instar larvae of *Aedes* spp. mosquitoes were sampled with ovitraps from June to November 2014 and adult mosquitoes with sweeping nets from May to November 2014. Other sampling details have been described in detail elsewhere (Chaves et al., 2015), and samples were collected at the exact same locations of the study by Zea Iriarte et al. (1991).

Presence/absence of immature (Fig. 8.3A) and adult (Fig. 8.3B) *Aedes albopictus* and *Aedes japonicus* records from 1989 were compared with data from 2014 (Fig. 8.3C,D) using logistic regression models (Kuhn & Johnson, 2013). Only presence/absence data were compared because it was not possible to obtain the raw abundance records from the 1989 study (Zea Iriarte et al., 1991). We built a "full" logistic model which considered that presence/absence of a mosquito species i (i could be *Ae albopictus* or *Ae japonicus*) around a tree was a function of the presence of other species k ($k \neq i$), elevation, sampling year and land cover at each sampling location. For land cover we used, respectively, vegetation survey data from 1984 (Fig. 8.3E) and 2005 (Fig. 8.3F) for the mosquito data from 1989 and 2014.

Changes in the distribution of *Ae albopictus* and *Ae japonicus* between 1989 and 2014 can be observed in Figure 8.3. Figures 8.3A and B show data for 1989, while Figures 8.3C and D show data for 2014. When comparing data for immatures (Fig. 8.3A and C), the distribution of *Ae japonicus* oviposition was very restricted and generally similar, with the exception that no immatures were observed high in the mountain in 2014, unlike 1989 when one trap was positive at around 300 m. In addition, two major changes are clear, first is the appearance of *Ae flavopictus* in 2014, which was able to oviposit all over the altitudinal gradient of Mt Konpira (Fig. 8.3C) and second is the contraction of the altitudinal range of *Ae albopictus* which was as high as 240 m in 1989 (Fig. 8.3A) but went below 160 m in 2014 (Fig. 8.3C). Regarding changes in the distribution of adults (Figs. 8.3B and D), *Ae. albopictus* was collected all over Mt Konpira in 2014 (Fig. 8.3D), in contrast with the 250 m limit observed in 1989 (Fig. 8.3C). In contrast, *Ae japonicus* adults were sampled along the altitudinal gradient in 1989 and 2014 (Figs. 8.3C and D) and in 2014 *Ae flavopictus* was also widely distributed across the altitudinal gradient (Fig. 8.3D). The statistical analysis for changes in the distribution of *Ae albopictus* and *Ae japonicus* in 1989 and 2014 (Table 8.1), suggests that immatures of the former were less likely to occur

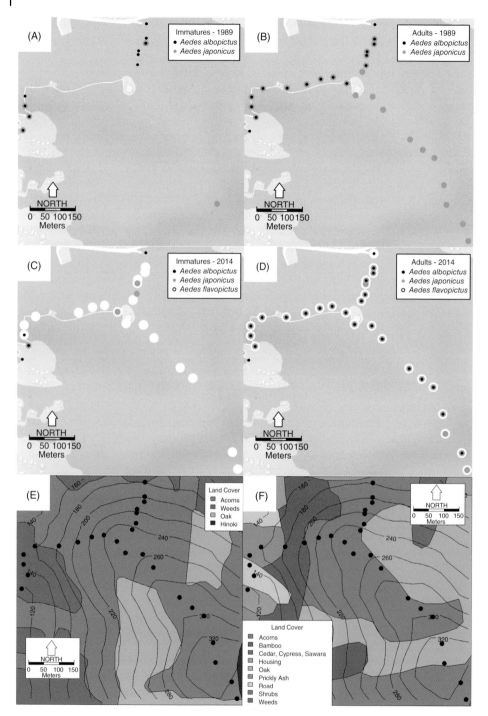

Figure 8.3 Land use change and presence of immature and adult *Aedes albopictus*, *Ae. flavopictus* and *Ae. japonicus* in/around trees with ovitraps at Mt Konpira in 1989 and 2014. (A) immatures 1989, (B) adults 1989, (C) immatures 2014, (D) adults 2014, (E) Mt Konpira land use and vegetation in 1984, (F) Mt Konpira land use and vegetation cover in 2005. In all panels further details are presented in the inset legends, contours in panels (E) and (F) are for altitude.

Table 8.1 Factors associated with *Aedes albopictus* and *Aedes japonicus* ovitrap colonization (or adult presence) at focal trees in 1989 and 2014 at Mt Konpira, Nagasaki, Japan. Parameter estimates are for the best logistic generalized linear model selected through a process of backward elimination. Moran's I indicates the Moran's I index of spatial autocorrelation estimated from model residuals through a 1000 replicates Monte Carlo. ΔAIC is the difference between the AIC from the "full" model, including all potential covariates, and the "best" model.

Model	Species	Parameter	Odds-ratio	Estimate	S.E.	z	P
Ovitrap colonization	*Aedes albopictus*	Intercept	–	9.418	3.058	3.08	0.00207
		Elevation	0.954	−0.0474	0.0149	−3.19	0.00143[a]
		Year – 2014	0.0896	−2.413	1.079	−2.24	0.0253[a]
		Moran's I	–	−0.0274	–	–	0.799
		ΔAIC	–	2.67	–	–	[b]
	Aedes japonicus	Intercept	–	−2.20	0.527	−4.17	0.0000306[a]
		Presence of *Ae. albopictus*	4.99	1.61	0.767	2.10	0.036[a]
		Moran's I	–	−0.0184	–	–	0.437
		ΔAIC	–	10.03	–	–	[b]
Adult presence	*Aedes albopictus*	Intercept	–	11.212	3.501	3.20	0.00136[a]
		Elevation	0.955	−0.0460	0.0146	−3.15	0.00163[a]
		Year – 2014	37.35	3.621	1.390	2.64	0.00922[a]
		Moran's I	–	−0.015	–	–	0.328
		ΔAIC	–	11.77	–	–	[b]
	Aedes japonicus	Intercept	–	−4.639	2.799	−1.66	0.0975
		Elevation	1.041	0.0400	0.0180	2.23	0.0261[a]
		Moran's I	–	−0.031	–	–	0.963
		ΔAIC	–	10.35	–	–	[b]

a) Statistically significant (P<0.05),
b) The "best" model significantly reduces the number of parameters and/or log likelihood.

at increasing altitudes (5% less for an additional metre of altitude) and less likely to be recorded in 2014 (82%), while in both years immatures of *Ae japonicus* were more likely to co-occur (around 500% or an odds ratio of nearly 5) with *Ae albopictus*, suggesting the lack of antagonistic interactions between the two species. Meanwhile, changes in the distribution of adults also showed that *Ae albopictus* was less likely to occur at high altitudes (5% less for an additional metre of altitude) but that changed radically in 2014 (when adults were 37 times more likely to be found in the mountain irrespective of altitude) and that *Ae japonicus* was more likely to be present at higher altitudes (4% more for an additional metre of altitude). Interestingly, the process of model selection suggests that observed changes are unlikely attributable to fine scale changes in the dominant vegetation, since this variable was eliminated as uninformative by the process of model selection, suggesting that other changes might have influenced the widespread distribution of *Ae albopictus* adults in 2014. These results illustrate how a third mosquito species, *Aedes flavopictus*, a competent vector of dengue virus (Eshita et al., 1982) and formerly a rare species (Omori et al., 1952), has now become the dominant *Aedes* species in

Mt Konpira, in a fashion that illustrates how predictions about vector species distribution need to consider the role that other species can play on the expansion of a focal species, highlighting the need to understand the role that climate change has on the co-existence of co-occurring vector species (Chaves & Añez, 2004; Lounibos et al., 2010; Hoshi et al., 2014b).

8.6 Vector Ecology and Evolution in Changing Environments

A summary of the patterns of climate change impacts on insect vectors of disease is presented in Table 8.2, specially linking them to vectorial capacity, a parameter that quantifies the ability of a vector to propagate a disease (Garrett-Jones, 1964). It should be noted that impacts of climate change do not always necessarily increase vectorial capacity, and in some cases, there might be trade-offs, since, for example, a decrease in EIP might be attenuated by a concomitant increase in mosquito mortality and survival reduction. Similarly, as illustrated by several examples in this chapter, there are exceptions to the general patterns presented in Table 8.2 that might emerge because of the evolutionary trajectories

Table 8.2 Summary of observed and expected patterns of climate change impacts on the different components of vectorial capacity, and other factors that might influence the ability of bloodsucking insects to transmit pathogens. Vectorial capacity (VC) is a parameter that quantifies the expected number of secondary inoculations on vertebrate hosts per infective vertebrate per time unit and is defined by the following formula (Garrett-Jones, 1964): $VC = \frac{Na^2 e^{-\mu n}}{\mu}$, where N is the abundance of vector per vertebrate host, a is the biting rate, μ is the mortality rate per unit time, $e^{-\mu}$ is the survival rate per unit time and n is the duration of the extrinsic incubation period. Climate change impacts are presented regarding both an increase in temperature (global warming) and increased weather variability.

VC Component	Global Warming	Increased Weather Variability
N	Outbreaks more likely to occur where populations are forced by temperature	Outbreaks more likely to occur where changes occur in the most important forcing factor for vector abundance
a	Expected increase per unit time, based on observations about the shortening of the gonotrophic cycle	Observed increase in more humid environments, and increase with more frequent rainfall.
μ	Observed increase with high temperature	Observed increase, especially at early stages of development
$e^{-\mu}$	Observed decrease with temperature	Expected decrease against a background of high average temperature
n	Observed decrease until reaching a threshold	Expected decrease against a background of low average temperature
Other Factors		
Dispersal of vectors	Expected reduction in areas that become less humid Expected increase in areas that become more humid	Expected decrease in humid areas where rain becomes more frequent Expected increase in dry areas where rain becomes more frequent
Evolutionary changes	Observed reduction in genetic diversity, which is expected if there is a strong selection for traits associated with surviving in a new extreme environment	Expected increased genetic diversity when the adaptation to changes is finely grained, same diversity when the adaptation to changes is coarsely grained

of specific vectors and pathogens, illustrating the realization that phenotypes always emerge from the interaction of genotypes and environments (Levins & Lewontin, 1985).

Schmalhausen's law states that any organism subjected to its limits of tolerance along any of its dimensions of existence is prone to be more sensitive to "normal" changes in other variables relevant for its existence (Chaves & Koenraadt, 2010). This can be illustrated by the observation that *Ae aegypti* abundance in Thailand, a place with a strong seasonality in rainfall, was associated with temperature fluctuations. In contrast, the opposite was observed in Puerto Rico, where abundance changes where associated with rainfall, yet seasonality is stronger for temperature (Scott et al., 2000; Chaves et al., 2012b; 2014b). Thus, Schmalhausen's law is extremely relevant to understand the impacts of climate change on both insect vectors and the diseases they transmit (Table 8.2). Some of the changes and responses to climate are conditioned in other factors, and vectorial capacity being an abstraction (Levins, 2006) might miss some key elements for disease transmission, such as the movement/dispersal of vectors (Reiner et al., 2013), which is indeed known to be sensitive to rainfall, decreasing when high rainfall occurs (Higa et al., 2000) or facilitated when humidity is high (Day & Curtis, 1994). Similarly, a major source of contingency is that pathogens and vectors can evolve in response to climate change. For example, it has been observed that allelic diversity decreased with increasing temperature and that genetic population distance increased in neighboring populations of *Ae japonicus* occurring at different temperatures (Egizi et al., 2015). These patterns pose questions about whether vectors will respond to climate change following the strategy of organisms that deal with the environment as finely grained, where genetic diversification is a likely outcome, or coarsely grained, where no major genetic changes are expected, unless there is a strong selective pressure (Levins, 1968), such as the one global warming imposes in ectothermic organisms with defined tolerance boundaries to extreme high temperatures.

Vectors belong to communities which tend to be ignored as a whole (Chaves et al., 2011a; Hoshi et al., 2014b; Chaves & Añez, 2016). Research tends to be narrowly focused on "dominant vector species", yet over and over cases like the one presented in the case study, where suddenly a "new" species be it invasive or native, can change the scenario for disease transmission. For example, following the earthquake and tsunami of 2011 in Eastern Japan it was observed that *Cx inatomii*, formerly a rare species, became dominant in the affected region probably displacing major JEV vectors (Watanabe et al., 2012). Likewise, *Cx stigmatosa*, a species difficult to collect as an adult, has otherwise all traits necessary to be an efficient vector for SLEV and WNV and co-occurs with *Cx pipiens* and *Cx restuans* in California, United States, where the latter two species are recognized as the dominant vectors of SLEV and WNV (Reisen, 2012). In that sense, a better incorporation of phylogenetic information has the potential to aid predictions about the impact of climate change via the incorporation of evolutionary knowledge about vectors. This approach has been used for predicting the threat of emerging arboviruses such as Chikungunya and Zika (Weaver & Reisen, 2010), well before they became public health major problems as they have recently widely invaded new habitats worldwide. Finally, there are plenty of exciting research questions needing answers to better understand the impacts of climate change on vectors of human pathogens, all this knowledge being necessary to better control the diseases they transmit and to ultimately alleviate or eliminate their burden from society.

Acknowledgements

The case study was funded by the Sumitomo Foundation grant No. 153107 to LFC. Nagasaki City direction of Parks and Recreational kindly provided us with all relevant permits to perform this study. Tomonori Hoshi and Nozomi Imanishi performed the net sweeping sampling and helped with mosquito identification. Trang T. T. Huynh kindly helped measuring diverse micro-environmental characteristics of the sampling sites and with the larval surveys. Philip Smith is thanked for proof-reading this chapter. Finally, Ms. Junko Sakemoto provided valuable administrative support.

References

Afrane, Y.A., Lawson, B.W., Githeko, A.K. & Yan, G. (2005). Effects of microclimatic changes caused by land use and land cover on duration of gonotrophic cycles of *Anopheles gambiae* (Diptera: Culicidae) in Western Kenya Highlands. *Journal of Medical Entomology*, **42**, 974–980.

Afrane, Y.A., Zhou, G., Lawson, B.W., Githeko, A.K. & Yan, G. (2007). Life-table analysis of *Anopheles arabiensis* in Western Kenya Highlands: effects of land covers on larval and adult survivorship. *American Journal of Tropical Medicine and Hygiene*, **77**, 660–666.

Altizer, S., Ostfeld, R.S., Johnson, P.T.J., Kutz, S. & Harvell, C.D. (2013). Climate change and infectious diseases: from evidence to a predictive framework. *Science*, **341**, 514–519.

Alto, B.W. & Bettinardi, D. (2013). Temperature and dengue virus infection in mosquitoes: independent effects on the immature and adult stages. *American Journal of Tropical Medicine and Hygiene*, **88**, 497–505.

Armbruster, P. & Conn, J.E. (2006). Geographic variation of larval growth in North American *Aedes albopictus* (Diptera : Culicidae). *Annals of the Entomological Society of America*, **99**, 1234–1243.

Asin, S. & Catalá, S. (1995). Development of *Trypanosoma cruzi* in *Triatoma infestans*: influence of temperature and blood consumption. *The Journal of Parasitology*, **81**, 1–7.

Bar-Zeev, M. (1958a). The effect of extreme temperatures on different stages of *Aëdes aegypti* (L.). *Bulletin of Entomological Research*, **48**, 593–599.

Bar-Zeev, M. (1958b). The effect of temperature on the growth rate and survival of the immature stages of *Aëdes aegypti* (L.). *Bulletin of Entomological Research*, **49**, 157–163.

Barker, C.M., Eldridge, B.F. & Reisen, W.K. (2010). Seasonal abundance of *Culex tarsalis* and *Culex pipiens* complex mosquitoes (Diptera: Culicidae) in California. *Journal of Medical Entomology*, **47**, 759–768.

Bates, M. & Roca-García, M. (1946). The development of the virus of yellow fever in haemagogus mosquitoes. *American Journal of Tropical Medicine and Hygiene*, **s1–26**, 585–605.

Campbell, L.P., Luther, C., Moo-Llanes, D., Ramsey, J.M., Danis-Lozano, R. & Peterson, A.T. (2015). Climate change influences on global distributions of dengue and chikungunya virus vectors. *Philosophical Transactions of the Royal Society of London B: Biological Sciences*, **370**, 20140135.

Carrington, L.B., Armijos, M.V., Lambrechts, L., Barker, C.M. & Scott, T.W. (2013a). Effects of fluctuating daily temperatures at critical thermal extremes on *Aedes aegypti* life-history traits. *PLOS ONE*, **8**, e58824.

Carrington, L.B., Armijos, M.V., Lambrechts, L. & Scott, T.W. (2013b). Fluctuations at a Low Mean Temperature Accelerate Dengue Virus Transmission by *Aedes aegypti*. *PLOS Neglected Tropical Diseases*, **7**, 8.

Carrington, L.B., Seifert, S.N., Willits, N.H., Lambrechts, L. & Scott, T.W. (2013c). Large diurnal temperature fluctuations negatively influence *Aedes aegypti* (Diptera: Culicidae) life-history traits. *Journal of Medical Entomology*, **50**, 43–51.

Chamberlain, R.W. & Sudia, W.D. (1955). The effects of temperature upon the extrinsic incubation period of Eastern Equine Encephalitis in mosquitoes. *American Journal of Epidemiology*, **62**, 295–305.

Chaniotis, B.N., Neely, J.M., Correa, M.A., Tesh, R.B. & Johnson, K.M. (1971). Natural population dynamics of Phlebotomine Sandflies in Panama. *Journal of Medical Entomology*, **8**, 339–352.

Chaves, L.F. & Añez, N. (2004). Species co-occurrence and feeding behavior in sand fly transmission of American cutaneous leishmaniasis in western Venezuela. *Acta Tropica*, **92**, 219–224.

Chaves, L.F. & Añez, N. (2016). Nestedness patterns of sand fly (Diptera: Psychodidae) species in a neotropical semi-arid environment. *Acta Tropica*, **153**, 7–13.

Chaves, L.F., Calzada, J.E., Rigg, C., Valderrama, A., Gottdenker, N.L. & Saldaña, A. (2013a). Leishmaniasis sand fly vector density reduction is less marked in destitute housing after insecticide thermal fogging. *Parasites & Vectors*, **6**, 164.

Chaves, L.F., Calzada, J.E., Valderama, A. & Saldaña, A. (2014a). Cutaneous Leishmaniasis and Sand Fly fluctuations are associated with El Niño in Panamá. *PLOS Neglected Tropical Diseases*, **8**, e3210.

Chaves, L.F., Hamer, G.L., Walker, E.D., Brown, W.M., Ruiz, M.O. & Kitron, U.D. (2011a). Climatic variability and landscape heterogeneity impact urban mosquito diversity and vector abundance and infection. *Ecosphere*, **2**, art70.

Chaves, L.F., Harrington, L.C., Keogh, C.L., Nguyen, A.M. & Kitron, U.D. (2010). Blood feeding patterns of mosquitoes: random or structured? *Frontiers in Zoology*, **7**, 3.

Chaves, L.F., Hashizume, M., Satake, A. & Minakawa, N. (2012a). Regime shifts and heterogeneous trends in malaria time series from Western Kenya Highlands. *Parasitology*, **139**, 14–25.

Chaves, L.F., Higa, Y., Lee, S.H., Jeong, J.Y., Heo, S.T., Kim, M, Minakawa, N. & Lee, K.H. (2013b). Environmental forcing shapes regional house mosquito synchrony in a warming temperate island. *Environmental Entomology*, **42**, 605–613.

Chaves, L.F., Imanishi, N. & Hoshi, T. (2015). Population dynamics of *Armigeres subalbatus* (Diptera: Culicidae) across a temperate altitudinal gradient. *Bulletin of Entomological Research*, **105**, 589–597.

Chaves, L.F., Keogh, C.L., Nguyen, A.M., Decker, G.M., Vazquez-Prokopec, G.M. & Kitron, U.D. (2011b). Combined sewage overflow accelerates immature development and increases body size in the urban mosquito *Culex quinquefasciatus*. *Journal of Applied Entomology*, **135**, 611–620.

Chaves, L.F. & Kitron, U.D. (2011). Weather variability impacts on oviposition dynamics of the southern house mosquito at intermediate time scales. *Bulletin of Entomological Research*, **101**, 633–641.

Chaves, L.F. & Koenraadt, C.J.M. (2010). Climate change and highland malaria: fresh air for a hot debate. *The Quarterly Review of Biology*, **85**, 27–55.

Chaves, L.F., Morrison, A.C., Kitron, U.D. & Scott, T.W. (2012b). Nonlinear impacts of climatic variability on the density-dependent regulation of an insect vector of disease. *Global Change Biology*, **18**, 457–468.

Chaves, L.F., Satake, A., Hashizume, M. & Minakawa, N. (2012c). Indian Ocean dipole and rainfall drive a Moran effect in East Africa malaria transmission. *Journal of Infectious Diseases*, **205**, 1885–1891.

Chaves, L.F., Scott, T.W., Morrison, A.C. & Takada, T. (2014b). Hot temperatures can force delayed mosquito outbreaks via sequential changes in *Aedes aegypti* demographic parameters in autocorrelated environments. *Acta Tropica*, **129**, 15–24.

Cheke, R.A., Basáñez, M.-G., Perry, M., White, M.T., Garms, R., Obuobie, E., Lamberton, P.H.L., Young, S., Osei-Atweneboana, M.Y., Intsiful, J., Shen, M., Boakye, D.A. & Wilson, M.D. (2015). Potential effects of warmer worms and vectors on onchocerciasis transmission in West Africa. *Philosophical Transactions of the Royal Society of London B: Biological Sciences*, **370**, 20130559.

Chuang, Y.-Y., Lin, C.-S., Wang, C.-H. & Yeh, C.-C. (2000). Distribution and seasonal occurrence of *Forcipomyia taiwana* (Diptera: Ceratopogonidae) in the Nantou area in Taiwan. *Journal of Medical Entomology*, **37**, 205–209.

Colinet, H., Sinclair, B.J., Vernon, P. & Renault, D. (2015). Insects in fluctuating thermal environments. *Annual Review of Entomology*, **60**, 123–140.

Couret, J. (2013). Meta-analysis of factors affecting ontogenetic development rate in the *Culex pipiens* (Diptera: Culicidae) complex. *Environmental Entomology*, **42**, 614–626.

Couret, J. & Benedict, M.Q. (2014). A meta-analysis of the factors influencing development rate variation in *Aedes aegypti* (Diptera: Culicidae). *BMC Ecology*, **14**, 3.

Cupo, P. (2015). Bites and stings from venomous animals: a neglected Brazilian tropical disease. *Revista da Sociedade Brasileira de Medicina Tropical*, **48**, 639–641.

Day, J.F. & Curtis, G.A. (1989). Influence of rainfall on *Culex nigripalpus* (Diptera: Culicidae) blood-feeding behavior in Indian River County, Florida. *Annals of the Entomological Society of America*, **82**, 32–37.

Day, J.F. & Curtis, G.A. (1994). When it rains they soar – and that makes *Culex nigripalpus* a dangerous mosquito. *American Entomologist*, **40**, 162–167.

Day, J.F., Curtis, G.A. & Edman, J.D. (1990a). Rainfall-directed oviposition behavior of *Culex nigripalpus* (Diptera: Culicidae) and its influence on St. Louis encephalitis virus transmission in Indian River County, Florida. *Journal of Medical Entomology*, **27**, 43–50.

Day, J.F., Ramsey, A.M. & Zhang, J.-T. (1990b). Environmentally mediated seasonal variation in mosquito body size. *Environmental Entomology*, **19**, 469–473.

Dohm, D.J., O'Guinn, M.L. & Turell, M.J. (2002). Effect of environmental temperature on the ability of *Culex pipiens* (Diptera: Culicidae) to transmit West Nile virus. *Journal of Medical Entomology*, **39**, 221–225.

Edman, J.D. (1988). Disease control through manipulation of vector-host interaction: some historical and evolutionary perspectives. *Proceedings of a Symposium: The Role of vector-host interactions in disease transmission* (eds T.W. Scott & J. Grumstrup-Scott). Entomological Society of America, Washington, D.C., pp. 43–50.

Egizi, A., Fefferman, N.H. & Fonseca, D.M. (2015). Evidence that implicit assumptions of 'no evolution' of disease vectors in changing environments can be violated on a rapid timescale. *Philosophical Transactions of the Royal Society of London B: Biological Sciences*, **370**, 20140136.

Eshita, Y., Kurihara, T., Ogata, T. & Oya, A. (1982). Studies on the susceptibility of mosquitoes to dengue virus: I. Susceptibility of Japanese mosquitoes to the virus. *Japanese Journal of Sanitary Zoology*, **33**, 61–64.

Futami, K., Valderrama, A., Baldi, M., Minakawa, N., Marín Rodríguez, R. & Chaves, L.F. (2015). New and common haplotypes shape genetic diversity in Asian tiger mosquito populations from Costa Rica and Panamá. *Journal of Economic Entomology*, **108**, doi:10.1093/jee/tou1028.

Garrett-Jones, C. (1964). Prognosis for interruption of malaria transmission through assessment of mosquito's vectorial capacity. *Nature*, **204**, 1173–1175.

Gubler, D.J. (2001). Prevention and control of tropical diseases in the 21st century: Back to the field. *American Journal of Tropical Medicine and Hygiene*, **65**, V–Xi.

Hartley, D.M., Barker, C.M., Le Menach, A., Niu, T., Gaff, H.D. & Reisen, W.K. (2012). Effects of temperature on emergence and seasonality of West Nile virus in California. *American Journal of Tropical Medicine and Hygiene*, **86**, 884–894.

Higa, Y., Tsuda, Y., Tuno, N. & Takagi, M. (2000). Tempo-spatial variation in feeding activity and density of *Aedes albopictus* (Diptera: Culicidae) at peridomestic habitat in Nagasaki, Japan. *Medical Entomology and Zoology*, **51**, 205–209.

Hlavacova, J., Votypka, J. & Volf, P. (2013). The effect of temperature on *Leishmania* (Kinetoplastida: Trypanosomatidae) development in sand flies. *Journal of Medical Entomology*, **50**, 955–958.

Hoshi, T., Higa, Y. & Chaves, L.F. (2014a). *Uranotaenia novobscura ryukyuana* (Diptera: Culicidae) population dynamics are denso-dependent and autonomous from weather fluctuations. *Annals of the Entomological Society of America*, **107**, 136–142.

Hoshi, T., Imanishi, N., Higa, Y. & Chaves, L.F. (2014b). Mosquito biodiversity patterns around urban environments in South-Central Okinawa island, Japan. *Journal of the American Mosquito Control Association*, **30**, 260–267.

Huey, R.B. & Berrigan, D. (2001). Temperature, demography, and ectotherm fitness. *American Naturalist*, **158**, 204–210.

Huffaker, C.B. (1944). The temperature relations of the immature stages of the malarial mosquito, *Anopheles quadrimaculatus* Say, with a comparison of the developmental power of constant and variable temperatures in insect metabolism. *Annals of the Entomological Society of America*, **37**, 1–27.

Hurlbut, H.S. (1973). The effect of environmental temperature upon the transmission of St. Louis encephalitis virus by *Culex pipiens quinquefasciatus*. *Journal of Medical Entomology*, **10**, 1–12.

Hurtado, L.A., Cáceres, L., Chaves, L.F. & Calzada, J.E. (2014). When climate change couples social neglect: malaria dynamics in Panamá. *Emerging Microbes & Infections*, **3**, e28.

Isida, R. (1969). *Geography of Japan*. Kokusai Bunka Shinkokai, Tokyo.

Janisch, E. (1932). The influence of temperature on the life history of insects. *Transactions of the Royal Entomological Society of London*, **80**, 137–168.

Juliano, S.A., Lounibos, L.P. & O'Meara, G.F. (2004). A field test for competitive effects of *Aedes albopictus* on *A. aegypti* in South Florida: differences between sites of coexistence and exclusion? *Oecologia*, **139**, 583–593.

Juliano, S.A. (2009). Species interactions among larval mosquitoes: context dependence across habitat gradients. *Annual Review of Entomology*, **54**, 37–56.

Juliano, S.A. & Lounibos, L.P. (2005). Ecology of invasive mosquitoes: effects on resident species and on human health. *Ecology Letters*, **8**, 558–574.

Kaufman, M.G. & Fonseca, D.M. (2014). Invasion biology of *Aedes japonicus japonicus* (Diptera: Culicidae). *Annual Review of Entomology*, **59**, 31–49.

Kay, B.H., Fanning, I.D. & Mottram, P. (1989). Rearing temperature influences flavivirus vector competence of mosquitoes. *Medical and Veterinary Entomology*, **3**, 415–422.

Kay, B.H. & Jennings, C.D. (2002). Enhancement or modulation of the vector competence of *Ochlerotatus vigilax* (Diptera: Culicidae) for Ross River virus by temperature. *Journal of Medical Entomology*, **39**, 99–105.

Kilpatrick, A.M., Meola, M.A., Moudy, R.M. & Kramer, L.D. (2008). Temperature, viral genetics, and the transmission of West Nile virus by *Culex pipiens* Mosquitoes. *PLOS Pathogens*, **4**, e1000092.

Kobayashi, M., Nihei, N. & Kurihara, T. (2002). Analysis of Northern distribution of *Aedes albopictus* (Diptera:Culicidae) in Japan by geographical information system. *Journal of Medical Entomology*, **39**, 4–11.

Kuhn, M. & Johnson, K. (2013). *Applied Predictive Modeling*. Springer, New York.

Lambrechts, L., Paaijmans, K.P., Fansiri, T., Carrington, L.B., Kramer, L.D., Thomas, M.B. & Scott, T.M. (2011). Impact of daily temperature fluctuations on dengue virus transmission by *Aedes aegypti*. *Proceedings of the National Academy of Sciences of the United States of America*, **108**, 7460–7465.

Lehane, M.J. (2005). *The Biology of Blood-sucking in Insects*. Cambridge University Press, Cambridge.

Levins, R. (1968). *Evolution in Changing Environments. Some theoretical explorations*. Princeton University Press, Princeton.

Levins, R. (1995). Toward an integrated epidemiology. *Trends in Ecology & Evolution*, **10**, 304.

Levins, R. (2006). Strategies of abstraction. *Biology & Philosophy*, **21**, 741–755.

Levins, R., Awerbuch, T., Brinkmann, U., Eckardt, I., Epstein, P., Makhoul, N., de Possas, C.A., Puccia, C., Spielman, A. & Wilson, M.E. (1994). The emergence of new diseases. *American Scientist*, **82**, 52–60.

Levins, R. & Lewontin, R. (1985). *The dialectical biologist*. Harvard University Press, Cambridge, MA.

Levins, R. & Lopez, C. (1999). Toward an ecosocial view of health. *International Journal of Health Services*, **29**, 261–293.

Lindsay, S.W. & Birley, M.H. (1996). Climate change and malaria transmission. *Annals of Tropical Medicine and Parasitology*, **90**, 573–588.

Lounibos, L.P. (2002). Invasions by insect vectors of human disease. *Annual Review of Entomology*, **47**, 233–266.

Lounibos, L.P., O'Meara, G.F., Juliano, S.A., Nishimura, N., Escher, R.L., Reiskind, M.H., Cutwa, M. & Greene, K. (2010). Differential survivorship of invasive mosquito species in South Florida cemeteries: do site-specific microclimates explain patterns of coexistence and exclusion? *Annals of the Entomological Society of America*, **103**, 757–770.

Luz, C., Fargues, J. & Grunewald, J. (1999). Development of *Rhodnius prolixus* (Hemiptera: Reduviidae) under constant and cyclic conditions of temperature and humidity. *Memórias do Instituto Oswaldo Cruz*, **94**, 403–409.

Lyimo, E.O., Takken, W. & Koella, J.C. (1992). Effect of rearing temperature and larval density on larval survival, age at pupation and adult size of *Anopheles gambiae*. *Entomologia Experimentalis Et Applicata*, **63**, 265–271.

McMichael, A.J., Woodruff, R.E. & Hales, S. (2006). Climate change and human health: present and future risks. *The Lancet*, **367**, 859–869.

Medone, P., Ceccarelli, S., Parham, P.E., Figuera, A. & Rabinovich, J.E. (2015). The impact of climate change on the geographical distribution of two vectors of Chagas disease: implications for the force of infection. *Philosophical Transactions of the Royal Society of London B: Biological Sciences*, **370**, 20130560.

Minakawa, N., Sonye, G., Mogi, M., Githeko, A. & Yan, G.Y. (2002). The effects of climatic factors on the distribution and abundance of malaria vectors in Kenya. *Journal of Medical Entomology*, **39**, 833–841.

Mogi, M., Armbruster, P., Tuno, N., Campos, R. & Eritja, R. (2015). Simple indices provide insight to climate attributes delineating the geographic range of *Aedes albopictus* (Diptera: Culicidae) prior to worldwide invasion. *Journal of Medical Entomology*, **52**, 647–657.

Mogi, M. & Tuno, N. (2014). Impact of climate change on the distribution of *Aedes albopictus* (Diptera: Culicidae) in Northern Japan: retrospective analyses. *Journal of Medical Entomology*, **51**, 572–579.

Moo-Llanes, D., Ibarra-Cerdeña, C.N., Rebollar-Téllez, E.A., Ibáñez-Bernal, S., González, C. & Ramsey, J.M. (2013). Current and future niche of North and Central American sand flies (Diptera: Psychodidae) in climate change scenarios. *PLOS Neglected Tropical Diseases*, **7**, e2421.

Morrison, A.C., Ferro, C., Pardo, R., Torres, M., Devlin, B., Wilson, M.L. & Tesh, R.B. (1995). Seasonal abundance of *Lutzomyia longipalpis* (Dipteral Psychodidae) at an endemic focus of visceral leishmaniasis in Colombia. *Journal of Medical Entomology*, **32**, 538–548.

Nguyen, A.T., Williams-Newkirk, A.J., Kitron, U.D. & Chaves, L.F. (2012). Seasonal weather, nutrients and conspecific presence impacts on the southern house mosquito oviposition dynamics in combined sewage overflows. *Journal of Medical Entomology*, **49**, 1328–1338.

Nisbet, R.M. & Gurney, W.S.C. (1976). Population dynamics in a periodically varying environment. *Journal of Theoretical Biology*, **56**, 459–475.

Oliva, C.F., Benedict, M.Q., Soliban, S.M., Lemperiere, G., Balestrino, F. & Gilles, J.R.L. (2012). Comparisons of life-history characteristics of a genetic sexing strain with laboratory strains of *Anopheles arabiensis* (Diptera: Culicidae) from Northern Sudan. *Journal of Medical Entomology*, **49**, 1045–1051.

Omori, N., Osima, M., Bekku, H. & Fujisaki, K. (1952). On the mosquitoes found in Nagasaki prefecture. *Contributions from the Research Institute of Endemics at Nagasaki University*, **27**, 281–284.

Paaijmans, K.P., Blanford, S., Bell, A.S., Blanford, J.I., Read, A.F. & Thomas, M.B. (2010). Influence of climate on malaria transmission depends on daily temperature variation. *Proceedings of the National Academy of Sciences of the United States of America*, **107**, 15135–15139.

Paaijmans, K.P., Read, A.F. & Thomas, M.B. (2009). Understanding the link between malaria risk and climate. *Proceedings of the National Academy of Sciences of the United States of America*, **106**, 13844–13849.

Parham, P.E., Waldock, J., Christophides, G.K. & Michael, E. (2015). Climate change and vector-borne diseases of humans. *Philosophical Transactions of the Royal Society of London B: Biological Sciences*, **370**, 20140377.

Parmesan, C. (2006). Ecological and evolutionary responses to recent climate change. *Annual Review of Ecology, Evolution, and Systematics*, **37**, 637–669.

Pascual, M., Ahumada, J.A., Chaves, L.F., Rodó, X. & Bouma, M. (2006). Malaria resurgence in the East African highlands: temperature trends revisited. *Proceedings of the National Academy of Sciences of the United States of America*, **103**, 5829–5834.

Patz, J.A. & Olson, S.H. (2006a). Climate change and health: global to local influences on disease risk. *Annals of Tropical Medicine & Parasitology*, **100**, 535–549.

Patz, J.A. & Olson, S.H. (2006b). Malaria risk and temperature: influences from global climate change and local land use practices. *Proceedings of the National Academy of Sciences of the United States of America*, **103**, 5635–5636.

Proestos, Y., Christophides, G.K., Ergüler, K., Tanarhte, M., Waldock, J. & Lelieveld, J. (2015). Present and future projections of habitat suitability of the Asian tiger mosquito, a vector of viral pathogens, from global climate simulation. *Philosophical Transactions of the Royal Society of London B: Biological Sciences*, **370**, 20130554.

Rabinovich, J.E., Kitron, U.D., Obed, Y., Yoshioka, M., Gottdenker, N. & Chaves, L.F. (2011). Ecological patterns of blood-feeding by kissing-bugs (Hemiptera: Reduviidae: Triatominae). *Memórias do Instituto Oswaldo Cruz*, **106**, 479–494.

Reiner, R.C., Perkins, T.A., Barker, C.M., Niu, T., Chaves, L.F., Ellis, A.M., George, D.B., Le Menach, A., Pulliam, J.R.C., Bisanzio, D., Buckee, C., Chiyaka, C., Cummings, D.A.T., Garcia, A.J., Gatton, M.L., Gething, P.W., Hartley, D.M., Johnston, G., Klein, E.Y., Michael, E., Lindsay, S.W., Lloyd, A.L., Pigott, D.M., Reisen, W.K., Ruktanonchai, N., Singh, B.K., Tatem, A.J., Kitron, U., Hay, S.I., Scott, T.W. & Smith, D.L. (2013). A systematic review of mathematical models of mosquito-borne pathogen transmission: 1970–2010. *Journal of The Royal Society Interface*, **10**, 20120921.

Reisen, W.K. (2012). The contrasting bionomics of *Culex* mosquitoes in western North America. *Journal of the American Mosquito Control Association*, **28**, 82–91.

Reisen, W.K., Fang, Y. & Martinez, V.M. (2006). Effects of temperature on the transmission of West Nile virus by *Culex tarsalis* (Diptera: Culicidae). *Journal of Medical Entomology*, **43**, 309–317.

Reisen, W.K., Meyer, R.P., Presser, S.B. & Hardy, J.L. (1993). Effect of temperature on the transmission of western equine encephalomyelitis and St. Louis encephalitis viruses by *Culex tarsalis* (Diptera: Culicidae). *Journal of Medical Entomology*, **30**, 151–160.

Reisen, W.K., Milby, M.M. & Bock, M.E. (1984). The effects of immature stress on selected events in the life history of *Culex tarsalis*. *Mosquito News*, **44**, 385–395.

Reisen, W.K., Milby, M.M. & Meyer, R.P. (1992). Population dynamics of adult *Culex* mosquitoes (Diptera: Culicidae) along the Kern River, Kern County, California, in 1990. *Journal of Medical Entomology*, **29**, 531–543.

Reisen, W.K., Thiemann, T., Barker, C.M., Lu, H., Carroll, B., Fang, Y. & Lothrop, H.D. (2010). Effects of warm winter temperature on the abundance and gonotrophic activity of *Culex* (Diptera: Culicidae) in California. *Journal of Medical Entomology*, **47**, 230–237.

Reiter, P. (2001). Climate change and mosquito-borne disease. *Environmental Health Perspectives*, **109**, 141–161.

Rueda, L.M., Patel, K.J., Axtell, R.C. & Stinner, R.E. (1990). Temperature-dependent development and survival rates of *Culex quinquefasciatus* and *Aedes aegypti* (Diptera: Culicidae). *Journal of Medical Entomology*, **27**, 892–898.

Ruiz, M.O., Chaves, L.F., Hamer, G.L., Sun, T., Brown, W.M., Walker, E.D., Haramis, L., Goldberg, T.L. & Kitron, U.D. (2010). Local impact of temperature and precipitation on West Nile virus infection in *Culex* species mosquitoes in northeast Illinois, USA. *Parasites & Vectors*, **3**, 19.

Salazar, M.I., Richardson, J.H., Sánchez-Vargas, I., Olson, K.E. & Beaty, B.J. (2007). Dengue virus type 2: replication and tropisms in orally infected *Aedes aegypti* mosquitoes. *BMC Microbiology*, **7**, 9.

Saldaña, A., Calzada, J.E., Pineda, V., Perea, M., Rigg, C., González, K., Santamaria, A.M., Gottdenker, N.L. & Chaves, L.F. (2015). Risk factors associated with *Trypanosoma cruzi* exposure in domestic dogs from a rural community in Panama. *Memórias do Instituto Oswaldo Cruz*, **110**, 936–944.

Scott, T.W., Morrison, A.C., Lorenz, L.H., Clark, G.G., Strickman, D., Kittayapong, P., Zhou, H. & Edman, J.D. (2000). Longitudinal studies of *Aedes aegypti* (Diptera : Culicidae) in Thailand and Puerto Rico: Population dynamics. *Journal of Medical Entomology*, **37**, 77–88.

Shaman, J. & Day, J.F. (2007). Reproductive phase locking of mosquito populations in response to rainfall frequency. *PLOS ONE*, **2**, e331.

Shaman, J., Spiegelman, M., Cane, M. & Stieglitz, M. (2006). A hydrologically driven model of swamp water mosquito population dynamics. *Ecological Modelling*, **194**, 395–404.

Smith, D.L., Drakeley, C.J., Chiyaka, C. & Hay, S.I. (2010). A quantitative analysis of transmission efficiency versus intensity for malaria. *Nature Communications*, **1**, 108.

Suwonkerd, W., Tsuda, Y., Takagi, M. & Wada, Y. (1996). Seasonal ocurrence of *Aedes aegypti* and *Ae. albopictus* in used tires in 1992–1994, Chiangmai, Thailand. *Tropical Medicine*, **38**, 101–105.

Suzuki, A., Tsuda, Y., Takagi, M. & Wada, Y. (1993). Seasonal observation on some population attributes of *Aedes albopictus* females in Nagasaki, Japan, with emphasis on the relation between the body size and the survival. *Tropical Medicine*, **35**, 91–99.

Takahashi, M. (1976). The effects of environmental and physiological conditions of *Culex tritaeniorhynchus* on the pattern of transmission of Japanese Encephalitis Virus. *Journal of Medical Entomology*, **13**, 275–284.

Taylor, F. (1981). Ecology and evolution of physiological time in insects. *American Naturalist*, **117**, 1–23.

Tsuda, Y., Takagi, M. & Wada, Y. (1994). Ecological study on mosquito communities in tree holes in Nagasaki, Japan, with special reference to *Aedes albopictus* (Diptera: Culicidae). *Japanese Journal of Sanitary Zoology*, **45**, 103–111.

Tsunoda, T., Chaves, L.F., Nguyen, G.T.T., Nguyen, Y.T. & Takagi, M. (2015). Winter activity and diapause of *Aedes albopictus* (Diptera: Culicidae) in Hanoi, Northern Vietnam. *Journal of Medical Entomology*, **52**, 1203–1212.

Uvarov, B.P. (1931). Insects and climate. *Transactions of the Royal Entomological Society of London*, **79**, 1–247.

Watanabe, M., Watanabe, H., Tabaru, Y., Hirao, M., Roychoudhury, S., Sawabe, K., Ishikawa, Y., Kawabata, T. & Kanno, K. (2012). Occurrence of vector mosquitoes at Tsunami disaster areas of the Great East Japan Earthquake. *Medical Entomology and Zoology*, **63**, 31–43.

Watts, D.M., Burke, D.S., Harrison, B.A., Whitmire, R.E. & Nisalak, A. (1986). Effect of temperature on the vector efficiency of *Aedes aegypti* for dengue 2 virus. *American Journal of Tropical Medicine & Hygiene*, **36**, 143–152.

Weaver, S.C. & Reisen, W.K. (2010). Present and future arboviral threats. *Antiviral Research*, **85**, 328–345.

Whitehorn, J., Kien, D.T.H., Nguyen, N.M., Nguyen, H.L., Kyrylos, P.P., Carrington, L.B., Tran, C.N.B., Quyen, N.T.H., Thi, L.V., Thi, D.L., Truong, N.T., Luong, T.T.H., Nguyen, C.V.V., Wills, B., Wolbers, M., & Simmons, C.P. (2015). Comparative susceptibility of *Aedes albopictus* and *Aedes aegypti* to dengue virus infection after feeding on blood of viremic humans: implications for public health. *Journal of Infectious Diseases*, **212**, 1182–1190.

WHO (2016). Vector-borne Diseases. http://www.who.int/mediacentre/factsheets/fs387/en/ Last accessed 24 October 2016.

Wittmann, E.J., Mellor, P.S. & Baylis, M. (2002). Effect of temperature on the transmission of orbiviruses by the biting midge, *Culicoides sonorensis*. *Medical and Veterinary Entomology*, **16**, 147–156.

Yang, G.-J., Bradshaw, C.J.A., Whelan, P.I. & Brook, B.W. (2008a). Importance of endogenous feedback controlling the long-term abundance of tropical mosquito species. *Population Ecology*, **50**, 293–305.

Yang, G.-J., Brook, B.W., Whelan, P.I., Cleland, S. & Bradshaw, C.J.A. (2008b). Endogenous and exogenous factors controlling temporal abundance patterns of tropical mosquitoes. *Ecological Applications*, **18**, 2028–2040.

Zea Iriarte, W.L., Tsuda, Y., Wada, Y. & Takagi, M. (1991). Distribution of mosquitoes on a hill of Nagasaki city, with emphasis to the distance from human dwellings. *Tropical Medicine*, **33**, 55–60.

Zouache, K., Fontaine, A., Vega-Rua, A., Mousson, L., Thiberge, J.-M., Lourenco-De-Oliveira, R., Caro, V., Lambrechts, L. & Failloux, A.-B. (2014). Three-way interactions between mosquito population, viral strain and temperature underlying chikungunya virus transmission potential. *Proceedings of the Royal Society B-Biological Sciences*, **281**, 20141078.

9

Climate and Atmospheric Change Impacts on Aphids as Vectors of Plant Diseases

James M. W. Ryalls[1] and Richard Harrington[2]

[1] Hawkesbury Institute for the Environment, Western Sydney University, Australia, NSW
[2] Rothamsted Insect Survey, Rothamsted Research, UK

Summary

Transmission of viruses is an important component of plant losses due to aphids. Viruses, plants and aphids interact in a range of ways which affect transmission. These interactions can be modified by climatic and atmospheric variables, and the effects of changes in these may scale up to play important roles in governing community productivity and ecosystem function. This chapter demonstrates how multiple environmental variables can interact to affect plant and aphid populations and, subsequently, the spread of viruses. We highlight gaps in the literature that are necessary for predicting how viruses, aphids and their host-plants will respond in the future. Realistic predictions about the extent of damage to plants will only be obtained if more studies incorporate multiple abiotic and biotic factors simultaneously.

9.1   The Disease Pyramid

Aphids are renowned pests of agriculture, horticulture and forestry. There are around 6000 species worldwide, most having developed an intimate relationship with a narrow range of closely-related host-plants. They are sap feeders, with mandibles and maxillae adapted to form stylets for probing between cells and penetrating the phloem tubes, into which they inject saliva and from which they withdraw sap. Whilst removal of sap inevitably causes some damage, especially when aphid populations are dense and the plant is weakened by other stresses, it is through transmission of viruses that the greatest economic losses usually occur. Just as aphids have evolved specialist relationships with plants, for example by overcoming their chemical defences, so have plant viruses evolved specialist relationships with aphids to effect transmission. The viruses have likewise evolved specialist relationships with their host-plants. There is thus a "disease triangle" whereby interactions between aphid and plant, plant and virus, and virus and aphid are each important in determining the rate of virus spread. Climate and atmospheric gases affect each component of the triangle and each interaction,

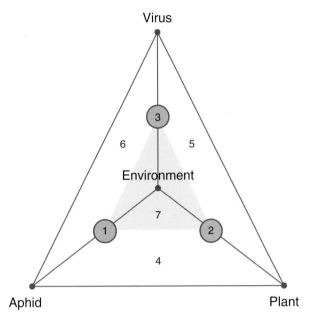

Figure 9.1 Disease pyramid. Numbers refer to sections within this chapter. Sections 9.1.1, 9.1.2 and 9.1.3 highlight the impacts of global environmental change on aphids, plants and viruses, respectively. Sections 9.2.1, 9.2.2 and 9.2.3 highlight the effects of environmental change on the two-way aphid–host-plant, plant–virus and virus–aphid interactions, respectively. Section 9.2.4 combines all four corners of the disease pyramid to highlight the effects of environmental change on the three-way aphid–host-plant–virus interaction.

thus forming the fourth node of a "disease pyramid". This chapter will consider, at the physiological and population levels, the implications of climatic and atmospheric changes on aphids, viruses and host plants, on the two-way and three-way interactions between them and hence on virus spread (Fig. 9.1).

9.1.1 Aphids

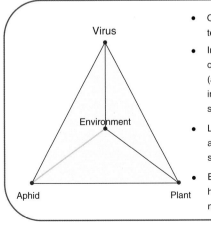

- Of climatic and atmospheric factors, changes in temperature have the biggest direct effects on aphids.

- In regions where temperatures are generally below optima, rising temperatures increase aphid growth rate (associated with an increase in fecundity and decrease in development time and mortality) and movement speed, although thresholds vary with species.

- Low temperatures tend to suppress wing development, and temperature thresholds for flight vary between species, genotype and life-cycle phases.

- Endosymbiotic bacteria can induce aphid resistance to heat stress but the presence of other symbiotic bacteria may alleviate this effect.

Climate change can have direct individual- and population-level impacts on aphids. Individual-level impacts include changes in behaviour, metabolism or life-history traits, namely development time, longevity and fecundity. Population-level impacts include genotypic variation within populations or changes in population dynamics (e.g., phenology, abundance, mortality rate and distribution). While any significant effects of CO_2 on aphids are likely to be plant-mediated (Sun & Ge, 2011), direct effects on aphids could occur through changes in gas exchange dynamics within their tracheal system (Ryan et al., 2015). Direct effects of wind or rainfall may also cause mortality by preventing aphid movement or through drowning (Menon & Christudas, 1967; Walters & Dixon, 1984; Harrington & Taylor, 1990). Temperature, however, is considered the major factor driving the direct responses of aphids to climate change. Like most insects, the intrinsic rate of increase (r_m) of aphids tends to increase with temperature up to a certain threshold, after which it decreases abruptly (van Baaren et al., 2010). The shape and magnitude of the curve is region-, aphid species- and genotype-dependent (Awmack & Leather, 2007). The lower developmental temperature thresholds of *Sitobion avenae* (grain aphid) and *Brevicoryne brassicae* (mealy cabbage aphid), for example, have been reported as −3.6°C and 7.1°C, respectively (Campbell et al., 1974; Carter et al., 1982), although thresholds often depend on the analytical methods used and may vary between aphid genotypes that have adapted to different regions (Bale et al., 2007). *Rhopalosiphum padi* (bird cherry–oat aphid) is particularly tolerant of high temperatures, although Mediterranean populations are more tolerant than British populations (de Barro, 1992; Asín & Pons, 2001).

Higher aphid numbers are usually associated with a decrease in development time and an increase in fecundity. *Aphis gossypii* (cotton aphid or melon aphid), for example, developed 33% faster and had a 20% higher r_m at 30°C compared with 20°C, although r_m increased by 31% at the optimum temperatures of 25°C compared with 20°C (van Steenis & El-Khawass, 1995). High temperatures may decrease aphid fitness by reducing embryo development or maturation and inhibiting aphid growth or size (Leather & Dixon, 1982; Carroll & Hoyt, 1986; Collins & Leather, 2001). High temperatures also inhibit aphid movement, which would be expected to become erratic as a result of gradual failure of physiological processes as temperature increases above an optimum, although such optima have rarely been investigated. Walking ceases altogether (heat coma) at around 40°C (Hazell et al., 2008). The lower temperature at which movement ceases has been shown to be dependent on acclimatisation temperature (Alford et al., 2012). For example, clones that were acclimatised to 20 and 25°C stopped walking at 7.5 and 12.5°C, respectively, whereas those acclimatised to 10°C remained mobile at 0°C. Alatae (winged adult aphid morphs) develop in all host-alternating species to effect movement between winter and summer hosts and vice versa and, in most species, when conditions become unfavourable (e.g., overcrowding or poor nutritional quality). Low temperatures have been shown to suppress wing development in aphids, with the proportion of *S. avenae* nymphs developing as wingless morphs increasing from < 20% in control conditions (20°C) to > 40% when subjected to a −5°C cold-pulse for one hour (Parish & Bale, 1990). Alatae usually disperse when atmospheric conditions favour take-off, sometimes travelling long distances in high winds (Irwin et al., 2007; Lawrence, 2009), with important implications for virus spread. Like most other physiological characteristics, temperature thresholds for aphid flight vary between and within species, and with season and region (Bale et al., 2007). Flight thresholds have also been reported to vary between life-cycle phases, with *Aphis fabae* (black bean aphid) taking off at 17°C and above, increasing altitude at 15°C and above, flying horizontally at 13°C and above and beating their wings to stay aloft at 6.5°C and above (Johnson & Taylor, 1957; Cockbain, 1961).

In warm temperate climates many aphid species reproduce parthenogenetically throughout the year, producing female clones to expand their populations rapidly. In cooler climates, however, aphids need to avoid overwintering in the mobile stages as these are not very tolerant of low temperatures. This is achieved by producing sexual males and females in autumn, which mate and produce cold-hardy eggs that hatch in spring (Dixon, 1998). Many species have both options open, with some clones genetically determined to produce winter eggs and some to continue parthenogenesis throughout the year. The minimum temperature threshold for survival of eggs tends to be considerably lower than the minimum temperature of their winter environments and thus overwintering success of eggs is largely independent of annual variations in winter minimum temperatures (Bale, 1999; Bale et al., 2007). As winter temperatures increase, so does survival of those aphids overwintering in the mobile stages and, over time, it is likely that selection will lead to a larger proportion overwintering in the mobile, rather than the egg, stage in species that have both options. This is likely to lead to increased damage from virus in both winter- and spring-sown crops. For example, an important vector of barley yellow dwarf virus (BYDV) in the UK is *R. padi* and secondary spread in autumn-sown crops is only by individuals from continuously parthenogenetic clones. For a range of spring-sown crops, warmer winters lead to continuously parthenogenetic aphids appearing earlier in the crop growth stage when they are more susceptible to damage from viruses (Harrington et al., 2007). In cold conditions, mobile aphids that are not killed may suffer a range of deleterious effects that determine their population dynamics, including longer development time, and reduced fecundity and longevity (Carter & Nichols, 1989; Parish & Bale, 1993; Hutchinson & Bale, 1994). In addition to producing overwintering eggs, however, aphids may adopt a number of other mechanisms to cope with low temperatures, including earlier-born nymphs in a progeny sequence being more cold-hardy than later-born nymphs, and acclimation (e.g., cold hardening) (Bale et al., 2007).

Most aphids form an obligate association with the primary endosymbiont *Buchnera aphidicola*, which synthesises essential amino acids and other nutrients that aphids cannot acquire from their phloem diet (Buchner, 1965; Baumann et al., 2013). Aphids may also harbour one or more secondary (or facultative) endosymbionts, which exert diverse effects on their aphid hosts, including changes in body colour, protection against parasitoids or resistance to heat stress (van Baaren et al., 2010; Ryalls et al., 2013a). The effects of heat stress on *Acyrthosiphon pisum* (pea aphid) fecundity and the number of bacteriocytes (in which *B. aphidicola* resides), for example, were reduced by the presence of the pea aphid secondary symbiont 'Candidatus Serratia symbiotica', although the presence of another secondary symbiont, pea aphid *Rickettsia*, negated the beneficial effects of *S. symbiotica* to heat-stressed aphids (Montllor et al., 2002). Dunbar et al. (2007) identified a point mutation in *B. aphidicola* that governs the thermal tolerance of *A. pisum*. At low temperatures, aphid fecundity increased when they contained the mutated symbiont. When aphids were exposed to heat, however, the fecundity of those containing the mutated symbiont (present in ~20% of field populations) dramatically decreased compared with non-mutants. Symbiotic associations play a fundamental role in aphid biology, with implications for predicting aphid responses to environmental change and incorporating novel biological control strategies (Goggin, 2007).

While this chapter does not detail predator or parasitoid impacts on aphids, it is necessary to note the impact of other trophic levels on aphid population responses to climate change (Barton & Ives, 2014). For example, thermal optima often vary between species and rising temperatures may release aphids from natural enemies by minimising the overlap between their distributions, or *vice versa*. Moreover, elevated temperature (eT) may influence other population parameters, including fecundity, longevity, and predator or parasitoid attack rate. At temperatures below 11°C, for example, *A. pisum* reproduces faster than the predator *Coccinella septempunctata* can consume it, but the reverse occurs above 11°C (Dunn, 1952). Understanding how climate change impacts aphid biology is key to identifying the underlying mechanisms that influence population-level responses, including phenology, abundance, mortality, distribution and genetic variability. Knowledge at the individual-level can be used in models, allowing better supported predictions of long-term aphid population responses to climate change. Brabec et al. (2014) used time-series data over a 24-year period to model the regulation by temperature of *Metopolophium dirhodum* (rose–grain aphid), *S. avenae* and *R. padi* on winter wheat. They concluded that wheat can escape aphid attack if temperatures range from 0 to 5°C in winter and noted the important role of pre-season termination of egg dormancy and hatching. Zhou et al. (1995) used suction-traps throughout Great Britain between 1964 and 1991 to study how temperature influences the migration phenologies of five aphid species: *Brachycaudus helichrysi* (peach leaf curl aphid or leaf-curling plum aphid), *Elatobium abietinum* (green spruce aphid), *M. dirhodum*, *M. persicae* and *S. avenae*. They identified the importance of winter temperature on aphid phenology, with a 1°C rise in average winter temperature advancing the migration phenology by 4–19 days depending on species. Harrington et al. (2007) compiled 40 years of flying aphid abundance data throughout Europe to determine the effects of climate change on aphid phenology. They predicted that the date of first flight record of aphids would advance by an average of 8 days over the next 50 years and promoted the value of trait-based groupings in predicting responses to environmental change. While data on land use change or plant host phenology may improve predictions of aphid migration phenology, the general predicted advance in aphid first flight record as temperatures rise (Harrington et al., 1995; Zhou et al., 1995; Harrington et al., 2007) is likely to increase crop damage and virus spread, especially if aphids arrive when crops are younger and more susceptible to attack.

9.1.2 Host-Plants

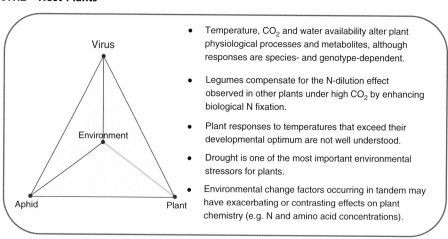

- Temperature, CO_2 and water availability alter plant physiological processes and metabolites, although responses are species- and genotype-dependent.
- Legumes compensate for the N-dilution effect observed in other plants under high CO_2 by enhancing biological N fixation.
- Plant responses to temperatures that exceed their developmental optimum are not well understood.
- Drought is one of the most important environmental stressors for plants.
- Environmental change factors occurring in tandem may have exacerbating or contrasting effects on plant chemistry (e.g. N and amino acid concentrations).

Various abiotic environmental factors, including temperature, CO_2 concentration and water availability can alter plant physiological processes, together with the concentration and composition of plant secondary metabolites (PSMs) and primary metabolites in the phloem (e.g. amino acids), which are important for aphid performance (Docherty et al., 1997; Douglas, 2003; Aslam et al., 2012). Plants generally respond to CO_2 enrichment by increasing rates of photosynthesis and growth. This dilutes limiting nutrients such as nitrogen, usually in the form of free amino acids and soluble protein, and increases the carbon:nitrogen ratio (C:N) in whole-plant tissues (Robinson et al., 2012). Plant type-, species-, and genotype-specific responses, however, are prevalent (Goverde et al., 2004; Ziska et al., 2012; Porter et al., 2014). Rice yield enhancement under elevated CO_2 concentrations (eCO_2), for example, ranged from 3 to 35% among different cultivars (Hasegawa et al., 2013). C_3 plants (e.g., cotton, potato, rice, soybean, sugar beet and wheat) tend to be more responsive to changes in

CO_2 concentration than C_4 plants (e.g., corn, sorghum and sugarcane) (Leakey, 2009; Ainsworth & McGrath, 2010; DaMatta et al., 2010) and tuber crops, with their capacity to store extra carbohydrates in belowground organs, often show the highest fertilisation responses to eCO_2 (Fleisher et al., 2008; Högy & Fangmeier, 2009). While eCO_2 decreases N availability in most plants, legumes are able to compensate for this N-dilution effect under eCO_2 by enhancing biological nitrogen fixation (Rogers et al., 2009), which may subsequently increase concentrations of phloem amino acids (Ryalls et al., 2015).

Plant responses to elevated temperature (eT) are more difficult to generalise. Temperature optima, as with plant responses to CO_2 concentration, can vary according to plant species or genotype within species (Bita & Gerats, 2013). At low temperatures (14/12°C day/night), for example, the photosynthetic activities of chilling-tolerant maize (*Zea mays*) genotypes were significantly higher than those of chilling-sensitive genotypes (Haldimann, 1999). Potato productivity is highly sensitive to temperature regime, with yield increasing when high daytime temperatures (25–30°C) are followed by low night-time temperatures (c. 15°C). This enhances the transport of assimilates in the phloem to the tubers and reduces respiration (Loebenstein et al., 2001).

While plant responses to temperature are generally well understood for temperatures up to the optimum temperature for development, plant responses to temperatures that exceed their developmental optimum are not as well understood (Craufurd & Wheeler, 2009). Wheat plants, for example, senesce more rapidly after flowering when subjected to temperatures above 32–34°C (Asseng et al., 2011; Lobell et al., 2012). Earlier flowering and maturity have been observed in a number of perennial horticultural crops throughout the world (Glenn et al., 2013), including grapes (Duchêne et al., 2010; Sadras & Petrie, 2011; Webb et al., 2011) and apples (Fujisawa & Kobayashi, 2010; Grab & Craparo, 2011). Climate variability, including changes in temperature, precipitation or their interaction, accounts for roughly a third (~32–39%) of the observed crop yield variability worldwide (Ray et al., 2015). Both flooding and drought are major drivers of food insecurity but the impact of drought on plant productivity is more widely studied than the impact of floods (Porter et al., 2014). Drought is one of the most important environmental stressors for plants, often reducing plant abundance, ground cover, plant vigour and yield (Cook & Sims, 1975) and altering the allocation of resources and profile of primary and secondary metabolites (Bray, 1997; Garg et al., 2001; Khan et al., 2011).

Interactions between multiple global change factors, predicted to occur in tandem in the future, can affect plants in complex ways. The majority of studies, however, tend to consider the effects of individual abiotic factors, which may lead to important oversights for management planning to prevent the spread of viruses. While eT may negate any increases in leaf carbohydrates caused by eCO_2, both eT and eCO_2, in combination, are predicted to amplify decreases in plant N and further increase C:N (Zvereva & Kozlov, 2006; DeLucia et al., 2012) in most plants. In legumes, however, eT and eCO_2 are likely to have a contrasting effect on plant N and, subsequently, phloem amino acid concentrations (Ryalls et al., 2013b; Ryalls et al., 2015). Elevated temperatures may exacerbate the impacts of drought on plants by creating a higher demand for water and reducing their nutrient uptake efficiency (Gregory, 2006; Staley & Johnson, 2008; Lobell et al., 2013). This is likely to impact phloem nutrient compositions by increasing demands for phloem solutes. Plants under eCO_2, however, may alleviate these effects by closing their stomata under drought conditions to minimise water loss and increase water use efficiency (Pritchard et al., 2007; Vanuytrecht et al., 2012; Trębicki et al., 2015). Stomatal closure may subsequently increase plant canopy temperatures by reducing transpiration and plant cooling (Kimball et al., 2002).

9.1.3 Viruses

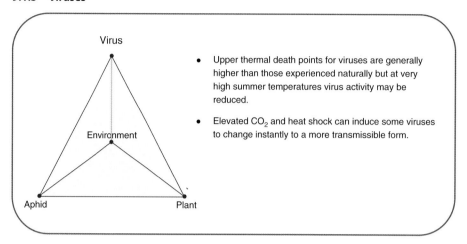

It seems intuitively unlikely that climate change will have significant direct impacts on the viruses, which cannot survive outside the plant or vector. Upper thermal death points for viruses are generally higher than those experienced naturally (Matthews, 1981), but viruses can become inactivated at temperatures which do not kill plants. Holding plants artificially at high temperatures is sometimes used to obtain virus-free stocks. For example, tobacco necrosis virus (TNV) (which is transmitted by fungi, not aphids) can be inactivated at 32–35°C so, if the same applies to aphid-borne viruses, warmer summers could potentially lead to reduced virus activity. The rate of replication of a virus species in plants varies with temperature, plant species, leaf age and virus strain. For example, for the first day after inoculation, 23°C is optimal for TNV replication in French bean leaves (Matthews, 1981).

It has recently been discovered that the non-circulative cauliflower mosaic virus (CaMV) can exist in different forms within the infected cell, some more transmissible than others (Martinière et al., 2013). The authors have shown that the virus reacts instantly (within seconds) to the presence of the vector, i.e., to the stylet punctures and/or saliva components and rapidly (and reversibly) transmutes into the more transmissible form. This ability to sense and react to the vector presence was also demonstrated for turnip mosaic virus (TuMV) (Drucker, 2015) and may therefore be widespread. Presumably, as with their insect vectors, there are costs associated with being dispersive and, when no vectors are present, it pays to remain non-dispersive so that more resources can be put into multiplication. The change from a less to more transmissible form appears to involve the rapid loading of soluble tubulin into the inclusion body (transmission body) of the virus, followed by relocation of the virus particle and helper component onto the microtubule network, which make the virus far more accessible to the vector. It is possible that mechanical stress due to disturbance by aphid stylets and/or compound(s) in the aphid saliva trigger the response (Gutiérrez et al., 2013). Importantly, in relation to this section, certain abiotic stresses such as elevated CO_2 or heat shock can induce the same changes, probably because of partial overlap in the signalling pathways triggered by the aphid and these abiotic stresses. Thus environmental conditions can potentially have a direct impact on the transmissibility of viruses. It has also been shown that stresses (salt, osmotic and wounding), probably via Ca^{2+} signalling, can influence rate and site of viral accumulation in plants (Suntio & Mäkinen, 2012). Relationships between plant stresses and the accumulation and epidemiology of viruses merit much further study (Suntio & Mäkinen, 2012; Gutiérrez et al., 2013).

9.2 Interactions with the Pyramid

9.2.1 Aphid–Host-Plant Interactions

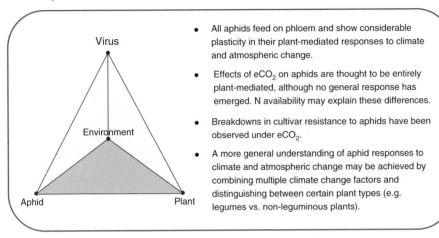

- All aphids feed on phloem and show considerable plasticity in their plant-mediated responses to climate and atmospheric change.

- Effects of eCO$_2$ on aphids are thought to be entirely plant-mediated, although no general response has emerged. N availability may explain these differences.

- Breakdowns in cultivar resistance to aphids have been observed under eCO$_2$.

- A more general understanding of aphid responses to climate and atmospheric change may be achieved by combining multiple climate change factors and distinguishing between certain plant types (e.g. legumes vs. non-leguminous plants).

Aphids form intricate relationships with their specific host-plants, often feeding from a single phloem sieve element cell for hours or days, while potentially evading and manipulating host-plant defences and nutritional compounds, respectively (Jiang & Miles, 1993; Will et al., 2007; Edwards et al., 2008; Gao et al., 2008b). Just as aphids can compromise host plant fitness (Dixon, 1998; Hussain et al., 2014), bottom-up effects of host plants, including nutritional quality, affect the performance of aphids (Awmack & Leather, 2002), which tend to favour plants with higher N and amino acid concentrations (Ponder et al., 2000; Nowak & Komor, 2010). Seasonal variation and environmental change can alter host plant quality, inducing chemical or nutritional changes that may benefit or deleteriously affect aphids. Sap-sucking aphids, as opposed to chewing insect guilds, show considerable plasticity in their plant-mediated responses to climate and, especially, atmospheric change, making long-term predictions challenging (Pritchard et al., 2007; Sun & Ge, 2011).

Direct effects of temperature on aphids and plants are prevalent as described above, although eT may indirectly impact aphids via changes in plant quality, morphology, phenology or defensive compounds (Holopainen & Kainulainen, 2004; Adler et al., 2007). In particular, aphid development rates and the subsequent number of generations in a given year may increase if rising temperatures increase the length of the host-plant growing season (Bale et al., 2007; Ziter et al., 2012). Additionally, leafy plant morphs can, through shading, provide protection against the negative effects of direct sunlight, high temperatures or other adverse weather conditions on aphids (Soroka & Mackay, 1991; Dunbar et al., 2007; Buchman & Cuddington, 2009). Plant genotype resistance to aphids may also alter the effects of eT on aphid success. For example, eT reduced *Acyrthosiphon pisum* (pea aphid) fecundity on both an aphid-resistant and -susceptible lucerne (*Medicago sativa*) cultivar but to a lesser degree on the resistant cultivar (Karner & Manglitz, 1985). While short-term studies are important for identifying mechanisms linking climate change with plants and aphids, long-term studies instil realism into climate change predictions. The effects of long-term experimental warming on aphids (*Obtusicauda coweni*) feeding on big sagebrush (*Artemisia tridentata*) showed that warming tended to decrease aphid density, although no significant relationship between aphids and plant C:N ratio or size was observed (Adler et al., 2007). Empirically, disentangling the direct and plant-mediated effects of temperature on aphids proves logistically challenging. In glasshouse studies, for example, only by growing plants at different temperatures and inoculating the plants with insects under the same temperature conditions (e.g., Murray et al., 2013) could direct and plant-mediated effects of temperature on aphids and other insects be differentiated.

Predicted increases in temperature should be considered alongside changes in rainfall patterns since global warming effects on aphids often interact with soil moisture conditions (Romo & Tylianakis, 2013), although few studies have considered both simultaneously. Sustained water stress (e.g., drought) in plants is likely to impact aphids negatively via changes in phloem physiology (e.g., increases in osmotic pressure or sap viscosity and decreases in turgor), which may outweigh the impact of any increases in N (Johnson et al., 2011) or amino acid concentrations (Hale et al., 2003) under drought conditions. Studies have observed positive (Wearing, 1972; Miles et al., 1982), negative (Kennedy et al., 1958; Xing et al., 2003) or no (Pons & Tatchell, 1995; Bethke et al., 1998; Aslam et al., 2012) effects of drought on aphid abundance, although the plant-mediated mechanisms are poorly understood. Additionally, some aphid species may be more resilient to water stress than others. Population growth rates of *A. pisum*, for example, decreased on water-stressed *M. sativa* plants, whereas *Therioaphis maculata* (spotted alfalfa aphid) was insensitive across a wide range of moisture levels (Forbes et al., 2005). Similarly, drought stress increased populations of the generalist aphid *Myzus persicae* (peach–potato aphid) but had no effect on the crucifer specialist *Brevicoryne brassicae* (mealy cabbage aphid) (Mewis et al., 2012). In this study, increases in *M. persicae* under drought were attributed to increases in plant (*Arabidopsis thaliana*) sugars and specific amino acids (namely, glutamate, proline, isoleucine and lysine), while water-logged conditions decreased these amino acids. Other studies have observed changes in aphid population demography with, for example, ratios of nymphs to adults increasing on drought-stressed plants (Aslam et al., 2012), which may reflect intraspecific variation in aphid developmental rates (Sandström & Pettersson, 1994; Xing et al., 2003).

The majority of studies consider sustained water stress events, such as drought or elevated precipitation. Realistically, however, stress-related fluctuations in precipitation are discontinuous, resulting in pulsed stress events (Huberty & Denno, 2004) or changes in drought intensity (English-Loeb et al., 1997) that may periodically alter plant nutritional quality or hydraulic properties. Plants recovering from drought, for example, may regain phloem turgor and increase plant nutritional compounds (e.g., N and amino acids), which aphids may capitalise on. However, if plants remain stressed due to other environmental pressures or if aphids remain on the plant, stress-related increases in plant quality are less likely (Huberty & Denno, 2004; Ryalls et al., 2015). Tariq et al. (2012) identified an increase in aphid performance on medium drought-stressed plants (50% of ambient water) compared with unstressed (ambient water) and highly drought-stressed plants (25% of ambient water), although no effect of pulsed drought stress was observed. Multiple theorems, including the plant stress (White, 1969, 1984), plant vigour (Price, 1991) or pulsed stress (Huberty & Denno, 2004) hypotheses, are used to help explain these conflicting trends, which demonstrate the complexity of these plant–aphid interactions in water-stressed environments (Pritchard et al., 2007). The field would benefit from more studies incorporating a range of predicted precipitation patterns, including pulsed stress, as well as seasonal or temporal differences that may contribute to the variation in plant quality and aphid performance. In the Western Australia wheatbelt, aphid flights into crops in winter correlate with summer rainfall events, which support the growth of alternative host plants on which aphids proliferate before flying to crops (Thackray et al., 2004; Knight & Thackray, 2007). While controlling these neighbouring plants, especially during months with higher rainfall, may serve as a way to limit aphid expansion and damage to crops, this strategy would almost certainly prove too expensive. Understanding the mechanisms underpinning crop susceptibility to aphids and building resistance into new crop genotypes may serve as the best way to mitigate the damage caused by aphids.

Compared with other factors associated with climate change, a wealth of knowledge has been accumulated on the effects of eCO_2 on aphids, which are thought to be entirely plant-mediated, yet few general trends have emerged (Sun & Ge, 2011). Positive, negative and neutral responses of aphids to eCO_2 have been observed, suggesting that aphids respond idiosyncratically to eCO_2 (Bezemer et al., 1999; Hughes & Bazzaz, 2001; Pritchard et al., 2007). The plant-mediated mechanisms underpinning responses to eCO_2 remain uncertain but are most likely driven by changes in plant amino acids (Weibull, 1987; Dixon et al., 1993; Ponder et al., 2000; Karley et al., 2002). Recent studies by Ryan et al. (2014b, 2015), for example, suggest that aphid responses to eCO_2 are linked to changes in individual plant amino acids. *R. padi* abundance on tall fescue (*Festuca arundinacea*) decreased under eCO_2, associated with a decrease in some, mostly essential, amino acids (Ryan et al., 2014b). In contrast, amino acid-mediated increases in *R. padi* abundance were observed on barley (*Hordeum vulgare*) under eCO_2 (Ryan et al., 2015), suggesting that aphids are able to overcome the disadvantageous decrease in plant N concentrations under eCO_2, potentially by reducing developmental times and increasing fecundity (Sun & Ge, 2011). Theoretical models by Newman and colleagues investigating the population dynamics of cereal aphids in Great Britain (Newman, 2003; Newman et al., 2003; Hoover & Newman, 2004; Newman, 2005) suggested that eCO_2 will have positive effects on aphids when nitrogen is not limited. They also proposed that N availability and environmental conditions would trump any differences between aphid species and predicted an outlook whereby generalisations may indeed be associated with the plant-mediated effects of climate and atmospheric change on aphids.

While it currently remains difficult to predict the effects of eCO_2 on aphids feeding on most plants, aphid performance is likely to improve on leguminous plants, which are able to fix atmospheric nitrogen and compensate for the N-diluting effect of eCO_2 experienced by other plant types (Rogers et al., 2009; Sulieman et al., 2015). For example, eCO_2 increased the nutritional quality (N and amino acid content) and decreased *A. pisum* resistance of *Medicago truncatula*, associated with the down-regulation of genes in the ethylene signalling pathway (Guo et al., 2014). Furthermore, *A. pisum* enhanced amino acid metabolism in the host plant (*M. truncatula*) to increase the aphid's fitness (population abundance) under eCO_2 (Guo et al., 2013). While plant nitrogen and amino acid concentrations often determine the performance of aphids, it remains fruitful to consider other plant-mediated mechanisms that may contribute to the variety of aphid responses observed under eCO_2, including other defensive or chemical mechanisms. For example, Goławska et al. (2006, 2008, 2012, 2014) noted a reduction in *A. pisum* activity (e.g., probing behaviour) and performance (e.g., reproduction) on lucerne plants with higher saponin concentrations and identified higher concentrations of saponins in plants infested with *A. pisum*, although no effects of climate change on saponin concentrations have yet been observed. Plants may be able to compensate for aphid infestation under eCO_2 by reallocating resources towards shoot tissue maintenance, as suggested by a reduction in flowering of broad beans (*Vicia faba*) caused by *A. pisum* under eCO_2 (Awmack & Harrington, 2000). Additionally, as predicted by the Carbon Nutrient Balance Hypothesis, eCO_2 conditions may allow plants to allocate more resources to defensive compounds, such as condensed tannin concentrations, and alter enzyme composition in aphids (Hartley et al., 2000; Wu et al., 2011).

Some plant–aphid interaction studies have incorporated multiple genotypes of an individual plant species and identified a breakdown in aphid resistance of certain cultivars under eCO_2 (Martin & Johnson, 2011; Hentley et al., 2014; Johnson et al., 2014), whereas others have observed changes in aphid performance under eCO_2 that have been consistent across host plant genotypes (Gao et al., 2008a; Ryalls et al., 2013b). Contrasting differences have also been observed across studies comparing aphid genotypes under eCO_2. For example, Mondor et al. (2005) identified neutral and positive effects of eCO_2 in pink and green genotypes of *A. pisum*, respectively, whereas Ryan et al. (2014a) identified no interaction between eCO_2 and *R. padi* genotype.

Studies into the effects of individual climate change factors remain important, although future increases in temperature and CO_2 concentrations are likely to occur in tandem. Only a handful of empirical studies (e.g., Himanen et al., 2008; Murray et al., 2013) have incorporated the effects of both eT and eCO_2 on plants and insects, and fewer still have examined their effects on aphids. Flynn et al. (2006) suggested that the combined effects of eT and eCO_2 will exacerbate *Macrosiphum euphorbiae* (potato aphid) damage to certain plants, including the C_3 perennial *Solanum dulcamara*. Ryalls et al. (2015) examined the effects of both factors on a leguminous plant (*M. sativa*) and demonstrated the ability of eT to negate the positive effects of eCO_2 on both plant amino acids and *A. pisum*. They also identified specific groups of individual amino acids responsible for driving aphid responses to eT and eCO_2 and plant genotypic resistance. Future studies should incorporate these factors simultaneously, especially in light of the exacerbating or contrasting effects of eCO_2 and eT on certain plants and aphids.

In summary, variation in host plant quality attributed to global climate change can shape the extent to which plants are attacked by aphids, yet marked differences have been observed between both aphid and plant species and genotypes. From this standpoint, it remains difficult to extrapolate general trends from such a variety of individual responses. However, a more holistic approach combining the effects of multiple climate change factors on a multi-trophic scale, while extrapolating across plant types (e.g., leguminous and non-leguminous plants) may pave the way towards a general understanding of these responses.

9.2.2 Host-Plant–Virus Interactions

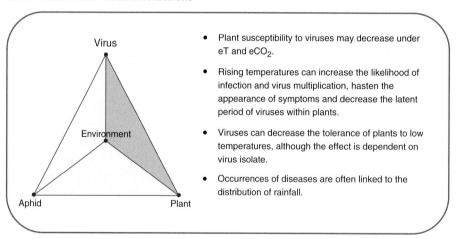

- Plant susceptibility to viruses may decrease under eT and eCO_2.
- Rising temperatures can increase the likelihood of infection and virus multiplication, hasten the appearance of symptoms and decrease the latent period of viruses within plants.
- Viruses can decrease the tolerance of plants to low temperatures, although the effect is dependent on virus isolate.
- Occurrences of diseases are often linked to the distribution of rainfall.

Climate and atmospheric conditions can modify virus biology and plant growth individually, although few studies have linked virus titre with plant growth under expected future conditions. Nancarrow et al. (2014) investigated the effects of eT on BYDV in wheat using real time quantitative PCR. BYDV titre and plant growth increased under eT, with wheat expressing earlier symptoms of BYDV infection. While virus symptoms may be expressed sooner under eT, they may be less severe than at lower temperatures (Broadbent, 1957). A number of studies have identified a decrease in virus titre and virulence in plants subjected to eT (Velázquez et al., 2010; Aguilar et al., 2015; Ma et al., 2016) and eCO_2 (Matros et al., 2006; Huang et al., 2012; Zhang et al., 2015). Trębicki et al. (2015) determined the effects of eCO_2 on BYDV in wheat. They identified a 36.8% increase in virus titre under eCO_2

compared with ambient CO_2, although the mechanisms underpinning this response were unclear, with no correlations observed between virus titre and plant growth. Considering that plant canopy temperatures can increase under eCO_2 by reducing transpiration via stomatal closure (Kimball et al., 2002), they speculated that an increase in wheat canopy temperatures under eCO_2 may have contributed to the observed increase in BYDV titre. In this case, eT is likely to exacerbate the effects of eCO_2 on virus titre, although these conditions may also decrease plant susceptibility to viruses via changes in plant resistance mechanisms, presenting a potential barrier to virus spread. For example, tobacco (*Nicotiana tabacum*) resistance to potato virus Y (PVY) increased in response to eCO_2, associated with changes in plant secondary metabolites (Matros et al., 2006). eT has also been shown to increase antiviral resistance by promoting RNA-mediated processes (i.e., RNA interference) responsible for silencing viral RNAs (Chellappan et al., 2005; Canto et al., 2009; Velázquez et al., 2010; Del Toro et al., 2015; Ma et al., 2016). For example, RNA-based antiviral resistance of *Nicotiana benthamiana* against Cymbidium ringspot virus increased at 24–27°C compared with 15°C, which reduced virus titre and symptom expression (Szittya et al., 2003). Temperature dependency has also been observed in protein-induced dominant resistances against viruses (e.g., tobacco mosaic virus), whereby hosts activate resistance by recognising viral genes (Schoelz, 2006). Specifically, and in contrast to the response of gene silencing-based resistance, some protein-elicited dominant resistances to tobacco mosaic virus and tomato mosaic virus have been shown to diminish under eT (Fraser & Loughlin, 1982; Dinesh-Kumar et al., 1995; Canto & Palukaitis, 2002; Pfitzner, 2006).

Temperature can determine both the ease with which plants become infected with a virus as well as the development (i.e., whether or not the virus multiplies within the plant), incubation (i.e., the interval between the establishment of an infection in the plant and the first appearance of symptoms) and latent period of the virus within the plant (i.e., the interval between the plant becoming infected and the virus being available for acquisition by an aphid) period of the virus within the plant. Many plants become more easily infected by viruses at higher temperatures (36°C) than at lower temperatures (20°C) and the incubation period often decreases at higher temperatures (Kassanis 1952). Increased plant growth observed under moderate temperature increases, however, may reduce the infection susceptibility period of the plant. Temperature can also determine virus symptom severity, with tobacco plants infected with tobacco mosaic virus (TMV), for example, producing necrotic local lesions at 20°C but only systemic mottling at 36°C. Continuous temperatures above 36°C, however, may render the virus inactive within the plant (Agrios, 2005). Stress tolerance in plants, including their ability to tolerate sub-zero temperatures, may be modified by the virus development period within the plant. For example, wheat plants with long disease (BYDV) development periods of ~34 days failed to tolerate low temperature conditions compared with plants with short (~18 days) development periods of BYDV (Andrews & Paliwal, 1986). Different isolates of the virus may also differ in their modifying effect on low temperature or ice tolerance of plants (Paliwal & Andrews, 1990). Furthermore, infecting plants with BYDV prior to cold-hardening and flooding them for two weeks after the hardening period may further reduce their low-temperature tolerance (Andrews & Sinha, 1991).

Moisture, like temperature, influences the development of the virus and the susceptibility of the plant to certain pathogens, with the occurrence of disease often linked with the amount and distribution of rainfall within the year. Moisture is often linked with host plant nutrition, which may affect the rate of growth and the ability of plants to defend themselves against virus infection. An abundance or shortage of nitrogen, for example, may delay plant maturity or weaken the plant, respectively, making it more susceptible to infection (Agrios, 2005).

Xu et al. (2008) have shown that virus infection can improve tolerance to abiotic stress, possibly as a result of increased osmoprotectant and antioxidant levels in infected plants. For example, infection with the aphid-transmitted cucumber mosaic virus (CMV) delayed the onset of drought symptoms and improved tolerance to freezing in beet plants.

9.2.3 Virus–Aphid Interactions

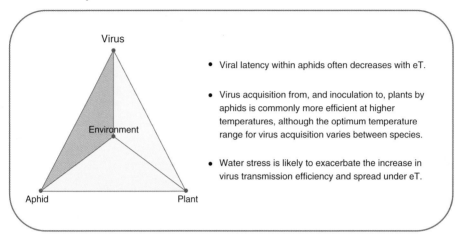

- Viral latency within aphids often decreases with eT.

- Virus acquisition from, and inoculation to, plants by aphids is commonly more efficient at higher temperatures, although the optimum temperature range for virus acquisition varies between species.

- Water stress is likely to exacerbate the increase in virus transmission efficiency and spread under eT.

Aphids can transmit viruses in a non-persistent, semi-persistent or persistent manner, depending on how long the aphid retains the virus, which ranges from minutes or hours (non-persistent) to days or weeks (persistent). The proportion of viruses transmitted non-persistently by aphids (c. 75%) far outweighs those transmitted semi-persistently or persistently due to the probing behaviour of aphids during host seeking, which facilitates acquisition from, and inoculation to, the epidermal cells in which non-persistent viruses reside (Katis et al., 2007). Only aphids can transmit viruses in the non-persistent manner. The viability of non-persistently transmitted viruses is short-lived, with aphids readily acquiring virions (i.e., virus particles) from infected plants and inoculating healthy plants within minutes. At the other end of the spectrum, persistently transmitted viruses have a long retention time in the aphid and a latent period prior to inoculation, during which time the virus passes into the haemolymph and thence to the accessory salivary glands from where it is inoculated during salivation. These viruses are retained during moulting of the aphid. Any given persistent virus is generally transmitted by fewer aphid species than a non-persistent virus since persistent viruses reside in the phloem and transmission can only occur during long feeding periods (12–24 hours). Midway along the spectrum lie semi-persistently transmitted viruses, which share some of the properties of both non-persistently and persistently transmitted viruses, but typically remain transmittable for hours. Non-circulative viruses, consisting of those transmitted both non-persistently and semi-persistently, are carried at the tips of the stylet or in the foregut, whereas circulative viruses (i.e., those transmitted persistently) are carried through the haemolymph to the salivary glands prior to inoculation (Pirone & Harris, 1977; Katis et al., 2007).

Climate and atmospheric changes have the potential to alter various characteristics associated with the virus–vector interaction, including acquisition efficiency, latency of persistently transmitted viruses or inoculation efficiency. Temperature effects on viral latency within aphids have been tested since the 1960s when Duffus (1963) demonstrated that the average latent period of sowthistle yellow vein virus in *Hyperomyzus lactucae* (currant–sowthistle aphid) was halved for each 10°C rise in the 5–25°C range. Latencies of three isolates of BYDV in two aphid species were shown to decrease by an average of over 20 hours (~30%) at 25°C compared with 20°C. Moreover, the percentage of aphids that transmitted two of the three isolates increased by 9.6 and 5%, suggesting that the virulence of BYDV may increase under eT by decreasing the latent period and increasing the number of infective aphids (Van der Broek & Gill, 1980). Other studies have similarly identified reductions in the latent period at increased temperatures (Tanaka & Shiota, 1970; Tamada & Harrison, 1981). In contrast, Sylvester and Osler (1977) found no difference between latent periods of filaree red-leaf virus in *Acyrthosiphon malvae* at 20 and 25°C, which may reflect the idiosyncratic characteristics of different viruses or isolates of the same virus, with temperature having variable effects on virus recognition efficiency and rates of transport through aphid tissues (Gildow, 1990).

Virus acquisition and inoculation are commonly more efficient at higher temperatures (Webb, 1956; Robert & Rouzé-Jouan, 1971; Singh et al., 1988; Lucio-Zavaleta et al., 2001), often associated with changes in aphid behaviour or the availability of infected plants (Syller, 1994). Wu and Su (1990) demonstrated an increase in transmission efficiency of banana bunchy top virus (BBTV) in *Pentalonia nigronervosa* (banana aphid) as temperatures increased from 16 to 27°C. Similarly, BBTV was transmitted more efficiently by *P. nigronervosa* at 25 and 30°C compared with 20°C, although no differences were observed in aphid nymphs (Anhalt & Almeida, 2008). Some virus–aphid interactions (e.g., potato leaf roll virus (PLRV) transmitted by *M. persicae*) suggest that temperature affects virus acquisition more than inoculation (Sylvester & Richardson, 1966; Syller, 1987), whereas others (e.g., BBTV and *P. nigronervosa*) suggest the opposite (Anhalt & Almeida, 2008). In warmer regions, predicted temperature increases may diminish virus transmission rates by consistently exceeding the optimum temperature range for virus acquisition by aphid vectors. For example, *Aulacorthum solani* (foxglove or glasshouse-potato aphid) acquired and inoculated soybean dwarf virus more efficiently at 20–22°C than at 5, 10–11 or 29°C (Damsteegt & Hewings, 1987). In cooler climates, however, transmission rates are likely to increase as temperatures increase towards the optimum, although optimum conditions and transmission rates are likely to vary between viruses, isolates, aphid species and clones (Selvarajan & Balasubramanian, 2014). For example, transmission efficiency has been shown to differ between clones of *R. padi* at 5 and 10°C but not at 15°C, whereas transmission efficiency did not vary between clones of *Sitobion avenae* (grain aphid) at any temperature (Smyrnioudis et al., 2001). Additionally, *M. persicae* has been found to transmit zucchini yellow mosaic potyvirus (ZYMV) over a broader range of temperatures and humidities than *A. gossypii* (Fereres et al., 1992).

On a population scale, environmental conditions can influence the virus transmission process, which can drive seasonal variation in aphid abundance. For example, the warmer the January to August period in France, the higher the percentage of viruliferous (virus-carrying) *R. padi* aphids the following autumn, most likely associated with increased rates of population growth and virus (BYDV) transmission. Similarly, the incidence of BYDV in the UK between 1995 and 1998 was associated with the incidence of aphids in the preceding autumn. Relative to temperature, few studies have demonstrated the effects of other environmental conditions on virus–aphid interactions. Humid environments have been shown to favour transmission of PVY and PLRV by *A. gossypi* and *M. persicae* (Singh et al., 1988). Fereres et al. (1992) observed a reduction in ZYMV transmission efficiency of *A. gossypii* in low-humidity conditions (40% RH). In contrast, ZYMV transmission efficiency of *M. persicae* was favoured by low humidity conditions at 8°C, whereas at 21°C transmission by *M. persicae* was favoured by a relative humidity of 65% rather than 40 and 85%. Predicted increases in temperature coupled with water stress, including more intense rainfall events or longer drought periods are likely to exacerbate the spread of viruses through increased transmission efficiency (Smyrnioudis et al., 2000).

9.2.4 Aphid–Host-Plant–Virus Interactions

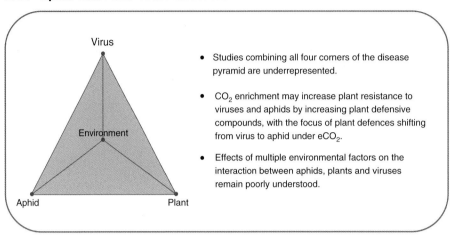

- Studies combining all four corners of the disease pyramid are underrepresented.

- CO_2 enrichment may increase plant resistance to viruses and aphids by increasing plant defensive compounds, with the focus of plant defences shifting from virus to aphid under eCO_2.

- Effects of multiple environmental factors on the interaction between aphids, plants and viruses remain poorly understood.

While the effects of climate and atmospheric changes on plants, insects and viruses are often inter-linked, few studies have incorporated interactions between all four corners of the disease pyramid. Viruses have been shown to alter plant volatile organic compounds to attract non-viruliferous aphids (Jiménez-Martínez & Bosque-Pérez, 2004). Specifically, BYDV and PLRV infection altered the relative composition of volatile organic compounds in wheat, which attracted both *R. padi* and *M. persicae* (Bosque-Pérez & Eigenbrode, 2011). Acquisition of BYDV has also been found to alter directly the host selection behaviour of *R. padi* (Ingwell et al., 2012). These virus–host-plant–aphid interactions emphasise the complexity of considering multiple factors simultaneously but are, nonetheless, important to consider when studying the effects of climate change on virus spread. Considering the ability of viruses to attract aphids and the direct effects of temperature and CO_2 on viruses, plants and aphids individually, climate change may exacerbate virus spread. For example, an increase in virus titre associated with eCO_2 combined with a preference by aphids for virus-infected plants may increase virus transmission and spread (Trębicki et al., 2015). Plant resistance to viruses and aphids may, however, increase under eCO_2. For example, Fu et al. (2010) suggested that plants increase defence against viruses and aphids under eCO_2. Specifically, tobacco plants produced more defensive secondary metabolites when inoculated with both aphids and cucumber mosaic virus (CMV) and *M. persicae* density increased on CMV-infected tobacco plants compared with uninfected plants, but only at ambient CO_2. They also observed an increase in plant defensive compounds on CMV-infected plants compared with aphid-infested plants under ambient CO_2 and a contrasting response at eCO_2, suggesting that eCO_2 shifts the focus of plant defences from virus to aphid. This highlights the importance of considering multiple biotic factors, although effects of multiple climate change factors combining plants, aphids and viruses are yet to be considered. This may be an important oversight considering the contrasting effects that climate and atmospheric change can have on aphids and plants (Newman, 2003; Ryalls et al., 2015) and potentially virus spread.

9.3 Conclusions and Future Perspectives

Plants, viruses and aphids do not experience climate change alone, but as part of a wider ecosystem combining multiple direct and indirect interlinking mechanisms. Environmental variables can clearly interact and govern plant and aphid population dynamics, and subsequently virus transmission (Newman, 2003; Hoover & Newman, 2004; van Baaren et al., 2010). Gaps in our knowledge of basic biological mechanisms, however, make it difficult to decipher the potential impacts of environmental change on the complex relationships between host plant, pest, pathogen and environment (Finlay & Luck, 2011). The effects of climate change on aphids are determined by the relationships between temperature and development, reproduction, movement and mortality. Direct effects of temperature on host plant and virus survival (Syller, 1987) and aphid behaviour during acquisition and/or inoculation (Robert et al., 2000; Katis et al., 2007) are prevalent, although uncertainty exists as to whether aphids will be able to acclimatise to higher temperatures over time. Impacts of eCO_2 on aphids are thought to be entirely plant-mediated, with many studies identifying a breakdown in plant resistance to aphids in some cultivars under eCO_2 (McMenemy et al., 2009; Hentley et al., 2014). This may force growers to revert back to using insecticide applications to suppress disease impacts on crops, which are associated with adverse environmental effects and high production costs (Jones, 2004; Oerke & Dehne, 2004). Climate change studies incorporating multiple cultivars are key to identifying the underlying mechanisms driving plant resistance to aphids and developing effective cultivars with long-lasting resistance. Moreover, impacts of climate and atmospheric change on viruses and aphids are often genotype-specific (Katis et al., 2007), which warrants further study combining additional virus isolates and aphid genotypes to establish generalisations and improve predictions. Interactions with neighbouring plants that can act as alternative aphid and/or virus hosts would also be beneficial to include as they can act as virus reservoirs and alter virus dynamics within a community (Malmstrom et al., 2005; Malmstrom et al., 2011).

Aphid, virus and host plant responses to climate change are likely to exacerbate each other. Earlier increases in virus titre coinciding with early aphid movement and greater dispersal under eT, for example, would increase virus transmission and enhance crop losses (Trębicki et al., 2015). Increased temperatures coupled with water stress (i.e., drought and increased precipitation) are likely to enhance the spread of viruses (Smyrnioudis et al., 2000). However, interactions between other climatic variables may counteract each other (Zvereva & Kozlov, 2006; Ryalls et al., 2015) and somewhat limit the spread of viruses, stressing the importance of considering multiple climate change factors simultaneously. The prospect of increasing temperatures and drought conditions worldwide raises the need for breeding additional drought-tolerant crop varieties (Naylor et al., 2007; Tao & Zhang, 2013) and including efficient plant mixtures (e.g., legume–grass mixtures) in cropping systems to maximise water use (Humphries, 2012), as well as optimised irrigation regimes, including efficient water delivery systems and improved irrigation technologies, among others (Porter et al., 2014).

Surprisingly few studies incorporate multiple plant, virus and aphid interactions into climate change studies. No studies to date have extended this to include higher trophic levels, including predators and parasitoids. Combining multiple environmental variables and multitrophic factors presents logistical constraints, yet realistic predictions about the spread of viruses will only be obtained if this challenge is met. Incorporating physiological and behavioural aphid, plant and virus responses into climate change models will elucidate the potential impacts of climate variability on interactions between all four corners of the disease pyramid, allowing more informed pest management strategies to defend our food resources against aphids and limit the spread of viruses.

Acknowledgements

We are grateful to Philip Smith for proofreading this chapter.

References

Adler, L.S., de Valpine, P., Harte, J. & Call, J. (2007) Effects of long-term experimental warming on aphid density in the field. *Journal of the Kansas Entomological Society*, **80**, 156–168.

Agrios, G.N. (2005) Chapter seven – environmental effects on the development of infectious plant disease. *Plant Pathology (Fifth Edition)* (G.N. Agrios), pp. 249–263. Academic Press, San Diego.

Aguilar, E., Allende, L., del Toro, F.J., Chung, B.-N., Canto, T. & Tenllado, F. (2015) Effects of elevated CO_2 and temperature on pathogenicity determinants and virulence of *Potato virus X*/potyvirus-associated synergism. *Molecular Plant–Microbe Interactions*, **28**, 1364–1373.

Ainsworth, E.A. & McGrath, J.M. (2010) Direct effects of rising atmospheric carbon dioxide and ozone on crop yields. *Climate Change and Food Security : Adapting Agriculture to a Warmer World* (eds D. Lobell & M. Burke), pp. 109–130. Springer Netherlands, Dordrecht.

Alford, L., Hughes, G.E., Blackburn, T.M. & Bale, J.S. (2012) Walking speed adaptation ability of *Myzus persicae* to different temperature conditions. *Bulletin of Entomological Research*, **102**, 303–313.

Andrews, C.J. & Sinha, R.C. (1991) Interactions between barley yellow dwarf virus infection and winter-stress tolerance in cereals. *Advances in Disease Vector Research* (ed. K.F. Harris), pp. 73–101. Springer New York, New York.

Andrews, C.J. & Paliwal, Y.C. (1986) Effects of barley yellow dwarf virus infection and low temperature flooding on cold stress tolerance of winter cereals. *Canadian Journal of Plant Pathology*, **8**, 311–316.

Anhalt, M.D. & Almeida, R.P.P. (2008) Effect of temperature, vector life stage, and plant access period on transmission of *Banana bunchy top virus* to banana. *Phytopathology*, **98**, 743–748.

Asín, L. & Pons, X. (2001) Effect of high temperature on the growth and reproduction of corn aphids (Homoptera: Aphididae) and implications for their population dynamics on the northeastern Iberian peninsula. *Environmental Entomology*, **30**, 1127–1134.

Aslam, T.J., Johnson, S.N. & Karley, A.J. (2012) Plant-mediated effects of drought on aphid population structure and parasitoid attack. *Journal of Applied Entomology*, **137**, 136–145.

Asseng, S., Foster, I. & Turner, N.C. (2011) The impact of temperature variability on wheat yields. *Global Change Biology*, **17**, 997–1012.

Awmack, C.S. & Harrington, R. (2000) Elevated CO_2 affects the interactions between aphid pests and host plant flowering. *Agricultural and Forest Entomology*, **2**, 57–61.

Awmack, C.S. & Leather, S.R. (2002) Host plant quality and fecundity in herbivorous insects. *Annual Review of Entomology*, **47**, 817–844.

Awmack, C.S. & Leather, S.R. (2007) Growth and development. *Aphids as Crop Pests* (eds H.F. van Emden & R. Harrington), pp. 135–151. CABI, Wallingford, UK.

Bale, J.S. (1999) Impacts of climate warming on arctic aphids: a comparative analysis. *Ecological Bulletins*, **47**, 38–47.

Bale, J.S., Masters, G.J., Hodkinson, I.D., Awmack, C., Bezemer, T.M., Brown, V.K., Butterfield, J., Buse, A., Coulson, J.C., Farrar, J., Good, J.E.G., Harrington, R., Hartley, S., Jones, T.H., Lindroth, R.L., Press, M.C., Symrnioudis, I., Watt, A.D. & Whittaker, J.B. (2002) Herbivory in global climate change research: direct effects of rising temperature on insect herbivores. *Global Change Biology*, **8**, 1–16.

Bale, J.S., Ponder, K.L. & Pritchard, J. (2007) Coping with stress. *Aphids as Crop Pests* (eds H.F. van Emden & R. Harrington), pp. 287–309. CABI, Wallingford, UK.

Barton, B.T. & Ives, A.R. (2014) Species interactions and a chain of indirect effects driven by reduced precipitation. *Ecology*, **95**, 486–494.

Baumann, P., Moran, N.A. & Baumann, L. (2013) Bacteriocyte-associated endosymbionts of insects. *The Prokaryotes* (eds M. Dworkin, S. Falkow, E. Rosenberg, K.-H. Schleifer & E. Stackebrandt), pp. 465–496. Springer, New York.

Bethke, J.A., Redak, R.A. & Schuch, U.K. (1998) Melon aphid performance on chrysanthemum as mediated by cultivar, and differential levels of fertilization and irrigation. *Entomologia Experimentalis et Applicata*, **88**, 41–47.

Bezemer, T.M., Knight, K.J., Newington, J.E. & Jones, T.H. (1999) How general are aphid responses to elevated atmospheric CO_2? *Annals of the Entomological Society of America*, **92**, 724–730.

Bita, C.E. & Gerats, T. (2013) Plant tolerance to high temperature in a changing environment: scientific fundamentals and production of heat stress-tolerant crops. *Frontiers in Plant Science*, **4**, 273.

Bosque-Pérez, N.A. & Eigenbrode, S.D. (2011) The influence of virus-induced changes in plants on aphid vectors: Insights from luteovirus pathosystems. *Virus Research*, **159**, 201–205.

Brabec, M., Honěk, A., Pekár, S. & Martinková, Z. (2014) Population dynamics of aphids on cereals: digging in the time-series data to reveal population regulation caused by temperature. *PLOS ONE*, **9**, e106228.

Bray, E.A. (1997) Plant responses to water deficit. *Trends in Plant Science*, **2**, 48–54.

Broadbent, L. (1957) *Investigation of virus diseases of brassica crops*. ARC Report Series 14. Cambridge University Press, London.

Buchman, N. & Cuddington, K. (2009) Influences of pea morphology and interacting factors on pea aphid (Homoptera: Aphididae) reproduction. *Environmental Entomology*, **38**, 962–970.

Buchner, P. (1965) *Endosymbiosis of Animals with Plant Microorganisms*. Interscience, New York.

Campbell, A., Frazer, B.D., Gilbert, N., Gutierrez, A.P. & Mackauer, M. (1974) Temperature requirements of some aphids and their parasites. *Journal of Applied Ecology*, **11**, 431–438.

Canto, T., Aranda, M.A. & Fereres, A. (2009) Climate change effects on physiology and population processes of hosts and vectors that influence the spread of hemipteran-borne plant viruses. *Global Change Biology*, **15**, 1884–1894.

Canto, T. & Palukaitis, P. (2002) Novel *N* gene-associated, temperature-independent resistance to the movement of *Tobacco mosaic virus* vectors neutralized by a *Cucumber mosaic virus* RNA1 transgene. *Journal of Virology*, **76**, 12908–12916.

Carroll, D.P. & Hoyt, S.C. (1986) Some effects of parental rearing conditions and age on progeny birth weight, growth, development, and reproduction in the apple aphid, *Aphis pomi* (Homoptera: Aphididae). *Environmental Entomology*, **15**, 614–619.

Carter, C.I. & Nichols, J.F.A. (1989) Winter survival of the lupin aphid *Macrosiphum albifrons* Essig. *Journal of Applied Entomology*, **108**, 213–216.

Carter, N., Dixon, A.F.G. & Rabbinge, R. (1982) *Cereal Aphid Populations: Biology, Simulation and Prediction*. Centre for Agricultural Publishing and Documentation (PUDOC), Wageningen.

Chellappan, P., Vanitharani, R., Ogbe, F. & Fauquet, C.M. (2005) Effect of temperature on geminivirus-induced RNA silencing in plants. *Plant Physiology*, **138**, 1828–1841.

Cockbain, A.J. (1961) Low temperature thresholds for flight in *Aphis fabae* Scop. *Entomologia Experimentalis et Applicata*, **4**, 211–219.

Collins, C.M. & Leather, S.R. (2001) Effect of temperature on fecundity and development of the giant willow aphid, *Tuberolachnus salignus* (Sternorrhyncha: Aphididae). *European Journal of Entomology*, **98**, 177–182.

Cook, C.W. & Sims, P.L. (1975) Drought and its relationship to dynamics of primary productivity and production of grazing animals. *Evaluation and Mapping of Tropical African Rangelands*, pp. 163-168. Proceedings of a seminar. International Livestock Centre for Africa, Addis Ababa, Ethiopia.

Craufurd, P.Q. & Wheeler, T.R. (2009) Climate change and the flowering time of annual crops. *Journal of Experimental Botany*, **60**, 2529–2539.

DaMatta, F.M., Grandis, A., Arenque, B.C. & Buckeridge, M.S. (2010) Impacts of climate changes on crop physiology and food quality. *Food Research International*, **43**, 1814–1823.

Damsteegt, V.D. & Hewings, A.D. (1987) Relationships between *Aulacorthum solani* and soybean dwarf virus: effect of temperature on transmission. *Phytopathology*, **77**, 515–518.

de Barro, P.J. (1992) The role of temperature, photoperiod, crowding and plant quality on the production of alate viviparous females of the bird cherry-oat aphid, *Rhopalosiphum padi*. *Entomologia Experimentalis et Applicata*, **65**, 205–214.

Del Toro, F.J., Aguilar, E., Hernández-Walias, F.J., Tenllado, F., Chung, B.-N. & Canto, T. (2015) High temperature, high ambient CO_2 affect the interactions between three positive-sense RNA viruses and a compatible host differentially, but not their silencing suppression efficiencies. *PLOS ONE*, **10**, e0136062.

DeLucia, E.H., Nabity, P.D., Zavala, J.A. & Berenbaum, M.R. (2012) Climate change: resetting plant-insect interactions. *Plant Physiology*, **160**, 1677–1685.

Dinesh-Kumar, S.P., Whitham, S., Choi, D., Hehl, R., Corr, C. & Baker, B. (1995) Transposon tagging of tobacco mosaic virus resistance gene *N*: its possible role in the TMV-*N*-mediated signal transduction pathway. *Proceedings of the National Academy of Sciences of the United States of America*, **92**, 4175-4180.

Dixon, A.F.G., Wellings, P.W., Carter, C. & Nichols, J.F.A. (1993) The role of food quality and competition in shaping the seasonal cycle in the reproductive activity of the sycamore aphid. *Oecologia*, **95**, 89–92.

Dixon, A.F.G. (1998) *Aphid Ecology (Second Edition)*. Chapman & Hall, London, UK.

Docherty, M., Wade, F., Hurst, D., Whittaker, J. & Lea, P. (1997) Responses of tree sap-feeding herbivores to elevated CO_2. *Global Change Biology*, **3**, 51–59.

Douglas, A.E. (2003) The nutritional physiology of aphids. *Advances in Insect Physiology*, **31**, 73–140.

Drucker, M. (2015) Turnip mosaic virus is a second virus that responds to the presence of aphid vectors and activates transmission. *Abstracts of UK - French Joint Meeting on Aphids*. 5-6 November 2015, Paris.

Duchêne, E., Huard, F., Dumas, V., Schneider, C. & Merdinoglu, D. (2010) The challenge of adapting grapevine varieties to climate change. *Climate Research*, **41**, 193–204.

Duffus, J.E. (1963) Possible multiplication in the aphid vector of sowthistle yellow vein virus, a virus with an extremely long insect latent period. *Virology*, **21**, 194–202.

Dunbar, H.E., Wilson, A.C.C., Ferguson, N.R. & Moran, N.A. (2007) Aphid thermal tolerance is governed by a point mutation in bacterial symbionts. *PLOS Biology*, **5**, 96.

Dunn, J.A. (1952) The effect of temperature on the pea aphid–ladybird relationship. Annual Report of the National Vegetable Research Station for 1951. National Vegetable Research Station. pp. 21-23. Wellesbourne, Warwick.

Edwards, O.R., Franzmann, B., Thackray, D. & Micic, S. (2008) Insecticide resistance and implications for future aphid management in Australian grains and pastures: a review. *Australian Journal of Experimental Agriculture*, **48**, 1523–1530.

English-Loeb, G., Stout, M.J. & Duffey, S.S. (1997) Drought stress in tomatoes: changes in plant chemistry and potential nonlinear consequences for insect herbivores. *Oikos*, **79**, 456–468.

Fereres, A., Blua, M.J. & Perring, T.M. (1992) Retention and transmission characteristics of zucchini yellow mosaic virus by *Aphis gossypii* and *Myzus persicae* (Homoptera: Aphididae). *Journal of Economic Entomology*, **85**, 759–765.

Finlay, K.J. & Luck, J.E. (2011) Response of the bird cherry-oat aphid (*Rhopalosiphum padi*) to climate change in relation to its pest status, vectoring potential and function in a crop–vector–virus pathosystem. *Agriculture, Ecosystems & Environment*, **144**, 405–421.

Fleisher, D.H., Timlin, D.J. & Reddy, V.R. (2008) Interactive effects of carbon dioxide and water stress on potato canopy growth and development. *Agronomy Journal*, **100**, 711–719.

Flynn, D.F.B., Sudderth, E.A. & Bazzaz, F.A. (2006) Effects of aphid herbivory on biomass and leaf-level physiology of *Solanum dulcamara* under elevated temperature and CO_2. *Environmental and Experimental Botany*, **56**, 10–18.

Forbes, A.E., Harvey, C.T. & Tilmon, K.J. (2005) Variation in the responses of spotted alfalfa aphids, *Therioaphis maculata* Buckton (Homoptera: Aphididae) and pea aphids, *Acythosiphon pisum* Harris (Homoptera: Aphididae) to drought conditions in alfalfa (*Medicago sativa* L., Fabaceae). *Journal of the Kansas Entomological Society*, **78**, 387–389.

Fraser, R.S.S. & Loughlin, S.A.R. (1982) Effects of temperature on the *Tm-1* gene for resistance to tobacco mosaic virus in tomato. *Physiological Plant Pathology*, **20**, 109–117.

Fu, X., Ye, L., Kang, L. & Ge, F. (2010) Elevated CO_2 shifts the focus of tobacco plant defences from cucumber mosaic virus to the green peach aphid. *Plant, Cell & Environment*, **33**, 2056–2064.

Fujisawa, M. & Kobayashi, K. (2010) Apple (*Malus pumila* var. domestica) phenology is advancing due to rising air temperature in northern Japan. *Global Change Biology*, **16**, 2651–2660.

Gao, F., Zhu, S.-R., Sun, Y.-C., Du, L., Parajulee, M., Kang, L. & Ge, F. (2008a) Interactive effects of elevated CO_2 and cotton cultivar on tri-trophic interaction of *Gossypium hirsutum*, *Aphis gossyppii*, and *Propylaea japonica*. *Environmental Entomology*, **37**, 29–37.

Gao, L.-L., Klingler, J.P., Anderson, J.P., Edwards, O.R. & Singh, K.B. (2008b) Characterization of pea aphid resistance in *Medicago truncatula*. *Plant Physiology*, **146**, 996–1009.

Garg, B.K., Kathju, S. & Burman, U. (2001) Influence of water stress on water relations, photosynthetic parameters and nitrogen metabolism of moth bean genotypes. *Biologia Plantarum*, **44**, 289–292.

Gildow, F. (1990) Barley yellow dwarf virus-aphid vector interactions associated with virus transmission and vector specificity. *World Perspectives on Barley Yellow Dwarf International Workshop, Udine, Italy* (ed. P. Burnett).CIMMYT, México, DF, Mexico.

Glenn, D.M., Kim, S.-H., Ramirez-Villegas, J. & Läderach, P. (2013) Response of perennial horticultural crops to climate change. *Horticultural Reviews* (ed. J. Janick), pp. 47–130. Wiley-Blackwell, Hoboken, NJ, USA.

Goggin, F.L. (2007) Plant–aphid interactions: molecular and ecological perspectives. *Current Opinion in Plant Biology*, **10**, 399–408.

Gołgawska, S., Bogumil, L. & Oleszek, W. (2006) Effect of low and high-saponin lines of alfalfa on pea aphid. *Journal of Insect Physiology*, **52**, 737–743.

Gołgawska, S., Łukasik, I. & Leszczyński, B. (2008) Effect of alfalfa saponins and flavonoids on pea aphid. *Entomologia Experimentalis et Applicata*, **128**, 147–153.

Gołgawska, S., Łukasik, I., Wójcicka, A. & Sytykiewicz, H. (2012) Relationship between saponin content in alfalfa and aphid development. *Acta Biologica Cracoviensia Series Botanica*, **54**, 39–46.

Gołgawska, S., Sprawka, I. & Łukasik, I. (2014) Effect of saponins and apigenin mixtures on feeding behavior of the pea aphid, *Acyrthosiphon pisum* Harris. *Biochemical Systematics and Ecology*, **55**, 137–144.

Goverde, M., Erhardt, A. & Stöcklin, J. (2004) Genotype-specific response of a lycaenid herbivore to elevated carbon dioxide and phosphorus availability in calcareous grassland. *Oecologia*, **139**, 383–391.

Grab, S. & Craparo, A. (2011) Advance of apple and pear tree full bloom dates in response to climate change in the southwestern Cape, South Africa: 1973–2009. *Agricultural and Forest Meteorology*, **151**, 406–413.

Gregory, P.J. (2006) *Plant Roots: Growth, Activity and Interaction with Soils (First Edition)*. Blackwell, Oxford.

Guo, H., Sun, Y., Li, Y., Liu, X., Zhang, W. & Ge, F. (2014) Elevated CO_2 decreases the response of the ethylene signaling pathway in *Medicago truncatula* and increases the abundance of the pea aphid. *New Phytologist*, **201**, 279–291.

Guo, H., Yucheng, S., Li, Y., Tong, B., Harris, M., Zhu-Salzman, K. & Ge, F. (2013) Pea aphid promotes amino acid metabolism both in *Medicago truncatula* and bacteriocytes to favor aphid population growth under elevated CO_2. *Global Change Biology*, **19**, 3210–3223.

Gutiérrez, S., Michalakis, Y., Van Munster, M. & Blanc, S. (2013) Plant feeding by insect vectors can affect life cycle, population genetics and evolution of plant viruses. *Functional Ecology*, **27**, 610–622.

Haldimann, P. (1999) How do changes in temperature during growth affect leaf pigment composition and photosynthesis in *Zea mays* genotypes differing in sensitivity to low temperature? *Journal of Experimental Botany*, **50**, 543–550.

Hale, B.K., Bale, J.S., Pritchard, J., Masters, G.J. & Brown, V.K. (2003) Effects of host plant drought stress on the performance of the bird cherry-oat aphid, *Rhopalosiphum padi* (L.): a mechanistic analysis. *Ecological Entomology*, **28**, 666–677.

Harrington, R., Bale, J.S. & Tatchell, G.M. (1995) Aphids in a changing climate. *Insects in a Changing Environment* (eds R. Harrington & N.E. Stork), pp. 126–155. Academic Press, London, UK.

Harrington, R., Clark, S.J., Welham, S.J., Verrier, P.J., Denholm, C.H., Hullé, M., Maurice, D., Rounsevell, M.D., Cocu, N. & European Union Examine Consortium (2007) Environmental change and the phenology of European aphids. *Global Change Biology*, **13**, 1550–1564.

Harrington, R. & Taylor, L.R. (1990) Migration for survival: fine-scale population redistribution in an aphid, *Myzus persicae*. *Journal of Animal Ecology*, **59**, 1177–1193.

Hazell, S.P., Pedersen, B.P., Worland, M.R., Blackburn, T.M. & Bale, J.S. (2008) A method for the rapid measurement of thermal tolerance traits in studies of small insects. *Physiological Entomology*, **33**, 389–394.

Hartley, S.E., Jones, C.G., Couper, G.C. & Jones, T.H. (2000) Biosynthesis of plant phenolic compounds in elevated atmospheric CO_2. *Global Change Biology*, **6**, 497–506.

Hasegawa, T., Sakai, H., Tokida, T., Nakamura, H., Zhu, C., Usui, Y., Yoshimoto, M., Fukuoka, M., Wakatsuki, H., Katayanagi, N., Matsunami, T., Kaneta, Y., Sato, T., Takakai, F., Sameshima, R., Okada, M., Mae, T. & Makino, A. (2013) Rice cultivar responses to elevated CO_2 at two free-air CO_2 enrichment (FACE) sites in Japan. *Functional Plant Biology*, **40**, 148–159.

Hentley, W.T., Hails, R.S., Johnson, S.N., Jones, T.H. & Vanbergen, A.J. (2014) Top-down control by *Harmonia axyridis* mitigates the impact of elevated atmospheric CO_2 on a plant–aphid interaction. *Agricultural and Forest Entomology*, **16**, 350–358.

Himanen, S.J., Nissinen, A., Dong, W.-X., Nerg, A.-M., Stewart, C.N., Jr., Poppy, G.M. & Holopainen, J.K. (2008) Interactions of elevated carbon dioxide and temperature with aphid feeding on transgenic oilseed rape: are *Bacillus thuringiensis* (Bt) plants more susceptible to nontarget herbivores in future climate? *Global Change Biology*, **14**, 1437–1454.

Högy, P. & Fangmeier, A. (2009) Atmospheric CO_2 enrichment affects potatoes: 1. Aboveground biomass production and tuber yield. *European Journal of Agronomy*, **30**, 78–84.

Holopainen, J.K. & Kainulainen, P. (2004) Reproductive capacity of the grey pine aphid and allocation response of Scots pine seedlings across temperature gradients: a test of hypotheses predicting outcomes of global warming. *Canadian Journal of Forest Research*, **34**, 94–102.

Hoover, J.K. & Newman, J.A. (2004) Tritrophic interactions in the context of climate change: a model of grasses, cereal Aphids and their parasitoids. *Global Change Biology*, **10**, 1197–1208.

Huang, L., Ren, Q., Sun, Y., Ye, L., Cao, H. & Ge, F. (2012) Lower incidence and severity of tomato virus in elevated CO_2 is accompanied by modulated plant induced defence in tomato. *Plant Biology*, **14**, 905–913.

Huberty, A.F. & Denno, R.F. (2004) Plant water stress and its consequences for herbivorous insects: a new synthesis. *Ecology*, **85**, 1383–1398.

Hughes, L. & Bazzaz, F.A. (2001) Effects of elevated CO_2 on five plant-aphid interactions. *Entomologia Experimentalis et Applicata*, **99**, 87–96.

Humphries, A.W. (2012) Future applications of lucerne for efficient livestock production in southern Australia. *Crop and Pasture Science*, **63**, 909–917.

Hussain, A., Razaq, M., Shahzad, W., Mahmood, K. & Ahmad Khan, F.Z. (2014) Influence of aphid herbivory on the photosynthetic parameters of *Brassica campestris* at multan, Punjab, Pakistan. *Journal of Biodiversity and Environmental Sciences*, **5**, 410–416.

Hutchinson, L.A. & Bale, J.S. (1994) Effects of sublethal cold stress on the aphid *Rhopalosiphum padi*. *Journal of Applied Ecology*, **31**, 102–108.

Ingwell, L.L., Eigenbrode, S.D. & Bosque-Pérez, N.A. (2012) Plant viruses alter insect behavior to enhance their spread. *Scientific Reports*, **2**, 578.

Irwin, M.E., Kampmeier, G.E. & Weisser, W.W. (2007) Aphid movement: process and consequences. *Aphids as Crop Pests* (eds H.F. van Emden & R. Harrington), pp. 153–186. CABI, Wallingford, UK.

Jiang, Y. & Miles, P.W. (1993) Responses of a compatible lucerne variety to attack by spotted alfalfa aphid: changes in the redox balance in affected tissues. *Entomologia Experimentalis et Applicata*, **67**, 263–274.

Jiménez-Martínez, E.S. & Bosque-Pérez, N.A. (2004) Variation in barley yellow dwarf virus transmission efficiency by *Rhopalosiphum padi* (Homoptera: Aphididae) after acquisition from transgenic and nontransformed wheat genotypes. *Journal of Economic Entomology*, **97**, 1790–1796.

Johnson, C.G. & Taylor, L.R. (1957) Periodism and energy summation with special reference to flight rhythms in aphids. *Journal of Experimental Biology*, **34**, 209–221.

Johnson, S.N., Ryalls, J.M.W. & Karley, A.J. (2014) Global climate change and crop resistance to aphids: contrasting responses of lucerne genotypes to elevated atmospheric carbon dioxide. *Annals of Applied Biology*, **165**, 62–72.

Johnson, S.N., Staley, J.T., McLeod, F.A.L. & Hartley, S.E. (2011) Plant-mediated effects of soil invertebrates and summer drought on above-ground multitrophic interactions. *Journal of Ecology*, **99**, 57–65.

Jones, R.A.C. (2004) Using epidemiological information to develop effective integrated virus disease management strategies. *Virus Research*, **100**, 5–30.

Karley, A.J., Douglas, A.E. & Parker, W.E. (2002) Amino acid composition and nutritional quality of potato leaf phloem sap for aphids. *Journal of Experimental Biology*, **205**, 3009–3018.

Karner, M.A. & Manglitz, G.R. (1985) Effects of temperature and alfalfa cultivar on pea aphid (Homoptera: Aphididae) fecundity and feeding activity of convergent lady beetle (Coleoptera: Coccinellidae). *Journal of the Kansas Entomological Society*, **58**, 131–136.

Kassanis, B. (1952) Some effects of high temperature on the susceptibility of plants to infection with viruses. *Annals of Applied Biology*, **39**, 358–369.

Katis, N.I., Tsitsipis, J.A., Stevens, M. & Powell, G. (2007) Transmission of plant viruses. *Aphids as Crop Pests* (eds H.F. van Emden & R. Harrington), pp. 353–377. CABI, Wallingford, UK.

Kennedy, J.S., Lamb, K.P. & Booth, C.O. (1958) Responses of *Aphis fabae* Scop. to water shortage in host plants in pots. *Entomologia Experimentalis et Applicata*, **1**, 274–290.

Khan, M.A.M., Ulrichs, C. & Mewis, I. (2011) Water stress alters aphid-induced glucosinolate response in *Brassica oleracea* var. *italica* differently. *Chemoecology*, **21**, 235–242.

Kimball, B.A., Kobayashi, K. & Bindi, M. (2002) Responses of agricultural crops to free-air CO_2 enrichment. *Advances in Agronomy* (ed. L.S. Donald), pp. 293–368. Academic Press, San Diego.

Knight, J.D. & Thackray, D.J. (2007) Decision support system. *Aphids as Crop Pests* (eds H.F. van Emden & R. Harrington), pp. 677–688. CABI, Wallingford, UK.

Lawrence, L. (2009) The future for aphids in Australian grain crops and pastures. *Outlooks on Pest Management*, **20**, 285–288.

Leakey, A.D.B. (2009) Rising atmospheric carbon dioxide concentration and the future of C_4 crops for food and fuel. *Proceedings of the Royal Society B: Biological Sciences*, **276**, 2333–2343.

Leather, S.R. & Dixon, A.F.G. (1982) Secondary host preferences and reproductive activity of the bird cherry-oat aphid, *Rhopalosiphum padi*. *Annals of Applied Biology*, **101**, 219–228.

Lobell, D.B., Hammer, G.L., McLean, G., Messina, C., Roberts, M.J. & Schlenker, W. (2013) The critical role of extreme heat for maize production in the United States. *Nature Climate Change*, **3**, 497–501.

Lobell, D.B., Sibley, A. & Ortiz-Monasterio, J.I. (2012) Extreme heat effects on wheat senescence in India. *Nature Climate Change*, **2**, 186–189.

Loebenstein, G., Berger, P.H., Brunt, A.A. & Lawson, R.H. (eds) (2001) *Virus and Virus-like Diseases of Potatoes and Production of Seed-potatoes*. Kluwer Academic Publishers, Dordrecht, Netherlands.

Lucio-Zavaleta, E., Smith, D.M. & Gray, S.M. (2001) Variation in transmission efficiency among *Barley yellow dwarf virus*-RMV Isolates and clones of the normally inefficient aphid vector, *Rhopalosiphum padi*. *Phytopathology*, **91**, 792–796.

Ma, L., Huang, X., Yu, R., Jing, X.L., Xu, J., Wu, C.A., Zhu, C.X. & Liu, H.M. (2016) Elevated ambient temperature differentially affects virus resistance in two tobacco species. *Phytopathology*, **106**, 94–100.

Malmstrom, C.M., Hughes, C.C., Newton, L.A. & Stoner, C.J. (2005) Virus infection in remnant native bunchgrasses from invaded California grasslands. *New Phytologist*, **168**, 217–230.

Malmstrom, C.M., Melcher, U. & Bosque-Pérez, N.A. (2011) The expanding field of plant virus ecology: historical foundations, knowledge gaps, and research directions. *Virus Research*, **159**, 84–94.

Martin, P. & Johnson, S.N. (2011) Evidence that elevated CO_2 reduces resistance to the European large raspberry aphid in some raspberry cultivars. *Journal of Applied Entomology*, **135**, 237–240.

Martinière, A., Bak, A., Macia, J.-L., Lautredou, N., Gargani, D., Doumayrou, J., Garzo, E., Moreno, A., Fereres, A., Blanc, S. & Drucker, M. (2013) A virus responds instantly to the presence of the vector on the host and forms transmission morphs. *eLife*, **2**, doi: 10.7554/e00183.

Matros, A., Amme, S., Kettig, B., Buck-Sorlin, G.H., Sonnewald, U. & Mock, H.-P. (2006) Growth at elevated CO_2 concentrations leads to modified profiles of secondary metabolites in tobacco cv. SamsunNN and to increased resistance against infection with *potato virus Y*. *Plant, Cell & Environment*, **29**, 126–137.

Matthews, R.E.F. (1981) *Plant Virology (Second Edition)*. Academic Press Inc., New York.

McMenemy, L.S., Mitchell, C. & Johnson, S.N. (2009) Biology of the European large raspberry aphid (*Amphorophora idaei*): its role in virus transmission and resistance breakdown in red raspberry. *Agricultural and Forest Entomology*, **11**, 61–71.

Menon, M.R. & Christudas, S.P. (1967) Studies on the population of the aphid *Pentalonia nigronervosa* Coq. on banana plants in Kerala. *Agriculture Research Journal of Kerala*, **5**, 84–86.

Mewis, I., Khan, M.A.M., Glawischnig, E., Schreiner, M. & Ulrichs, C. (2012) Water stress and aphid feeding differentially influence metabolite composition in *Arabidopsis thaliana* (L.). *PLOS ONE*, **7**, e48661.

Miles, P.W., Aspinall, D. & Rosenberg, L. (1982) Performance of the cabbage aphid, *Brevicoryne brassicae* (L.), on water-stressed rape plants, in relation to changes in their chemical composition. *Australian Journal of Zoology*, **30**, 337–346.

Mondor, E.B., Tremblay, M.N., Awmack, C.S. & Lindroth, R.L. (2005) Altered genotypic and phenotypic frequencies of aphid populations under enriched CO_2 and O_3 atmospheres. *Global Change Biology*, **11**, 1990–1996.

Montllor, C.B., Maxmen, A. & Purcell, A.H. (2002) Facultative bacterial endosymbionts benefit pea aphids *Acyrthosiphon pisum* under heat stress. *Ecological Entomology*, **27**, 189–195.

Murray, T.J., Ellsworth, D.S., Tissue, D.T. & Riegler, M. (2013) Interactive direct and plant-mediated effects of elevated atmospheric [CO_2] and temperature on a eucalypt-feeding insect herbivore. *Global Change Biology*, **19**, 1407–1416.

Nancarrow, N., Constable, F.E., Finlay, K.J., Freeman, A.J., Rodoni, B.C., Trebicki, P., Vassiliadis, S., Yen, A.L. & Luck, J.E. (2014) The effect of elevated temperature on Barley yellow dwarf virus-PAV in wheat. *Virus Research*, **186**, 97–103.

Naylor, R.L., Battisti, D.S., Vimont, D.J., Falcon, W.P. & Burke, M.B. (2007) Assessing risks of climate variability and climate change for Indonesian rice agriculture. *Proceedings of the National Academy of Sciences of the United States of America*, **104**, 7752–7757.

Newman, J. (2005) Climate change and the fate of cereal aphids in Southern Britain. *Global Change Biology*, **11**, 940–944.

Newman, J.A. (2003) Climate change and cereal aphids: the relative effects of increasing CO_2 and temperature on aphid population dynamics. *Global Change Biology*, **10**, 5–15.

Newman, J.A., Gibson, D.J., Parsons, A.J. & Thornley, J.H.M. (2003) How predictable are aphid population responses to elevated CO_2? *Journal of Animal Ecology*, **72**, 556–566.

Nowak, H. & Komor, E. (2010) How aphids decide what is good for them: experiments to test aphid feeding behaviour on *Tanacetum vulgare* (L.) using different nitrogen regimes. *Oecologia*, **163**, 973–984.

Oerke, E.C. & Dehne, H.W. (2004) Safeguarding production – losses in major crops and the role of crop protection. *Crop Protection*, **23**, 275–285.

Paliwal, Y. & Andrews, C. (1990) Barley yellow dwarf virus–host plant interactions affecting winter stress tolerance in cereals. *World Perspectives on Barley Yellow Dwarf International Workshop, Udine (Italy), 6-11 Jul 1987*. CIMMYT, México, DF, Mexico.

Parish, W.E.G. & Bale, J.S. (1993) Effects of brief exposures to low temperature on the development, longevity and fecundity of the grain aphid *Sitobion auenae* (Hemiptera: Aphididae). *Annals of Applied Biology*, **122**, 9–21.

Parish, W.E.G. & Bale, J.S. (1990) Effects of short term exposure to low temperature on wing development in the grain aphid *Sitobion avenae* (F.) (Hem., Aphididae). *Journal of Applied Entomology*, **109**, 175–181.

Pfitzner, A.J.P. (2006) Resistance to tobacco mosaic virus and tomato mosaic virus in tomato. *Natural Resistance Mechanisms of Plants to Viruses* (eds G. Loebenstein & J.P. Carr), pp. 399–413. Springer Netherlands, Dordrecht, The Netherlands.

Pirone, T.P. & Harris, K.F. (1977) Nonpersistent transmission of plant viruses by aphids. *Annual Review of Phytopathology*, **15**, 55–73.

Ponder, K.L., Pritchard, J., Harrington, R. & Bale, J.S. (2000) Difficulties in location and acceptance of phloem sap combined with reduced concentration of phloem amino acids explain lowered performance of the aphid *Rhopalosiphum padi* on nitrogen deficient barley (*Hordeum vulgare*) seedlings. *Entomologia Experimentalis et Applicata*, **97**, 203–210.

Pons, X. & Tatchell, G.M. (1995) Drought stress and cereal aphid performance. *Annals of Applied Biology*, **126**, 19–31.

Porter, J.R., Xie, L., Challinor, A.J., Cochrane, K., Howden, S.M., Iqbal, M.M., Lobell, D.B. & Travasso, M.I. (2014) Food security and food production systems. *Climate Change 2014: Impacts, Adaptation, and Vulnerability. Part A: Global and Sectoral Aspects. Contribution of Working Group II to the Fifth Assessment Report of the Intergovernmental Panel on Climate Change* (eds C.B. Field, V.R. Barros, D.J. Dokken, K.J. Mach, M.D. Mastrandrea, T.E. Bilir, M. Chatterjee, K.L. Ebi, Y.O. Estrada, R.C. Genova, B. Girma, E.S. Kissel, A.N. Levy, S. MacCracken, P.R. Mastrandrea & L.L. White), pp. 485–533. Cambridge University Press, Cambridge, UK and New York, USA.

Price, P.W. (1991) The plant vigor hypothesis and herbivore attack. *Oikos*, **62**, 244–251.

Pritchard, J., Griffiths, B. & Hunt, E.J. (2007) Can the plant-mediated impacts on aphids of elevated CO_2 and drought be predicted? *Global Change Biology*, **13**, 1616–1629.

Ray, D.K., Gerber, J.S., MacDonald, G.K. & West, P.C. (2015) Climate variation explains a third of global crop yield variability. *Nature Communications*, **6**, 5989.

Robert, Y. & Rouzé-Jouan, J. (1971) Premières observations sur le rôle de la température au moment de la transmission de l'Enroulement par *Aulacorthum solani* Kltb. *Macrosiphum euphorbiae* Thomas et *Myzus persicae* Sulz. *Potato Research*, **14**, 154–157.

Robert, Y., Woodford, J.A.T. & Ducray-Bourdin, D.G. (2000) Some epidemiological approaches to the control of aphid-borne virus diseases in seed potato crops in northern Europe. *Virus Research*, **71**, 33–47.

Robinson, E.A., Ryan, G.D. & Newman, J.A. (2012) A meta-analytical review of the effects of elevated CO_2 on plant–arthropod interactions highlights the importance of interacting environmental and biological variables. *New Phytologist*, **194**, 321–336.

Rogers, A., Ainsworth, E.A. & Leakey, A.D.B. (2009) Will elevated carbon dioxide concentration amplify the benefits of nitrogen fixation in legumes? *Plant Physiology*, **151**, 1009–1016.

Romo, C.M. & Tylianakis, J.M. (2013) Elevated temperature and drought interact to reduce parasitoid effectiveness in suppressing hosts. *PLOS ONE*, **8**, e58136.

Ryalls, J.M.W., Moore, B.D., Riegler, M., Gherlenda, A.N. & Johnson, S.N. (2015) Amino acid-mediated impacts of elevated carbon dioxide and simulated root herbivory on aphids are neutralized by increased air temperatures. *Journal of Experimental Botany*, **66**, 613–623.

Ryalls, J.M.W., Riegler, M., Moore, B.D. & Johnson, S.N. (2013a) Biology and trophic interactions of lucerne aphids. *Agricultural and Forest Entomology*, **15**, 335–350.

Ryalls, J.M.W., Riegler, M., Moore, B.D., Lopaticki, G. & Johnson, S.N. (2013b) Effects of elevated temperature and CO_2 on aboveground–belowground systems: a case study with plants, their mutualistic bacteria and root/shoot herbivores. *Frontiers in Plant Science*, **4**, 445.

Ryan, G.D., Emiljanowicz, L., Härri, S.A. & Newman, J.A. (2014a) Aphid and host-plant genotype × genotype interactions under elevated CO_2. *Ecological Entomology*, **39**, 309–315.

Ryan, G.D., Shukla, K., Rasmussen, S., Shelp, B.J. & Newman, J.A. (2014b) Phloem phytochemistry and aphid responses to elevated CO_2, nitrogen fertilization and endophyte infection. *Agricultural and Forest Entomology*, **16**, 273–283.

Ryan, G.D., Sylvester, E.V.A., Shelp, B.J. & Newman, J.A. (2015) Towards an understanding of how phloem amino acid composition shapes elevated CO_2-induced changes in aphid population dynamics. *Ecological Entomology*, **40**, 247–257.

Sadras, V.O. & Petrie, P.R. (2011) Climate shifts in south-eastern Australia: early maturity of Chardonnay, Shiraz and Cabernet Sauvignon is associated with early onset rather than faster ripening. *Australian Journal of Grape and Wine Research*, **17**, 199–205.

Sandström, J. & Pettersson, J. (1994) Amino acid composition of phloem sap and the relation to intraspecific variation in pea aphid (*Acyrthosiphon pisum*) performance. *Journal of Insect Physiology*, **40**, 947–955.

Schoelz, J.E. (2006) Viral determinants of resistance versus susceptibility. *Natural Resistance Mechanisms of Plants to Viruses* (eds G. Loebenstein & J.P. Carr), pp. 13–43. Springer Netherlands, Dordrecht, The Netherlands.

Selvarajan, R. & Balasubramanian, V. (2014) Host–virus interactions in banana-infecting viruses. *Plant Virus–Host Interaction: Molecular Approaches and Viral Evolution* (eds R.K. Gaur, T. Hohn & P. Sharma), pp. 57–72. Academic Press, Oxford, UK.

Singh, M.N., Paul Khurana, S.M., Nagaich, B.B. & Agrawal, H.O. (1988) Environmental factors influencing aphid transmission of potato virus Y and potato leafroll virus. *Potato Research*, **31**, 501–509.

Smyrnioudis, I.N., Harrington, R., Hall, M., Katis, N. & Clark, S.J. (2001) The effect of temperature on variation in transmission of a BYDV PAV-like isolate by clones of *Rhopalosiphum padi* and *Sitobion avenae*. *European Journal of Plant Pathology*, **107**, 167–173.

Smyrnioudis, I.N., Harrington, R., Katis, N. & Clark, S.J. (2000) The effect of drought stress and temperature on spread of barley yellow dwarf virus (BYDV). *Agricultural and Forest Entomology*, **2**, 161–166.

Soroka, J.J. & Mackay, P.A. (1991) Antibiosis and antixenosis to pea aphid (Homoptera: Aphididae) in cultivars of field peas. *Journal of Economic Entomology*, **84**, 1951–1956.

Staley, J.T. & Johnson, S.N. (2008) Climate change impacts on root herbivores. *Root Feeders – an ecosystem perspective* (eds S.N. Johnson & P.J. Murray), pp. 192–213. CABI, Wallingford, UK.

Sulieman, S., Thao, N.P. & Tran, L.-S. (2015) Does elevated CO_2 provide real benefits for N_2-fixing leguminous symbioses? *Legume Nitrogen Fixation in a Changing Environment* (eds S. Sulieman & L.-S. Tran), pp. 89–112. Springer International Publishing, New York.

Sun, Y.C. & Ge, F. (2011) How do aphids respond to elevated CO_2? *Journal of Asia-Pacific Entomology*, **14**, 217–220.

Suntio, T. & Mäkinen, K. (2012) Abiotic stress responses promote *Potato virus A* infection in *Nicotiana benthamiana*. *Molecular Plant Pathology*, **13**, 775–784.

Syller, J. (1987) The influence of temperature on transmission of potato leaf roll virus by *Myzus persicae* Sulz. *Potato Research*, **30**, 47–58.

Syller, J. (1994) The effects of temperature on the availability and acquisition of potato leafroll luteovirus by *Myzus persicae*. *Annals of Applied Biology*, **125**, 141–145.

Sylvester, E.S. & Osler, R. (1977) Further studies on the transmission of the filaree red-leaf virus by the aphid *Acyrthosiphon pelargonii zerozalphum*. *Environmental Entomology*, **6**, 39–42.

Sylvester, E.S. & Richardson, J. (1966) Some effects of temperature on the transmission of pea enation mosaic virus and on the biology of the pea aphid vector. *Journal of Economic Entomology*, **59**, 255–261.

Szittya, G., Silhavy, D., Molnár, A., Havelda, Z., Lovas, Á., Lakatos, L., Bánfalvi, Z. & Burgyán, J. (2003) Low temperature inhibits RNA silencing-mediated defence by the control of siRNA generation. *The EMBO Journal*, **22**, 633–640.

Tamada, T. & Harrison, B.D. (1981) Quantitative studies on the uptake and retention of potato leafroll virus by aphids in laboratory and field conditions. *Annals of Applied Biology*, **98**, 261–276.

Tanaka, S. & Shiota, H. (1970) Latent period of potato leaf roll virus in the green peach aphid (*Myzus persicae* Sulzer). *Japanese Journal of Phytopathology*, **36**, 106–111.

Tao, F. & Zhang, Z. (2013) Climate change, wheat productivity and water use in the North China Plain: a new super-ensemble-based probabilistic projection. *Agricultural and Forest Meteorology*, **170**, 146–165.

Tariq, M., Wright, D.J., Rossiter, J.T. & Staley, J.T. (2012) Aphids in a changing world: testing the plant stress, plant vigour and pulsed stress hypotheses. *Agricultural and Forest Entomology*, **14**, 177–185.

Thackray, D.J., Diggle, A.J., Berlandier, F.A. & Jones, R.A.C. (2004) Forecasting aphid outbreaks and epidemics of cucumber mosaic virus in lupin crops in a Mediterranean-type environment. *Virus Research*, **100**, 67–82.

Trębicki, P., Nancarrow, N., Cole, E., Bosque-Pérez, N.A., Constable, F.E., Freeman, A.J., Rodoni, B., Yen, A.L., Luck, J.E. & Fitzgerald, G.J. (2015) Virus disease in wheat predicted to increase with a changing climate. *Global Change Biology*, **21**, 3511–3519.

van Baaren, J., Le Lann, C. & van Alphen, J.J.M. (2010) Consequences of climate change for aphid-based multi-trophic systems. *Aphid Biodiversity under Environmental Change* (eds P. Kindlmann, A.F.G. Dixon & J.P. Michaud), pp. 55–68. Springer, Dordrecht, The Netherlands.

Van der Broek, L.J. & Gill, C.C. (1980) The median latent periods for three isolates of barley yellow dwarf virus in aphid vectors. *Phytopathology*, **70**, 644–646.

van Steenis, M.J. & El-Khawass, K.A.M.H. (1995) Life history of *Aphis gossypii* on cucumber: influence of temperature, host plant and parasitism. *Entomologia Experimentalis et Applicata*, **76**, 121–131.

Vanuytrecht, E., Raes, D., Willems, P. & Geerts, S. (2012) Quantifying field-scale effects of elevated carbon dioxide concentration on crops. *Climate Research*, **54**, 35–47.

Velázquez, K., Renovell, A., Comellas, M., Serra, P., García, M.L., Pina, J.A., Navarro, L., Moreno, P. & Guerri, J. (2010) Effect of temperature on RNA silencing of a negative-stranded RNA plant virus: *Citrus psorosis virus*. *Plant Pathology*, **59**, 982–990.

Walters, K.F.A. & Dixon, A.F.G. (1984) The effect of temperature and wind on the flight activity of cereal aphids. *Annals of Applied Biology*, **104**, 17–26.

Wearing, C.H. (1972) Responses of *Myzus persicae* and *Brevicoryne brassicae* to leaf age and water stress in brussels sprouts grown in pots. *Entomologia Experimentalis et Applicata*, **15**, 61–80.

Webb, L.B., Whetton, P.H. & Barlow, E.W.R. (2011) Observed trends in winegrape maturity in Australia. *Global Change Biology*, **17**, 2707–2719.

Webb, R.E. (1956) Relation of temperature to transmission of the potato leafroll virus. *Phytopathology*, **46**, 470.

Weibull, J. (1987) Seasonal changes in the free amino acids of oat and barley phloem sap in relation to plant growth stage and growth of *Rhopalosiphum padi*. *Annals of Applied Biology*, **111**, 729–737.

White, T.C.R. (1984) The abundance of invertebrate herbivores in relation to the availability of nitrogen in stressed food plants. *Oecologia*, **63**, 90–105.

White, T.C.R. (1969) An index to measure weather-induced stress of trees associated with outbreaks of psyllids in Australia. *Ecology*, **50**, 905–909.

Will, T., Tjallingii, W.F., Thönnessen, A. & van Bel, A.J.E. (2007) Molecular sabotage of plant defense by aphid saliva. *Proceedings of the National Academy of Sciences of the United States of America*, **104**, 10536–10541.

Wu, G., Chen, F.-J., Xiao, N.-W. & Ge, F. (2011) Influences of elevated CO_2 and pest damage on the allocation of plant defense compounds in Bt-transgenic cotton and enzymatic activity of cotton aphid. *Insect Science*, **18**, 401–408.

Wu, R.-Y. & Su, H.-J. (1990) Transmission of banana bunchy top virus by aphids to banana plantlets from tissue culture. *Botanical Bulletin of Academia Sinica*, **31**, 7–10.

Xing, G., Zhang, J., Liu, J., Zhang, X., Wang, G. & Wang, Y. (2003) Impacts of atmospheric CO_2 concentrations and soil water on the population dynamics, fecundity and development of the bird cherry–oat aphid *Rhopalosiphum padi*. *Phytoparasitica*, **31**, 499–514.

Xu, P., Chen, F., Mannas, J.P., Feldman, T., Sumner, L.W. & Roossinck, M.J. (2008) Virus infection improves drought tolerance. *New Phytologist*, **180**, 911–921.

Zhang, S., Li, X., Sun, Z., Shao, S., Hu, L., Ye, M., Zhou, Y., Xia, X., Yu, J. & Shi, K. (2015) Antagonism between phytohormone signalling underlies the variation in disease susceptibility of tomato plants under elevated CO_2. *Journal of Experimental Botany*, **66**, 1951–1963.

Zhou, X., Harrington, R., Woiwod, I.P., Perry, J.N., Bale, J.S. & Clark, S.J. (1995) Effects of temperature on aphid phenology. *Global Change Biology*, **1**, 303–313.

Ziska, L.H., Bunce, J.A., Shimono, H., Gealy, D.R., Baker, J.T., Newton, P.C.D., Reynolds, M.P., Jagadish, K.S.V., Zhu, C., Howden, M. & Wilson, L.T. (2012) Food security and climate change: on the potential to adapt global crop production by active selection to rising atmospheric carbon dioxide. *Proceedings of the Royal Society B: Biological Sciences*, **279**, 4097–4105.

Ziter, C., Robinson, E.A. & Newman, J.A. (2012) Climate change and voltinism in Californian insect pest species: sensitivity to location, scenario and climate model choice. *Global Change Biology*, **18**, 2771–2780.

Zvereva, E.L. & Kozlov, M.V. (2006) Consequences of simultaneous elevation of carbon dioxide and temperature for plant–herbivore interactions: a meta-analysis. *Global Change Biology*, **12**, 27–41.

Part III

Multi-Trophic Interactions and Invertebrate Communities

10

Global Change, Herbivores and Their Natural Enemies

William T. Hentley and Ruth N. Wade

The Department of Animal and Plant Sciences, The University of Sheffield, Sheffield, U.K.

Summary

Global change threatens the interaction between insect herbivores and their natural enemies, potentially destabilising terrestrial ecosystems through changes in insect growth, development and population demography. Understanding the impact of global change on trophic interactions is vitally important to provide information for future management, modelling and functioning of ecosystems. Major changes to the abiotic environment are predicted, such as increased concentrations of atmospheric carbon dioxide (CO_2) and ground-level ozone (O_3), increased temperature, altered precipitation patterns and more extreme weather events. These changes to the abiotic environment can impact herbivores and their natural enemies both directly and indirectly through changes in habitat, food quality and quantity, defence chemicals and developmental timing potentially creating phenological mismatch between interacting species. In this chapter we summarise the latest research investigating the effects of climate change on the interactions between insect herbivores and their natural enemies.

Published research demonstrates that climate change can have significant impacts on trophic interactions with potential consequences for agroecosystems and pest management. However, there are many different and interacting factors which can influence the precise outcomes such as insect feeding guild, the developmental time of the insect and plant at which the stress occurs, experimental design and severity of the stress. Therefore, generalisations regarding the overall impact of climate change on herbivores and their natural enemies cannot currently be made. Most of the current research also focuses on one aspect of the abiotic environment, whereas global change is multifaceted and the interactions between multiple environmental perturbations will have very different impacts on herbivores and their natural enemies. Further research in this area is crucial to elucidate the impact of multiple aspects of global change including extreme weather events on the response of herbivores and their natural enemies.

Global Climate Change and Terrestrial Invertebrates, First Edition. Edited by Scott N. Johnson and T. Hefin Jones.
© 2017 John Wiley & Sons, Ltd. Published 2017 by John Wiley & Sons, Ltd.

10.1   Introduction

This chapter will focus on the impact of global change on invertebrate herbivores and their natural enemies within agro-ecosystems. We define agro-ecosystem as a multi-trophic community that is managed to maximise the output of commodities for human consumption (e.g., food production) (Moonen & Bàrberi, 2008). This includes managed grassland and forests highlighting the diversity of habitat types in agro-ecosystems (see Chapter 13 for more detail on the response of forest communities). Over 40% of the earth's landmass is therefore classified as agro-ecosystems (FAOSTAT, 2014). The human population is predicted to reach nine billion by 2050 (United Nations, 2015), doubling in only 75 years. Food security provided by agro-ecosystems is critical for meeting the estimated 70% rise in global food demand over the next 50 years (Godfray et al., 2010). Insect herbivores ("pests" in agro-ecosystems) are one of the most detrimental factors affecting output from agro-ecosystems (Rusch et al., 2010). Arguably, one of the biggest threats to the control of these insect herbivores over the next century will be global change (McLean et al., 2011).

The major changes to the abiotic environment discussed in this chapter will be projected increases in the concentration of atmospheric carbon dioxide (CO_2) and ground-level ozone (O_3), increased temperature, altered precipitation patterns and extreme weather events. The cause of these changes and the environmental effects are summarised in Table 10.1. Elevated atmospheric CO_2 (eCO_2), increased temperature and altered precipitation are all intrinsically linked. Higher concentrations of atmospheric CO_2 increase downwelling of infrared radiation to the earth's surface, increasing global temperature (IPCC, 2013). Elevated atmospheric temperature can then significantly impact the amount, intensity and frequency of global precipitation (Trenberth et al., 2007) and the occurence of extreme weather events.

Changes to the abiotic environment are likely to impact ecosystem functioning, with some communities more susceptible than others. The "diversity–stability" hypothesis suggests that greater species diversity (i.e., more species) within a biological community will reduce the risk of ecological collapse such as a complete failure in ecosystem functioning (MacArthur, 1955). Compared to un-managed natural systems, agro-ecosystems have relatively low species diversity, making them more susceptible to the detrimental impacts of global change. Understanding the impact of global change is crucial to maintain the services agro-ecosystems provide. As a result, agro-ecosystems have been the focus for the majority of global change research. For example, to date, a Web of Science™ search for climate/global change, invertebrates and either grassland, forest or agro-ecosystem produces 900, 4,450 and 14,377 hits respectively. Similarly, much of the research understanding trophic interactions in

Table 10.1 Summary of major aspects of global change, the causes and the likely impacts on organisms.

Environmental change	Linked to:	Predicted outcomes
Elevated CO_2	Anthropogenic activities *e.g. burning fossil fuels, industrial processes*	Increased photosynthesis and biomass in plants
		Driver of global warming
Elevated O_3	Burning of fossil fuels	Damage to plant tissue reducing growth
Elevated Temperature	Greenhouse gasses accumulating in atmosphere *e.g. CO_2 and methane*	Affects all life – increased metabolism, growth rate, senescence
Altered precipitation	Global temperature increase	Changes in rates of photosynthesis and resource availability

agro-ecosystems has been focused on limiting the damage caused by herbivory in agro-ecosystems. As a result, research into the impact of global change has followed a similar pattern with a focus on natural enemies, particularly parasitoids, as biological control agents. Controlling invertebrate herbivores involves a combination of selective plant breeding to confer resistance (bottom-up processes), biological control by natural enemies (top-down processes) and pesticides (Kogan, 1998). Research has found that the latter are having severe detrimental impacts on the wider ecological community (Carson, 2002) and even human health (Blair et al., 2014). As a result, their use in controlling invertebrate herbivores is becoming limited (e.g., EU legislation (1107/2009) and U.S. legislation (Federal Insecticide, Fungicide and Rodenticide Act)) and are unlikely to play a key role in the future of controlling herbivore abundance in agro-ecosystems. Controlling or limiting the abundance of herbivores will therefore rely on bottom-up and top-down processes, both of which are directly and indirectly susceptible to global change.

10.2   Global Climate Change and Insect Herbivores

The influence of global climate change, on insect herbivores, directly and indirectly (i.e., plant-mediated), has been relatively well studied, and several general patterns of response identified (Table 10.2), despite certain perturbations (e.g., atmospheric CO_2 concentration and mean temperature) receiving more attention than others (e.g., UV-B, precipitation). In the absence of predation, herbivore responses appear to be feeding guild specific (Table 10.2). For example, under eCO_2, the reduced nutritional value (i.e., lowered concentration of leaf protein) of plant tissue (Lincoln et al., 1986) often results in compensatory feeding, especially by free-living chewing herbivores. High concentrations of carbon-based defence compounds in plant tissue can, however, confound such compensatory feeding by reducing digestion efficiency of plant material (Stiling & Cornelissen, 2007). Sap-feeding herbivores can circumvent many of the defence compounds and therefore show little response to eCO_2 (Robinson et al., 2012). When trophic complexity increases to include natural enemies, generalisations on the response of herbivores and natural enemies are currently limited by the lack of published studies (Table 10.3). Current research mainly focuses on individual aspects of global change with a particular emphasis on eCO_2 and elevated temperature.

Table 10.2 Effects of abiotic stressors associated with global change (increased temperature, CO_2, UV-B radiation and altered precipitation) on the performance (e.g., reduced developmental rate or increased abundance) of aboveground herbivore feeding guilds. General trends identified as "+" positive response, "−" negative response and "?" unknown. References citing examples of prevailing trends are given where possible.

Environmental change		Chewer	Phloem	Galling and mining	Reference
Temperature		+	+	+	Bale et al., 2002
CO_2		−	+	−	Stiling & Cornelissen, 2007; Robinson et al. 2012
Precipitation	Increase	?	+	?	Huberty & Denno, 2004
	Decrease	?	+	?	
UV-B		−	?	−	Ballaré et al., 2011
Ozone		+	+	?	Valkama et al., 2007

Table 10.3 The effect of component aspects of global change on interspecific interactions between herbivores and their natural enemies in agro-ecosystems. Net effect on prey performance in the presence of an antagonist; '+' positive response, '–' negative response, '=' no effect. Only studies that gave a clear indication of herbivore performance were included. (* Studies from forest or grassland ecosystems).

Environmental change	Predator/ Parasitoid	Herbivore guild	Natural enemy	Herbivore	Effect on natural enemy	Net effect herbivore	Cite
Elevated CO_2	Parasitoid	Aphid	Diaeretiella rapae	Myzus persicae, Brevicoryne brassicae	No change in parasitism	=	1
			Diaeretiella rapae	Brevicoryne brassicae	Reduced performance	+	2
			Aphidius picipes	Sitobion avenae	Increased parasitism	–	3
			Lysiphlebia japonica	Aphis gossypii	Increased parasitism	–	4
		Chewer	*Compsilura concinnata	Malacosoma disstria	No change in predation	=	5
			Cotesia vestalis	Plutella xylostella	No effect on orientation to host	=	6
			*Cotesia melanoscela	Lymantria dispar	No change in parasitism	=	7
			Cotesia plutellae	Plutella xylostella	Reduced orientation to host	+	8
			Microplitis mediator	Heliocoverpa armigera	Negligible effect	=	9
		Miner	*Multiple species	Multiple species	Increased parasitism	–	10
	Predator	Aphid	Harmonia axyridis	Sitobion avenae	No change in predation	=	3
			Hippodamia convergens	Myzus persicae, Brevicoryne brassicae	No change in predation	=	1
			*Multiple species	Cepegillettea betulaefoliae	Increased abundance	=	11
			Harmonia axyridis	Amphorophora idaei	No change in predation	=	44
			Leis axyridis	Aphis gossypii	Increased performance	–	43
			Chrysopa sinica	Aphis gossypii	Reduced predation	+	12
			Propylea japonica	Aphis gossypii	Longer development	+	13
		Chewer	Oechalia schellenbergii	Helicoverpa armigera	Increased predation	–	14
			Podisus maculiventris	Plutella xylostella	Reduced orientation to prey	+	8
Elevated temperature	Parasitoid	Aphid	Aphidius ervi, Aphidius eadyi	Acyrthosiphon pisum	No change in parasitism	=	15
			Aphidius matricariae	Myzus persicae	No change in parasitism	=	16

			Natural enemy	Host/prey	Effect		Ref
			Lysiphlebus fabarum	Aphis fabae	No effect on host resistance	=	17
			Lysiphlebus fabarum	Aphis fabae	No effect of heat shock	=	18
			Aphidius ervi	Acyrthosiphon pisum	Reduced resistance to parasitism	–	19
			Aphidius ervi	Acyrthosiphon pisum	Reduced resistance to parasitism	–	20
			Aphidius matricariae	Myzus persicae	Increased parasitism	–	21
			Multiple species	Sitobion avenae	Increased parasitism	–	22
			Diaeretiella rapae	Brevicoryne brassicae	Increased parasitism	–	23
			Aphidius rhopalosiphi	Sitobion avenae	Reduced parasitism, increased host defence	+	24
			Aphidius avenae	Sitobion avenae	Reduced oviposition after heat shock	+	25
		Phloem	*Multiple species	Parthenolecanium quercifex	Phenological mismatch	+	26
		Chewer	Perilitus brevicollis	Phratora vulgatissima	Increased performance	–	27
			Cotesia marginiventris	Spodoptera exigua	Reduced survival at constant high	+	28
			Tetrastichus julis	Oulema melanopus	Phenological miss-match	+	29
			Cotesia bignellii	Euphydryas aurinia	Remained in synchrony	=	31
	Predator	Aphid	Hippodamia convergens	Acyrthosiphon pisum	No change in predation	=	15
			Coleomegilla maculata	Myzus persicae	No change in predation	=	30
Elevated O₃	Parasitoid	Aphid	*Multiple species	Cepegillettea betulaefoliae	No significant effect	=	11
		Chewer	Cotesia plutellae	Plutella xylostella	No significant effect	=	32
			Phytoseiulus persimilis	Tetranychus urticae	No significant effect	=	45
			Cotesia vestalis	Plutella xylostella	Reduced orientation to host	+	6
			*Compsilura concinnata	Malacosoma disstria	Reduced survivorship	+	33

(Continued)

Table 10.3 (Continued)

Environmental change	Predator/ Parasitoid	Natural enemy	Herbivore guild	Herbivore	Effect on natural enemy	Net effect herbivore	Cite
	Predator	*Multiple species	Aphid	Cepegillettea betulaefoliae	No change in predation	=	11
UV - radiation	Parasitoid	Aphidius ervi	Aphid	Acyrthosiphon pisum	No significant effect	=	19
		Cotesia plutellae	Chewer	Plutella xylostella	Increased orientation to host	−	34
Reduced precipitation	Parasitoid	Aphis colemani, Diaeretiella rapae	Aphid	Myzus persicae, Brevicoryne brassicae	Reduced parasitism	+	35
		Aphidius ervi		Rhopalosiphum padi	Reduced parasitism	+	36
		Multiple species	Phloem	Phenacoccus herreni	Reduced abundance	+	37
	Predator	Multiple species	Multiple	Multiple species	No change in abundance	−	38
		*Multiple species		Multiple Species	Reduced abundance	+	39
		Heterorhabditis marelatus	Chewer	Hepialus californicus	Reduced predation	+	40
Variable Precipitation	Parasitoid	Multiple species	Chewer	Multiple species	Variable precipitation reduced parasitism	+	41
Elevated CO₂ + elevated temp	Parasitoid	Cotesia marginiventris	Chewer	Spodoptera exigua	Reduced parasitism	+	42
Elevated CO₂ + elevated O₃	Parasitoid	*Compsilura concinnata	Chewer	Malacosoma disstria	Negligible effect on parasitism	=	33
	Predator	Multiple species	Aphid	Cepegillettea betulaefoliae	No effect on predation	=	11

1) Stacey & Fellowes, 2002a; 2) Klaiber et al., 2013; 3) Chen et al., 2007b; 4) Sun et al., 2011; 5) Holton et al., 2003; 6) Himanen et al., 2009; 7) Roth & Lindroth, 1995; 8) Vuorinen et al., 2004a; 9) Yin et al., 2009; 10) Stiling et al., 1999; 11) Awmack et al., 2004; 12) Gao et al., 2010; 13) Gao et al., 2008; 14) Coll & Hughes, 2008; 15) Stacey & Fellows, 2002b; 16) Bannerman et al., 2011; 17) Cayetano & Vorburger, 2013a; 18) Cayetano & Vorburger, 2013b; 19) Guay et al., 2009; 20) Bensadia et al., 2006; 21) Bezemer et al., 1998; 22) Dong et al., 2013; 23) Romo & Tylianakis, 2013; 24) Le Lann et al. 2014; 25) Roux et al., 2010; 26) Meineke et al., 2014; 27) Baffoe et al., 2012; 28) Butler & Trumble, 2010; 29) Evans et al., 2013; 30) Sentis et al., 2013; 31) Klapwijk et al., 2010; 32) Pinto et al., 2008; 33) Holton et al., 2003; 34) Foggo et al., 2007; 35) Tariq et al., 2013; 36) Aslam et al., 2013; 37) Calatayud et al., 2002; 38) Zhu et al., 2014; 39) Trotter et al., 2008; 40) Preisser & Strong, 2004; 41) Stireman et al., 2005; 42) Dyer et al., 2013; 43) Chen et al., 2005; 44) Hentley et al., 2014b; 45) Vuorinen et al., 2004b

The finding of this research has been summarised in Table 10.3. Natural enemies have been divided into predator and parasitoid to reflect the very different methods by which these groups attack their prey.

10.3   Global Climate Change and Natural Enemies of Insect Herbivores

10.3.1   Elevated Atmospheric CO$_2$

The impact of eCO$_2$ on natural enemies is one of the most intensely studied aspects of global change. The most significant impacts of eCO$_2$ on natural enemies are likely to be indirect, either from changes to the plant, herbivore or both (Bezemer & Jones, 1998; Chen et al., 2005). As a result, research into the effect of global change is somewhat limited compared to other trophic levels (i.e., herbivores – see Table 10.2). Although some generalisations of plant–herbivore responses are emerging (Table 10.2), as a whole, there remains a high degree of variability in the response of natural enemies. Of the 17 examples in Table 10.3, nine found no effect of eCO$_2$, four found a negative effect on natural enemies and four found a positive effect. It has become a cliché within global change research to state "the response is species and context specific, generalisations are impossible", but the evidence presented in Table 10.3 supports this. For example, parasitism by *Diaeretiella rapae* on the aphid *Brevicoryne brassicae* can be unresponsive (Stacey & Fellowes, 2002a) and negatively affected (Klaiber et al., 2013) to eCO$_2$. Looking closer at these two examples, both studies used the same plant species *Brassica oleracea*, but methodology differed significantly. Stacey and Fellows used 650 ppm of CO$_2$ for the eCO$_2$ treatment and allowed a single, mated female parasitoid 6 hours to attack 20 same-age aphids. In contrast, Klaiber et al. used 800 ppm as the eCO$_2$ treatment and allowed a mated female parasitoid 4.5 hours to attack 60 mixed age aphids. Differences in experimental approach may therefore be playing a significant role in the variability of responses to global change. Although generalisations are not possible, two important characteristics are crucial for natural enemies: the ability of the natural enemy to locate their prey and the quality of the prey as a resource. Both of these characteristics may be differentially impacted by eCO$_2$ and considering them separately may offer some insight into the responses of herbivores and their natural enemies under eCO$_2$.

10.3.1.1   Prey Location

Herbivore induced plant volatiles (HIPVs) have been identified as one of the most important cues used by natural enemies to locate their prey (Mumm & Dicke, 2010). The emissions of HIPVs are triggered by herbivore damage, which then induces a signalling pathway within the plant to produce HIPVs. Herbivore specific stimuli (e.g., saliva) facilitate the production of plant hormones, such as jasmonic and salicylic acid (Van Poecke & Dicke, 2004). The blend of these chemicals is specific to the mode of herbivory, for example, chewing insects such as lepidoptera larvae and cell-sucking organisms such as mites, induce a wound response stimulating the jasmonic acid-signalling pathway (Thaler, 1999). Phloem feeding insects such as aphids are perceived by the plant as pathogens, stimulating the salicylic acid signalling pathway (Walling, 2000). As a result of these feeding guild specific signalling pathways, HIPVs can be a reliable indicator to natural enemies for the location of their prey. Elevated CO$_2$ conditions are known to suppress the production of some plant signalling hormones (e.g., jasmonates and ethylene), but also stimulate others such as salicylic acid (Zavala, Nabity, & DeLucia, 2013). The mechanisms by which HIPVs are altered under eCO$_2$ remains unknown (Ode et al., 2014), but whether HIPVs increase or decrease are system specific (Vuorinen et al., 2004). Parasitoids appear to have herbivore-specific feeding guild responses to eCO$_2$ (Table 10.3).

For example, parasitism of phloem feeding aphids increased under eCO_2 compared to ambient conditions, whereas parasitism on chewing herbivores decreased (Table 10.3). Reduced parasitism on chewing herbivores may be linked to a down regulation of the jasmonic signalling pathway seen under eCO_2 (Zavala et al., 2013). Damage caused by the chewing herbivore *Plutella xylostella*, for example, was not detected by the parasitoid *Cotesia plutellae* under eCO_2 (Vuorinen et al., 2004b), suggesting herbivore induced volatile cues used by the parasitoid are altered under eCO_2. In contrast, host location by the parasitoid *Cotesia vestalis* on the same herbivore species, *P. xylostella*, was not altered under eCO_2 despite both studies using the same plant species, *Brassica oleracea* (Himanen et al., 2009). In stark contrast, when an effect has been shown, eCO_2 increases parasitism on aphids. This, again, may be linked to the plant-signalling pathway in the production of HIPVs. Aphid herbivory stimulates the salicylic acid signalling pathway (Walling, 2000), which is up-regulated under eCO_2 (Zavala et al., 2013). This may result in a significant increase of parasitoid abundance, which was seen by the parasitoid *Aphidius picipes* parasitising the aphid *Sitobion avenae* (Chen et al., 2007a).

10.3.1.2 Prey Quality

Once suitable prey has been found, eCO_2 conditions may also alter prey quality, further modifying the response of natural enemies. Plants grown under eCO_2 are generally nutritionally poorer than plants from ambient conditions (Ainsworth & Long, 2005; Robinson et al., 2012). The impact of eCO_2 on the plant nutritional quality may also transfer to herbivores feeding on the plant. Aphids feeding on host plants with low C:N ratios (i.e., high N input environment) may have a high nutritional value for predators (Couture et al., 2010), therefore in a high CO_2 environment, where the C:N ratio is increased, predators may require greater numbers of prey to fulfil their physiological demands. This is analogous to compensatory feeding seen in herbivores (e.g., Watt et al., 1995) and detritivores (e.g., Dray et al., 2014).

Natural enemies may also benefit from the slow development of prey feeding on nutritionally poor plants. A poor diet may increase herbivore development time, which in turn creates a longer window for predators and parasitoids to attack (Stiling & Cornelissen, 2007). The parasitoid *Aphidius ervi*, for example, will only oviposit in first and second instar *Amphorophora idaei* nymphs (Mitchell et al., 2010), therefore, slower aphid development would significantly expand the window of opportunity for this parasitoid attack. Elevated CO_2 has also been shown to affect the avoidance behaviour of herbivores. It has been reported in multiple different aphid species that responsiveness to conspecific alarm pheromones is significantly diminished under eCO_2 (Awmack et al., 1997; Hentley et al., 2014b). For example, Hentley et al. (2014b) conducted an experiment where both plants and aphids reared under eCO_2 were reciprocally crossed with plants and aphids from ambient conditions (Fig. 10.1A). It was found that, compared to aphids and plants reared under ambient CO_2 (Fig. 10.1B I), aphid escape responses to ladybird predation were significantly reduced when either the plants or aphids were reared under eCO_2 (Fig. 10.1B II-IV). Substituting ladybird predation with conspecific alarm pheromone E-β-farnesene elicited a similar pattern (Fig. 10.1C), suggesting a down-regulation of aphid alarm response under eCO_2.

10.3.2 Temperature Change

Arthropods generally have limited abilities to thermoregulate and as such are likely to be directly impacted by increases in global temperatures. The majority of published research reports that rises in temperatures can increase the rate of arthropod growth and development, benefitting many insect herbivores and thus increasing the number of insect generations per year (Porter et al., 1991; Bale

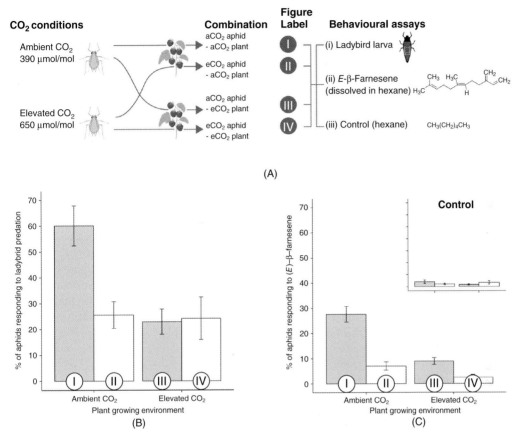

Figure 10.1 Summary of the findings from Hentley et al. (2014b) where a reciprocally crossed experiment (A) tested the effect of eCO_2 on aphid escape responses to either ladybird predation (B) or conspecific alarm pheromone (C). Grey bars correspond to ambient CO_2 and white bars eCO_2. Figure and full details of the methodology and results can be found in Hentley et al. (2014b).

et al., 2002; Musolin, 2007; Lemoine et al., 2014). However, other studies report that the relationship between rearing temperature and insect body size is complex. For example average consumption rates of 21 herbivore–plant pairs increased between 20°C to 30°C; however under extreme temperatures there was greater variation in the responses of herbivore species with both increasing and declining consumption rates reported (Lemoine et al., 2014). It has been suggested that body size increases as temperatures rise from very low temperatures to optimum. However, once maximum body size and optimum temperature are reached, any further rises in temperature result in a decline in insect body size (Kingsolver & Huey, 2008).

Rise in temperature also has the potential to generate asynchrony between herbivores and their natural enemies through changes in insect phenology, timing of emergence and migration (Bale et al., 2002; Klapwijk et al., 2010; Dyer et al., 2013; Evans et al., 2013; Facey et al., 2014). An increase in temperature may allow insect herbivores and their natural enemies to disperse into new regions that were previously unsuitable due to low temperatures (Bebber et al., 2013; Delava et al., 2014). Certain insect species may also have a greater ability to adapt to the rapidly increasing temperature than others, further creating phenological mismatches (DeLucia et al., 2012; McCluney et al., 2012;

Facey et al., 2014) and allowing opportunities for invasive species to prosper (Olesen et al., 2011). The behaviour of insect herbivores and natural enemies has been shown to be affected by elevated temperatures, with aphid and parasitoid wasp host preference recorded to change under warmer conditions (Hance et al., 2007). *Sitobion avenae* dropping off behaviour, considered to be an important anti-predator behaviour for aphids, was recorded to be influenced by temperature and time exposed to high temperatures (Ma & Ma, 2012). Under warmer temperatures aphids had a lowered 'dropping off' temperature, which is likely to have negative consequences for aphid development and population growth under global warming due to the time and energy spent re-locating a host plant and increased likelihood of starvation.

Global warming also has the potential to affect insect herbivores indirectly through changes in host plant quality and defensive capabilities (Gregory et al., 2009; Falik et al., 2012; Facey et al., 2014). Plant growth, development and chemical composition can all be influenced by rises in atmospheric temperatures, however, the level of impact will depend on the crop species and growing region of the crop (McCarty, 2001; Bale et al., 2002). Warmer temperatures in northern regions may enhance crop growth rates, which could benefit insect herbivores through increased movement of nutrients within the plant and reduced competition for space on larger, rapidly growing plants (Price, 1991). However, warmer temperatures may also prolong the length of the growing season, allowing earlier sowing and increasing the range of potential growing areas, creating spatial and temporal asynchrony between plants and herbivores (Gregory et al., 2009; IPCC, 2014). Particularly in warmer, lower latitudes, where plant growth and development are already constrained by high temperatures, further rises in temperature will increase plant respiration. This will create sub-optimal growing conditions resulting in a plant stress response and reduced plant growth and development (Coakley et al., 1999; Bale et al., 2002). In these regions herbivores may benefit from reduced plant defences and changes in the nutritional quality of host plants exposed to high temperatures (Porter et al., 1991; Lemoine et al., 2014). However, reduced plant growth and development may negatively affect herbivores due to reductions in available food sources (Lemoine et al., 2014). The response of the herbivore to heat stressed plants will depend on the feeding guild of the herbivore, the type of plant tissue the herbivore attacks and the nutritional requirements of the herbivore (Porter et al., 1991; Lemoine et al., 2014). In contrast to northern regions, phloem feeding herbivores in lower latitudes may be negatively affected due to restricted plant growth and loss of phloem turgor reducing the movement of nutrients through the plant (Inbar et al., 2001). However, chewing feeding herbivores may avoid any negative impacts due to changes in the movement of nutrients through the plant and instead benefit from reduced plant defence under high temperatures (Ayres, 1993).

10.3.3 Reduction in Mean Precipitation

Changes in precipitation will affect insect herbivores mostly through alterations in the quality of their food source as a consequence of changes in the host plant (Johnson et al., 2009). There are a number of hypotheses underpinning predictions about how insect herbivores will respond to feeding on drought stressed plants. The Plant Stress Hypothesis suggests that drought stressed plants are a better quality food resource for insect herbivores (White, 1984) as plant defence is often compromised under drought conditions (White, 1984; Abebe et al., 2010). Environmental stress that reduces plant growth has the potential to shift the allocation of resources to or away from defence strategies. As drought severity increases carbon becomes limited and plant secondary metabolite content declines. Increased damage to leaf discs collected from drought stressed *Alliaria petiolata* by Lepidoptera (specialist *Pieris brassicae* and generalist *Spodoptera littoralis*) correlated with decreased tissue concentrations of defensive metabolites (Gutbrodt et al., 2011). Other defence strategies such as silicon (Si) deposition as abrasive structures, a physical defence in many grasses to deter herbivores (Massey et al., 2007) is also reduced in drought stressed plants (Wade, 2016). Silicon is mainly taken up by plants passively through aquaporin-type transporters in the root (McNaughton & Tarrants, 1983; Hartley, 2015) therefore reduced transpiration under drought conditions could result in reduced Si uptake and weaker physical defence. Increases in tissue concentrations of free amino acids common in drought stressed plants also benefits insect herbivores with nitrogen often limiting insect growth (White, 1984; Huberty & Denno, 2004). Furthermore, drought stressed plants may also appear more attractive to insects with a larger number of yellowing leaves and an increase in the release of stress related cues (Chown et al., 2011).

In contrast, other studies report that herbivorous insects perform better on healthy, rapidly growing plants and in response Price (1991) developed the Plant Vigour Hypothesis. Rapidly growing plants move larger quantities of plant nutrients through the phloem which provides higher concentrations of nutrients, particularly for sap feeding insects (Inbar et al., 2001). Fast-growing plants provide a larger resource for feeding insects, minimising competition with other organisms attacking the plant (Huberty & Denno, 2004). It is has also been suggested that plants experiencing intermediate levels of water stress may provide optimal food quality for insect herbivores (Mattson & Haack, 1987; Huberty & Denno, 2004) through increased nutritional quality (e.g., free amino acid concentrations) without the negative effects to plant growth and composition caused by severe and prolonged drought.

There are numerous studies reporting both negative and positive effects of drought on the performance of insect herbivores. These contradictions in findings may be due to differences in watering regimes and the severity and/or timing of the drought stress. Interactions between drought and other environmental stresses such as shading and soil nutrient availability may also influence the experimental outcomes, as well as the methods used to impose and measure drought stress (Huberty & Denno, 2004). There will also be temporal and spatial differences in the impact of drought on many insect herbivores. The severity of reductions in mean precipitation will depend upon the season and location; reduced precipitation will have larger effects on crop growth and chemical composition and thus insect herbivores in regions with very low rainfall (Bates et al., 2008; Gregory et al., 2009; Newton et al., 2011; IPCC, 2014). Furthermore, the level of stress caused by changes in water availability and the plant response to water stress will differ between different plant species and varieties, influencing the quality of the herbivore's food source. Different insect herbivore feeding guilds will also be differentially affected by drought (Huberty & Denno, 2004) depending on the type of plant tissue they attack and their nutritional requirements. Phloem feeders are thought to be affected by changes in plant water status to a greater extent than chewing insects due to the significant effects of water stress on phloem sap viscosity, nutrient and allelochemical composition (Larsson, 1989; Huberty & Denno, 2004; Mody et al., 2009). However, chewing herbivores may benefit from reductions in Si deposition reducing the abrasiveness of leaves and increasing the ability of the insect to absorb nutrients (Massey et al., 2007). Soil dwelling insect herbivores may be particularly sensitive to changes in soil moisture availability (Erb & Lu, 2013). Belowground insect herbivore abundance and vertical distribution has been reported to be sensitive to soil moisture, with soil dwelling insects becoming less mobile and moving deeper in the soil profile under drought conditions reducing herbivory (Briones et al., 1997; Sinka et al., 2007; Staley & Johnson, 2008).

Herbivore–natural enemy interactions can be affected by reductions in rainfall through changes in the outcome of plant-mediated interactions between insect herbivores, altering the number and quality of the prey or host of the natural enemy (Giles et al., 2002). It has been reported that higher trophic levels may be more sensitive to changes in climate (Voigt et al., 2003). There are few studies investigating the effect of water stress on trophic interactions. Johnson et al. (2011) reported a larger effect of drought stress on parasitoid wasps than on their aphid prey. The rate of attack by a parasitoid wasp (*Aphidius ervi*) was lower on drought stressed plants compared to ambient plants (Aslam et al., 2013). This was thought to be due to changes in aphid population demography, with a greater number of adult aphids recorded on drought stressed plants which are more difficult to parasitize (Dixon, 1958). Under ambient watering conditions, *Agriotes* larvae were recorded to reduce the abundance and performance of *Stephensia brunnichella* and reduce the rate of parasitism of a leaf-miner when sharing the same host (Staley et al., 2007). Drought conditions removed the interaction between the *Agriotes* larvae, leaf-mining (*S. brunnichella)* larvae and its associated parasitoid. Furthermore, caterpillar parasitism levels by parasitoid wasps (Hymenoptera) and tachinid flies (Diptera) decreased as climate variability increased (Stireman et al., 2005). Very few studies have investigated the effects of changes in water availability on predators in agroecosystems. Invasive ladybird species (*Harmonia axyridis*) had increased weight when feeding on aphids collected from plants grown under a watering regime that reduced the quantity and frequency of precipitation events (Wade, 2016). However, ladybird choice of prey in this study did not match prey quality, suggesting that the effect of future precipitation patterns on ladybirds may be influenced by other factors such as prey location, which requires further investigation. Therefore it is very difficult to make generalisations as to how all herbivores and natural enemies will respond to changes in precipitation due to the many complicating factors which can influence the precise outcome.

10.3.4 Extreme Events

Climate models predict significant increase in the frequency and intensity of extreme weather events (IPCC, 2014). In contrast to gradual or continuous stress events, extreme climate change events represent a distinctive stress for insect herbivores and their natural enemies. The lack of warning, short duration and unpredictable nature of these extreme events can reduce the ability of plants, herbivores and natural enemies to prepare, prime or adapt (Hance et al., 2007). The impact of these extreme events will differ depending on the length and severity of the event. There is a significant lack of research investigating the impact of extreme events on multi-trophic interactions, despite the potential consequences for managing and predicting the impacts of global change at the ecosystem level. Severe drought and flooding events are predicted to increase under climate change with times of drought followed by heavy rainfall, resulting in periods of stress and recovery (or pulsed stress) for the plant and associated organisms, both above- and belowground. This is likely to have different effects on insect herbivore growth, abundance and survival than continuous water stress. Tariq et al. (2013) reported that fecundity of *Myzus persicae* feeding on *Brassica* species was reduced on pulsed stressed plants compared to those feeding on unstressed and moderately drought stressed plants. However, aphids increased in mass when feeding on plants subjected to a 40% reduction in quantity of water as well as a 75% reduction in watering frequency, simulating future precipitation pulses; this watering regime also benefitted the ladybirds feeding on these aphids (Wade, 2016). High temperatures associated with severe drought events may also significantly affect insect herbivores and their natural enemies. For example, aphids and their symbionts are particularly sensitive to high temperatures. Under extreme temperatures, aphid populations may reduce as a result of damage to their obligate and facultative symbionts (Guay et al., 2009). Reduced numbers of aphids will negatively affect natural enemies through a reduction in their prey availability (Giles et al., 2002; Schmitz & Barton, 2014).

10.3.5 Ozone and UV-B

Projected increases to tropospheric ozone and UV-B radiation are likely to have significant impacts on herbivores and their natural enemies, but current research is lacking. Similar to eCO_2, elevated

O$_3$ (eO$_3$) at the levels projected for the next century will have little or no direct effect on insects (Alstad et al., 1982). Elevated O$_3$ can, however, have a significant detrimental impact on plants such as reduced plant growth and crop yield (Ashmore, 2005). Concentrations of phenolics and terpenes, both associated with plant defence against herbivory, also increase under eO$_3$ (Valkama et al., 2007). For example, eO$_3$ treatment has been shown to increase enzyme activity controlling phenylpropanoid and flavonoid biosynthesis pathways (Kangasjärvi et al., 1994) linked to plant defences. A meta-analysis of the literature concerning eO$_3$ and herbivores found they were generally positively affected (decreased development time, increased pupal mass) by the changes to the host plant under eO$_3$, despite the increase in some plant defence compounds (Valkama et al., 2007). Studies investigating the impacts of eO$_3$ on natural enemies are currently limited to five studies (Table 10.3). Only two of these studies show any impact of eO$_3$ on natural enemies, both of which are negative. Exposure of plants to eO$_3$ can trigger the emission of volatile compounds known to be attractive to natural enemies (Pinto et al., 2010), but also reduce the emission of others, such as terpenoids, potentially altering the ability of natural enemies to locate their prey/hosts (Himanen et al., 2009). Under eCO$_2$, small, undetectable alterations to the volatile blend released from herbivore damaged white cabbage (*Brassica oleracea*) resulted in the parasitoid wasp *Cotesia plutellae* being unable to locate their host (Vuorinen et al., 2004b). In contrast, eO$_3$ generally has no effect on host-location abilities of natural enemies (Table 10.3), despite significant changes to the amount and composition of volatile blends (Mumm & Dicke, 2010). This suggests that the volatiles important for host location by natural enemies are not sensitive to altered O$_3$ concentrations.

In contrast to ground level O$_3$, stratospheric O$_3$, or the "ozone layer", is beneficial to life, absorbing ultraviolet radiation before reaching the earth. Anthropogenic emission of compounds such as chlorofluorocarbons (CFCs) and nitrous oxide is responsible for the depletion of the ozone layer (Farman et al., 1985) and subsequent increases in UV-B radiation. There are currently very few studies that have investigated the impact of increased UV-B radiation on herbivores and their natural enemies (Table 10.3). These are limited to studies which investigate parasitism on phloem feeding and chewing herbivores. The results of these studies are variable demonstrating a potential increase in parasitism on chewing herbivores (Foggo et al., 2007) but no change in parasitism on aphids (Guay et al., 2009).

Recent evidence suggests that future levels of UV-B radiation may remain constant and will therefore not be considered under global change scenarios. An international treaty (The Montreal Protocol, 1987) resulted in significant global reductions in the emissions of compounds depleting the ozone layer. A recent report by NASA suggests that ozone levels will return to pre-industrial levels by 2040 (Strahan et al., 2014).

10.4 Multiple Abiotic Factors

Overall, research has focused on the effects of individual components of climate change on multi-trophic interactions. Such a reductionist approach is crucial for our understanding of how climate-driven variation, particularly in the abiotic environment, affects plant–herbivore–natural enemy interactions. However, global change will result in the simultaneous alteration of multiple abiotic factors. Over the last decade there has been an increased emphasis on the study of multiple climate change impacts on trophic interactions (Bezemer et al., 1998; Veteli et al., 2002; Riikonen et al., 2010), with a focus on plant–herbivore interactions. Still, this complexity is rarely represented for natural enemies, with very few studies that have looked at the impact of multiple abiotic factors on plant–herbivore–natural enemy interactions. To manipulate several abiotic variables within the same experiment represents a technical and logistical challenge. Consequently, most studies have only assessed the impacts of pairwise combinations of eCO_2 with another climate variable (temperature, nitrogen, UV radiation, O_3 and soil moisture) on trophic interactions. For example, Dyer et al. found that elevated temperature benefited (faster development time) the herbivorous beet army worm (*Spodoptera exigua* (Hübner)) but this was then removed when CO_2 was simultaneously elevated. The faster development of the beet army worm under elevated temperature meant that it pupated before parasitoid eclosion, significantly reducing parasitism. Under eCO_2 development time of the beet army worm increased due to reduced plant quality, but this was also associated with reduced parasitism. The combined effects of elevated temperature and eCO_2 resulted in zero successful parasitism being recorded compared to 29% under ambient conditions.

The direct and indirect effects of global change on trophic interactions are often hard to distinguish. For example, projected increases in atmospheric CO_2 can positively affect plants by increasing photosynthetic rate and water use efficiency. Within plant tissue, there is a dilution of nitrogen-based compounds (defence chemicals containing nitrogen) as a result of the increased growth and additional carbon available to the plant. This causes plant tissue to be of lower nutritional quality for the herbivore. In addition, carbon based defence compounds within plant tissue increase, thus making the plant harder to digest in addition to being less nutritious. In this example, the herbivore is not directly affected by elevated CO_2, therefore, the net effect of elevated CO_2 on the herbivore is observed as negative due to changes in the host plant. When the combined effect of two modified abiotic factors are acting on the same system, the net outcome can change significantly. In this example, elevated temperature generally negates the increase in carbon defence compounds seen under eCO_2 (Veteli et al., 2002), but the reduction in nutritional quality remains. Whereas eCO_2 has no direct effect on herbivores, under elevated temperature, insect metabolism increases along with growth rate (Bale et al., 2002). Therefore, the net effect for herbivores on this system is neutral; although plant material is of poorer nutritional quality due to eCO_2, the increase in temperature reduces generation time and also increases digestion efficiency to compensate for the extra food intake that is required. Although, low food quality (reduced N) and increased temperature can result in reduced mass gain during larval stages (Yang & Joern, 1994). It is clear that the simultaneous manipulation of multiple abiotic factors can have unexpected impacts on herbivores and their natural enemies. Further research is required to understand the impact of simultaneously altered biotic factors, but there is a limit to the complexity of empirical studies. For example, altering temperature, CO_2, O_3 and precipitation in the same experiment would represent a major logistical challenge. New techniques that combine empirical and theoretical techniques may therefore be required to explore this area further.

10.5   Conclusions

In this chapter we have reviewed the evidence for global change impacts on insect herbivores and their natural enemies. It is clear that global change will significantly affect herbivores and their natural enemies, but discerning a pattern in these changes is very difficult. For example, under eCO_2, 53% of the studies documented in this chapter (Table 10.3) found there was no impact on natural enemies. Of the studies that found an effect, natural enemies benefited 50% of the time. Similarly, under increased temperature, 38% of the studies found no response from natural enemies. Within the studies that found an effect, natural enemies benefited 54% of the time. These figures highlight the lack of a discernible pattern; the responses of herbivores and their natural enemies are apparently species and context specific. Is this really true? Weather patterns and indeed global change are seemingly random, but are predictable. The Chaos Theory, proposed by meteorologist Edward Lorenz in the 1960s, refers to an apparent randomness resulting from interactions between and within complex systems that ultimately obey particular laws or rules (Lorenz, 1963). Research to date on the impact of global change on insect herbivores and their natural enemies has focused on observed changes in species abundance or behaviour. The mechanisms underpinning species responses to global change still remain largely unknown – these could be the key to making the unpredictable responses of herbivores and their natural enemies predictable.

In addition to understanding mechanisms, it is clear that research into the response of herbivores and their natural enemies to global change needs to become much more complicated. To date, the majority of studies have focused on one abiotic factor and one or two herbivore–natural enemy interactions (Table 10.3). Global change is the simultaneous alteration of multiple abiotic factors such as temperature, precipitation, CO_2 and O_3. Currently the logistical challenges of integrating multiple abiotic factors and interspecific interactions within a network of herbivores and natural enemies make it impossible to quantitatively test realistic global change scenarios. Theoretical approaches combining our mechanistic understanding with both historical and experimental observations of species responses to global change may prove an invaluable tool for predicting global change impacts on herbivores and their natural enemies. Subsequently integrating these predicted responses into crop models will allow for more realistic estimates of future crop production allowing for robust food security policies to be created (Gregory et al., 2009).

Acknowledgements

The authors are grateful to Philip Smith for proofreading this chapter.

References

Abebe, T., Melmaiee, K., Berg, V. & Wise, R. (2010) Drought response in the spikes of barley: gene expression in the lemma, palea, awn, and seed. *Functional & Integrative Genomics*, **10**, 191–205.

Ainsworth, E.A. & Long, S.P. (2005) What have we learned from 15 years of free-air CO_2 enrichment (FACE)? A meta-analytic review of the responses of photosynthesis, canopy properties and plant production to rising CO_2. *New Phytologist*, **165**, 351–372.

Alstad, D.N., Edmunds Jr, G.F. & Weinstein, L.H. (1982) Effects of air pollutants on insect populations. *Annual Review of Entomology*, **27**, 369–384.

Ashmore, M.R. (2005) Assessing the future global impacts of ozone on vegetation. *Plant, Cell and Environment*, **28**, 949–964.

Aslam, T.J., Johnson, S.N. & Karley, A.J. (2013) Plant-mediated effects of drought on aphid population structure and parasitoid attack. *Journal of Applied Entomology*, **137**, 136–145.

Awmack, C., Harrington, R. & Leather, S. (1997) Host plant effects on the performance of the aphid *Aulacorthum solani* (Kalt.) (Homoptera : Aphididae) at ambient and elevated CO_2. *Global Change Biology*, **3**, 545–549.

Awmack, C.S., Harrington, R. & Lindroth, R.L. (2004) Aphid individual performance may not predict population responses to elevated CO_2 or O_3. *Global Change Biology*, **10**, 1414–1423.

Ayres, M.P. (1993) Plant defense, herbivory, and climate change. *Biotic Interactions and Global Change* (eds P.M. Karieva, J.G. Kingsolver & R.B. Huey), pp. 75–94. Sinauer, Sunderland, MA.

Baffoe, K.O., Dalin, P., Nordlander, G. & Stenberg, J.A. (2012) Importance of temperature for the performance and biocontrol efficiency of the parasitoid *Perilitus brevicollis* (Hymenoptera: Braconidae) on *Salix*. *BioControl*, **57**, 611–618.

Bale, J.S., Masters, G.J., Hodkinson, I.D., Awmack, C., Bezemer, T.M., Brown, V.K., Butterfield, J., Buse, A., Coulson, J.C., Farrar, J., Good, J.E.G., Harrington, R., Hartley, S., Jones, T.H., Lindroth, R.L., Press, M.C., Symrnioudis, I., Watt, A.D. & Whittaker, J.B. (2002) Herbivory in global climate change research: direct effects of rising temperature on insect herbivores. *Global Change Biology*, **8**, 1–16.

Ballaré, C.L., Caldwell, M.M., Flint, S.D., Robinson, S.A. & Bornman, J.F. (2011). Effects of solar ultraviolet radiation on terrestrial ecosystems. Patterns, mechanisms, and interactions with climate change. *Photochemical & Photobiological Sciences*, **10**, 226–241.

Bannerman, J.A., Gillespie, D.R. & Roitberg, B.D. (2011) The impacts of extreme and fluctuating temperatures on trait-mediated indirect aphid–parasitoid interactions. *Ecological Entomology*, **36**, 490–498.

Bates, B.C., Kundzewicz, Z.W., Wu, S. & Palutikof, J.S. (2008) *Climate change and water: Technical paper VI*. Intergovernmental Panel on Climate Change (IPCC). IPCC Secretariat, Geneva.

Bebber, D.P., Ramotowski, M.A.T. & Gurr, S.J. (2013) Crop pests and pathogens move polewards in a warming world. *Nature Climate Change*, **3**, 985–988.

Bensadia, F., Boudreault, S., Guay, J.F., Michaud, D. & Cloutier, C. (2006) Aphid clonal resistance to a parasitoid fails under heat stress. *Journal of Insect Physiology*, **52**, 146–157.

Bezemer, T.M., Jones, T.H. & Knight, K.J. (1998) Long-term effects of elevated CO_2 and temperature on populations of the peach potato aphid *Myzus persicae* and its parasitoid *Aphidius matricariae*. *Oecologia*, **116**, 128–135.

Blair, A., Ritz, B., Wesseling, C. & Beane Freeman, L. (2014) Pesticides and human health. *Occupational and Environmental Medicine*, **72**, 81–82.

Briones, M.J.I., Ineson, P. & Piearce, T.G. (1997) Effects of climate change on soil fauna; responses of enchytraeids, Diptera larvae and tardigrades in a transplant experiment. *Applied Soil Ecology*, **6**, 117–134.

Butler, C.D. & Trumble, J.T. (2010) Predicting population dynamics of the parasitoid *Cotesia marginiventris* (Hymenoptera: Braconidae) resulting from novel interactions of temperature and selenium. *Biocontrol Science and Technology*, **20**, 391–406.

Calatayud, P.A., Polania, M.A., Seligmann, C.D. & Bellotti, A.C. (2002) Influence of water-stressed cassava on *Phenacoccus herreni* and three associated parasitoids. *Entomologia Experimentalis Et Applicata*, **102**, 163–175.

Carson, R. (2002) *Silent Spring*. Houghton Mifflin Harcourt, Boston.

Cayetano, L. & Vorburger, C. (2013a) Effects of heat shock on resistance to parasitoids and on life history traits in an aphid/endosymbiont system. *PLOS ONE*, **8**, e75966.

Cayetano, L. & Vorburger, C. (2013b) Genotype-by-genotype specificity remains robust to average temperature variation in an aphid/endosymbiont/parasitoid system. *Journal of Evolutionary Biology*, **26**, 1603–1610.

Chen, F., Wu, G., Parajulee, M.N. & Ge, F. (2007a) Long-term impacts of elevated carbon dioxide and transgenic Bt cotton on performance and feeding of three generations of cotton bollworm. *Entomologia Experimentalis et Applicata*, **124**, 27–35.

Chen, F.J., Ge, F. & Parajulee, M.N. (2005) Impact of elevated CO_2 on tri-trophic interaction of *Gossypium hirsutum*, *Aphis gossypii*, and *Leis axyridis*. *Environmental Entomology*, **34**, 37–46.

Chen, F.J., Wu, G., Parajulee, M.N. & Ge, F. (2007b) Impact of elevated CO_2 on the third trophic level: A predator *Harmonia axyridis* and a parasitoid *Aphidius picipes*. *Biocontrol Science and Technology*, **17**, 313–324.

Chown, S.L., Sørensen, J.G. & Terblanche, J.S. (2011) Water loss in insects: An environmental change perspective. *Journal of Insect Physiology*, **57**, 1070–1084.

Coakley, S.M., Scherm, H. & Chakraborty, S. (1999) Climate change and plant disease management. *Annual Review of Phytopathology*, **37**, 399–426.

Coll, M. & Hughes, L. (2008) Effects of elevated CO_2 on an insect omnivore: A test for nutritional effects mediated by host plants and prey. *Agriculture, Ecosystems & Environment*, **123**, 271–279.

Couture, J.J., Servi, J.S. & Lindroth, R.L. (2010) Increased nitrogen availability influences predator–prey interactions by altering host-plant quality. *Chemoecology*, **20**, 277–284.

Delava, E., Allemand, R., Léger, L., Fleury, F. & Gibert, P. (2014) The rapid northward shift of the range margin of a Mediterranean parasitoid insect (Hymenoptera) associated with regional climate warming. *Journal of Biogeography*, **41**, 1379–1389.

DeLucia, E.H., Nabity, P.D., Zavala, J.A. & Berenbaum, M.R. (2012) Climate change: resetting plant-insect interactions. *Plant Physiology*, **160**, 1677–1685.

Dixon, A.F.G., (1958) The escape responses shown by certain aphids to the presence of the coccinellid *Adalia decempunctata* (L.). *Transactions of the Royal Entomological Society of London*, **110**, 319–334.

Dong, Z.K., Hou, R.X., Ouyang, Z. & Zhang, R.Z. (2013) Tritrophic interaction influenced by warming and tillage: A field study on winter wheat, aphids and parasitoids. *Agriculture, Ecosystems & Environment*, **181**, 144–148.

Dyer, L.A., Richards, L.A., Short, S.A. & Dodson, C.D. (2013) Effects of CO_2 and temperature on tritrophic interactions. *PLOS ONE*, **8**, e62528.

Erb, M. & Lu, J. (2013) Soil abiotic factors influence interactions between belowground herbivores and plant roots. *Journal of Experimental Botany*, **64**, 1295–1303.

Evans, E.W., Carlile, N.R., Innes, M.B. & Pitigala, N. (2013) Warm springs reduce parasitism of the cereal leaf beetle through phenological mismatch. *Journal of Applied Entomology*, **137**, 383–391.

Facey, S.L., Ellsworth, D.S., Staley, J.T., Wright, D.J. & Johnson, S.N. (2014) Upsetting the order: how atmospheric and climate change affects herbivore–enemy interactions. *Current Opinion in Insect Science*, **5**, 66–74.

FAOSTAT (2014) Food and Agriculture Organization of the United Nations - Production Data Archives.

Farman, J.C., Gardiner, B.G. & Shanklin, J.D. (1985) Large losses of total ozone in Antarctica reveal seasonal ClO_x/NO_x interaction. *Nature*, **315**, 207–210.

Foggo, A., Higgins, S., Wargent, J.J. & Coleman, R.A. (2007) Tri-trophic consequences of UV-B exposure: plants, herbivores and parasitoids. *Oecologia*, **154**, 505–512.

Gao, F., Chen, F. & Ge, F. (2010) Elevated CO_2 lessens predation of *Chrysopa sinica* on *Aphis gossypii*. *Entomologia Experimentalis Et Applicata*, **135**, 135–140.

Gao, F., Zhu, S.R., Sun, Y.C., Du, L., Parajulee, M., Kang, L. & Ge, F. (2008) Interactive effects of elevated CO_2 and cotton cultivar on tri-trophic interaction of *Gossypium hirsutum*, *Aphis gossyppii*, and *Propylaea japonica*. *Environmental Entomology*, **37**, 29–37.

Giles, K.L., Madden, R.D., Stockland, R., Payton, M.E. & Dillwith, J.W. (2002) Host plants affect predator fitness via the nutritional value of herbivore prey: Investigation of a plant-aphid-ladybeetle system. *Biocontrol*, **47**, 1–21.

Godfray, H.C.J., Beddington, J.R., Crute, I.R., Haddad, L., Lawrence, D., Muir, J.F., Pretty, J., Robinson, S., Thomas, S.M. & Toulmin, C. (2010) Food security: the challenge of feeding 9 billion people. *Science*, **327**, 812–818.

Gregory, P.J., Johnson, S.N., Newton, A.C. & Ingram, J.S.I. (2009) Integrating pests and pathogens into the climate change/food security debate. *Journal of Experimental Botany*, **60**, 2827–2838.

Guay, J.-F., Boudreault, S., Michaud, D. & Cloutier, C. (2009) Impact of environmental stress on aphid clonal resistance to parasitoids: Role of *Hamiltonella defensa* bacterial symbiosis in association with a new facultative symbiont of the pea aphid. *Journal of Insect Physiology*, **55**, 919–926.

Gutbrodt, B., Mody, K. & Dorn, S. (2011) Drought changes plant chemistry and causes contrasting responses in lepidopteran herbivores. *Oikos*, **120**, 1732–1740.

Hance, T., van Baaren, J., Vernon, P. & Boivin, G. (2007) Impact of extreme temperatures on parasitoids in a climate change perspective. *Annual Review of Entomology*, **52**, 107–126.

Hartley, S.E. (2015) Round and round in cycles? Silicon-based plant defences and vole population dynamics. *Functional Ecology*, **29**, 151–153.

Hentley, W.T., Hails, R.S., Johnson, S.N., Jones, T.H. & Vanbergen, A.J. (2014a) Top-down control by *Harmonia axyridis* mitigates the impact of elevated atmospheric CO_2 on a plant–aphid interaction. *Agricultural and Forest Entomology*, **16**, 350–358.

Hentley, W.T., Vanbergen, A.J., Hails, R.S., Jones, T.H. & Johnson, S.N. (2014b) Elevated atmospheric CO_2 impairs aphid escape responses to predators and conspecific alarm signals. *Journal of Chemical Ecology*, **40**, 1110–1114.

Himanen, S.J., Nerg, A.-M., Nissinen, A., Pinto, D.M., Stewart, C.N., Jr, Poppy, G.M. & Holopainen, J.K. (2009) Effects of elevated carbon dioxide and ozone on volatile terpenoid emissions and multitrophic communication of transgenic insecticidal oilseed rape (*Brassica napus*). *New Phytologist*, **181**, 174–186.

Holton, M.K., Lindroth, R.L. & Nordheim, E.V. (2003) Foliar quality influences tree-herbivore-parasitoid interactions: effects of elevated CO_2, O_3, and plant genotype. *Oecologia*, **137**, 233–244.

Huberty, A.F. & Denno, R.F. (2004) Plant water stress and its consequences for herbivorous insects: A new synthesis. *Ecology*, **85**, 1383–1398.

Inbar, M., Doostdar, H. & Mayer, R.T. (2001) Suitability of stressed and vigorous plants to various insect herbivores. *Oikos*, **94**, 228–235.

IPCC (2013) Summary for Policymakers. *Climate Change 2013: The Physical Science Basis. Contribution of Working Group I to the Fifth Assessment Report of the Intergovernmental Panel on Climate Change* (eds T.F. Stocker, D. Qin, G.-K. Plattner, M. Tignor, S. K. Allen, J. Boschung, A. Nauels, Y. Xia, V. Bex & P.M. Midgley). Cambridge University Press, Cambridge and New York.

IPCC (2014) Summary for Policymakers. *Climate Change 2014: Impacts, Adaptation, and Vulnerability. Part A: Global and Sectoral Aspects. Contribution of Working Group II to the Fifth Assessment Report of the Intergovernmental Panel on Climate Change* (eds C.B. Field, V.R. Barros, D.J. Dokken, K.J. Mach, M.D. Mastrandrea, T.E. Bilir, M. Chatterjee, K.L. Ebi, Y.O. Estrada, R.C. Genova, B. Girma, E.S. Kissel, A.N. Levy, S. MacCracken, P.R. Mastrandrea & L.L. White), pp. 1–32. Cambridge University Press, Cambridge and New York.

Johnson, S.N., Hawes, C. & Karley, A.J. (2009) Reappraising the role of plant nutrients as mediators of interactions between root- and foliar-feeding insects. *Functional Ecology*, **23**, 699–706.

Johnson, S.N., Staley, J.T., McLeod, F.A.L. & Hartley, S.E. (2011) Plant-mediated effects of soil invertebrates and summer drought on above-ground multitrophic interactions. *Journal of Ecology*, **99**, 57–65.

Kangasjärvi, J., Talvinen, J., Utriainen, M. & Karjalainen, R. (1994). Plant defence systems induced by ozone. *Plant, Cell & Environment*, **17**, 783–794.

Kingsolver, J.G., Huey, R,B. (2008) Size, temperature, and fitness: three rules. *Evolutionary Ecology Research*, **10**, 251–268.

Klaiber, J., Najar-Rodriguez, A.J., Dialer, E. & Dorn, S. (2013) Elevated carbon dioxide impairs the performance of a specialized parasitoid of an aphid host feeding on *Brassica* plants. *Biological Control*, **66**, 49–55.

Klapwijk, M.J., Gröbler, B.C., Ward, K., Wheeler, D. & Lewis, O.T. (2010) Influence of experimental warming and shading on host–parasitoid synchrony. *Global Change Biology*, **16**, 102–112.

Kogan, M. (1998) Integrated pest management: historical perspectives and contemporary developments. *Annual Review of Entomology*, **43**, 243–270.

Larsson, S. (1989) Stressful times for the plant stress–insect performance hypothesis. *Oikos*, **56**, 277–283.

Le Lann, C., Lodi, M. & Ellers, J. (2014) Thermal change alters the outcome of behavioural interactions between antagonistic partners. *Ecological Entomology*, **39**, 578–588.

Lemoine, N.P., Burkepile, D.E. & Parker, J.D. (2014) Variable effects of temperature on insect herbivory. *PeerJ*, **2**, e376.

Lincoln, D.E., Couvet, D. & Sionit, N. (1986) Response of an insect herbivore to host plants grown in carbon dioxide enriched atmospheres. *Oecologia*, **69**, 556–560.

Lorenz, E.N. (1963) Deterministic nonperiodic flow. *Journal of the Atmospheric Sciences*, **20**, 130–141.

Ma, G. & Ma, C.S. (2012) Climate warming may increase aphids' dropping probabilities in response to high temperatures. *Journal of Insect Physiology*, **58**, 1456–1462.

MacArthur, R. (1955) Fluctuations of animal populations and a measure of community stability. *Ecology*, **36**, 533–536.

Massey, F.P., Ennos, A.R. & Hartley, S.E. (2007) Herbivore specific induction of silica-based plant defences. *Oecologia*, **152**, 677–683.

Mattson, W.J. & Haack, R.A. (1987) The role of drought in outbreaks of plant-eating insects. *Bioscience*, **37**, 110–118.

McCarty, J.P. (2001) Ecological consequences of recent climate change. *Conservation Biology*, **15**, 320–331.

McCluney, K.E., Belnap, J., Collins, S.L., González, A.L., Hagen, E.M., Holland, J.N., Kotler, B.P., Maestre, F.T., Smith, S.D. & Wolf, B.O. (2012) Shifting species interactions in terrestrial dryland ecosystems under altered water availability and climate change. *Biological Reviews*, **87**, 563–582.

McLean, A.H.C., van Asch, M., Ferrari, J. & Godfray, H.C.J. (2011) Effects of bacterial secondary symbionts on host plant use in pea aphids. *Proceedings of the Royal Society B: Biological Sciences*, **278**, 760–766.

McNaughton, S.J. & Tarrants, J.L. (1983) Grass leaf silicification: Natural selection for an inducible defense against herbivores. *Proceedings of the National Academy of Sciences of the United States of America*, **80**, 790–791.

Meineke, E.K., Dunn, R.R. & Frank, S.D. (2014) Early pest development and loss of biological control are associated with urban warming. *Biology Letters*, **10**, 20140586.

Mitchell, C., Johnson, S.N., Gordon, S.C., Birch, A.N.E. & Hubbard, S.F. (2010) Combining plant resistance and a natural enemy to control *Amphorophora idaei*. *Biocontrol*, **55**, 321–327.

Mody, K., Eichenberger, D. & Dorn, S. (2009) Stress magnitude matters: different intensities of pulsed water stress produce non-monotonic resistance responses of host plants to insect herbivores. *Ecological Entomology*, **34**, 133–143.

Moonen, A.-C. & Bàrberi, P. (2008) Functional biodiversity: An agroecosystem approach. *Agriculture, Ecosystems & Environment*, **127**, 7–21.

Mumm, R. & Dicke, M. (2010) Variation in natural plant products and the attraction of bodyguards involved in indirect plant defense. *Canadian Journal of Zoology – Revue Canadienne De Zoologie*, **88**, 628–667.

Musolin, D.L. (2007) Insects in a warmer world: ecological, physiological and life-history responses of true bugs (Heteroptera) to climate change. *Global Change Biology*, **13**, 1565–1585.

Newton, A.C., Johnson, S.N. & Gregory, P.J. (2011) Implications of climate change for diseases, crop yields and food security. *Euphytica*, **179**, 3–18.

Ode, P.J., Johnson, S.N. & Moore, B.D. (2014). Atmospheric change and induced plant secondary metabolites — are we reshaping the building blocks of multi-trophic interactions? *Current Opinion in Insect Science*, **5**, 57–65.

Olesen, J.E., Trnka, M., Kersebaum, K.C., Skjelvåg, A.O., Seguin, B., Peltonen-Sainio, P., Rossi, F., Kozyra, J. & Micale, F. (2011) Impacts and adaptation of European crop production systems to climate change. *European Journal of Agronomy*, **34**, 96–112.

Pinto, D.M., Blande, J.D, Souza, S.R., Nerg, A.-M. & Holopainen, J.K. (2010) Plant volatile organic compounds (VOCs) in ozone (O_3) polluted atmospheres: the ecological effects. *Journal of Chemical Ecology*, **36**, 22–34.

Pinto, D.M., Himanen, S.J., Nissinen, A., Nerg, A.-M. & Holopainen, J.K. (2008) Host location behavior of *Cotesia plutellae* Kurdjumov (Hymenoptera: Braconidae) in ambient and moderately elevated ozone in field conditions. *Environmental Pollution*, **156**, 227–231.

Porter, J.H., Parry, M.L. & Carter, T.R. (1991) The potential effects of climatic-change on agricultural insect pests. *Agricultural and Forest Meteorology*, **57**, 221–240.

Preisser, E.L. & Strong, D.R. (2004) Climate affects predator control of an herbivore outbreak. *American Naturalist*, **163**, 754–762.

Price, P.W. (1991) The Plant Vigor Hypothesis and Herbivore Attack. *Oikos*, **62**, 244–251.

Riikonen, J., Percy, K.E., Kivimäenpää, M., Kubiske, M.E., Nelson, N.D., Vapaavuori, E. & Karnosky, D.F. (2010) Leaf size and surface characteristics of *Betula papyrifera* exposed to elevated CO_2 and O_3. *Environmental Pollution*, **158**, 1029–1035.

Robinson, E.A., Ryan, G.D. & Newman, J.A. (2012) A meta-analytical review of the effects of elevated CO_2 on plant–arthropod interactions highlights the importance of interacting environmental and biological variables. *New Phytologist*, **194**, 321–336.

Romo, C.M. & Tylianakis, J.M. (2013) Elevated temperature and drought interact to reduce parasitoid effectiveness in suppressing hosts. *PLOS ONE*, **8**, e58136.

Roth, S.K. & Lindroth, R.L. (1995) Elevated atmospheric CO_2: effects on phytochemistry, insect performance and insect–parasitoid interactions. *Global Change Biology*, **1**, 173–182.

Roux, O., Le Lann, C., van Alphen, J.J.M. & van Baaren, J. (2010) How does heat shock affect the life history traits of adults and progeny of the aphid parasitoid *Aphidius avenae* (Hymenoptera: Aphidiidae)? *Bulletin of Entomological Research*, **100**, 543–549.

Rusch, A., Valantin-Morison, M., Sarthou, J.-P. & Roger-Estrade, J. (2010) Biological control of insect pests in agroecosystems: effects of crop management, farming systems, and seminatural habitats at the landscape scale: a review. *Advances in Agronomy*, Vol **109**, pp. 219–259. Elsevier Academic Press Inc, San Diego.

Schmitz, O.J. & Barton, B.T. (2014) Climate change effects on behavioral and physiological ecology of predator–prey interactions: Implications for conservation biological control. *Biological Control*, **75**, 87–96.

Sentis, A., Hemptinne, J.-L. & Brodeur, J. (2013) Effects of simulated heat waves on an experimental plant–herbivore–predator food chain. *Global Change Biology*, **19**, 833–842.

Sinka, M., Jones, T.H. & Hartley, S.E. (2007) The indirect effect of above-ground herbivory on collembola populations is not mediated by changes in soil water content. *Applied Soil Ecology*, **36**, 92–99.

Stacey, D.A. & Fellowes, M.D.E. (2002a) Influence of elevated CO_2 on interspecific interactions at higher trophic levels. *Global Change Biology*, **8**, 668–678.

Stacey, D.A. & Fellowes, M.D.E. (2002b) Influence of temperature on pea aphid *Acyrthosiphon pisum* (Hemiptera: Aphididae) resistance to natural enemy attack. *Bulletin of Entomological Research*, **92**, 351–357.

Staley, J.T. & Johnson, S.N. (2008) Climate change impacts on root herbivores. *Root Feeders - an ecosystem perspective* (eds S.N. Johnson & P.J. Murray), pp. 192–213. CABI, Wallingford, UK.

Staley, J.T., Mortimer, S.R., Morecroft, M.D., Brown, V.K. & Masters, G.J. (2007) Summer drought alters plant-mediated competition between foliar- and root-feeding insects. *Global Change Biology*, **13**, 866–877.

Stiling, P. & Cornelissen, T. (2007) How does elevated carbon dioxide (CO_2) affect plant–herbivore interactions? A field experiment and meta-analysis of CO_2-mediated changes on plant chemistry and herbivore performance. *Global Change Biology*, **13**, 1823–1842.

Stiling, P., Rossi, A.M., Hungate, B., Dijkstra, P., Hinkle, C.R., Knott, W.M. III & Drake, B. (1999) Decreased leaf-miner abundance in elevated CO_2: Reduced leaf quality and increased parasitoid attack. *Ecological Applications*, **9**, 240–244.

Stireman, J.O. III, Dyer, L.A., Janzen, D.H., Singer, M.S., Lill, J.T., Marquis, R.J., Ricklefs, R.E., Gentry, G.L., Hallwachs, W., Coley, P.D., Barone, J.A., Greeney, H.F., Connahs, H., Barbosa, P., Morais, H.C. & Diniz, I.R. (2005) Climatic unpredictability and parasitism of caterpillars: Implications of global warming. *Proceedings of the National Academy of Sciences of the United States of America*, **102**, 17384–17387.

Strahan, S.E., Douglass, A.R., Newman, P.A. & Steenrod, S.D. (2014) Inorganic chlorine variability in the Antarctic vortex and implications for ozone recovery. *Journal of Geophysical Research: Atmospheres*, **119**, 14,098–014,109.

Sun, Y.C., Feng, L., Gao, F. & Ge, F. (2011) Effects of elevated CO_2 and plant genotype on interactions among cotton, aphids and parasitoids. *Insect Science*, **18**, 451–461.

Tariq, M., Wright, D.J., Bruce, T.J.A. & Staley, J.T. (2013) Drought and root herbivory interact to alter the response of above-ground parasitoids to aphid infested plants and associated plant volatile signals. *PLOS ONE*, **8**, e69013.

Thaler, J.S. (1999) Jasmonate-inducible plant defences cause increased parasitism of herbivores. *Nature*, **399**, 686–688.

Trenberth, K.E., Jones, P.D., Ambenje, P., Bojariu, R., Easterling, D., Klein Tank, A., Parker, D., Rahimzadeh, F., Renwick, J.A., Rusticucci, M., Soden, B. & Zhai, P. (2007) Observations: Surface and Atmospheric Climate Change. *Climate Change 2007: The Physical Science Basis. Contribution of Working Group I to the Fourth Assessment Report of the Intergovernmental Panel on Climate Change* (eds S. Solomon, D. Qin, M. Manning, Z. Chen, M. Marquis, K.B. Averyt, M. Tignor & H.L. Miller). Cambridge University Press, Cambridge and New York.

Trotter, R.T. III, Cobb, N.S. & Whitham, T.G. (2008) Arthropod community diversity and trophic structure: a comparison between extremes of plant stress. *Ecological Entomology*, **33**, 1–11.

United Nations (2015) World Population Prospects: The 2015 Revision, Key Findings and Advance Tables. *Department of Economic and Social Affairs, Population Division*, Working Paper No. ESA/P/WP.241.

Valkama, E., Koricheva, J. & Oksanen, E. (2007) Effects of elevated O_3, alone and in combination with elevated CO_2, on tree leaf chemistry and insect herbivore performance: a meta-analysis. *Global Change Biology*, **13**, 184–201.

Van Poecke, R.M.P. & Dicke, M. (2004) Indirect defence of plants against herbivores: Using *Arabidopsis thaliana* as a model plant. *Plant Biology*, **6**, 387–401.

Veteli, T.O., Kuokkanen, K., Julkunen-Tiitto, R., Roininen, H. & Tahvanainen, J. (2002) Effects of elevated CO_2 and temperature on plant growth and herbivore defensive chemistry. *Global Change Biology*, **8**, 1240–1252.

Voigt, W., Perner, J., Davis, A.J., Eggers, T., Schumacher, J., Bährmann, R., Fabian, B., Heinrich, W., Köhler, G., Lichter, D., Marstaller, R. & Sander, F.W. (2003) Trophic levels are differentially sensitive to climate. *Ecology*, **84**, 2444–2453.

Vuorinen, T., Nerg, A.-M. & Holopainen, J.K. (2004a) Ozone exposure triggers the emission of herbivore-induced plant volatiles, but does not disturb tritrophic signalling. *Environmental Pollution*, **131**, 305–311.

Vuorinen, T., Nerg, A.-M., Ibrahim, M.A., Reddy, G.V.P. & Holopainen, J.K. (2004b) Emission of *Plutella xylostella*-induced compounds from cabbages grown at elevated CO_2 and orientation behavior of the natural enemies. *Plant Physiology*, **135**, 1984–1992.

Wade, RN. (2016) The Effect of Simulated Precipitation Change on Multi-trophic Interactions in a Cereal Crop. PhD thesis, University of York, UK.

Walling, L.L. (2000) The myriad plant responses to herbivores. *Journal of Plant Growth Regulation*, **19**, 195–216.

White, T.C.R. (1984) The abundance of invertebrate herbivores in relation to the availability of nitrogen in stressed food plants. *Oecologia*, **63**, 90–105.

Yang, Y. & Joern, A. (1994) Influence of diet quality, developmental stage, and temperature on food residence time in the grasshopper *Melanoplus differentialis*. *Physiological Zoology*, **67**, 598–616.

Yin, J., Sun, Y., Wu, G., Parajulee, M.N. & Ge, F. (2009) No effects of elevated CO_2 on the population relationship between cotton bollworm, *Helicoverpa armigera* Hübner (Lepidoptera: Noctuidae), and its parasitoid, *Microplitis mediator* Haliday (Hymenoptera: Braconidae). *Agriculture, Ecosystems & Environment*, **132**, 267–275.

Zavala, J.A., Nabity, P.D. & DeLucia, E.H. (2013) An emerging understanding of mechanisms governing insect herbivory under elevated CO_2. *Annual Review of Entomology*, **58**, 79–97.

Zhu, H., Wang, D.L., Wang, L., Fang, J., Sun, W. & Ren, B.Z. (2014) Effects of altered precipitation on insect community composition and structure in a meadow steppe. *Ecological Entomology*, **39**, 453–461.

11

Climate Change in the Underworld: Impacts for Soil-Dwelling Invertebrates

Ivan Hiltpold[1,2], Scott N. Johnson[1], Renée-Claire Le Bayon[3] and Uffe N. Nielsen[1]

[1] *Hawkesbury Institute for the Environment, Western Sydney University, Australia*
[2] *Department of Entomology and Wildlife Ecology, University of Delaware, USA*
[3] *Functional Ecology Laboratory, University of Neuchâtel, Switzerland*

Summary

Increasing concentrations of atmospheric CO_2 and associated changes in climate are undoubtedly impacting communities and ecosystem functioning. Despite the importance of soil ecosystem services, little is known on how these changes will affect soil-dwelling communities. In this chapter, we review and discuss the impact of elevated atmospheric CO_2 and climatic changes on three of the functionally most important invertebrate taxa in soil ecosystems: nematodes, insects, and earthworms. Elevated atmospheric CO_2 concentrations are mostly impacting these three groups indirectly (e.g., plant-mediated mechanisms), whereas climatic changes (elevated temperature and altered precipitation) are both directly and indirectly affecting soil invertebrates. Earthworms are mainly positively affected by elevated atmospheric CO_2 and climate change; these effects are however mostly indirect and more subtle than those observed in the other two taxa discussed. Chemical ecology underpins biotic interactions belowground. Plant secondary metabolites patterns and their diffusion in soil are influenced by climate change; therefore belowground chemical interactions are likely to be influenced too. Soils are at high risk and better comprehending the effect of the rapid changes occurring will help to protect these highly complex ecosystems.

11.1 Introduction

Climate fluctuations have occurred over millions of years and have shaped ecosystems throughout this period. The greenhouse gases carbon dioxide (CO_2), methane (CH_4), and nitrous oxide (N_2O) concentrations have increased by 40%, 150%, and 20%, respectively, from 1970 to 2010 (IPCC, 2013b). The combined average temperature of the atmosphere and oceans is increasing together with changes in both the amount and patterns of precipitation (IPCC, 2013b). The rapid pace of the recent changes is alarming and likely to have serious impacts on ecosystems and biological communities therein (Newman et al., 2011).

Most contemporary research on climate change has focused on aboveground communities to the comparative neglect of belowground ecosystems. This is paradoxical since soils are at high risk with, for example, ca. 11 ha of arable soil sealed (e.g., concreted or paved upon) every hour in Europe

Global Climate Change and Terrestrial Invertebrates, First Edition. Edited by Scott N. Johnson and T. Hefin Jones.
© 2017 John Wiley & Sons, Ltd. Published 2017 by John Wiley & Sons, Ltd.

(FAO, 2015; Wall & Six, 2015). It is therefore essential to take up the challenge and assess the impact of climate change on soil communities before irremediable damage is done to the wealth of ecological services that soils confer.

The effects of climate and atmospheric CO_2 change (from here on, climate change encompasses atmospheric changes and variations in CO_2, warming and altered precipitations) on soil abiotic conditions vary and numerous biotic feedbacks occur. For example, elevated atmospheric CO_2 concentrations (eCO_2) decrease plant stomatal conductance and consequently increase soil moisture (Cowan & Farquhar, 1977), whereas higher temperatures reduce soil moisture by increasing plant evapotranspiration (Norby & Luo, 2004; Dermody et al., 2007). These variations in soil abiotic conditions arising from climate change are likely to shape soil invertebrate communities (e.g., Blankinship et al., 2011; Newman et al., 2011; Lang et al., 2014; García-Palacios et al., 2015; Nielsen & Ball, 2015), which will have impacts at a global scale (Lavelle et al., 1997).

11.1.1   Soil Community Responses to Climate Change

The response of a diverse group of organisms, such as soil invertebrates, to eCO_2 cannot be easily generalised and will vary for different members of the community. For example, in a greenhouse experiment, earthworm biomass was ca. 50% greater at 700 ppm than a 350 ppm CO_2, whereas eCO_2 treatment reduced some nematode communities (Yeates et al., 1997). In the field, eCO_2 did not affect the abundance of invertebrate grazers (i.e., nematodes, mites and springtails) but large omnivorous and predacious nematodes were fewer, possibly because of the changes in the soil structure and particularly in the aggregate and pore sizes (Yeates et al., 2002; Niklaus et al., 2003).

Similarly, the consequences of warming and altered precipitation on soil invertebrates are also highly variable, though changes in precipitation patterns are likely to have the strongest effect on the abundance of soil biota (Blankinship et al., 2011) since arthropods are highly dependent on soil moisture (e.g., Chikoski et al., 2006; Kardol et al., 2011; Sylvain et al., 2014; Williams et al., 2014). The response of invertebrate communities appears to be largely idiosyncratic (Blankinship et al., 2011). Yet, there is evidence that taxonomy, ecosystem type and the direction of the environmental change are determinant factors that influence the response of soil-dwelling invertebrates to climate change. The response of soil biota to climate change may also depend on the body size of the organisms as well as on their trophic or functional grouping (Blankinship et al., 2011; A'Bear et al., 2014). In addition, as suggested by the trophic dynamic theory (Lindeman, 1942), trophic levels differ in limitations and thus climate and atmospheric change affect them differently. Moreover, the consequences of climate change at any trophic level will have cascading effects up and down the food web (e.g., Scheu, 2002; Xiao et al., 2005; Lang et al., 2014).

11.1.2   Scope of the Chapter

In this chapter we discuss in further detail the impact of eCO_2, warming and altered precipitation on three broad taxonomic and functional groups mentioned in a community context above: (1) nematodes, (2) insects, and (3) earthworms. These groups represent major components of the soil ecosystem. Nematodes are the most abundant soil invertebrates, represent several trophic groups (i.e., fungivores, bacterivores, herbivores and predators) and play an essential role in C and nutrient cycling (Yeates et al., 2009). Many soil-dwelling insects are herbivores and devastate crops, which impact human societies through yield decreases (Johnson & Murray, 2008); therefore an understanding of how climate change will affect their pest status is essential. Some insects are like earthworms in that they too are ecosystem engineers (Jones et al., 1994) and provide ecosystem services, and might be used in climate change adaptation (Johnson et al., 2016b). As ecosystem engineers, earthworms are pivotal in belowground processes such as bioturbation and the incorporation of organic matter to the mineral fraction of soil. Mitigating the impact of climatic changes on this group is therefore likely to be essential to ensure a sustainable use of soils as a natural resource. In closing, we discuss aspects of belowground ecology in the context of climate change.

11.2 Effect of Climate Change on Nematodes: Omnipresent Soil Invertebrates

Nematodes are ubiquitous in soils and the most numerous multicellular soil-dwelling animals in most ecosystems (Fierer et al., 2009). They represent at least five distinct trophic groups and a multitude of feeding guilds. Plant parasitic nematodes and microbial grazers, however, typically dominate soil nematode communities (Nielsen & Philipsen, 2004; Nielsen et al., 2014). The plant parasites affect carbon (C) and nutrient cycling directly by feeding on root biomass, and they can have substantial negative impacts on plant biomass production in both natural and managed ecosystems, including agriculture and horticulture. Microbial grazers stimulate the turnover of microbial biomass and moderate microbial activities, which in turn influence ecosystem processes including decomposition, mineralization and nutrient cycling. Other nematode trophic groups, including omnivores and predators, similarly influence C and nutrient cycling, but their impacts occur mostly indirectly through predation or control of plant pathogens. The true extent of the roles of nematodes in ecosystem functioning is not well quantified (Bardgett et al., 1999; Hunt & Wall, 2002; Nielsen et al., 2015), but given their known impacts on ecosystems, soil nematode community responses to climate change may have far-reaching effects.

In this section we discuss the observed and potential soil nematode community responses to climate change (Fig. 11.1) and how this might influence ecosystem functioning more broadly. Evidence from field surveys indicates that nematode community composition is strongly related to climate suggesting that predicted changes in temperature and precipitation might have substantial long-term effects (Nielsen et al., 2014). There is less evidence to suggest that eCO_2 should influence soil nematode communities directly but indirect effects may theoretically be expected. These observations are now being supported by in situ climate change manipulations, and we provide an overview of the observed effects under field conditions. We focus first on the effects of eCO_2, which is expected to influence nematode communities mostly mediated by vegetation responses (Nielsen et al., 2015), and then discuss the effects of altered precipitation regimes and increased temperatures (warming) as these two global change drivers are likely to have the strongest direct impact on soil nematode communities. It is worth noting that the realized effects of climate change will be moderated by interactions between eCO_2, warming, and altered precipitation regimes as these rarely change in isolation. Yet, very few studies have investigated the interactive effects of these global change drivers on soil nematode communities. Potential broader impacts of soil nematode community responses to climate change on ecosystems are briefly discussed at the end of the section.

11.2.1 Nematode Responses to eCO_2

Carbon dioxide concentrations are much higher in soil than in the atmosphere, and therefore, eCO_2 is unlikely to substantially influence soil nematode communities directly. However, indirect effects are likely to occur due to changes in root exudates and litter input associated with altered vegetation structure and composition (Nielsen et al., 2015). For instance, C:N ratio in plant tissue has been shown to increase under eCO_2 (Pendall et al., 2004b; Milchunas et al., 2005), which may influence their quality as a food source for belowground herbivores including plant parasitic nematodes with

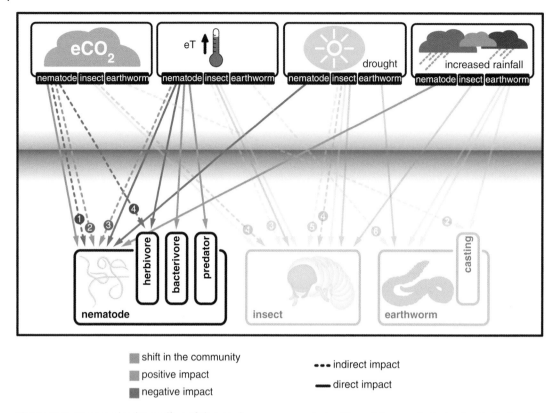

Figure 11.1 Direct and indirect effect of climate change on nematodes. The solid arrows indicate direct effects, while dashed arrows indicate indirect effects. Blue arrows represent shift in the community whereas green arrows indicate a positive impact and red arrows a negative impact on the community. Indirect effects are mediated by (1) changes in the soil structure, (2) increased soil moisture resulting from lower stomatal conductance, (3) shifts in the plant community, (4) a reduction of plant quality (often reflected by higher C:N ratio), and (5) decreased population of predators and/or parasitoids, and (6) a higher N rhizodeposition. Details and references in the text. (Drawings: I. Hiltpold, WSU, Australia)

cascading impacts on the soil food web more broadly. Along the same lines, Frederiksen et al. (2001) found that wheat litter grown under eCO_2 showed reduced decomposition rates and supported lower abundances of bacterial grazers including nematodes in situ indicating indirect effects of eCO_2 on ecosystem processes. Moreover, eCO_2 has been shown to enhance fine root growth and distribution (e.g., Rogers et al., 1996; Curtis & Wang, 1998; Pendall et al., 2004a; Prior et al., 2012; Madhu & Hatfeld, 2013; Pacholski et al., 2015). Root architecture plays a role in the ability of entomopathogenic nematodes to find and infect insect hosts in the rhizosphere (Demarta et al., 2014). Indeed, entomopathogenic nematodes appeared to be more infectious in the presence of root systems or structure mimicking root systems (Ennis et al., 2010; Demarta et al., 2014), but increased root complexity results in a lower infectiousness of entomopathogenic nematodes (Demarta et al., 2014). In this context, eCO_2 can potentially lessen successful implementation of entomopathogenic nematodes in pest control strategies. More broadly, it is often observed that eCO_2 reduces evapotranspiration with cascading effects on soil moisture, which could benefit nematode communities as these are limited by soil moisture availability at least in some ecosystem types. Furthermore, there is some evidence that suggests that eCO_2 may influence soil aggregation, which might in turn influence nematode communities through a reduction in soil pore neck diameters (Niklaus et al., 2003).

A recent meta-analysis by A'Bear et al. (2014) found an overall negative effect of eCO_2 on nematode abundances except for plant parasitic nematodes. However, plant parasitic nematodes may still respond to eCO_2. For instance, a study investigating the effects of eCO_2 at three grassland sites found no effect on total abundance of plant parasites, but the abundance of Anguinidae increased at one site and Hoplolaimidae decreased at another (Ayres et al., 2008) indicating changes in community structure. In another study, eCO_2 increased the abundance of the plant parasitic Trichodoridae and reduced the abundance of bacterial feeding Rhabditidae in low nitrogen (N) soil, whereas the abundance of predators/omnivores increased in high N soil due to an increase in Dorylaimida and the abundance of bacterial feeding Cephalobidae decreased (Hoeksema et al., 2000). Hence, it is clear that eCO_2 can influence both nematode abundances and community composition although some studies have found limited effects (Allen et al., 2005; Ayres et al., 2008; Eisenhauer et al., 2013). However, ecosystem level implications of these changes are not well established.

11.2.2 Nematode Responses to Warming

Nematodes are likely to respond to warming through both direct and indirect effects. Most regions are predicted to become warmer under future climates (IPCC, 2013b). While soil temperatures will be buffered to some degree compared to surface air temperatures, even moderate increases in soil temperatures are likely to enhance nematode activity and metabolic rates, and through this reproduction, provided that other environmental variables such as soil moisture are not limiting. Moreover, indirect effects are likely through shifts in plant community composition, microbial biomass and composition, and edaphic variables including potential increased stress associated with decreased soil moisture under elevated temperature (e.g., Nielsen et al., 2015). However, the number of studies reporting the effects of warming on soil nematode communities is still surprisingly limited and results are often idiosyncratic. For instance, two years of increased soil temperature ($+1.7°C$ and $+3.4°C$) had no effect on total or trophic group abundances in a temperate-boreal forest ecotone, although there was an increase in microbial feeders relative to plant parasites (Thakur et al., 2014) suggesting an overall shift in community composition. By contrast, $1°C$ soil warming in a heath was found to reduce the abundance of nematodes including root herbivores and interestingly there was shift towards longer-lived species perhaps due to longer growing seasons (Stevnbak et al., 2012). Similarly, in a subarctic heath and a fell field, summer warming was shown to influence nematode community structure with an observed increase in abundance of *Aphelenchoides* and *Filenchus*, both species with rapid life cycles and high fecundity, whereas the abundance of species with slower reproductive cycles, specifically *Eudorylaimus* and *Teratocephalus*, decreased (Ruess et al., 1999). Finally, warming has been observed to influence soil nematode community composition at least in some vegetation types and bare soil in temperate semiarid ecosystems (Bakonyi et al., 2007) as well as nematode diversity in maize fields (Dong et al., 2013). Collectively, these results suggest that warming can have strong effects on soil nematode community composition even when total nematode abundance is not affected.

A couple of meta-analyses aimed to synthesize the effects of warming on soil nematode communities. One of the first meta-analyses of studies that investigated belowground effects of climate change suggested that increased temperature consistently increase nematode abundances (Blankinship et al., 2011), but this was not confirmed by a more recent meta-analysis (A'Bear et al., 2014). The latter paper found no strong directional effect of increased temperature on total nematode abundance, but instead showed that plant parasitic nematodes respond negatively, and fungal feeding and predatory nematodes respond positively, to increased temperatures (A'Bear et al., 2014). Such divergent trophic level responses to warming indicate whole soil food web level responses to warming and suggest broader impacts on ecosystem functioning. However, it is clear that responses are context dependent and further research is required to establish general patterns and to further our understanding of potential feedbacks of warming driven nematode community responses to the ecosystem.

11.2.3 Nematode Responses to Altered Precipitation Regimes

Soil nematodes are effectively aquatic animals that rely on water films for movement and feeding (Vandegehuchte et al., 2015). Thus, any climate change that influences soil moisture availability is likely to affect soil nematode communities. Water is a limiting resource in many ecosystems and increased precipitation is theoretically expected to have a positive effect on soil nematodes. By extension, reduced precipitation could be expected to have negative effects on soil nematodes, whereas the effects of changes in seasonality or event size and frequency are more difficult to predict (Nielsen & Ball, 2015). Most field studies investigating the effect of altered precipitation regimes to date have focused on increased or reduced precipitation, the latter often in terms of a drought event, while very few studies have manipulated event size and frequency.

As discussed above, the effect of altered precipitation regimes has been summarized in two meta-analyses. The earlier meta-analysis found no consistent effect of reduced precipitation on soil nematode abundances (Blankinship et al., 2011), but given new data a more recent paper found that drought significantly reduces the abundance of plant parasites and bacterial feeders (A'Bear et al., 2014). For instance, summer drought imposed for eight years in a heath negatively affected nematode abundances and all other decomposer biota measured (Stevnbak et al., 2012). These results suggested a substantial reduction in C flow through the decomposer community, which in turn indicate significant ecosystem responses to summer drought. Also, reduced precipitation has been found to influence soil nematode community composition with more pronounced effects in bare soils than under poplar, and no effects in fescue grassland and a temperate semi-arid shrubland (Bakonyi et al., 2007). Another study found that nematode families responded differentially to a one-year imposed drought in pine woodland with Plectidae being highly sensitive to drought conditions whereas drought had no effect on Cephalobidae and Qudsianematidae (Landesman et al., 2011). Hence, there may be taxonomically specific responses to reduced precipitation and responses are often context-dependent (Sylvain et al., 2014; Vandegehuchte et al., 2015). In particular, nematode communities in dryland soils appear to be very robust to altered precipitation regimes with limited observed responses to reduced precipitation (Vandegehuchte et al., 2015), at least in the short term. The capacity of nematodes to tolerate reduced precipitation in dryland ecosystems is likely due to their adaptation to already low soil moistures, and their capacity to enter stress tolerant life stages such as anhydrobiosis (Nielsen & Ball, 2015). However, longer-term exposure to reduced precipitation may have stronger effects.

The meta-analysis by A'Bear et al. (2014) further showed that increased precipitation has a strong positive effect on soil nematode abundances, although these results are based on a relatively small number of studies (n=7). For instance, irrigation to simulate a 100% increase in precipitation had a strong positive effect on nematode abundances in a pineland after a year's manipulation (Landesman et al., 2011), and another study, that investigated the effect of rainfall event size (2 mm versus 25 mm weekly), eCO_2, and warming in an artificial grassland community, showed that the effect of increased soil moisture availability was much greater than that of eCO_2 or warming (Kardol et al., 2010). However, increased water availability has been observed to have no effect on total and trophic group abundances in desert ecosystems (Darby et al., 2011; Sylvain et al., 2014), although increased abundance of plant parasites (Freckman et al., 1987; Vandegehuchte et al., 2015) and reduced abundances (Sylvain et al., 2014) have been observed elsewhere. Studies investigating belowground effects of changes in precipitation are increasing; however, more research is needed to establish general patterns, and the effects of changes in event size and frequency need to be considered more explicitly (Nielsen & Ball, 2015).

It is worth noting here that while rainfall is predicted to increase in some areas around the world, these increases are often predicted to be accompanied by warmer temperatures that may cause increased evapotranspiration and therefore no net increase in soil moisture (IPCC, 2013a). This interaction between precipitation regime and warming is crucial in understanding climate change effects. Moreover, changes in precipitation may also interact with other global change drivers. For instance, a recent study found that increased precipitation had a slight positive effect on total nematode densities in a temperate forest but when N was added to simulate increased N deposition, a strong negative effect was observed (Sun et al., 2013). Finally, even though relatively small effects of altered precipitation regimes are observed in trophic or total nematode abundances in some studies there may be ecologically relevant shifts in community composition. This is illustrated by recent findings that suggest that functional guild abundances are more sensitive to climate changes than trophic groups and therefore a better indication of potential shifts in food web structure and ecosystem functioning (Cesarz et al., 2015). It is therefore important that responses are quantified at appropriate taxonomic levels.

11.2.4 Ecosystem Level Effects of Nematode Responses to Climate Change

As discussed above, there is strong evidence that soil nematode communities show climate change responses that are moderated by ecosystem characteristics such as vegetation type and nutrient levels (Smolik & Dodd, 1983; Klironomos et al., 1996; Niklaus et al., 2003). However, we need more experimental data to build a robust framework to make generalizations, particularly if we want to understand potential feedbacks on ecosystem functioning. This is very important given the substantial impact trophic groups such as plant parasites might have on plant productivity. For instance, Yeates & Newton (2009) found that nine years of eCO_2 substantially increased the abundance of a dominant plant parasite in a grazed pasture (*Longidorus elongatus* increased 3.48-fold) and a follow-up pot experiment indicated that eCO_2 could result in increased belowground root herbivory by nematodes. This, in turn, could have cascading effects on the soil food web more broadly by inducing plant chemical responses. Similarly, *Phleum pratense* and *Poa pratensis* showed greater reduction in root biomass due to the plant parasite *Pratylenchus penetrans* when grown at eCO_2 (650 ppm) compared with plants grown at 350 ppm (Wilsey, 2001), indicating that global changes may moderate belowground interactions between plants and herbivorous nematodes. While such data are very insightful there is a need for more in-depth studies on how useful nematode communities are as bioindicators of ecosystem state and food web complexity, and to more accurately measure how great a role they play in C and nutrient cycling. In particular, developing new techniques that can assess the degree to which the activity of soil nematodes moderate microbial activity and functioning would be insightful in understanding ecosystem functioning and potential responses to global changes.

11.3 Effect of Climate Change on Insect Root Herbivores, the Grazers of the Dark

Soil-dwelling insects that feed on roots are usually the juvenile stages of insects that live aboveground as adults (Brown & Gange, 1990). These soil invertebrates can reach astonishing densities, with root-feeding cicadas of deciduous forests of North America having the largest collective biomass per unit area of any terrestrial animal (Karban, 1980). While the agricultural importance of root-feeding

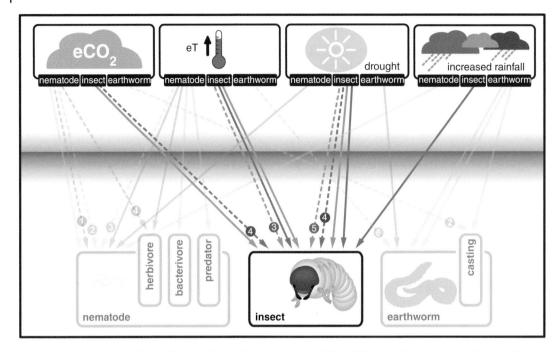

Figure 11.2 Direct and indirect effect of climate change on insects. The solid arrows indicate direct effects, while dashed arrows indicate indirect effects. Blue arrows represent shift in the community whereas green arrows indicate a positive impact and red arrows a negative impact on the community. See Figure 11.2 for nature of the indirect effects. Details and references in the text.

insects has long been known (Blackshaw & Kerry, 2008), it is increasingly recognised that they also affect numerous above- (Johnson et al., 2012a; Soler et al., 2012; Johnson et al., 2013a) and belowground organisms (Johnson & Rasmann, 2015). Root-feeding insects therefore represent important components of many ecosystems, and while far less is understood about how climate change affects this trophic group than their aboveground counterparts (Fig. 11.2), research in the field is undoubtedly increasing (Staley & Johnson, 2008; McKenzie et al., 2013).

11.3.1 Insect Root Herbivore Responses to eCO$_2$

To our knowledge, CO$_2$ is the only greenhouse gas to be investigated in the context of its impact on root-feeding insects (Staley & Johnson, 2008). Moreover, changes in CO$_2$ are likely to affect root-herbivores purely via indirect (e.g., plant-mediated, Fig. 11.2) mechanisms since, like nematodes and earthworms, they are already exposed to much higher concentrations in the rhizosphere than is the case aboveground (Payne & Gregory, 1988). While the effects of eCO$_2$ on aboveground insect herbivores are relatively well studied, and several reviews and meta-analyses are now available (e.g., Stiling & Cornelissen, 2007; DeLucia et al., 2012; Robinson et al., 2012; Zavala et al., 2013), studies involving root herbivores are scarce, by comparison, and seem to reach contrasting conclusions.

In an early study, Salt et al. (1996) reported that eCO$_2$ had no effect on the root-feeding aphid, *Pemphigus populi-transversus*, despite increasing root biomass in aphid-free plants. eCO$_2$ often increases root biomass relative to shoot mass (Rogers et al., 1994; Rogers et al., 1996) and also leads to more branched root systems that are generally shallower (Gregory, 2006). Staley & Johnson (2008) speculated that this increase in root biomass and architecture could promote the performance of some root herbivores, but it was more likely to affect root chewing herbivores that are limited by the quantity rather than quality of their host roots. Indeed, root quality might be anticipated to decrease in response to eCO_2; a meta-analysis of over 100 studies reported that while both C and N increased in both roots and shoots, C increased far more than N leading to an average increase in C:N ratios of 11%, effectively reducing N concentrations in both roots and shoots (Luo et al., 2006), either by dilution or reallocation (DeLucia et al., 2012). Given the importance of N in insect herbivore diets (Mattson, 1980), we might reasonably assume this would negatively affect root herbivores. There is recent evidence for this, with the body mass of root-feeding Argentine scarab (*Sericesthis nigrolineata*) decreasing by 24% under eCO_2, despite consuming 118% more root tissue when feeding on the grass *Microlaena stipoides* (Johnson et al., 2014a). Compensatory feeding by folivores is a common response to increased C:N in the foliage of plants grown under eCO_2 (Stiling & Cornelissen, 2007; DeLucia et al., 2012), so it seems probable that such tactics might be employed belowground too. Like folivores, which typically fail to redress this deterioration in host plant quality (Stiling & Cornelissen, 2007), it seems that the *S. nigrolineata* could not maintain performance levels even with compensatory feeding (Johnson et al., 2014a). Interestingly, when *S. nigrolineata* fed on a C_4 grass, *Cymbopogon refractus*, its performance was typically lower than when feeding on *M. stipoides*, but performance did not decline further under eCO_2. This was likely due to the fact that the C:N ratio in root tissue did not increase under eCO_2 (Johnson et al., 2014a). Unchanged C:N ratios in C_4 plants grown under eCO_2 is often reported (Wand et al., 1999) because Rubisco, the initial enzyme facilitating CO_2 assimilation into carbohydrates, operates close to its maximum capacity at ambient atmospheric concentration of CO_2 (aCO_2), so C_4 plants have less capacity to respond to eCO_2 than C_3 plants (Ainsworth et al. 2004; DeLucia et al., 2012). Hence, belowground responses to eCO_2 will be moderated by plant species functional type.

Secondary metabolites in shoots are frequently altered in response to eCO_2, and this too can affect herbivores aboveground (Ode et al., 2014). While it usually follows that eCO_2-induced increases in plant secondary metabolite concentrations diminish herbivore performance (Zavala et al., 2013; Ode et al., 2014), this may not always be the case for root herbivores which appear to respond quite differently to plant defences expressed belowground (Johnson et al., 2016a). For instance, vine weevils (*Otiorhynchus sulcatus*) feeding on blackcurrant grown under eCO_2 performed much worse (33% and 23% declines in abundance and body mass, respectively) than when feeding on plants grown at aCO_2 concentrations. Plants had 16% lower root growth under eCO_2 and this most likely drove this decline in herbivore performance but, surprisingly, root phenolic compounds were positively correlated with abundance and body mass. Vine weevils induced increases in root phenolic compounds under aCO_2 concentrations but not at eCO_2 (Johnson et al., 2011).

While there is a general trend for host plant quality for herbivores to decline under eCO_2 (Robinson & Jaffee, 1996; Stiling & Cornelissen, 2007; Ode et al., 2014), this isn't necessarily true for all plant taxa. In particular, leguminous plants will often increase rates of root nodulation (Ryle et al., 1992) and biological nitrogen fixation under eCO_2 (Soussana & Hartwig, 1996), which is accomplished via their association with nitrogen fixing bacteria. While this increase in N availability can promote populations of shoot herbivores, especially aphids (Guo et al., 2013; Guo et al., 2014; Johnson et al., 2014b; Ryalls et al., 2015), it has also been shown to affect root herbivores (Johnson & McNicol, 2010). In particular, the clover root weevil (*Sitona lepidus*) were 38% more abundant under eCO_2, which the authors linked to the doubling in root nodules in white clover (*Trifolium repens*) under these conditions (Johnson & McNicol, 2010). The larvae of *S. lepidus* are particularly dependent on root nodules (Gerard, 2001) and it appears that the increased abundance of root nodules under eCO_2 promoted weevil populations. Warmer air temperatures, which are anticipated to increase in tandem with eCO_2, tend to have opposite effects on nodulation in legumes because higher temperatures are inhibitory to many N fixing bacteria (Zahran, 1999). In a study combining eCO_2 and air temperature, root nodulation in lucerne (*Medicago sativa*) increased and decreased, respectively, under these treatments. Emergence success of *S. lineatus* was positively correlated with these patterns of nodulation, so eCO_2 promoted emergence whereas elevated temperatures diminished it (Ryalls et al., 2013). Further studies, preferably those involving multiple environmental changes, are clearly needed before we can determine any patterns in root herbivore responses to atmospheric change.

11.3.2   Insect Root Herbivore Responses to Warming

The soil is buffered, to some extent, from fluctuating temperatures and so soil dwelling invertebrates might be affected less by predicted rises in air temperature than those living aboveground (Bale et al., 2002). However, many root herbivores are affected by soil temperature (see Barnett & Johnson, 2013 for a full account), and especially those feeding close to the soil surface. Leatherjacket larvae (Tipulidae) represent shallow feeders and reciprocal soil core transfer experiments using these insects showed that warmer temperatures increased rates of development such as time to pupation (Coulson et al., 1976). Using historical records, it was shown that UK spring (specifically May) temperatures were highly correlated with emergence of Tipulids and for every 1°C increase, Tipulids would emerge, on average, 7 days earlier (Pearce-Higgins et al., 2005). Tipulid emergence was predicted to advance by 12 days by 2100 (Pearce-Higgins et al., 2005), which could affect the numerous bird species that depend on them to feed their young (see Thomas et al., this volume). The cabbage root fly (*Delia radicum*) is another root herbivore to receive attention in respect to global warming; simulation models predict that an increase of 3°C in soil temperatures would advance and prolong emergence in Spring and create a larger third generation in late summer. A 5°C increase in soil temperature would even result in a fourth generation in southern UK (Collier et al., 1991).

In addition to these direct effects, rising temperatures are likely to impact root herbivores via plant-mediated mechanisms (Staley & Johnson, 2008; Barnett & Johnson, 2013). Plants respond in extraordinarily diverse ways to increased temperatures, which cannot be easily summarised or generalised. Extrapolating from the meta-analysis of Zvereva et al. (2006), which focuses on *foliar* responses to elevated temperature and eCO_2, we might anticipate a general decline in root quality for herbivores. When elevated temperature and CO_2 acted in tandem, elevated temperature negated the eCO_2 induced increase in leaf carbohydrates, but amplified decreases in leaf N, causing the leaf C:N ratio to increase overall (Zvereva et al., 2006). Decreases in root quality can trigger compensatory feeding in root herbivores (Johnson et al., 2014a), and given that warmer temperatures directly accelerate insect metabolism we could see sharp increases in rates of root herbivory and damage in the future.

11.3.3   Insect Root Herbivore Responses to Altered Precipitation

Altered patterns of precipitation are highly likely to affect root-feeding insects since soil moisture is amongst the most important abiotic factors affecting their physiology and behaviour (Barnett & Johnson, 2013; Erb & Lu, 2013). Numerous studies characterise how water availability affects root-feeding insects, and several reviews cover this in some depth (Brown & Gange, 1990; Villani & Wright, 1990; Barnett & Johnson, 2013; Erb et al., 2013). In this section, we therefore focus primarily on extended periods of drought, since climate change models predict this will occur more frequently in many parts of the world.

Dry conditions usually affect soil-dwelling herbivores adversely because they often possess permeable cuticles that make them prone to desiccation (Brown & Gange, 1990). Indeed, there are several examples of rainfall manipulation experiments that link reduced precipitation to population decline (see Staley & Johnson, 2008 for details). Despite this, many soil-dwelling insects have evolved physiological and behavioural strategies for coping with decreased water availability (Barnett & Johnson, 2013). Physiological adaptations include the production of metabolic water, a by-product of the oxidative catabolism of fats and carbohydrates (Wharton, 1985) and the formation of physical spiracle plastrons and hydrofugic hairs for water retention (Villani et al., 1999). Behavioural adaptations are probably more effective and at the simplest level include movement down the soil profile from dry regions to damper soils (Villani & Wright, 1990). In other cases, root-feeding scarabs manipulate the microclimate by constructing earthen chambers in which they encase themselves, allowing them to endure periods of drought (Villani et al., 1999). In contrast, seasonal declines in rainfall can benefit some root-feeding insects. African black beetle (*Heteronychus arator*) populations, for instance, are often suppressed by early summer rainfall (Matthiessen & Ridsdill-Smith, 1991), because first instar larvae cope badly with high soil moistures (King et al., 1981). In periods of seasonal drought, the larval populations are no longer suppressed by the normally high moisture content, resulting in damaging outbreaks (Matthiessen & Ridsdill-Smith, 1991).

In addition to the direct impacts of water availability on root-feeding insects, plant-mediated effects are likely to be very important (Barnett & Johnson, 2013; Erb & Lu, 2013). Generally speaking, periods of drought decrease growth and nutritional quality of the roots (Erb et al., 2013) and there is evidence for it increasing concentrations of root defensive compounds too (Zhang et al., 2008). It might reasonably be expected that many root herbivores will be adversely affected via such plant-mediated impacts. While periods of drought might adversely affect root herbivores via bottom-up (i.e., plant-mediated) drivers, they could benefit from less effective top-down (i.e., natural enemies) control. Preisser & Strong (2004), for instance, showed that lupine ghost moth (*Hepialus californicus*) larvae escaped control by entomopathogenic nematodes under dry conditions, which resulted in population outbreaks of the moth. Seasonal shifts in rainfall, or increased frequency of extreme events, could therefore cause temporal mismatches in such predator–prey cycles simply because nematodes are so sensitive to prevailing soil water conditions.

11.3.4 Soil-Dwelling Insects as Modifiers of Climate Change Effects

The plant-mediated effects of environmental change on invertebrates is well represented in the literature, but the effects of invertebrates on plant responses to environmental change, has received relatively little attention. Two groups of soil-dwelling insect, in particular, have the capacity to modify how plants respond to climatic and atmospheric change: herbivores and detritivores.

Root herbivores have a distinctive ecology that makes them particularly damaging to plants (Zvereva et al., 2012; Johnson et al., 2016a). Root herbivores can: (i) decrease nutrient and water uptake, (ii) cause disproportionate resource losses by severing roots, (iii) divert assimilates belowground for root re-growth and (iv) impair photosynthesis by imposing water deficits (Johnson & Murray, 2008; Zvereva et al., 2012). Plant biomass of C_3 plants is predicted to increase by 10–20% under eCO_2 and C_4 plants by 0–10% (Newman et al., 2011), yet this crucially depends on plants increasing rates of photosynthesis to accomplish this growth. While some foliar chewing herbivores stimulate rates of photosynthesis (Nykänen & Koricheva, 2004), presumably as the plant attempts compensatory growth to tolerate attack, root herbivory has the exact opposite effect causing significant declines of around 12% (Zvereva et al., 2012). So while any kind of herbivory has the potential to reduce the positive effects of eCO_2, it seems that soil-dwelling herbivores have the capacity to completely negate these effects. Indeed, this is exactly what happened when root herbivores attacked a brassica (*Cardamine pratensis*) (Salt et al., 1996) and seedling eucalypts (Johnson & Riegler, 2013b). In both cases, plant growth was restrained and plants grown under eCO_2 had similar biomass to those grown under aCO_2. Legumes possibly cope better and root herbivory did not diminish the fertilising impacts of eCO_2 (Johnson & McNicol, 2010; Ryalls et al., 2013).

In contrast, soil-dwelling detritivores such as dung beetles may mitigate the negative impacts of environmental change. In addition to consuming animal faeces, dung beetles disperse and incorporate it into the soil via burrowing activities. Most recently, dung beetles have been shown to mitigate the effects of simulated drought on plants by increasing soil permeability to water and reducing surface runoff (Johnson et al., 2016b). Considered as ecosystem engineers, those species of dung beetle that build networks of tunnels in the soil beneath the dung pad (the paracoprids) are most likely to be beneficial in this respect (Johnson et al., 2016), and field trials in Southern Australia showed that dung beetles improved pasture productivity for up to a decade due, in large part, to improved access to water (Doube & Marshall, 2014). Moreover, Doube & Marshall (2014) make a convincing argument that dung beetles promote carbon sequestration, so these invertebrates may reduce atmospheric change (and therefore its impact) directly, at least as much as other carbon sequestration measures (e.g., conservation tillage).

11.4 Effect of Climate Change on Earthworms: the Crawling Engineers of Soil

In 1881, Charles Darwin wrote: "It may be doubted whether there are many other animals which have played so important part of the history of the world, as these lowly organized creatures [earthworms]." (Darwin, 1881). This quotation embraces the importance of these ecosystem engineers in our contemporary natural and agro-ecosystems. Indeed, these ubiquitous soil invertebrates incessantly burrow galleries, mix organic and mineral matter, and produce casts. Earthworms are essential to the formation of humic matter, are key players in nutrient cycling, alter soil hydrology and drainage, or still affect plant population dynamic and community composition (i.e., dissemination of seeds) (Lal, 1991; Thompson et al., 1993; Jones et al., 1994). In addition to modifying soil structure (Shipitalo & Le Bayon, 2004; Amossé et al., 2015), earthworms impact biota composition and soil processes (Edwards & Bohlen, 1996; Eisenhauer, 2010; Blouin et al., 2013). Because they are a crucial component of soil fertility and promote agricultural productivity (van Groenigen et al., 2014), understanding the effect of climate change on earthworms will help to ensure food security. Yet, knowledge on the impact of climate change on earthworm communities is still very scarce (Fig. 11.3).

11.4.1 Earthworm Responses to eCO_2

Few studies report the effects of eCO_2 on earthworm density, biomass, or diversity. Milcu et al. (2011) observed an increased earthworm biomass of *Lumbricus terrestris* in microcosms with high plant diversity (eight species) under eCO_2, probably due to higher N rhizodeposition. Conversely, Chevallier et al. (2006) did not record any effect of eCO_2 on earthworm biomass (*Lumbricus rubellus* and *Aporrectodea caliginosa*) in grazed pastures in New Zealand. The effect of eCO_2 on earthworm fitness, especially on cocoon production, incubation and hatching, are neglected (Butt et al., 1992). Earthworms may even contribute modestly to global warming (Lubbers et al., 2013); while being widely beneficial, they also increase soil CO_2 emissions by 33% through aerobic respiration and N_2O emission by 37% via their intestinal bacteria, as compared to emissions from soil where they are absent (Lubbers et al., 2013).

Moving into the soil matrix, mixing, ingesting and excreting soil and organic matter particles (bio-turbation), earthworms create biogenic structures such as burrows and casts (Lavelle et al., 1997). These newly formed habitats are enriched in organic matter and nutrients and act as hotspots of biological activity (Fonte et al., 2007; Le Bayon & Milleret, 2009; Chapuis-Lardy et al., 2011). On a pasture

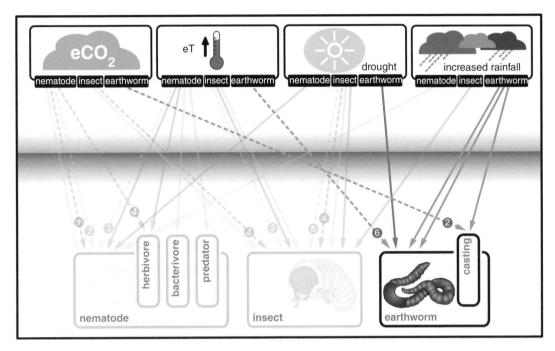

Figure 11.3 Direct and indirect effect of climate change on earthworms. The solid arrows indicate direct effects, while dashed arrows indicate indirect effects. Blue arrows represent shift in the community whereas green arrows indicate a positive impact and red arrows a negative impact on the community. See Figure 11.2 for nature of the indirect effects. Details and references in the text.

in Switzerland, annual production of earthworm surface-casts was 35% greater under eCO_2 (610 µl l^{-1}) due to a 10% increase in soil moisture (Arnone & Zaller, 1997). In a second study, the stimulatory effect of eCO_2 on cast production was not observed when plant species diversity decreased from 31 to 12 and five species (Arnone et al., 2013). Conversely, Chevallier et al. (2006) did not detect any direct effect of eCO_2 on earthworm (*L. rubellus* and *A. caliginosa*) casting in a grazed pasture in New Zealand. However, cast available N was lower under eCO_2 suggesting that N concentrations in casts reflect N cycle processes (Chevallier et al., 2006). This impact of earthworms on nutrient biogeochemical cycles in a context of climate change is reinforced in a review by Blouin et al. (2013) which highlighted the crucial role of earthworms in C sequestration both in their burrow linings and in casts.

The variable distribution of organic matter patchiness in soil determines how earthworms influence plant response to nutrient heterogeneity. Thus, under eCO_2, lower shoot and root biomass of *Lolium perenne* were found in heterogeneous *vs.* homogeneous treatments when epigeic earthworms (*Eisenia fetida*) were present (Garcia-Palacios et al., 2014). The authors suggested that earthworms foraging from patches of *L. perenne* shoots stimulate microbial N mobilization thus decreasing N availability for plants. Under eCO_2, earthworms may therefore mediate plant biomass responses to nutrient patchiness by affecting N capture (Garcia-Palacios et al., 2014).

As reviewed by Norby et al. (2001), eCO_2 influence on litter quality is not yet entirely clarified and several experimental artefacts render a general interpretation of the refereed literature difficult. Because of resorption of nutrients to perennial tissues during senescence, the decline in leaf litter N concentration under eCO_2 is lower than N concentration measured in green leaves of plants grown in the similar atmospheric conditions. This may suggest that indirect effect of eCO_2 would be greater on soil herbivores, consuming live tissues, than on decomposers, such as earthworms, consuming mainly senescent material.

11.4.2   Earthworm Responses to Warming and Altered Precipitation

Hackenberger & Hackenberger (2014) demonstrated that earthworm species composition, the ratio of ecological categories (epigeics, endogeics and anecics) and juvenile:adult ratio changed along a transect of varying climate types and elevations. This suggests that temperature influences the species composition of communities, possibly impacting the ecosystem services provided by these same communities. As earthworm populations are regulated in a density-dependent manner, inter- and intra-specific interactions may strongly affect species response and thereby structure and functioning of lumbricid communities (Uvarov, 2009) with further consequence on soil processes.

Zaller & Arnone (1999b) noticed that elevated rainfall increased earthworm density, and consequently casting activities. However, they did not record any impact on aboveground net primary production (ANPP) of graminoids. Conversely, the same authors further demonstrated that increasing earthworm activities stimulates shoot growth (Zaller & Arnone, 1999a) and enhances ANPP (Zaller et al., 2013). To unravel these conflicting results, the authors conducted a third experiment focussing on earthworms' impact on plant roots (Arnone & Zaller, 2014). At low earthworms density (37 individuals m^{-2}), plants produced more deep roots to compensate for lower nutrient availability in shallow soils resulting from reduced casting activities, whereas at high earthworm densities (114 and 169 individuals m^{-2}), large amounts of casts and greater nutrient availability compensated for physical disruption of roots by worms and induced a shift in the higher carbon allocation toward ANPP (Arnone & Zaller, 2014). These successive studies underlined the essential and integrative role of earthworms on plant communities' aboveground *vs.* belowground biomass allocation under climate change, possibly impacting other functional and trophic groups in the ecosystem in addition to agricultural yield. Earthworm bioturbation activity depends on both soil temperature (Whalen et al., 2004) and moisture (Edwards & Bohlen, 1996), and is therefore likely to influenced by climate change.

11.4.3   Climate Change Modification of Earthworm–Plant–Microbe Interactions

Because earthworms impact soil seed banks (Forey et al., 2011), mainly by dispersing and feeding on seeds (Eisenhauer et al., 2010), changes in their community composition or behaviour caused by climate change will probably affect plant communities. Indeed, earthworms play a role in seed transport and translocation into deep soil layers, acceleration or deceleration of seed germination and seedling establishment. Eisenhauer et al. (2010) found that the selective ingestion of seeds depends on several variables: earthworm body size, species specific habits and mode of digestion (e.g., depending on gut enzyme activities and gizzard contraction) (Eisenhauer et al., 2010). Plants that are highly reliant on earthworms for propagation will either benefit or suffer from climate change impact on these organisms, depending on the direction of the effect. eCO_2 affects the quality of plants (C:N ratio) (Korner, 2000), as well as starch storage, lignin concentrations (Coûteaux & Bolger, 2000), and in allocation of photosynthates within the plant (Young, 1998). This effect on plants in turn results in an increased consumption of litter by earthworms to compensate for N (Coûteaux et al., 1991). Root biomass either increases (Zaller et al., 2013) or remains stable (Eisenhauer et al., 2009) in the presence of earthworms, therefore, disentangling the effects of climate change in this particular context is challenging.

An increasing number of studies focus on the influence of climate change on earthworm and soil microbe interactions (Zirbes et al., 2012). Niklaus et al. (2003) showed that soil microbial communities are not affected by eCO_2 while Cesarz et al. (2015) highlight that the decomposer community may switch from a bacterial-dominated to a fungal-dominated system at eCO_2, indicating shifts in the microbial community as well as the functional structure of belowground food-webs. Earthworms have a dramatic impact on microorganism communities and therefore on the ecological services provided by soil microbes. For instance, *A. caliginosa* mucus enhanced the mineralization and humification of plant residues through the activation of microorganisms (Bityutskii et al., 2012). Earthworms and mycorrhizal fungi, especially arbuscular (AMF), are commonly co-occurring and interacting. These interactions depend on earthworm and plant species and are, generally speaking, beneficial for soil ecosystems as they result in increased soil aggregate stability (Kohler-Milleret et al., 2013), dispersal of non-pathogenic fungal spores and increased AMF colonization of plant roots by 140% (Zaller et al., 2013; Trouvé et al., 2014). Earthworms may also moderate plant performance by disrupting the interactions between plants and their AMF symbionts (Grabmaier et al., 2014). For instance, anecic earthworms (i.e., *L. terrestris*) reduced seedling emergence and diversity in the presence of AMF (*Glomus* sp.) indicating that their feeding reduced the effect of AMF on the seeds and seedlings (Zaller et al., 2011).

These interactions are species-specific and/or context dependent and general patterns are hard to draw but it has to be emphasised that climate change may dramatically modify these systems with unpredictable consequences at the ecosystem level.

11.4.4 Influence of Climate Change on Earthworms in Belowground Food Webs

In the context of belowground multitrophic interactions, several studies focus on tritrophic interactions in the presence of earthworms. Earthworms increase the incorporation of plant residues deep in the soil, which generally stimulate microbial activity but could also cause changes in the microbial communities that induce plant defences (Coûteaux et al., 1991; Wurst, 2010). However, earthworms with disparate life history traits may differ in their effects on aboveground herbivore performance, with impacts mediated by plant responses, while earthworms and root herbivores interact either directly, over the trophic web, or indirectly through the plants or changes in soil characteristics (Wurst, 2010). Regarding root-feeding nematodes, earthworms have a direct trophic effect (digestion) or indirect effects through the modification of soil properties (structure, water regime, nutrient cycling) (Yeates, 1981). Thus earthworms counteract the negative effects of root-feeding nematodes on plant performance (Blouin et al., 2005; Lohmann et al., 2009; Wurst, 2010). However, these effects depend on the community composition (Wurst, 2010) that inevitably may vary under climatic changes. There is increasing evidence that earthworms also play a role in the dispersion of beneficial entomopathogenic nematodes (Poinar & Thomas, 1975; Shapiro et al., 1993; Shapiro et al., 1995; Campos-Herrera et al., 2006; Shapiro-Ilan & Brown, 2013). Whether climate change will have an impact on such interactions (and the dependent trophic cascades) remains unknown but it is likely that any impact of climate on earthworm behaviour will cascade to upper and lower levels of the complex soil food webs.

11.4.5 Influence of Climate Change on Earthworm Colonization of New Habitats

Climate change induces migration of species, and earthworms may become invasive species with severe above- and belowground consequences on native ecosystems. As revealed by Uvarov (2009), in regions rich in native earthworm communities (i.e., Australia, New Zealand), the invasion of peregrine earthworms may largely impact the functioning of natural and agroecosystems (Baker et al., 2006). In areas devoid of earthworms, lumbricid invasions may cause substantial changes in ecosystem structure and functioning (Frelich et al., 2006) with important implications for climate change. Suárez et al. (2006) found that the invasions follow a successional sequence (epigeic then anecic and finally endogeic species) and that the activities of the previous group facilitated the habitat exploration by the following one. Changes in plant communities may also be induced by earthworm invasive species as vectors of seeds (Forey et al., 2011). In North America, Frelich & Reich (2010) highlighted the role of exotic earthworms as drivers of "savannification" of the forest through increasing soil bulk density, decreasing N availability and removing the organic layer, thus inhibiting the establishment of tree seedlings. In addition to such shifts in plant communities, the extended range expansion of macro-detritivores, including earthworms (*L. rubellus*) into subarctic environments, will probably result in a positive ecosystem feedback mechanism for climate change via increased CO_2 release in the atmosphere (van Geffen et al., 2011).

11.5 Conclusions and Future Perspectives

Climate change impacts on soil invertebrate communities are mostly idiosyncratic although some general patterns can be hypothesised (Sections 11.2-11.4 and Table 11.1). Because several effects of climate change on soil-dwelling invertebrate communities are indirect, often mediated by plants,

Table 11.1 Summary of the Impact of Climate Change on Nematodes, Insects and Earthworms.

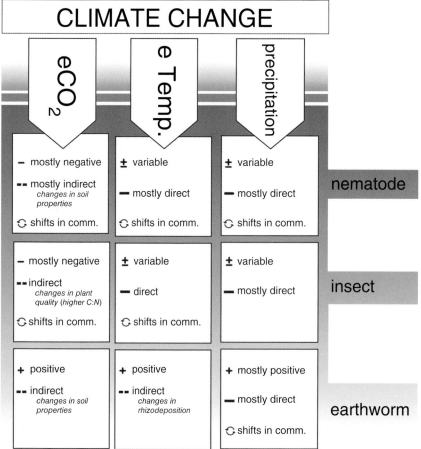

it is very difficult to encompass the consequences on soil invertebrate communities, and cascading effects on soil processes and ecological services. We have identified gaps in our current knowledge on the effect of climate change on soil invertebrate communities. Experimental time scale needs further attention given that most studies to date are relatively short-term. Such studies may be well equipped to elucidate the potential effects of extreme events such heavy rainfall, high temperatures or drought, but are unlikely to realistically reflect responses to longer-term impacts such as altered precipitation regimes, increased temperature or eCO_2. This is illustrated in studies that collect samples across years. For instance, Sonnemann & Wolters (2005) found that root hair feeding nematodes showed a strong response to moderate increases in atmospheric CO_2 after one year but this effect was not apparent in the second or third years, whereas predators were suppressed by eCO_2 in the first year and enhanced in the second. Consequently, different conclusions will often be made depending on the length of the study. There is a great need for studies that evaluate both short- and long-term responses to climate change drivers.

In addition, very few studies have experimentally investigated the effects of changes in precipitation, increased temperatures and eCO_2 simultaneously. This is not surprising given logistical constraints and substantial variability in climate change scenarios between regions and even between models. However, climate change drivers are unlikely to change in isolation and their interactions need to be considered to understand belowground community responses in detail. Moreover, the effect of climate change will be moderated by other global change drivers such as land use and N deposition further complicating the matter. On a positive note, there is evidence that an increasing number of studies are addressing this issue.

Chemical ecology plays a particularly important role in soil invertebrate trophic interactions since the soil environment precludes other sensory mechanisms (i.e., visual and acoustic cues, e.g., Rasmann et al., 2012b; Hiltpold et al., 2013); yet, to date this field of research is mostly neglected in the context of climate change. Insect root herbivores use plant secondary metabolites to identify and locate palatable root systems (e.g., Johnson & Gregory, 2006; Johnson & Nielsen, 2012b; Hiltpold et al., 2013). Plant parasitic nematodes also rely on root exudates to find hosts (Ali et al., 2011; Rasmann et al., 2012a) and entomopathogenic nematodes exploit volatiles emitted by insect damaged roots to locate their host (Boff et al., 2002; Rasmann et al., 2005; Ali et al., 2010; Hiltpold et al., 2010; Hiltpold et al., 2011; Rasmann et al., 2011; Laznik & Trdan, 2013). The impact of root exudates in soil food webs highly depends on the concentration of the released metabolites (e.g., Hiltpold et al., 2015) and changes in exudate quality and quantity at eCO_2 (e.g., Tarnawski & Aragno, 2006) will possibly severely impact invertebrate communities (e.g., Drigo et al., 2010). Changes in abiotic factors such as temperature, soil moisture, and rainfall patterns will also interfere with plant chemical signalling. Indeed, Hiltpold & Turlings (2008) showed a negative correlation between soil moisture and the diffusion of root volatiles in the ground. Despite this example, the impact of climate change on belowground chemical ecology has so far been largely overlooked.

Filling these gaps in our knowledge is likely to be the next boundary to expand and will provide us with a more holistic comprehension of the impact of climate change on belowground invertebrate communities and soil ecosystem services.

Acknowledgements

The authors are grateful to Philip Smith for proofreading this chapter.

References

A'Bear, A.D., Jones, T.H. & Boddy, L. (2014) Potential impacts of climate change on interactions among saprotrophic cord-forming fungal mycelia and grazing soil invertebrates. *Fungal Ecology*, **10**, 34–43.

Ainsworth, E.A., Rogers, A., Nelson, R. & Long, S.P. (2004) Testing the "source–sink" hypothesis of down-regulation of photosynthesis in elevated [CO_2] in the field with single gene substitutions in *Glycine max*. *Agricultural and Forest Meteorology*, **122**, 85–94.

Ali, J.G., Alborn, H.T. & Stelinski, L.L. (2010) Subterranean herbivore-induced volatiles released by citrus roots upon feeding by *Diaprepes abbreviatus* recruit entomopathogenic nematodes. *Journal of Chemical Ecology*, **36**, 361–368.

Ali, J.G., Alborn, H.T. & Stelinski, L.L. (2011) Constitutive and induced subterranean plant volatiles attract both entomopathogenic and plant parasitic nematodes. *Journal of Ecology*, **99**, 26–35.

Allen, M.F., Klironomos, J.N., Treseder, K.K. & Oechel, A.W.C. (2005) Responses of soil biota to elevated CO_2 in a chaparral ecosystem. *Ecological Applications*, **15**, 1701–1711.

Amossé, J., Le Bayon, R.-C. & Gobat, J.M. (2015) Are urban soils similar to natural soils of river valleys? *Journal of Soils and Sediments*, **15**, 1716–1724.

Arnone, J.A.I. & Zaller, J.G. (1997) Activity of surface-casting earthworms in a calcareous grassland under elevated atmospheric CO_2. *Oecologia*, **111**, 249–254.

Arnone, J.A.I. & Zaller, J.G. (2014) Earthworm effects on native grassland root system dynamics under natural and increased rainfall. *Frontiers in Plant Science*, **5**, 152.

Arnone, J.A.I., Zaller, J.G., Hofer, G., Schmid, B. & Körner, C. (2013) Loss of plant biodiversity eliminates stimulatory effect of elevated CO_2 on earthworm activity in grasslands. *Oecologia*, **171**, 613–622.

Ayres, E., Wall, D.H., Simmons, B.L., Field, C.B., Milchunas, D.G., Morgan, J.A. & Roy, J. (2008) Belowground nematode herbivores are resistant to elevated atmospheric CO_2 concentrations in grassland ecosystems. *Soil Biology and Biochemistry*, **40**, 978–985.

Baker, G.H., Brown, G., Butt, K., Curry, J.P. & Scullion, J. (2006) Introduced earthworms in agricultural and reclaimed land: their ecology and influences on soil properties, plant production and other soil biota. *Biological Invasions*, **8**, 1301–1316.

Bakonyi, G., Nagy, P., Kovács-Láng, E., Kovács, E., Barabás, S., Répási, V. & Seres, A. (2007) Soil nematode community structure as affected by temperature and moisture in a temperate semiarid shrubland. *Applied Soil Ecology*, **37**, 31–40.

Bale, J.S., Masters, G.J., Hodkinson, I.D., Awmack, C., Bezemer, T.M., Brown, V.K., Butterfield, J., Buse, A., Coulson, J.C., Farrar, J., Good, J.E.G., Harrington, R., Hartley, S., Jones, T.H., Lindroth, R.L., Press, M.C., Symrnioudis, I., Watt, A.D. & Whittaker, J.B. (2002) Herbivory in global climate change research: direct effects of rising temperature on insect herbivores. *Global Change Biology*, **8**, 1–16.

Bardgett, R.D., Denton, C.S. & Cook, R. (1999) Below-ground herbivory promotes soil nutrient transfer and root growth in grassland. *Ecology Letters*, **2**, 357–360.

Barnett, K. & Johnson, S.N. (2013) Living in the Soil Matrix: Abiotic Factors Affecting Root Herbivores. *Advances in Insect Physiology*, **45** (eds S.N. Johnson, I. Hiltpold & T.C.J. Turlings), pp. 1–52. Academic Press, Oxford, UK.

Bityutskii, N.P., Maiorov, E.I. & Orlova, N.E. (2012) The priming effects induced by earthworm mucus on mineralization and humification of plant residues. *European Journal of Soil Biology*, **50**, 1–6.

Blackshaw, R.P. & Kerry, B.R. (2008) Root herbivory in agricultural ecosystems. *Root Feeders – an ecosystem perspective* (eds S.N. Johnson & P.J. Murray), pp. 35–53. CABI, Wallingford, UK.

Blankinship, J.C., Niklaus, P.A. & Hungate, B.A. (2011) A meta-analysis of responses of soil biota to global change. *Oecologia*, **165**, 553–565.

Blouin, M., Hodson, M.E., Delgado, E.A., Baker, G., Brussaard, L., Butt, K.R., Dai, J., Dendooven, L., Peres, G., Tondoh, J.E., Cluzeau, D. & Brun, J.-J. (2013) A review of earthworm impact on soil function and ecosystem services. *European Journal of Soil Science*, **64**, 161–182.

Blouin, M., Zuily-Fodil, Y., Pham-Thi, A.-T., Laffray, D., Reversat, G., Pando, A., Tondoh, J. & Lavelle, P. (2005) Belowground organism activities affect plant aboveground phenotype, inducing plant tolerance to parasites. *Ecology Letters*, **8**, 202–208.

Boff, M.I.C., van Tol, R.H.W.M. & Smits, P.H. (2002) Behavioural response of *Heterorhabditis megidis* towards plant roots and insect larvae. *Biocontrol*, **47**, 67–83.

Brown, V.K. & Gange, A.C. (1990) Insect herbivory below ground. *Advances in Ecological Research*, **20**, 1–58.

Butt, K.R., Frederickson, J. & Morris, R.M. (1992) The intensive production of *Lumbricus terrestris* L. for soil amelioration. *Soil Biology & Biochemistry*, **24**, 1321–1325.

Campos-Herrera, R., Trigo, D. & Gutiérrez, C. (2006) Phoresy of the entomopathogenic nematode *Steinernema feltiae* by the earthworm *Eisenia fetida*. *Journal of Invertebrate Pathology*, **92**, 50–54.

Cesarz, S., Reich, P.B., Scheu, S., Ruess, L., Schaefer, M. & Eisenhauer, N. (2015) Nematode functional guilds, not trophic groups, reflect shifts in soil food webs and processes in response to interacting global change factors. *Pedobiologia*, **58**, 23–32.

Chapuis-Lardy, L., Le Bayon, R.-C., Brossard, M., López-Hernández, D. & Blanchart, E. (2011) Role of soil macrofauna in phosphorus cycling. *Phosphorus in Action*, **26** (eds E. Bünemann, A. Oberson & E. Frossard), pp. 199–213. Springer, Berlin Heidelberg.

Chevallier, A.J.T., Lieffering, M., Carran, R.A. & Newton, P.C.D. (2006) Mineral nitrogen cycling through earthworm casts in a grazed pasture under elevated atmospheric CO_2. *Global Change Biology*, **12**, 56–60.

Chikoski, J.M., Ferguson, S.H. & Meyer, L. (2006) Effects of water addition on soil arthropods and soil characteristics in a precipitation-limited environment. *Acta Oecologica*, **30**, 203–211.

Collier, R.H., Finch, S., Phelps, K. & Thompson, A.R. (1991) Possible impact of global warming on cabbage root fly (*Delia radicum*) activity in the UK. *Annals of Applied Biology*, **118**, 261–271.

Coulson, J.C., Horobin, J.C., Butterfield, J. & Smith, G.R.J. (1976) Maintenance of annual life-cycles in two species of *Tipulidae* (Diptera): field study relating development, temperature and altitude. *Journal of Animal Ecology*, **45**, 215–233.

Coûteaux, M.-M. & Bolger, T. (2000) Interactions between atmospheric CO_2 enrichment and soil fauna. *Plant and Soil*, **224**, 123–134.

Coûteaux, M.-M., Mousseau, M., Célérier, M.-L. & Bottner, P. (1991) Increased atmospheric CO_2 and litter quality: decomposition of sweet chestnut leaf litter with animal food webs of different complexities. *Oikos*, **61**, 54–64.

Cowan, I.R. & Farquhar, G.D. (1977) Stomatal function in relation to leaf metabolism and environment. *Symposia of the Society for Experimental Biology*, **31**, 471–505.

Curtis, P.S. & Wang, X. (1998) A meta-analysis of elevated CO_2 effects on woody plant mass, form, and physiology. *Oecologia*, **113**, 299–313.

Darby, B.J., Neher, D.A., Housman, D.C. & Belnap, J. (2011) Few apparent short-term effects of elevated soil temperature and increased frequency of summer precipitation on the abundance and taxonomic diversity of desert soil micro- and meso-fauna. *Soil Biology and Biochemistry*, **43**, 1474–1481.

Darwin, C.R. (1881) *The Formation of Vegetable Mould through the Action of Worms*. John Murray, London.

DeLucia, E.H., Nabity, P.D., Zavala, J.A. & Berenbaum, M.R. (2012) Climate change: resetting plant–insect interactions. *Plant Physiology*, **160**, 1677–1685.

Demarta, L., Hibbard, B.E., Bohn, M.O. & Hiltpold, I. (2014) The role of root architecture in foraging behavior of entomopathogenic nematodes. *Journal of Invertebrate Pathology*, **122**, 32–39.

Dermody, O., Weltzin, J.F., Engel, E.C., Allen, P. & Norby, R.J. (2007) How do elevated [CO_2], warming, and reduced precipitation interact to affect soil moisture and LAI in an old field ecosystem? *Plant and Soil*, **301**, 255–266.

Dong, Z., Hou, R., Chen, Q., Ouyang, Z. & Ge, F. (2013) Response of soil nematodes to elevated temperature in conventional and no-tillage cropland systems. *Plant and Soil*, **373**, 907–918.

Doube, B. & Marshall, T. (2014) *Dung down under: dung beetles for Australia*. Dung Beetle Solutions Australia, Bridgewater, South Australia, Australia.

Drigo, B., Pijl, A.S., Duyts, H., Kielak, A.M., Gamper, H.A., Houtekamer, M.J., Boschker, H.T.S., Bodelier, P.L.E., Whiteley, A.S., Van Veen, J.A. & Kowalchuk, G.A. (2010) Shifting carbon flow from roots into associated microbial communities in response to elevated atmospheric CO_2. *Proceedings of the National Academy of Sciences of the United States of America*, **107**, 10938–10942.

Edwards, C.A. & Bohlen, P.J. (1996) *The Biology and Ecology of Earthworms*. Springer, London.

Eisenhauer, N. (2010) The action of an animal ecosystem engineer: Identification of the main mechanisms of earthworm impacts on soil microarthropods. *Pedobiologia*, **53**, 343–352.

Eisenhauer, N., Butenschoen, O., Radsick, S. & Scheu, S. (2010) Earthworms as seedling predators: Importance of seeds and seedlings for earthworm nutrition. *Soil Biology & Biochemistry*, **42**, 1245–1252.

Eisenhauer, N., Dobies, T., Cesarz, S., Hobbie, S.E., Meyer, R.J., Worm, K. & Reich, P.B. (2013) Plant diversity effects on soil food webs are stronger than those of elevated CO_2 and N deposition in a long-term grassland experiment. *Proceedings of the National Academy of Sciences of the United States of America*, **110**, 6889–6894.

Eisenhauer, N., Schuy, M., Butenschoen, O. & Scheu, S. (2009) Direct and indirect effects of endogeic earthworms on plant seeds. *Pedobiologia*, **52**, 151–162.

Ennis, D.E., Dillon, A.B. & Griffin, C.T. (2010) Simulated roots and host feeding enhance infection of subterranean insects by the entomopathogenic nematode *Steinernema carpocapsae*. *Journal of Invertebrate Pathology*, **103**, 140–143.

Erb, M., Huber, M., Robert, C.A.M., Ferrieri, A.P., Machado, R.A.R. & Arce, C.C.M. (2013) The role of plant primary and secondary metabolites in root-herbivore behaviour, nutrition and physiology. *Advances in Insect Physiology*, **45** (eds S.N. Johnson, I. Hiltpold & T.C.J. Turlings), pp. 53–95. Academic Press, Oxford, UK.

Erb, M. & Lu, J. (2013) Soil abiotic factors influence interactions between belowground herbivores and plant roots. *Journal of Experimental Botany*, **64**, 1295–1303.

FAO (2015) Nothing dirty here: FAO kicks off International Year of Soils 2015. Food and Agriculture Organization of the United Nations, Rome. Retrieved from http://www.fao.org/news/story/en/item/270812/icode/

Fierer, N., Strickland, M.S., Liptzin, D., Bradford, M.A. & Cleveland, C.C. (2009) Global patterns in belowground communities. *Ecology Letters*, **12**, 1238–1249.

Fonte, S.J., Kong, A.Y.Y., van Kessel, C., Hendrix, P.F. & Six, J. (2007) Influence of earthworm activity on aggregate-associated carbon and nitrogen dynamics differs with agroecosystem management. *Soil Biology & Biochemistry*, **39**, 1014–1022.

Forey, E., Barot, S., Decaëns, T., Langlois, E., Laossi, K.-R., Margerie, P., Scheu, S. & Eisenhauer, N. (2011) Importance of earthworm–seed interactions for the composition and structure of plant communities: A review. *Acta Oecologica-International Journal of Ecology*, **37**, 594–603.

Freckman, D.W., Whitford, W.G. & Steinberger, Y. (1987) Effect of irrigation on nematode population dynamics and activity in desert soils. *Biology and Fertility of Soils*, **3**, 3–10.

Frederiksen, H.B., Rønn, R. & Christensen, S. (2001) Effect of elevated atmospheric CO_2 and vegetation type on microbiota associated with decomposing straw. *Global Change Biology*, 7, 313–321.

Frelich, L.E., Hale, C.M., Scheu, S., Holdsworth, A.R., Heneghan, L., Bohlen, P.J. & Reich, P.B. (2006) Earthworm invasion into previously earthworm-free temperate and boreal forests. *Biological Invasions*, **8**, 1235–1245.

Frelich, L.E. & Reich, P.B. (2010) Will environmental changes reinforce the impact of global warming on the prairie–forest border of central North America? *Frontiers in Ecology and the Environment*, **8**, 371–378.

Garcia-Palacios, P., Maestre, F.T., Bradford, M.A. & Reynolds, J.F. (2014) Earthworms modify plant biomass and nitrogen capture under conditions of soil nutrient heterogeneity and elevated atmospheric CO_2 concentrations. *Soil Biology & Biochemistry*, **78**, 182–188.

García-Palacios, P., Vandegehuchte, M.L., Shaw, E.A., Dam, M., Post, K.H., Ramirez, K.S., Sylvain, Z.A., de Tomasel, C.M. & Wall, D.H. (2015) Are there links between responses of soil microbes and ecosystem functioning to elevated CO_2, N deposition and warming? A global perspective. *Global Change Biology*, **21**, 1590–1600.

Gerard, P.J. (2001) Dependence of *Sitona lepidus* (Coleoptera: Curculionidae) larvae on abundance of white clover *Rhizobium* nodules. *Bulletin of Entomological Research*, **91**, 149–152.

Grabmaier, A., Heigl, F., Eisenhauer, N., van der Heijden, M.G.A. & Zaller, J.G. (2014) Stable isotope labelling of earthworms can help deciphering belowground–aboveground interactions involving earthworms, mycorrhizal fungi, plants and aphids. *Pedobiologia*, **57**, 197–203.

Gregory, P.J. (2006) *Plant Roots – Growth, Activity and Interaction with Soils*. Blackwell Publishing, Oxford, UK.

Guo, H., Sun, Y.C., Li, Y., Liu, X., Zhang, W. & Ge, F. (2014) Elevated CO_2 decreases the response of the ethylene signaling pathway in *Medicago truncatula* and increases the abundance of the pea aphid. *New Phytologist*, **201**, 279–291.

Guo, H., Sun, Y.C., Li, Y., Tong, B., Harris, M., Zhu-Salzman, K. & Ge, F. (2013) Pea aphid promotes amino acid metabolism both in *Medicago truncatula* and bacteriocytes to favor aphid population growth under elevated CO_2. *Global Change Biology*, **19**, 3210–3223.

Hackenberger, D.K. & Hackenberger, B.K. (2014) Earthworm community structure in grassland habitats differentiated by climate type during two consecutive seasons. *European Journal of Soil Biology*, **61**, 27–34.

Hiltpold, I., Bernklau, E., Bjostad, L.B., Alvarez, N., Miller-Struttmann, N.E., Lundgren, J.G. & Hibbard, B.E. (2013) Nature, evolution and characterisation of rhizospheric chemical exudates affecting root herbivores. *Advances in Insect Physiology*, **45** (eds S.N. Johnson, I. Hiltpold & T.C.J. Turlings), pp. 97–157. Academic Press, Oxford, UK.

Hiltpold, I., Erb, M., Robert, C.A.M. & Turlings, T.C.J. (2011) Systemic root signalling in a belowground, volatile-mediated tritrophic interaction. *Plant, Cell and Environment*, **34**, 1267–1275.

Hiltpold, I., Jaffuel, G. & Turlings, T.C.J. (2015) The dual effects of root-cap exudates on nematodes: from quiescence in plant-parasitic nematodes to frenzy in entomopathogenic nematodes. *Journal of Experimental Botany*, **66**, 603–611.

Hiltpold, I., Toepfer, S., Kuhlmann, U. & Turlings, T.C.J. (2010) How maize root volatiles influence the efficacy of entomopathogenic nematodes in controlling the western corn rootworm? *Chemoecology*, **20**, 155–162.

Hiltpold, I. & Turlings, T.C.J. (2008) Belowground chemical signalling in maize: when simplicity rhymes with efficiency. *Journal of Chemical Ecology*, **34**, 628–635.

Hoeksema, J.D., Lussenhop, J. & Teeri, J.A. (2000) Soil nematodes indicate food web responses to elevated atmospheric CO_2. *Pedobiologia*, **44**, 725–735.

Hunt, H.W. & Wall, D.H. (2002) Modelling the effects of loss of soil biodiversity on ecosystem function. *Global Change Biology*, **8**, 33–50.

IPCC (2013a) *Climate Change 2013: The Physical Science Basis. Contribution of Working Group I to the Fifth Assessment Report of the Intergovernmental Panel on Climate Change.* Cambridge University Press, Cambridge, United Kingdom and New York, NY, USA.

IPCC (2013b) Summary for Policymakers. *Climate change 2013: The Physical Science Basis. Summary for policymakers. Contribution of Working Group I to the Fifth Assessment Report of the Intergovernmental Panel on Climate Change* (eds T.F. Stocker, D. Qin, G.-K. Plattner, M.M.B. Tignor, S.K. Allen, A. Nauels, Y. Xia, V. Bex & P.M. Midgley). Cambridge University Press, Cambridge, United Kingdom and New York, NY, USA.

Johnson, S.N., Barton, A.T., Clark, K.E., Gregory, P.J., McMenemy, L.S. & Hancock, R.D. (2011) Elevated atmospheric carbon dioxide impairs the performance of root-feeding vine weevils by modifying root growth and secondary metabolites. *Global Change Biology*, **17**, 688–695.

Johnson, S.N., Clark, K.E., Hartley, S.E., Jones, T.H., McKenzie, S.W. & Koricheva, J. (2012a) Aboveground-belowground herbivore interactions: A meta-analysis. *Ecology*, **93**, 2208–2215.

Johnson, S.N., Erb, M. & Hartley, S.E. (2016a) Roots under attack: contrasting plant responses to below- and aboveground insect herbivory. *New Phytologist*, **210**, 413–418.

Johnson, S.N. & Gregory, P.J. (2006) Chemically-mediated host-plant location and selection by root-feeding insects. *Physiological Entomology*, **31**, 1–13.

Johnson, S.N., Lopaticki, G., Barnett, K., Facey, S.L., Powell, J.R. & Hartley, S.E. (2016b) An insect ecosystem engineer alleviates drought stress in plants without increasing plant susceptibility to an aboveground herbivore. *Functional Ecology*, **30**, 894–902.

Johnson, S.N., Lopaticki, G. & Hartley, S.E. (2014a) Elevated atmospheric CO_2 triggers compensatory feeding by root herbivores on a C_3 but not a C_4 grass. *PlOS ONE*, **9**, e90251.

Johnson, S.N. & McNicol, J.W. (2010) Elevated CO_2 and aboveground–belowground herbivory by the clover root weevil. *Oecologia*, **162**, 209–216.

Johnson, S.N., Mitchell, C., McNicol, J.W., Thompson, J. & Karley, A.J. (2013a) Downstairs drivers - root herbivores shape communities of above-ground herbivores and natural enemies via changes in plant nutrients. *Journal of Animal Ecology*, **82**, 1021–1030.

Johnson, S.N. & Murray, P.J. (2008) *Root Feeders – An Ecosystem Perspective.* CABI Publishing, Wallingford, UK.

Johnson, S.N. & Nielsen, U.N. (2012b) Foraging in the dark – chemically mediated host plant location by belowground insect herbivores. *Journal of Chemical Ecology*, **38**, 604–614.

Johnson, S.N. & Rasmann, S. (2015) Root-feeding insects and their interactions with organisms in the rhizosphere. *Annual Review of Entomology*, **60**, 517–535.

Johnson, S.N. & Riegler, M. (2013b) Root damage by insects reverses the effects of elevated atmospheric CO_2 on eucalypt seedlings. *PLOS ONE*, **8**, e79479.

Johnson, S.N., Ryalls, J.M.W. & Karley, A.J. (2014b) Global climate change and crop resistance to aphids: contrasting responses of lucerne genotypes to elevated atmospheric carbon dioxide. *Annals of Applied Biology*, **165**, 62–72.

Jones, C.G., Lawton, J.H. & Shachak, M. (1994) Organisms as ecosystem engineers. *Oikos*, **69**, 373–386.

Karban, R. (1980) Periodical cicada nymphs impose periodical oak tree wood accumulation. *Nature*, **287**, 326–327.

Kardol, P., Cregger, M.A., Campany, C.E. & Classen, A.T. (2010) Soil ecosystem functioning under climate change: Plant species and community effects. *Ecology*, **91**, 767–781.

Kardol, P., Reynolds, W.N., Norby, R.J. & Classen, A.T. (2011) Climate change effects on soil microarthropod abundance and community structure. *Applied Soil Ecology*, **47**, 37–44.

King, P.D., Mercer, C.F. & Meekings, J.S. (1981) Ecology of the black beetle, *Heteronychus arator* (Coleoptera, Scarabaeidae) – Influence of temperature on feeding, growth, and survival of the larvae. *New Zealand Journal of Zoology*, **8**, 113–117.

Klironomos, J.N., Rillig, M.C. & Allen, M.F. (1996) Below-ground microbial and microfaunal responses to *Artemisia tridentata* grown under elevated atmospheric CO_2. *Functional Ecology*, **10**, 527–534.

Kohler-Milleret, R., Le Bayon, R.C., Chenu, C., Gobat, J.-M. & Boivin, P. (2013) Impact of two root systems, earthworms and mycorrhizae on the physical properties of an unstable silt loam Luvisol and plant production. *Plant and Soil*, **370**, 251–265.

Korner, C. (2000) Biosphere responses to CO_2 enrichment. *Ecological Applications*, **10**, 1590–1619.

Lal, R. (1991) Soil conservation and biodiversity. *The Biodiversity of Microorganisms and Invertebrates: its role in Sustainable Agriculture* (ed. D.L. Hawksworth), pp. 89–103. CAB International, Wallingford.

Landesman, W.J., Treonis, A.M. & Dighton, J. (2011) Effects of a one-year rainfall manipulation on soil nematode abundances and community composition. *Pedobiologia*, **54**, 87–91.

Lang, B., Rall, B.C., Scheu, S. & Brose, U. (2014) Effects of environmental warming and drought on size-structured soil food webs. *Oikos*, **123**, 1224–1233.

Lavelle, P., Bignell, D., Lepage, M., Wolters, V., Roger, P., Ineson, P., Heal, O.W. & Dhillion, S. (1997) Soil function in a changing world: The role of invertebrate ecosystem engineers. *European Journal of Soil Biology*, **33**, 159–193.

Laznik, Ž. & Trdan, S. (2013) An investigation on the chemotactic responses of different entomopathogenic nematode strains to mechanically damaged maize root volatile compounds. *Experimental Parasitology*, **134**, 349–355.

Le Bayon, R.-C. & Milleret, R. (2009) Effects of earthworms on phosphorus dynamics – a review. *Dynamic Plant, Dynamic Soil*, **3**, 21–27.

Lindeman, R.L. (1942) The trophic-dynamic aspect of ecology. *Ecology*, **23**, 399–417.

Lohmann, M., Scheu, S. & Müller, C. (2009) Decomposers and root feeders interactively affect plant defence in *Sinapis alba*. *Oecologia*, **160**, 289–298.

Lubbers, I.M., van Groenigen, K.J., Fonte, S.J., Six, J., Brussaard, L. & van Groenigen, J.W. (2013) Greenhouse-gas emissions from soils increased by earthworms. *Nature Climate Change*, **3**, 187–194.

Luo, Y.Q., Hui, D.F. & Zhang, D.Q. (2006) Elevated CO_2 stimulates net accumulations of carbon and nitrogen in land ecosystems: A meta-analysis. *Ecology*, **87**, 53–63.

Madhu, M. & Hatfeld, J.L. (2013) Dynamics of plant root growth under increased atmospheric carbon dioxide. *Agronomy Journal*, **105**, 657–669.

Matthiessen, J.N. & Ridsdill-Smith, T.J. (1991) Populations of African Black Beetle, *Heteronychus arator* (Coleoptera, Scarabaeidae) in a Mediterranean climate region of Australia. *Bulletin of Entomological Research*, **81**, 85–91.

Mattson, W.J.J. (1980) Herbivory in relation to plant nitrogen content. *Annual Review of Ecology and Systematics*, **11**, 119–161.

McKenzie, S.W., Hentley, W.T., Hails, R.S., Jones, T.H., Vanbergen, A.J. & Johnson, S.N. (2013) Global climate change and above–belowground insect herbivore interactions. *Frontiers in Plant Science*, **4**, 412.

Milchunas, D.G., Mosier, A.R., Morgan, J.A., LeCain, D.R., King, J.Y. & Nelson, J.A. (2005) Root production and tissue quality in a shortgrass steppe exposed to elevated CO_2: Using a new ingrowth method. *Plant and Soil*, **268**, 111–122.

Milcu, A., Paul, S. & Lukac, M. (2011) Belowground interactive effects of elevated CO_2, plant diversity and earthworms in grassland microcosms. *Basic and Applied Ecology*, **12**, 600–608.

Newman, J.A., Anand, M., Henry, H.A.L., Hunt, S. & Gedalof, Z. (2011) *Climate Change Biology*. CABI, Walingford, UK.

Nielsen, O. & Philipsen, H. (2004) Seasonal population dynamics of inoculated and indigenous steinernematid nematodes in an organic cropping system. *Nematology*, **6**, 901–909.

Nielsen, U.N., Ayres, E., Wall, D.H., Li, G., Bardgett, R.D., Wu, T. & Garey, J.R. (2014) Global-scale patterns of assemblage structure of soil nematodes in relation to climate and ecosystem properties. *Global Ecology and Biogeography*, **23**, 968–978.

Nielsen, U.N. & Ball, B.A. (2015) Impacts of altered precipitation regimes on soil communities and biogeochemistry in arid and semi-arid ecosystems. *Global Change Biology*, **21**, 1407–1421.

Nielsen, U.N., Wall, D.H. & Six, J. (2015) Soil biodiversity and the environment. *Annual Review of Environment and Resources*, **40**, 63–90.

Niklaus, P.A., Alphei, J., Ebersberger, D., Kampichler, C., Kandeler, E. & Tscherko, D. (2003) Six years of in situ CO_2 enrichment evoke changes in soil structure and soil biota of nutrient-poor grassland. *Global Change Biology*, **9**, 585–600.

Norby, R.J., Cotrufo, M.F., Ineson, P., O'Neill, E.G. & Canadell, J.G. (2001) Elevated CO_2, litter chemistry, and decomposition: A synthesis. *Oecologia*, **127**, 153–165.

Norby, R.J. & Luo, Y. (2004) Evaluating ecosystem responses to rising atmospheric CO_2 and global warming in a multi-factor world. *New Phytologist*, **162**, 281–293.

Nykänen, H. & Koricheva, J. (2004) Damage-induced changes in woody plants and their effects on insect herbivore performance: a meta-analysis. *Oikos*, **104**, 247–268.

Ode, P.J., Johnson, S.N. & Moore, B.D. (2014) Atmospheric change and induced plant secondary metabolites — are we reshaping the building blocks of multi-trophic interactions? *Current Opinion in Insect Science*, **5**, 57–65.

Pacholski, A., Manderscheid, R. & Weigel, H.J. (2015) Effects of free air CO_2 enrichment on root growth of barley, sugar beet and wheat grown in a rotation under different nitrogen supply. *European Journal of Agronomy*, **63**, 36–46.

Payne, D. & Gregory, P.J. (1988) The soil atmosphere. *Russell's Soil Conditions and Plant Growth* (ed. A. Wild), pp. 298–314. Longman, Harlow, UK.

Pearce-Higgins, J.W., Yalden, D.W. & Whittingham, M.J. (2005) Warmer springs advance the breeding phenology of golden plovers *Pluvialis apricaria* and their prey (Tipulidae). *Oecologia*, **143**, 470–476.

Pendall, E., Bridgham, S., Hanson, P.J., Hungate, B., Kicklighter, D.W., Johnson, D.W., Law, B.E., Luo, Y., Megonigal, J.P., Olsrud, M., Ryan, M.G. & Wan, S. (2004a) Below-ground process responses to elevated CO_2 and temperature: A discussion of observations, measurement methods, and models. *New Phytologist*, **162**, 311–322.

Pendall, E., Mosier, A.R. & Morgan, J.A. (2004b) Rhizodeposition stimulated by elevated CO_2 in a semiarid grassland. *New Phytologist*, **162**, 447–458.

Poinar, G.O. & Thomas, G.M. (1975) *Rhabditis pellio* Schneider (Nematoda) from earthworm, *Aporrectodea trapezoides* Duges (Annelida). *Journal of Nematology*, **7**, 374–379.

Preisser, E.L. & Strong, D.R. (2004) Climate affects predator control of an herbivore outbreak. *The American Naturalist*, **163**, 754–762.

Prior, S.A., Runion, G.B., Torbert, H.A., Idso, S.B. & Kimball, B.A. (2012) Sour orange fine root distribution after seventeen years of atmospheric CO_2 enrichment. *Agricultural and Forest Meteorology*, **162-163**, 85–90.

Rasmann, S., Ali, J.G., Helder, J. & van der Putten, W.H. (2012a) Ecology and evolution of soil nematode chemotaxis. *Journal of Chemical Ecology*, **38**, 615–628.

Rasmann, S., Erwin, A.C., Halitschke, R. & Agrawal, A.A. (2011) Direct and indirect root defences of milkweed (*Asclepias syriaca*): trophic cascades, trade-offs and novel methods for studying subterranean herbivory. *Journal of Ecology*, **99**, 16–25.

Rasmann, S., Hiltpold, I. & Ali, J. (2012b) The role of root-produced volatile secondary metabolites in mediating soil interactions. *Advances in Selected Plant Physiology Aspects* (eds G. Montanaro & B. Dichio), pp. 269–290. InTech, Rijeka, Croatia.

Rasmann, S., Köllner, T.G., Degenhardt, J., Hiltpold, I., Toepfer, S., Kuhlmann, U., Gershenzon, J. & Turlings, T.C.J. (2005) Recruitment of entomopathogenic nematodes by insect-damaged maize roots. *Nature*, **434**, 732–737.

Robinson, A.F. & Jaffee, B.A. (1996) Repulsion of *Meloidogyne incognita* by alginate pellets containing hyphae of *Monacrosporium cionopagum*, *M. ellipsosporum*, or *Hirsutella rhossiliensis*. *Journal of Nematology*, **28**, 133–147.

Robinson, E.A., Ryan, G.D. & Newman, J.A. (2012) A meta-analytical review of the effects of elevated CO_2 on plant–arthropod interactions highlights the importance of interacting environmental and biological variables. *New Phytologist*, **194**, 321–336.

Rogers, H.H., Prior, S.A., Runion, G.B. & Mitchell, R.J. (1996) Root to shoot ratio of crops as influenced by CO_2. *Plant and Soil*, **187**, 229–248.

Rogers, H.H., Runion, G.B. & Krupa, S.V. (1994) Plant responses to atmospheric CO_2 enrichment with emphasis on roots and the rhizosphere. *Environmental Pollution*, **83**, 155–189.

Ruess, L., Michelsen, A. & Jonasson, S. (1999) Simulated climate change in subarctic soils: responses in nematode species composition and dominance structure. *Nematology*, **1**, 513–526.

Ryalls, J.M.W., Moore, B.D., Riegler, M., Gherlenda, A.N. & Johnson, S.N. (2015) Amino acid-mediated impacts of elevated carbon dioxide and simulated root herbivory on aphids are neutralised by increased air temperatures. *Journal of Experimental Botany*, **66**, 613–623.

Ryalls, J.M.W., Riegler, M., Moore, B.D., Lopaticki, G. & Johnson, S.N. (2013) Effects of elevated temperature and CO_2 on aboveground-belowground systems: a case study with plants, their mutualistic bacteria and root/shoot herbivores. *Frontiers in Plant Science*, **4**, 445.

Ryle, G.J.A., Powell, C.E. & Davidson, I.A. (1992) Growth of white clover, dependent on N_2 fixation, in elevated CO_2 and temperature. *Annals of Botany*, **70**, 221–228.

Salt, D.T., Fenwick, P. & Whittaker, J.B. (1996) Interspecific herbivore interactions in a high CO_2 environment: root and shoot aphids feeding on *Cardamine*. *Oikos*, **77**, 326–330.

Scheu, S. (2002) The soil food web: Structure and perspectives. *European Journal of Soil Biology*, **38**, 11–20.

Shapiro, D.I., Berry, E.C. & Lewis, L.C. (1993) Interactions between nematodes and earthworms – Enhanced dispersal of *Steinernema carpocapsae*. *Journal of Nematology*, **25**, 189–192.

Shapiro, D.I., Tylka, G.L., Berry, E.C. & Lewis, L.C. (1995) Effect of earthworms on the dispersal of *Steinernema* spp. *Journal of Nematology*, **27**, 21–28.

Shapiro-Ilan, D.I. & Brown, I. (2013) Earthworms as phoretic hosts for *Steinernema carpocapsae* and *Beauveria bassiana*: Implications for enhanced biological control. *Biological Control*, **66**, 41–48.

Shipitalo, M. & Le Bayon, R.C. (2004) Quantifying the effects of earthworms on soil aggregation and porosity. *Earthworm Ecology* (ed. C.A. Edward), pp. 183–200. CRC Press LLC, Boca Raton.

Smolik, J.D. & Dodd, J.L. (1983) Effect of Water and Nitrogen, and Grazing on Nematodes in a Shortgrass Prairie. *Journal of Range Management*, **36**, 744–748.

Soler, R., van der Putten, W.H., Harvey, J.A., Vet, L.E.M., Dicke, M. & Bezemer, T.M. (2012) Root herbivore effects on aboveground multitrophic interactions: patterns, processes and mechanisms. *Journal of Chemical Ecology*, **38**, 755–767.

Sonnemann, I. & Wolters, V. (2005) The microfood web of grassland soils responds to a moderate increase in atmospheric CO_2. *Global Change Biology*, **11**, 1148–1155.

Soussana, J.F. & Hartwig, U.A. (1996) The effects of elevated CO_2 on symbiotic N_2 fixation: a link between the carbon and nitrogen cycles in grassland ecosystems. *Plant and Soil*, **187**, 321–332.

Staley, J.T. & Johnson, S.N. (2008) Climate change impacts on root herbivores. *Root Feeders – An Ecosystem Perspective* (eds S.N. Johnson & P.J. Murray), pp. 192–213. CABI Publishing, Wallingford, UK.

Stevnbak, K., Maraldo, K., Georgieva, S., Bjørnlund, L., Beier, C., Schmidt, I.K. & Christensen, S. (2012) Suppression of soil decomposers and promotion of long-lived, root herbivorous nematodes by climate change. *European Journal of Soil Biology*, **52**, 1–7.

Stiling, P. & Cornelissen, T. (2007) How does elevated carbon dioxide (CO_2) affect plant–herbivore interactions? A field experiment and meta-analysis of CO_2-mediated changes on plant chemistry and herbivore performance. *Global Change Biology*, **13**, 1823–1842.

Suárez, E.R., Fahey, T.J., Groffman, P.M., Yavitt, J.B. & Bohlen, P.J. (2006) Spatial and temporal dynamics of exotic earthworm communities along invasion fronts in a temperate hardwood forest in south-central New York (USA). *Biological Invasions*, **8**, 553–564.

Sun, X., Zhang, X., Zhang, S., Dai, G., Han, S. & Liang, W. (2013) Soil nematode responses to increases in nitrogen deposition and precipitation in a temperate forest. *PLOS ONE*, **8**, e82468.

Sylvain, Z.A., Wall, D.H., Cherwin, K.L., Peters, D.P.C., Reichmann, L.G. & Sala, O.E. (2014) Soil animal responses to moisture availability are largely scale, not ecosystem dependent: Insight from a cross-site study. *Global Change Biology*, **20**, 2631–2643.

Tarnawski, S. & Aragno, M. (2006) The influence of elevated [CO_2] on diversity, activity and biogeochemical functions of rhizosphere and soil bacterial communities. *Managed Ecosystems and CO_2: Case Studies, Processes and Perspectives* **187** (eds J. Nösberger, S.P. Long, R.J. Norby, M. Stitt, G.R. Hendrey & H. Blum), pp. 393–412. Springer, Berlin Heidelberg.

Thakur, M.P., Reich, P.B., Fisichelli, N.A., Stefanski, A., Cesarz, S., Dobies, T., Rich, R.L., Hobbie, S.E. & Eisenhauer, N. (2014) Nematode community shifts in response to experimental warming and canopy conditions are associated with plant community changes in the temperate-boreal forest ecotone. *Oecologia*, **175**, 713–723.

Thompson, L., Thomas, C.D., Radley, J.M.A., Williamson, S. & Lawton, J.H. (1993) The effect of earthworms and snails in a simple plant community. *Oecologia*, **95**, 171–178.

Trouvé, R., Drapela, T., Frank, T., Hadacek, F. & Zaller, J.G. (2014) Herbivory of an invasive slug in a model grassland community can be affected by earthworms and mycorrhizal fungi. *Biology and Fertility of Soils*, **50**, 13–23.

Uvarov, A.V. (2009) Inter- and intraspecific interactions in lumbricid earthworms: Their role for earthworm performance and ecosystem functioning. *Pedobiologia*, **53**, 1–27.

van Geffen, K.G., Berg, M.P. & Aerts, R. (2011) Potential macro-detritivore range expansion into the subarctic stimulates litter decomposition: a new positive feedback mechanism to climate change? *Oecologia*, **167**, 1163–1175.

van Groenigen, J.W., Lubbers, I.M., Vos, H.M.J., Brown, G.G., De Deyn, G.B. & van Groenigen, K.J. (2014) Earthworms increase plant production: a meta-analysis. *Scientific Reports*, **4**, 6365.

Vandegehuchte, M.L., Sylvain, Z.A., Reichmann, L.G., De Tomasel, C.M., Nielsen, U.N., Wall, D.H. & Sala, O.E. (2015) Responses of a desert nematode community to changes in water availability. *Ecosphere*, **6**, 44.

Villani, M.G., Allee, L.L., Diaz, A. & Robbins, P.S. (1999) Adaptive strategies of edaphic arthropods. *Annual Review of Entomology*, **44**, 233–256.

Villani, M.G. & Wright, R.J. (1990) Environmental influences on soil macroarthropod behavior in agricultural systems. *Annual Review of Entomology*, **35**, 249–269.

Wall, D.H. & Six, J. (2015) Give soils their due. *Science*, **347**, 695.

Wand, S.J.E., Midgley, G.F., Jones, M.H. & Curtis, P.S. (1999) Responses of wild C4 and C3 grass (Poaceae) species to elevated atmospheric CO_2 concentration: a meta-analytic test of current theories and perceptions. *Global Change Biology*, **5**, 723–741.

Whalen, J.K., Sampedro, L. & Waheed, T. (2004) Quantifying surface and subsurface cast production by earthworms under controlled laboratory conditions. *Biology and Fertility of Soils*, **39**, 287–291.

Wharton, G.W. (1985) Water balance of insects. *Comprehensive Insect Physiology, Biochemistry and Pharmacology*, **4** (eds G.A. Kerkut & L.I. Gibert), pp. 565–601. Pergamon Press, Oxford, UK.

Williams, R.S., Marbert, B.S., Fisk, M.C. & Hanson, P.J. (2014) Ground-dwelling beetle responses to long-term precipitation alterations in a Hardwood forest. *Southeastern Naturalist*, **13**, 138–155.

Wilsey, B.J. (2001) Effects of elevated CO_2 on the response of *Phleum pratense* and *Poa pratensis* to aboveground defoliation and root-feeding nematodes. *International Journal of Plant Sciences*, **162**, 1275–1282.

Wurst, S. (2010) Effects of earthworms on above- and belowground herbivores. *Applied Soil Ecology*, **45**, 123–130.

Xiao, G., Lui, W., Xu, Q., Sun, Z. & Wang, J. (2005) Effects of temperature increase and elevated CO_2 concentration, with supplemental irrigation, on the yield of rain-fed spring wheat in a semiarid region of China. *Agricultural Water Management*, **74**, 243–255.

Yeates, G.W. (1981) Soil nematode populations depressed in the presence of earthworms. *Pedobiologia*, **22**, 191–195.

Yeates, G.W., Dando, J.L. & Shepherd, T.G. (2002) Pressure plate studies to determine how moisture affects access of bacterial-feeding nematodes to food in soil. *European Journal of Soil Science*, **53**, 355–365.

Yeates, G.W., Ferris, H., Moens, T. & van der Putten, W.H. (2009) The role of nematodes in ecosystems. *Nematodes as Environmental Indicators* (eds M.J. Wilson & T. Kakouli-Duarte), pp. 1–44. CABI Publishing, Walingford, UK.

Yeates, G.W. & Newton, P.C.D. (2009) Long-term changes in topsoil nematode populations in grazed pasture under elevated atmospheric carbon dioxide. *Biology and Fertility of Soils*, **45**, 799–808.

Yeates, G.W., Tate, K.R. & Newton, P.C.D. (1997) Response of the fauna of a grassland soil to doubling of atmospheric carbon dioxide concentration. *Biology and Fertility of Soils*, **25**, 307–315.

Young, I.M. (1998) Biophysical interactions at the root–soil interface: a review. *Journal of Agricultural Science*, **130**, 1–7.

Zahran, H.H. (1999) Rhizobium–legume symbiosis and nitrogen fixation under severe conditions and in an arid climate. *Microbiology and Molecular Biology Reviews*, **63**, 968–989.

Zaller, J.G. & Arnone, J.A.I. (1999a) Earthworm and soil moisture effects on the productivity and structure of grassland communities. *Soil Biology & Biochemistry*, **31**, 517–523.

Zaller, J.G. & Arnone, J.A.I. (1999b) Interactions between plant species and earthworm casts in a calcareous grassland under elevated CO_2. *Ecology*, **80**, 873–881.

Zaller, J.G., Heigl, F., Grabmaier, A., Lichtenegger, C., Piller, K., Allabashi, R., Frank, T. & Drapela, T. (2011) Earthworm-mycorrhiza interactions can affect the diversity, structure and functioning of establishing model grassland communities. *PLOS ONE*, **6**, e29293.

Zaller, J.G., Wechselberger, K.F., Gorfer, M., Hann, P., Frank, T., Wanek, W. & Drapela, T. (2013) Subsurface earthworm casts can be important soil microsites specifically influencing the growth of grassland plants. *Biology and Fertility of Soils*, **49**, 1097–1107.

Zavala, J.A., Nabity, P.D. & DeLucia, E.H. (2013) An emerging understanding of mechanisms governing insect herbivory under elevated CO_2. *Annual Review of Entomology*, **58**, 79–97.

Zhang, H., Schonhof, I., Krumbein, A., Gutezeit, B., Li, L., Stuezel, H. & Schreiner, M. (2008) Water supply and growing season influence glucosinolate concentration and composition in turnip root

(*Brassica rapa* ssp. *rapifera* L.). *Journal of Plant Nutrition and Soil Science-Zeitschrift Für Pflanzenernahrung und Bodenkunde*, **171**, 255–265.

Zirbes, L., Thonart, P. & Haubruge, E. (2012) Microscale interactions between earthworms and microorganisms: a review. *Biotechnologie Agronomie Societé et Environnement*, **16**, 125–131.

Zvereva, E.L. & Kozlov, M.V. (2006) Consequences of simultaneous elevation of carbon dioxide and temperature for plant–herbivore interactions: a metaanalysis. *Global Change Biology*, **12**, 27–41.

Zvereva, E.L. & Kozlov, M.V. (2012) Sources of variation in plant responses to belowground insect herbivory: a meta-analysis. *Oecologia*, **169**, 441–452.

12

Impacts of Atmospheric and Precipitation Change on Aboveground-Belowground Invertebrate Interactions

Scott N. Johnson[1], James M.W. Ryalls[1] and Joanna T. Staley[2]

[1] Hawkesbury Institute for the Environment, Western Sydney University, NSW, Australia
[2] Centre for Ecology and Hydrology, Wallingford, UK

Summary

Aboveground–belowground invertebrate interactions often operate when invertebrates (i) modify plant traits affecting behaviour and/or performance of other invertebrates, (ii) cause shifts in plant community composition that affect other invertebrates and (iii) alter patterns of deposition of plant-derived organic matter affecting soil-dwelling invertebrates. In this chapter, we show how atmospheric and climatic change has the capacity to shape both the magnitude and frequency of such interactions, often by affecting plants mediating the interaction. For example, elevated atmospheric carbon dioxide concentrations ($e[CO_2]$) caused a 36% increase in deposition of organic matter through increased herbivory of aspen and birch trees aboveground. This input most likely affected soil-dwelling invertebrates such as detritivores. Environmental change factors also interact with one another. For instance, increased air temperatures dampened the effects of $e[CO_2]$ on legume-mediated interactions between weevils and aphids. Changes in precipitation have the strongest effects on aboveground–belowground interactions, usually mediated via plant-stress related changes in plant chemistry. Our understanding of how atmospheric and climatic change might affect aboveground–belowground invertebrate interactions is limited by the scarcity of information available. We suggest three key areas of research that should be prioritised: (i) better characterisation of belowground components to assist hypothesis development, (ii) tests of multiple environmental factors simultaneously, particularly those involving warming, and (iii) expansion of study systems to include different plant functional groups and mixed plant communities.

12.1   Introduction

Ecological processes occurring above- and belowground interact with each other to affect the composition, structure and function of communities, but were rarely addressed together prior to the late 1990s (Van der Putten et al., 2001; Wardle et al., 2004). Since this time, there has been a number of influential studies (e.g., De Deyn et al., 2003; Bezemer et al., 2005) showing the strong linkages between above- and belowground organisms, and this has become a sub-discipline of ecology in its own right (Bardgett & Wardle, 2010; van Dam & Heil, 2010). Studies concerning interactions between invertebrates living above- and belowground have been prominent in the field from the outset (Moran & Whitham, 1990; Masters & Brown, 1992; Masters et al., 1993), and have grown in number to the extent that they have been subject to review (Blossey & Hunt-Joshi, 2003; Johnson et al., 2008) and meta-analysis (Johnson et al., 2012). For the most part, interactions between aboveground and belowground invertebrates are mediated by plants, which of course, span both sub-systems. Broadly speaking, these plant-mediated interactions can be divided into three types of mechanistic interaction which we summarise in Figure 12.1. These include (A) interactions between invertebrates which are mediated via modification of plant traits (e.g., induced primary and secondary metabolites); (B) shifts in plant community composition caused by invertebrates (e.g., grass–legume composition in pastures) that affect other invertebrates and (C) interactions based on aboveground invertebrates modifying plant-derived organic matter inputs to the soil directly (e.g., leaf litter, insect faecal matter).

In this chapter, we focus on the effects of predicted changes in atmosphere (increased concentrations of carbon dioxide, $e[CO_2]$ hereafter) and altered precipitation patterns on aboveground–belowground interactions. There are a limited number of studies that address how environmental changes affect aboveground–belowground invertebrate interactions (McKenzie et al., 2013a), and most of these focus on invertebrate-induced changes in plant quality (i.e., trait changes shown in Fig. 12.1A) as the principal mechanism underpinning the interaction. This route for interactions between above- and belowground invertebrates therefore features heavily in our

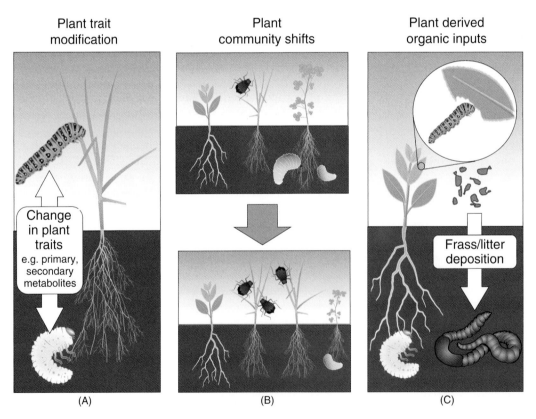

Figure 12.1 Aboveground–belowground invertebrate interactions can arise through three broad mechanisms, whereby invertebrates can affect each other by (A) Modifying plant traits that alter host plant suitability, (B) causing shifts in plant community composition that alter host plant availability or (C) Directly inputting plant-derived organic matter (e.g., Leaf litter or Insect faecal material or frass) into the soil.

chapter, though it must be acknowledged that plant community shifts (Fig. 12.1B) and altered patterns of how plant-derived organic matter enter the soil (Fig. 12.1C) will play an important role in how environmental change mediates aboveground–belowground interactions. In particular, at least three studies have shown that soil-dwelling detritivores interact with aboveground herbivores via changes in plant community structure rather than changes in plant traits *per se* (Thompson et al., 1993; Wurst & Rillig, 2011; Zaller et al., 2013).

Below, we summarise key aboveground–belowground interactions between invertebrates, focusing primarily on aboveground–belowground herbivore interactions. We go on to consider how these may be changed by $e[CO_2]$, operating alone and in tandem with increased air temperatures (Section 12.2), and altered precipitation patterns (Section 12.3). In closing (Section 12.4), we identify gaps in our knowledge and where future research directions might take us.

12.1.1 Interactions Between Shoot and Root Herbivores

It is well established that herbivores sharing a common host plant may indirectly affect one another via induced changes in their shared host plant (Lawton & Hassell, 1981; Denno et al., 1995; Kaplan & Denno, 2007). Herbivorous invertebrates may affect each other in this way even if they are temporally or spatially separated (Masters & Brown, 1997), and interactions between aboveground and belowground herbivores are now widely reported (see references contained in Johnson et al., 2008; Johnson et al., 2012). Based on the few studies available to them, Masters and Brown (1993) proposed a model whereby root herbivores would have beneficial impacts on shoot herbivores via a stress-induced response in plants that cause accumulation of nitrogen in the foliage. Essentially, root herbivores impair water uptake and induce a form of drought stress in plants (see Section 12.3) which it was hypothesised would be beneficial to some herbivores (White, 1984; Koricheva et al., 1998). The proposition that root herbivores promote the performance of shoot herbivores has some support, particularly when the aboveground herbivore is a sap-feeder such as an aphid (Johnson et al., 2012; Soler et al., 2013). This facilitation has been termed 'induced susceptibility' (van Dam & Heil, 2010). The mechanism for this interaction being a stress-induced change in nitrogen has received limited empirical support, however, and recent papers suggest defensive signalling in the plant may more commonly underpin such interactions (reviewed by Soler et al., 2013). In particular, simultaneous attack by above- and belowground herbivores with different feeding guilds may make plants more susceptible to both herbivores (Soler et al., 2013). This may arise because belowground herbivores (usually chewers) trigger the jasmonic acid pathway which interferes with the salicylic acid pathway triggered by the sap-feeders. This ultimately compromises plant resistance to either, or both, herbivores (Soler et al., 2013).

Examining the interaction in the opposite direction (i.e., impacts of shoot herbivory on root herbivores), Masters et al. (1993) proposed that shoot herbivores would detrimentally affect root herbivores because the plant would divert resources away from roots to allow for compensatory growth aboveground. This second proposition is now largely discounted as many plants translocate photoassimilates to the roots for storage after episodes of shoot herbivory (Schultz et al., 2013). Moreover, studies subsequent to the model of Masters et al. (1993) frequently show root herbivore performance is either unchanged (Blossey & Hunt-Joshi, 2003; Johnson et al., 2012 and references therein) or even improved in the presence of shoot herbivores (Johnson et al., 2009; Kaplan et al., 2009; McKenzie et al., 2013b).

The focus of research interest has moved towards how secondary metabolites mediate interactions between above- and belowground herbivores (Bezemer et al., 2003; Bezemer & van Dam, 2005). Systemic induction of plant defences by one herbivore that subsequently affects another has been reported in a number of cases (van Dam et al., 2003; Bezemer et al., 2004; Bezemer et al., 2005; Bezemer & van Dam, 2005; Soler et al., 2005), though it seems more common for root herbivores to induce defences systemically (potentially affecting shoot herbivores) than the other way around (Kaplan et al., 2008).

12.1.2 Interactions Between Herbivores and Non-Herbivorous Invertebrates

At present, the effects of climate and atmospheric change have been investigated mainly for interactions between above- and belowground herbivores, but it is likely that interactions involving other functional invertebrate groupings are shaped by such environmental changes and so we summarise the key features of these interactions below. While empirical evidence for how these interactions are affected by climate change is lacking, a recent review by A'Bear et al. (2014) attempts to predict some of these responses.

12.1.2.1 Detritivore–Shoot Herbivore Interactions

In the same way as herbivores can affect one another via changes in host plant quality and quantity (see Section 12.1.1), so too can soil-dwelling detritivores affect aboveground herbivores (A'Bear et al., 2014). Earthworms, for example often promote plant growth by increasing nutrient availability through a variety of mechanisms (reviewed by Scheu, 2003; Brown et al., 2004). Improved nutritional quality of plant tissue arising from such benefits has been shown to promote the performance of a leaf-chewing caterpillar (Newington et al., 2004) and aphids (Scheu et al., 1999; Bonkowski et al., 2001; Poveda et al., 2005; Eisenhauer et al., 2010). Conversely, decomposers can increase concentrations of plant secondary metabolites in the foliage (e.g., glucosinolates, phytosterols) which may adversely affect herbivores (A'Bear et al., 2014). For example, earthworms living in relatively nutrient rich soils promoted aphid performance on hairy bittercress (*Cardamine hirsuta*; Wurst & Jones, 2003), wild mustard (*Sinapis arvensis*; Poveda et al., 2005) and winter wheat (*Triticum aestivum*; Eisenhauer et al., 2010), but enhanced production of defensive secondary metabolites acting against aphids feeding on plantain (*Plantago lanceolata*; Wurst et al., 2004a; Wurst et al., 2004b).

12.1.2.2 Root Herbivore–Pollinator Interactions

Root herbivory usually imposes fitness costs on plants, resulting in trade-offs between re-growth, development and reproduction (Johnson et al., 2016a). We might reasonably expect root herbivory to reduce pollinator visitation due to plants not being able to invest in floral displays and production of pollinator-attracting Herbivore-Induced Plant Volatiles (HIPVs) (A'Bear et al., 2014). The impact of root herbivores on pollinators have only been considered in three systems to date; wild mustard (*Sinapis arvensis*), butternut squash (*Cucurbita moschata*) and cucumber (*Cucumis sativus*). Surprisingly, pollinator visitation increased in *S. arvensis* and *C. sativus* plants experiencing root herbivory (Poveda et al., 2003; Poveda et al., 2005; Barber et al., 2011). In these species, root herbivory either increased the frequency of pollinator visits (Poveda et al., 2003; 2005) or prolonged visitation time (Barber et al., 2011). The mechanisms underpinning these effects remain unclear but Poveda et al. (2003) suggested that root herbivory may have stimulated nutrient and water uptake via compensatory growth of root laterals. Compensatory growth in response to root herbivory is reported, but plants are less able to compensate for root damage, compared to aboveground herbivory (Johnson et al., 2016a) so this may be a short-term effect. It was mistakenly stated in A'Bear et al. (2014) that root herbivory stimulates rates of photosynthesis which may result in extra resources for floral displays, but this is rarely the case and root herbivory mostly decreases rates of photosynthesis (Zvereva & Kozlov, 2012; Johnson et al., 2016a).

12.2 Atmospheric Change – Elevated Carbon Dioxide Concentrations

12.2.1 Impacts of e[CO₂] on Interactions Mediated by Plant Trait Modification

The impacts of e[CO_2] on aboveground–belowground interactions have been investigated in a few systems. As discussed in Hiltpold et al. (this volume), one of the earliest studies examined the interaction between a root-feeding (*Pemphigus populitransversus*) and shoot-feeding (*Aphis fabae fabae*) aphid feeding on the brassica *Cardamine pratensis* (Salt et al., 1996). The baseline interaction (i.e. that occurring under ambient CO_2 conditions, a[CO_2]) was for the shoot-feeder to suppress populations of the root-feeder. Exposing the system to e[CO_2] did not alter the nature of the interaction and aphid populations remained at similar levels as those reared under a[CO_2] (Salt et al., 1996).

The responses of the clover root weevil (*Sitona lepidus*) feeding on white clover (*Trifolium repens*) grown under ambient carbon dioxide concentrations (a[CO_2]) and e[CO_2] have also been investigated (Staley & Johnson, 2008; Johnson & McNicol, 2010). As with many insect herbivores with root-feeding life-stages, the larval stages feed belowground whereas adults feed on foliage (Brown & Gange, 1990; Johnson et al., 2016a). Adult weevils consumed significantly more foliage under e[CO_2], perhaps indicative of compensatory feeding since foliage was of lower quality (higher C:N), and laid 23% fewer eggs into the soil. Despite fewer eggs being laid, larval populations grew by 38% under e[CO_2] which was directly correlated with increased levels of root nodulation (housing N-fixing bacteria) under e[CO_2] (Staley & Johnson, 2008; Johnson & McNicol, 2010). Larval *S. lepidus* performance is tightly linked with the availability of such root nodules (Gerard, 2001), as discussed in Hiltpold et al. (this volume), so larvae were able to take advantage of increased resource availability. Interestingly, the more than two-fold increase in root nodulation (and implicitly N-fixation) was not sufficient to stop e[CO_2] driving up foliar C:N by 9% (Johnson & McNicol, 2010). This might be explained by the fact that N-fixers generally show much larger increases in carbohydrate concentrations (+30%), compared to non-N-fixers (+20%), in response to e[CO_2] (Robinson et al., 2012). In this system, it was clear that e[CO_2] could decouple, or at least adjust, the relationship between adults aboveground and their belowground offspring. This raises interesting questions about whether e[CO_2] might introduce 'parent–offspring' conflict in species which have above- and belowground herbivorous life stages (see related discussion in Clark et al., 2011).

Also working with legumes and *Sitona* spp. weevils, Ryalls et al. (2013) found evidence that *S. discoideus* had a weakly negative impact on the pea aphid (*Acyrthosiphum pisum*)'s colonisation of lucerne (*Medicago sativa*). Like white clover and *S. lepidus*, e[CO_2] promoted nodulation and survival of *S. discoideus* which it might be assumed would strengthen the negative impacts on aphids aboveground. In fact, belowground herbivory stimulated nodulation and compensatory root growth, as was reported by Quinn and Hall (1992), which may have dampened the negative impacts on aphids aboveground (Ryalls et al., 2013). This was supported by a follow-up study, in which root damage by herbivores was simulated by cutting the roots (Ryalls et al., 2015). In particular, it was shown that cutting the roots early in the experiment allowed compensatory responses belowground, which promoted aphid performance, but late cutting of roots diminished aphid performance (Ryalls et al., 2015). Late cutting of roots did not allow the plant to sufficiently recover and compensate for the damage and as a result the negative interaction on aphids persisted. Moreover, the positive effects of e[CO_2] and negative effects of root damage on aphid performance were mirrored by their impacts on amino acid concentrations in the foliage; e[CO_2] increased concentrations whereas late root cutting caused significant decreases (Ryalls et al., 2015).

12.2.2 Impacts of e[CO₂] and Warming on Interactions Mediated by Plant Trait Modification

Surprisingly limited attention has been paid to the effects of warming on aboveground–belowground invertebrate interactions, which is why this chapter focuses on climate and precipitation impacts. One study (Stevnbak et al., 2012) focused on interactions via deposition (and is therefore considered in Section 12.2.3) found warming by itself had little impact. A second study (Lu et al., 2015) suggested that defoliation of sessile joyweed (*Alternanthera sessilis*) by several insect herbivores was unchanged by warming, but warming caused significant increases in populations of root-knot nematodes (*Meloidogyne incognita*). The mechanism was unclear, but previous glasshouse observations suggested that herbivory aboveground promoted nematode populations so warming has at least the capacity to shift the balance of this interaction.

In addition to manipulating e[CO₂], the studies by Ryalls et al. (2013; 2015) also manipulated air temperatures, which are predicted to increase in tandem with e[CO₂] (IPCC, 2014). The main conclusions of these studies was that higher temperatures reduced nodulation and performance of larval *S. discoideus*, which most probably reduced its impact aboveground (Ryalls et al., 2013). Reduced nodulation under higher temperatures led to lower concentrations of amino acids in the foliage and diminished the performance of aphids (Ryalls et al., 2015). Moreover, they identified a decrease in a specific group of individual amino acids (namely, arginine, aspartate, glutamate and histidine) associated with this response, which may have exacerbated any direct negative effects of high temperature on aphids. Temperature also interacted with root damage, whereby cutting early in the experiment under ambient temperature resulted in increased aphid performance, associated with higher concentrations of foliar amino acids (namely lysine, phenylalanine and tyrosine). The interactions of biotic and abiotic factors in these studies shows that there was a complex interplay of mechanisms operating in addition to the interaction between the above- and belowground subsystems (see summary in Fig. 12.2).

12.2.3 Impacts of Aboveground Herbivores on Belowground Invertebrates via Deposition Pathways

All of the studies above have principally focused on the effects of e[CO₂] aboveground–belowground interactions that are mediated via changes in plant traits (Fig. 12.1A), but there is increasing interest in whether climate change may moderate aboveground–belowground interactions via changes to deposition of plant-derived organic matter to the rhizosphere (Stevnbak et al., 2012). In particular, a large number of studies show that vertebrate herbivores can affect soil communities via deposition of plant derived organic matter, mostly in the form of animal wastes but also detached plant litter. This can promote nutrient cycling, microbial communities and plant nutrient uptake (Bardgett & Wardle, 2010). Invertebrate herbivores, while less studied, undeniably have similar impacts on nutrient cycling via deposition of faecal material (frass) and leaf litter (Belovsky & Slade, 2000; Frost & Hunter, 2004; Frost & Hunter, 2007; Frost & Hunter, 2008). Herbivory can also lead to changes in plant nutrient allocation, which cause short-term pulses of root exudates that affect soil microbial communities and their faunal consumers (Bardgett & Wardle, 2010).

While organic matter inputs via herbivory enhance nutrient cycling and are beneficial for some invertebrates, particularly decomposers, selective herbivory of more nutrient rich plants can have negative consequences on nutrient cycling. This arises because selective herbivory generally leads to the dominance of plants which are nutrient poor and/or better defended, which ultimately produces poorer quality and more recalcitrant litter (Bardgett & Wardle, 2010).

Given that e[CO₂] often has significant impacts on plant quality and patterns of herbivory (Robinson et al., 2012), how might this affect interactions between decomposers and aboveground

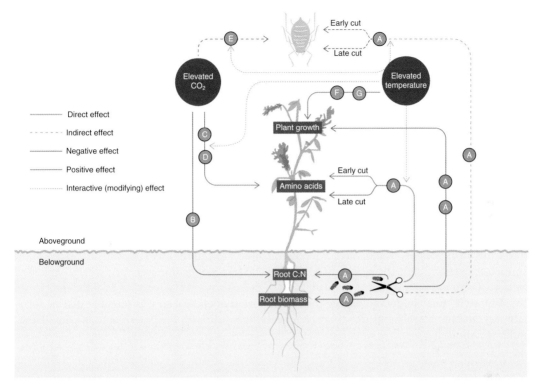

Figure 12.2 Simulated root herbivory (a) Reduced root C:N, root and shoot biomass, and affected foliar amino acids differently – roots cut early in the experiment stimulate amino acid production presumably through compensatory nodulation, whereas roots cut late in the experiment reduced foliar amino acid concentrations. Root herbivory therefore had positive and negative impacts on aphids, respectively. e[CO_2] reduced (b) root C:N, but increased concentrations of (c) essential and (d) non-essential amino acids in the foliage which stimulated (e) aphid population growth. Elevated air temperature increased plant growth, particular in terms of (f) plant height and (g) shoot biomass. Higher temperatures moderated both foliar amino acids and aphid performance, and therefore the aboveground–belowground interaction operating under ambient temperature conditions. Adapted from Ryalls et al. (2015).

herbivores (e.g., Fig. 12.1C)? In a landmark study, Stevnbak et al. (2012) investigated the impacts of e[CO_2], air temperature and drought on interactions between a folivore (grasshoppers) and soil communities (nematodes, protozoans and microbes). Herbivory greatly increased protozoan biomass and microbial activity, though did not affect nematodes. Grasshoppers ate less plant material under e[CO_2], however, which goes against the general prediction that herbivory increases deposition and nutrient cycling. Instead, the authors concluded that reduced levels of herbivory relaxed competition for nitrogen between plants and microbes and so some soil communities flourished (Stevnbak et al., 2012).

In addition to the lab study of Stevnbak et al. (2012), two Free Air Carbon dioxide Enrichment (FACE) experimental programmes in forest ecosystems have addressed insect defoliation and organic matter deposition belowground (see Facey and Gherlenda, Chapter 13). The first represents an aspen–birch stand in the United States (Hillstrom et al., 2010; Meehan et al., 2014; Couture et al., 2015) and the second is a mature *Eucalyptus* woodland in Australia (Duursma et al., 2015; Gherlenda et al., 2016a). In aspen and birch stands, e[CO_2] dramatically increased defoliation by

88%, deposition of organic matter by 36% and nitrogen by 38% (Couture et al., 2015). These changes in organic matter inputs were attributed to altered patterns of herbivory arising from the effects of e[CO_2] on plant traits including nutrients, structural compounds and secondary metabolites (Couture et al., 2015). The authors acknowledged that the proportion of nitrogen fluxed via insect excretion is small compared with that from litter inputs, but the increased nitrogen flux was at least comparable to the figure of 10–70% of atmospheric nitrogen deposition predicted for their study region by 2050 (Meehan et al., 2014). While not considered in the study conducted by Couture et al. (2015), it seems likely that such increased levels of organic input would affect soil-dwelling invertebrates, especially detritivores.

The second FACE study in a mature *Eucalyptus* woodland site found no evidence for increased levels of defoliation or frass deposition at e[CO_2] (Gherlenda et al., 2016a,b). However, patterns of precipitation altered levels of frass deposition (Gherlenda et al., 2016b) through stimulating the production of new leaves, the preferred food source of *Eucalyptus* feeding insects (Duursma et al., 2015; Gherlenda et al., 2016a,b). Alterations in the timing of frass deposition and leaf senescence as a result of changing precipitation patterns may alter the availability of nutrients or substrate for belowground microbes and invertebrates, which may have flow-on effects to community structure and population dynamics (see Section 12.3). The lack of e[CO_2] effects on frass deposition were principally explained by the lack of changes seen in plant chemistry and therefore patterns of herbivory (Gherlenda et al., 2016b). This study was conducted during the initial period of e[CO_2] so it is possible that patterns of herbivory, and therefore deposition, may change in the future with the potential to impact belowground invertebrates.

12.3   Altered Patterns of Precipitation

12.3.1   Precipitation Effects on the Outcome of Above–Belowground Interactions

As discussed in Section 12.1.2, root herbivory can impair water uptake in plants, which potentially promotes shoot herbivore performance via stress-induced accumulation of nitrogen compounds in the foliage (Masters et al., 1993). Under these circumstances, there is potential for interactions between above and belowground herbivores to be strengthened under drought, and ameliorated under increased rainfall under future climate scenarios (Kovats et al., 2014). In addition, defensive secondary metabolites often mediate interactions between herbivores (Section 12.1.2), and like primary metabolites (Girousse et al., 1996) these too can be modified by altered precipitation patterns (Hopkins et al., 2009; Tariq et al., 2013a).

A few studies have been conducted to experimentally manipulate soil water content to mimic the effects of altered precipitation patterns on interactions between above- and belowground invertebrates. The majority of these focus on drought, predicted to increase in severity and frequency in many regions under a future climate. Mean shallow soil moisture is predicted to decrease by between 1 and 4% by 2035 in most subtropical regions and central Europe (Kirtman et al., 2013). In the first study to investigate the effects of soil water content on interactions between phytophagous invertebrates feeding above- and belowground, Gange and Brown (1989) found that both a low water treatment and a root-feeding scarab beetle larvae (*Phyllopertha horticola*) increased the growth rate and weight of black-bean aphid (*Aphis fabae*) feeding aboveground on an annual Brassicaceae species (*Capsella bursa-pastoris*). Aphid fecundity and longevity were also increased by root herbivory, but not affected by the water treatment. The effect of root herbivores on aphid performance was reduced under a higher soil water content treatment (Gange & Brown, 1989).

Staley et al. (2007b) assessed the interaction between a leaf-mining micro-moth (*Stephensia brunnichella*) and root-feeding click beetle larva (*Agriotes* spp.) feeding on a perennial host plant (*Clinopodium vulgare*, wild basil) common to mesotrophic grasslands, under drought treatments in both field and laboratory experiments. Root herbivores reduced the weight of leaf-miner pupae and the abundance of leaf-miners on unstressed host plants grown under an enhanced precipitation treatment in the field. Percentage parasitism of leaf-miners in the field experiment was also reduced when root feeders were present, on unstressed host plants. Under summer drought treatment both the abundance and the pupal weight of the leaf-miners were substantially reduced, such that the root herbivore had no additional detrimental effect on either parameter. Summer drought reduced the survival of the root-feeding larvae, though this was unaffected by the presence of the leaf-miner (Staley et al., 2007b). In a laboratory study comparing the response of a leaf-mining Diptera (*Chromatomyia syngenesiae*) and the same click beetle species feeding on annual and perennial *Sonchus* species, the root feeders increased the pupal weight and reduced the development time of the leaf-mining species on one of the four plant species, but only under drought conditions (Staley et al., 2008). Leaf-miners reduced the relative growth rate of the root-feeding larvae on two of the four plant species (Staley et al., 2008). The laboratory studies show similar results, with a positive effect of root-feeding beetle larvae on aphids (Gange & Brown, 1989) and leaf-miners (Staley et al., 2008) under drought or low water treatments, which was partly or entirely removed under a high water treatment. The field-based study provides contrasting results, and is the only study to show that colonisation of the host plants by aboveground herbivores responds in a similar way to the performance of individual leaf-miners under drought and root herbivore treatments (Staley et al., 2007b).

Laboratory and field studies create different conditions in which to test the effects of precipitation on invertebrate interactions, as plants in the field suffer from multiple stressors including the presence of other herbivores. Different magnitudes of drought or low water treatments may also have been applied in different studies. For example, foliar relative water content was reduced by nearly 50% under a summer drought treatment designed to simulate climate change projections for 2080 in a field study (Staley et al., 2007b), while in one laboratory study relative water content was only reduced by about 6% in the low water compared to the high water treatment (Gange & Brown, 1989). Herbivorous invertebrates are predicted to respond differently to moderate and severe drought stress (Larsson, 1989; Tariq et al., 2012); Section 12.3.2), so experiments applying just one level of drought or low water may be an over-simplification. Drought and enhanced precipitation treatments that are designed to mimic predictions of future precipitation patterns, and cover a range of scenarios, are key to understanding how interactions between above- and belowground herbivores may be altered under climate change. Nonetheless, all of the studies discussed above showed that changes in water availability alter the outcome or strength of interactions between herbivores feeding above- and belowground.

12.3.1.1 Case Study – Impacts of Simulated Precipitation Changes on Aboveground–Belowground Interactions in the Brassicaceae

The Brassicaceae contains several thousand plant species, some of which have been used as models for unravelling the complexities of aboveground–belowground invertebrate interactions (van Dam et al., 2003). This family possesses a number of traits that make them amenable for such investigations, including a well characterised defensive chemistry, a broad range of comparative life-history traits across the family, the absence of mycorrhizae (which can represent a confounding experimental factor) and inclusion of the model plant *Arabidopsis thaliana* in the family (van Dam et al., 2003). Here, we describe a series of experiments that explored the impacts of predicted rainfall patterns on the interactions between belowground herbivores, aboveground herbivores and natural enemies of the aboveground herbivores.

The effects of drought treatments varying in severity, and root herbivores at low and high populations, on the performance of two aphid species and their parasitoids were investigated in detail (Tariq et al., 2012; Tariq et al., 2013a; Tariq et al., 2013b). Root-feeding fly larvae (*Delia radicum*) reduced the fecundity and increased the development time of both aphid species (*Myzus persicae* and *Brevicoryne brassicae*), with a stronger effect when more root-feeding larvae were present (three versus five larvae per plant, *Brassica oleracea* L. var. *gemmifera* cv. Oliver; Tariq et al., 2013a; Fig. 12.3). Glucosinolates are a group of secondary metabolites in Brassica species that can act both as defences against herbivores and as signalling compounds to specialist phytophages, depending on the insect species and compound (Hopkins et al., 2009). The increase in aphid development time correlated with the concentration of foliar glucosinolates, the concentration of which may be as important as nitrogen concentrations in determining herbivore responses to drought in these host plants (Gutbrodt et al., 2012). Severe drought stress reduced aphid performance further in the presence of root feeders, which was minimised under a combination of severe drought stress and high root feeder density for both aphid species. In contrast, moderate drought stress enhanced aphid performance, such that a combination of moderate drought stress and low root-feeder density maximised performance of *M. persicae*, whereas *B. brassicae*'s performance was greatest under medium drought stress without root herbivory (Tariq et al., 2013a). Although *M. persicae* is considered a generalist aphid species,

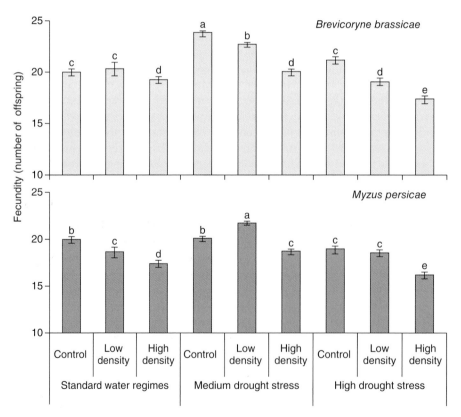

Figure 12.3 Fecundity (number of offspring produced in six days) of *Myzus persicae* and *Brevicoryne brassicae* (Mean ± S.E.M.) feeding on *Brassica oleracea* plants under different *Delia radicum* densities (control = no *D. radicum*) and drought-stress treatments. Within each aphid species, means with different letters are significantly different [ANOVA, post hoc Tukey honestly significant difference (HSD) test: P<0.05]. Reproduced from Tariq et al. (2013a).

and *B. brassicae* is a specialist on brassicas, the two aphid species responded in a similar way to root herbivory and drought stress (Tariq et al., 2013a).

Parasitoid wasps, feeding within *B. brassicae* and *M. persicae*, were also affected by a severe drought treatment and the presence of root-feeding fly larvae (Tariq et al., 2013b) (Fig. 12.4). Females of two wasp species (*Diaeretiella rapae* and *Aphidius colemani* feeding on *B. brassicae* and *M. persicae* respectively) parasitised fewer aphids and produced more male parasitoids (an indicator of poor host quality) when root herbivores were present or severe drought stress was applied. The volatile organic compound allyl isothiocyanate, an attractant of aphid parasitoids, was emitted at much lower concentrations by plants and aphids that were subjected to severe drought or root herbivory compared to plants without root feeders (Tariq et al., 2013b). This detailed laboratory study shows that percentage parasitism and parasitoid performance follow a similar response as their host invertebrates to both drought and root herbivory, possibly mediated by changes in plant signalling, and is supported by the only other study to assess rates of parasitism of aboveground herbivores feeding on plants subjected to root herbivory and drought treatments in the field (Staley et al., 2007b, discussed above).

12.3.2 Aboveground–Belowground Interactions in Mixed Plant Communities Under Altered Precipitation Scenarios

The majority of work exploring the impacts of altered rainfall on aboveground–belowground herbivore interactions has focussed on invertebrate induced changes in plant traits (Fig. 12.1A), but one recent study investigated whether a belowground herbivore could affect herbivores aboveground by changing plant community composition (e.g., Fig. 12.1B) in addition to changing plant traits (Ryalls, 2016). Using a model grass–legume system they investigated how simultaneous herbivory by belowground weevil (*Sitona discoideus*) larvae and aboveground aphids (*Acyrthosiphon pisum*) on a common legume host plant (lucerne or alfalfa, *Medicago sativa*) affected a neighbouring grass (*Phalaris aquatica*) under drought and elevated precipitation using mesocosms. Grass nitrogen

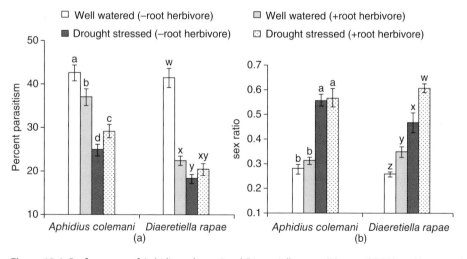

Figure 12.4 Performance of *Aphidius colemani* and *Diaeretiella rapae* (Mean ± S.E.M.) on *Myzus persicae* and *Brevicoryne brassicae* reared on *Brassica oleracea* plants under a well-watered regime (200 ml/pot/week; "Control") and a reduced water regime (100 ml/pot/week; "drought stressed") with/without *Delia radicum*. Within each parasitoid species, means with different letters are significantly different (P <0.05): (a) Percentage parasitism (b) sex ratio. Reproduced from Tariq et al. (2013b).

concentrations and productivity increased in response to root herbivory by weevil larvae, suggesting that nitrogen leached from lacerated lucerne root nodules was taken up by the neighbouring grass. Rates of nutrient mineralisation are particularly important in grass–legume systems, with one species often failing to persist under competition with the other if N dynamics are not maintained (Dear et al., 1999). In relation to aboveground–belowground herbivore interactions, root herbivory by *S. discoideus* negatively impacted aphid populations, associated with decreases in proline concentrations. Moreover, drought decreased the productivity of both plant species and reduced aphid populations on lucerne, most likely associated with reduced phloem turgor and increased sap viscosity (Hale et al., 2003). While elevated precipitation has the potential to alter plant competitive dynamics by causing the grass to dominate lucerne, no effect on aphids or their interaction with belowground weevils was observed. However, indirect negative effects on aphids are likely to occur if their host plant is suppressed by the neighbouring grass. In situations where aphids can feed on both plants, suppression of one species may cause aphids to migrate to the other. These results demonstrate how climate change and interactions between herbivores (both above- and belowground) can shape the competitive interactions between N-fixing legumes and non-N-fixing grasses, with consequences for plant community structure and productivity.

12.3.3   Altered Precipitation Impacts on Decomposer–Herbivore Interactions

Decomposers can also interact with aboveground invertebrates through their host plants, and two studies have assessed how these interactions might change under altered precipitation patterns. Summer drought (50% of ambient rainfall in line with predictions for the UK by 2080) decreased plant biomass while earthworms increased plant biomass in a field experiment using single and multi-species plant communities (Johnson et al., 2011). Earthworms strongly increased biomass of plants growing in monocultures but did not alter biomass in multi-species communities, while drought had a greater effect in multi-species communities, though some plant species were more responsive to drought than others. Aphid (*Rhopalosiphum padi*) density was reduced by drought and earthworms. Abundance of the parasitoid *Aphidius ervi* was correlated to aphid abundance, but drought had an additional negative effect on *A. ervi* abundance, indicative of a reduction in aphid host quality under drought (Johnson et al., 2011).

Dung beetles have also been shown to reverse the negative effects of drought on plant biomass in Brassica plants, through burrowing activity that promotes nutrient cycling and enhances water infiltration and soil porosity to reduce soil water runoff (Johnson et al., 2016b). Neither drought, increased precipitation nor the presence of dung beetles affected development time or pupal weight of the aboveground herbivore diamondback moth (*Plutella xylostella*). Diamondback moth is a cosmopolitan pest of Brassica species, and has previously been shown not to respond to changes in host plant nutritional content or secondary chemistry (Ratzka et al., 2002).These studies show the need for further work to address climate change and interactions between soil and aboveground invertebrates using multi-species assemblages of plants, and where possible, realistic plant communities. Moreover, the potential role for decomposers to ameliorate the negative effects of drought on plant biomass and nitrogen content make their inclusion in future above–belowground studies key.

12.3.4   Impacts of Increased Unpredictability and Variability of Precipitation Events on the Frequency of Above–Belowground Interactions

Atmospheric carbon dioxide concentrations and, to a lesser extent, air temperatures are increasing in a more or less predictable manner. Precipitation differs in the respect that increased unpredictability and variation in precipitation events is predicted (Parmesan et al., 2000). Changes in such precipitation patterns and events may affect the likelihood of interactions occurring between above- and belowground invertebrates, through changes to their abundance, population structure and behaviour. The potential for atypical patterns or quantities of precipitation to affect herbivorous invertebrate populations, through changes to their host plants, has been recognised for over fifty years (White, 1969; White, 1984). The 'Plant Stress Hypothesis' (PlSH) was proposed in response to observations that populations of the psyllid *Cardiaspina densitexta* reached outbreak levels when feeding on trees that were stressed by water shortage, postulated to be due to enhanced host plant quality caused by an increase in nitrogen concentration (White, 1969). The PlSH gained enough general acceptance for it to have 'paradigm status' (Waring & Cobb, 1992; Koricheva et al., 1998), although experimental tests of this hypothesis produced inconsistent results (Bultman & Faeth, 1987; Staley et al., 2006). Observations that some phytophagous insects choose vigorous, fast-growing parts of plants to feed on led to the 'Plant Vigour Hypothesis' (PVH), which has been interpreted as an alternative to the PlSH (Price, 1991). Attempts to reconcile the two hypotheses include the suggestion that feeding guilds respond differently to drought stress, with phloem-feeders most likely to be positively affected, and gall feeders negatively affected (Larsson, 1989). In addition the severity of drought stress may be important, with herbivores benefiting from plants that are under moderate stress, but performing poorly on those under severe drought stress (Larsson, 1989; Tariq et al., 2012).

Several reviews and meta-analyses have been conducted to identify general patterns in the response of herbivorous insects to changes in precipitation, usually in response to drought (Koricheva et al., 1998; Huberty & Denno, 2004; Jactel et al., 2012; Jamieson et al., 2012). Sap-feeding herbivores were found to respond negatively overall to continuously water-stressed host plants in terms of performance and population growth (Huberty & Denno, 2004), leading Huberty and Denno (2004) to put forward a 'Pulsed Stress Hypothesis' (PuSH) to explain observed outbreaks of phytophagous insects on plants under stress. According to this hypothesis periods without drought stress are required, to allow plants to recover their turgor pressure in order for sap-feeders to take advantage of the stress-induced increases in plant nitrogen (Huberty & Denno, 2004). Experimental tests of the PuSH have produced varying results (Gutbrodt et al., 2011; Simpson et al., 2012; Tariq et al., 2012). Two aphid species (*Myzus persicae* and *Brevicoryne brassicae*) feeding on a Brassica host plant (*Brassica oleracea* var. *capitata*) had reduced fecundity and potential population growth rate (intrinsic rate of increase) under an intermittent pulsed drought stress treatment, compared with those receiving the same drought stress through more regular applications of the same total amount of water. Performance of both aphid species was maximised on plants under a continuous, intermediate level of drought stress, and reduced on well-watered plants or those under constant, severe drought stress (Tariq et al., 2012), illustrating the importance of stress intensity in determining herbivore response. Gutbrodt et al. (2012) applied drought treatments described as 'pulsed' to *B. oleracea* plants, and found that drought stressed hosts were preferred by two chewing Lepidoptera herbivores, but as their high drought stress plants received water every 4–6 days these may be comparable to the continuously stressed Brassica plants in the previous study. A recent meta-analysis focussing on damage caused by forest herbivores found lower levels of damage on drought-stressed trees by insects feeding on woody organs, but more damage by herbivores on foliage (Jactel et al., 2012). Feeding guild was not found to be important in determining response to drought stress, but secondary herbivores that only colonise trees that are already severely stressed responded differently to primary herbivores, as they caused higher levels of damage on woody organs in trees under drought stress (Jactel et al., 2012).

While the PlSH and PuSH have historically emphasised changes in the concentration of foliar nitrogen compounds as the mechanism for phytophagous insects' responses to drought stress, more recent studies have considered changes in foliar secondary plant chemistry and other aspects of plant physiology. For example, population size and damage to sugar cane by a stalk-boring Lepidoptera larva was increased under drought stress, but this effect of drought was negated by an increase in concentration of the defensive compound silica, through the addition of calcium silicate (Kvedaras et al., 2007). In the study on *M. persicae* and *B. brassicae* above, five of six glucosinolate compounds responded to the drought treatments, though the effects of drought differed between compounds (Tariq et al., 2012). Changes to physiological parameters such as turgor pressure and stomatal conductance under drought stress can also affect herbivore feeding rates, and therefore their individual development and population growth. A drought stress treatment applied to *Brassica oleracea* var. *capitata*, similar in intensity to the severe drought stress discussed above (Tariq et al., 2012), resulted in reduced photosynthesis and water potential, and decreased stomatal conductance (Simpson et al., 2012). Changes to precipitation can alter the structure as well as the size of invertebrate herbivore populations, resulting in knock-on effects on subsequent trophic levels. Adults made up a much greater proportion of aphid (*Rhopalosiphum padi*) populations under drought treatments, resulting in reduced parasitism rates as adult aphids are harder to parasitise than nymphs (Aslam et al., 2013).

Fewer studies have addressed the response of soil invertebrates to changing precipitation patterns, compared to the response of aboveground invertebrates. Click beetle (*Agriotes* spp.) larvae, a major pest of root vegetable crops, were more abundant under an enhanced rainfall treatment than either an ambient control or a summer drought treatment in a long-term grassland field experiment, though the abundance of a second hemipteran root feeder (*Lecanopsis formicarum*) was not affected by the precipitation treatments (Staley et al., 2007a). Root feeders may alter their behaviour in response to changes in soil moisture; for example click beetle larvae feed lower in the soil profile under dry conditions (Lafrance, 1968; Seal et al., 1992; Parker & Howard, 2001). If this results in reduced root consumption, it could reduce the frequency of interactions with aboveground invertebrates. Decomposers and decomposition rates may also be altered by precipitation patterns (Riutta et al., 2012), potentially reducing nutrient availability for host plants shared with aboveground invertebrates.

Changes to the abundance, population structure and behaviour of invertebrates both above- and belowground in response to precipitation are likely to alter the frequency with which interactions between these groups occur under a future climate. While some aboveground herbivore populations may increase under moderate drought stress, severe drought stress is likely to reduce the abundance of many above- and belowground invertebrates, alter population structures and potentially drive soil invertebrates lower in the soil profile where fewer roots may be consumed, and less plant material may be available for decomposition. The order of arrival at a host plant can alter the outcome of interactions between above- and belowground herbivores, as a recent meta-analysis found that belowground feeders only increase the performance of aboveground herbivores if they arrive at the plant simultaneously (Johnson et al., 2012). Changes to the abundance and behaviour of invertebrates may thus alter the outcome of interactions through changing the order of arrival, as well as altering the frequency with which they occur (McKenzie et al., 2013a).

12.4   Conclusions and Future Directions

It is clear from the preceding sections that climate and atmospheric change have the capacity to modify aboveground–belowground invertebrate interactions, with altered patterns of precipitation seemingly being the most influential. Many research questions have yet to be answered, however, and aboveground–belowground ecology is notoriously complex without the added dimension of climate change. A good starting point may be to introduce environmental variation to those interactions that have been shown to be consistently present, preferably in both the laboratory and the field (Vandegehuchte et al., 2010; Johnson et al., 2013), and are thought to play an important role in shaping the invertebrate populations relative to other interactions (e.g., mutualism or predation). The danger with this approach, however, is that climate change might cause new interactions to arise in ecosystems where they had previously been absent. Below, we set out key issues that we consider should be at the forefront of future research in this area.

12.4.1 Redressing the Belowground Knowledge Gap

The impacts of $e[CO_2]$ on aboveground arthropods are relatively well characterised (Robinson et al., 2012), but comparatively less is known about the responses of belowground invertebrates (reviewed by Hiltpold et al., this volume). This limits our ability to infer or even hypothesise about how aboveground–belowground invertebrate interactions might be affected by atmospheric change (McKenzie et al., 2013a). Redressing this empirical imbalance, particularly about how root herbivores respond to climate change, will therefore be important for developing testable hypotheses in the future.

There are numerous examples of defoliators affecting microbial communities via deposition of plant-derived material, but very little is known about this in the context of global climate change (see Section 12.2.3; Fig. 12.1C). Given the diverse literature characterising how plant quality changes in response to climate change, and large responses in terms of defoliation and deposition (Couture et al., 2015), it is surely now timely to address how this is affecting belowground components of the ecosystem. Multiple hypotheses are possible. For example, increased defoliation, leaf litter input, organic matter and soil N under $e[CO_2]$ (as reported by Couture et al., 2015) could promote root herbivore performance when this increased nutrient input increased root growth, but might also reduce herbivore performance if this enabled plants to increase root defences (Barnett & Johnson, 2013). Increased or decreased levels of root herbivory would most likely affect defoliators and therefore litter input, particularly if they were moderating changes in foliar quality induced by $e[CO_2]$ as hypothesised by McKenzie et al. (2013a).

This particular hypothesis suggested that root herbivory may stop plants responding in the same way to $e[CO_2]$ because of impaired root function. In particular, many plant species increase rates of growth and photosynthesis in response to $e[CO_2]$, but this relies on regulating stomatal conductance and water use efficiency (Newman et al., 2011). If root herbivory impairs root function, then it is possible that typical plant responses to $e[CO_2]$ that affect foliar herbivores, such as increased C:N in the foliage negatively affecting foliar herbivores (Robinson et al., 2012), might not occur to the same extent, if at all. There is some support for this; specific leaf thickness and C:N both increased by about 30% in *Eucalyptus globulus* grown under $e[CO_2]$, but these changes were nullified in the presence of root damaging soil insects (Johnson & Riegler, 2013). While not tested in that study, the increased leaf thickness and C:N would likely be deleterious to foliar feeding insects, so in this instance the soil insects would have dampened the negative effects of $e[CO_2]$ on foliar feeders.

12.4.2 Testing Multiple Environmental Factors

Increases in temperature are likely to affect aboveground invertebrates much more than belowground invertebrates, since the latter are buffered from temperature changes to a greater extent (Barnett & Johnson, 2013). This discrepancy may affect one community more than another and therefore modify aboveground–belowground interactions, so it is surprising that so few warming studies have been conducted in this area. Moreover, it is apparent that testing single environmental factors can provide an incomplete (at best) or misleading (at worst) explanation of how climate change will affect plant–invertebrate interactions (Scherber et al., 2013; Facey et al., 2014). This appears to be true for aboveground–belowground interactions in all of the studies that have manipulated more than one environmental variable so far (Stevnbak et al., 2012; Ryalls et al., 2013; Ryalls et al., 2015).

Testing multiple factors presents a number of logistical and statistical difficulties, however, so researchers should first establish the mechanistic basis for particular aboveground–belowground interactions. Understanding the mechanics of the interaction (e.g., plant trait-mediated, community shifts or deposition) will allow researchers to identify which aspects of climate change might be most important to manipulate (see Lindroth and Raffa, this volume). In addition, as discussed above in relation to precipitation (Section 12.3.1), the magnitude of environmental factors may affect the outcome of aboveground–belowground interactions, and the relationship may not be linear. Testing environmental factors across a range of intensities that are likely to occur under climate change is needed to characterise these relationships.

12.4.3 New Study Systems

There are many benefits that come from concentrating on a small group of plants, such as the Brassicaceae (see Section 12.3.1.1), but broadening our understanding of the impacts of climate change on aboveground–belowground interactions will be important for making more generalised predictions. While the absence of mycorrhizae is empirically beneficial in terms of its confounding effect, the Brassicaceae are unlike most terrestrial plants in this regard and so this is a somewhat unrealistic scenario. Indeed, there is emerging recognition that soil microbes interact intimately with plants to shape their interaction with the biotic and abiotic environment (Edwards et al., 2015), so this clearly needs to be factored into future studies. Many studies on above–belowground interactions have been conducted on single-species plant systems in pots or small mesocosms due to the ease of manipulating invertebrate presence and abundance. The effects of belowground herbivores and detritivores on plant traits can differ between single and multi-species plant assemblages as discussed above, and changes to plant community structure may be more important in determining the effects of detritivores on aboveground invertebrates than changes to plant traits, making the use of more realistic plant assemblages key.

Broadening the range of plant species which are examined in such studies could assist the second hypothesis put forward by McKenzie et al. (2013a). Essentially, they hypothesised that plants of different functional groups will respond differently to atmospheric and climate changes, thereby affecting interactions between above- and belowground herbivores. In particular, C_3 plants are generally much more responsive to e[CO_2] than C_4 plants, for instance showing greater increases in C:N (Wand et al., 1999). This occurs because Rubisco, the initial carboxylating enzyme to facilitate the assimilation of CO_2 into carbohydrates, operates below its maximum capacity at current CO_2 concentrations in C_3 plants, so has the greater capacity to respond to eCO_2 (DeLucia et al., 2012). McKenzie et al. (2013a) therefore hypothesised that interactions occurring on C_3 plants would be more affected than those on C_4 plants. Higher foliar C:N often results in compensatory feeding by herbivores aboveground (Robinson et al., 2012), but has also been reported for a root herbivore (Johnson et al., 2014). In this instance, increased levels of root herbivory were seen under e[CO_2] on the C_3 grass, but not the C_4 grass. While not tested in that study, it could be hypothesised that e[CO_2] would therefore intensify any effects of root herbivory on aboveground herbivores to a greater extent on the C_3 grass than the C_4 grass.

12.4.4 Closing Remarks

Despite the knowledge gaps outlined above, research into interactions between above- and below-ground invertebrates has advanced considerably over the last 15–20 years, and an understanding of soil processes and linkages is now regarded integral to terrestrial ecology. The use of large-scale manipulative field experiments to assess climate change impacts, such as FACE and rainout shelters, are helping to address some of the research gaps discussed above. Nonetheless, the preceding sections show that precipitation and atmospheric change have strong effects on the occurrence, outcome and frequency of interactions between soil and aboveground herbivores, which can also cascade up to affect higher trophic levels.

Acknowledgements

We are grateful to Andrew Gherlenda for assistance with Section 12.2.3 and to Philip Smith for proofreading the chapter.

References

A'Bear, A.D., Johnson, S.N. & Jones, T.H. (2014) Putting the 'upstairs–downstairs' into ecosystem service: what can aboveground–belowground ecology tell us? *Biological Control*, **75**, 97–107.

Aslam, T.J., Johnson, S.N. & Karley, A.J. (2013) Plant-mediated effects of drought on aphid population structure and parasitoid attack rates. *Journal of Applied Entomology*, **137**, 136–145.

Barber, N.A., Adler, L.S. & Bernardo, H.L. (2011) Effects of above- and belowground herbivory on growth, pollination, and reproduction in cucumber. *Oecologia*, **165**, 377–386.

Bardgett, R.D. & Wardle, D.A. (2010) *Aboveground–belowground Linkages: Biotic Interactions, Ecosystem Processes, and Global Change*. Oxford University Press, New York, New York, USA.

Barnett, K. & Johnson, S.N. (2013) Living in the soil matrix: abiotic factors affecting root herbivores. *Advances in Insect Physiology*, **45**, 1–52.

Belovsky, G.E. & Slade, J.B. (2000) Insect herbivory accelerates nutrient cycling and increases plant production. *Proceedings of the National Academy of Sciences of the United States of America*, **97**, 14412–14417.

Bezemer, T.M., De Deyn, G.B., Bossinga, T.M., Van Dam, N.M., Harvey, J.A. & Van der Putten, W.H. (2005) Soil community composition drives aboveground plant–herbivore–parasitoid interactions. *Ecology Letters*, **8**, 652–661.

Bezemer, T.M. & van Dam, N.M. (2005) Linking aboveground and belowground interactions via induced plant defenses. *Trends in Ecology and Evolution*, **20**, 617–624.

Bezemer, T.M., Wagenaar, R., Van Dam, N.M., Van Der Putten, W.H. & Wäckers, F.L. (2004) Above- and below-ground terpenoid aldehyde induction in cotton, *Gossypium herbaceum*, following root and leaf injury. *Journal of Chemical Ecology*, **30**, 53–67.

Bezemer, T.M., Wagenaar, R., Van Dam, N.M. & Wäckers, F.L. (2003) Interactions between above- and belowground insect herbivores as mediated by the plant defense system. *Oikos*, **101**, 555–562.

Blossey, B. & Hunt-Joshi, T.R. (2003) Belowground herbivory by insects: influence on plants and aboveground herbivores. *Annual Review of Entomology*, **48**, 521–547.

Bonkowski, M., Geoghegan, I.E., Birch, A.N.E. & Griffiths, B.S. (2001) Effects of soil decomposer invertebrates (protozoa and earthworms) on an above-ground phytophagous insect (cereal aphid) mediated through changes in the host plant. *Oikos*, **95**, 441–450.

Brown, G.G., Edwards, C.A. & Brussaard, L. (2004) How earthworms affect plant growth: burrowing into the mechanisms. *Earthworm Ecology* (ed. C.A. Edwards). CRC Press, Boca Raton, Florida, USA.

Brown, V.K. & Gange, A.C. (1989) Herbivory by soil-dwelling insects depresses plant species richness. *Functional Ecology*, **3**, 667–671.

Brown, V.K. & Gange, A.C. (1990) Insect herbivory below ground. *Advances in Ecological Research*, **20**, 1–58.

Bultman, T.L. & Faeth, S.H. (1987) Impact of irrigation and experimental drought stress on leaf-mining insects of Emory oak. *Oikos*, **48**, 5–10.

Clark, K.E., Hartley, S.E. & Johnson, S.N. (2011) Does mother know best? The preference–performance hypothesis and parent–offspring conflict in aboveground–belowground herbivore life cycles. *Ecological Entomology*, **36**, 117–124.

Couture, J.J., Meehan, T.D., Kruger, E.L. & Lindroth, R.L. (2015) Insect herbivory alters impact of atmospheric change on northern temperate forests. *Nature Plants*, **1**, 15016.

De Deyn, G.B., Raaijmakers, C.E., Zoomer, H.R., Berg, M.P., de Ruiter, P.C., Verhoef, H.A., Bezemer, T.M. & van der Putten, W.H. (2003) Soil invertebrate fauna enhances grassland succession and diversity. *Nature*, **422**, 711–713.

Dear, B.S., Cocks, P.S., Peoples, M.B., Swan, A.D. & Smith, A.B. (1999) Nitrogen fixation by subterranean clover (*Trifolium subterraneum* L.) growing in pure culture and in mixtures with varying densities of lucerne (*Medicago sativa* L.) or phalaris (*Phalaris aquatica* L.). *Australian Journal of Agricultural Research*, **50**, 1047–1058.

DeLucia, E.H., Nabity, P.D., Zavala, J.A. & Berenbaum, M.R. (2012) Climate change: resetting plant–insect interactions. *Plant Physiology*, **160**, 1677–1685.

Denno, R.F., McClure, M.S. & Ott, J.R. (1995) Interspecific interactions in phytophagous insects: competition reexamined and resurrected. *Annual Review of Entomology*, **40**, 297–331.

Duursma, R.A., Gimeno, T.E., Boer, M.M., Crous, K.Y., Tjoelker, M.G. & Ellsworth, D.S. (2016) Canopy leaf area of a mature evergreen *Eucalyptus* woodland does not respond to elevated atmospheric [CO_2] but tracks water availability. *Global Change Biology*, **22**, 1666–1676.

Edwards, J., Johnson, C., Santos-Medellin, C., Lurie, E., Podishetty, N.K., Bhatnagar, S., Eisen, J.A. & Sundaresan, V. (2015) Structure, variation, and assembly of the root-associated microbiomes of rice. *Proceedings of the National Academy of Sciences of the United States of America*, **112**, E911–E920.

Eisenhauer, N., Hörsch, V., Moeser, J. & Scheu, S. (2010) Synergistic effects of microbial and animal decomposers on plant and herbivore performance. *Basic and Applied Ecology*, **11**, 23–34.

Facey, S.L., Ellsworth, D.S., Staley, J.T., Wright, D.J. & Johnson, S.N. (2014) Upsetting the order: how climate and atmospheric change affects herbivore–enemy interactions. *Current Opinion in Insect Science*, **5**, 66–74.

Frost, C.J. & Hunter, M.D. (2004) Insect canopy herbivory and frass deposition affect soil nutrient dynamics and export in oak mesocosms. *Ecology*, **85**, 3335–3347.

Frost, C.J. & Hunter, M.D. (2007) Recycling of nitrogen in herbivore feces: plant recovery, herbivore assimilation, soil retention, and leaching losses. *Oecologia*, **151**, 42–53.

Frost, C.J. & Hunter, M.D. (2008) Herbivore-induced shifts in carbon and nitrogen allocation in red oak seedlings. *New Phytologist*, **178**, 835–845.

Gange, A.C. & Brown, V.K. (1989) Effects of root herbivory by an insect on a foliar-feeding species, mediated through changes in the host plant. *Oecologia*, **81**, 38–42.

Gerard, P.J. (2001) Dependence of *Sitona lepidus* (Coleoptera: Curculionidae) larvae on abundance of white clover *Rhizobium* nodules. *Bulletin of Entomological Research*, **91**, 149–152.

Gherlenda, A., Moore, B.D., Haigh, A., Johnson, S.N. & Riegler, M. (2016a) Insect herbivory in a mature *Eucalyptus* woodland depends on rainfall and leaf phenology but not CO_2 enrichment. *BMC Ecology*, **16**, 47. DOI: 10.1186/s12898-016-0102-z.

Gherlenda, A.N., Crous, K.Y., Moore, B.D., Haigh, A.M., Johnson, S.N. & Riegler, M. (2016b) Precipitation, not CO_2 enrichment, drives insect herbivore frass deposition and subsequent nutrient dynamics in a mature *Eucalyptus* woodland. *Plant and Soil*, **399**, 29–39.

Girousse, C., Bournoville, R. & Bonnemain, J.-L. (1996) Water deficit-induced changes in concentrations in proline and some other amino acids in the phloem sap of alfalfa. *Plant Physiology*, **111**, 109–113.

Gutbrodt, B., Dorn, S., Unsicker, S.B. & Mody, K. (2012) Species-specific responses of herbivores to within-plant and environmentally mediated between-plant variability in plant chemistry. *Chemoecology*, **22**, 101–111.

Gutbrodt, B., Mody, K. & Dorn, S. (2011) Drought changes plant chemistry and causes contrasting responses in lepidopteran herbivores. *Oikos*, **120**, 1732–1740.

Hale, B.K., Bale, J.S., Pritchard, J., Masters, G.J. & Brown, V.K. (2003) Effects of host plant drought stress on the performance of the bird cherry-oat aphid, *Rhopalosiphum padi* (L.): a mechanistic analysis. *Ecological Entomology*, **28**, 666–677.

Hillstrom, M., Meehan, T.D., Kelly, K. & Lindroth, R.L. (2010) Soil carbon and nitrogen mineralization following deposition of insect frass and greenfall from forests under elevated CO_2 and O_3. *Plant and Soil*, **336**, 75–85.

Hopkins, R.J., van Dam, N.M. & van Loon, J.J.A. (2009) Role of glucosinolates in insect-plant relationships and multitrophic interactions. *Annual Review of Entomology*, **54**, 57–83.

Huberty, A.F. & Denno, R.F. (2004) Plant water stress and its consequences for herbivorous insects: a new synthesis. *Ecology*, **85**, 1383–1398.

IPCC (2014) Summary for policymakers. *Climate change 2014: impacts, adaptation, and vulnerability. Part A: Global and Sectoral Aspects. Contribution of Working Group II to the Fifth Assessment Report of the Intergovernmental Panel on Climate Change* (eds C.B. Field, V.R. Barros, D.J. Dokken, K.J. Mach, M.D. Mastrandrea, T.E. Bilir, M. Chatterjee, K.L. Ebi, Y.O. Estrada, R.C. Genova, B. Girma, E.S. Kissel, A.N. Levy, S. MacCracken, P.R. Mastrandrea & L.L. White), pp. 1–32. Cambridge University Press, Cambridge, United Kingdom and New York, NY, USA.

Jactel, H., Petit, J., Desprez-Loustau, M.-L., Delzon, S., Piou, D., Battisti, A. & Koricheva, J. (2012) Drought effects on damage by forest insects and pathogens: a meta-analysis. *Global Change Biology*, **18**, 267–276.

Jamieson, M.A., Trowbridge, A.M., Raffa, K.F. & Lindroth, R.L. (2012) Consequences of climate warming and altered precipitation patterns for plant-insect and multitrophic interactions. *Plant Physiology*, **160**, 1719–1727.

Johnson, S.N., Bezemer, T.M. & Jones, T.H. (2008) Linking aboveground and belowground herbivory. *Root Feeders - an ecosystem perspective* (eds S.N. Johnson & P.J. Murray), pp. 153–170. CABI, Wallingford, UK.

Johnson, S.N., Clark, K.E., Hartley, S.E., Jones, T.H., McKenzie, S.W. & Koricheva, J. (2012) Aboveground–belowground herbivore interactions: a meta-analysis. *Ecology*, **93**, 2208–2215.

Johnson, S.N., Erb, M. & Hartley, S.E. (2016a) Roots under attack: contrasting plant responses to below- and aboveground insect herbivory. *New Phytologist*, **210**, 413–418.

Johnson, S.N., Hawes, C. & Karley, A.J. (2009) Reappraising the role of plant nutrients as mediators of interactions between root- and foliar-feeding insects. *Functional Ecology*, **23**, 699–706.

Johnson, S.N., Lopaticki, G., Barnett, K., Facey, S.L., Powell, J.R. & Hartley, S.E. (2016b) An insect ecosystem engineer alleviates drought stress in plants without increasing plant susceptibility to an above-ground herbivore. *Functional Ecology*, **30**, 894–902.

Johnson, S.N., Lopaticki, G. & Hartley, S.E. (2014) Elevated atmospheric CO_2 triggers compensatory feeding by root herbivores on a C_3 but not a C_4 grass. *PLOS ONE*, **9**, e90251.

Johnson, S.N. & McNicol, J.W. (2010) Elevated CO_2 and aboveground–belowground herbivory by the clover root weevil. *Oecologia*, **162**, 209–216.

Johnson, S.N., Mitchell, C., McNicol, J.W., Thompson, J. & Karley, A.J. (2013) Downstairs drivers – root herbivores shape communities of above-ground herbivores and natural enemies via changes in plant nutrients. *Journal of Animal Ecology*, **82**, 1021–1030.

Johnson, S.N. & Riegler, M. (2013) Root damage by insects reverses the effects of elevated atmospheric CO_2 on eucalypt seedlings. *PLOS ONE*, **8**, e79479.

Johnson, S.N., Staley, J.T., McLeod, F.A.L. & Hartley, S.E. (2011) Plant-mediated effects of soil invertebrates and summer drought on above-ground multitrophic interactions. *Journal of Ecology*, **99**, 57–65.

Kaplan, I. & Denno, R.F. (2007) Interspecific interactions in phytophagous insects revisited: a quantitative assessment of competition theory. *Ecology Letters*, **10**, 977–994.

Kaplan, I., Halitschke, R., Kessler, A., Sardanelli, S. & Denno, R.F. (2008) Constitutive and induced defenses to herbivory in above- and belowground plant tissues. *Ecology*, **89**, 392–406.

Kaplan, I., Sardanelli, S. & Denno, R.F. (2009) Field evidence for indirect interactions between foliar-feeding insect and root-feeding nematode communities on *Nicotiana tabacum*. *Ecological Entomology*, **34**, 262–270.

Kirtman, B., Power, S.B., Adedoyin, J.A., Boer, G.J., Bojariu, R., Camilloni, I., Doblas-Reyes, F.J., Fiore, A.M., Kimoto, M., Meehl, G.A., Prather, M., Sarr, A., Schär, C., Sutton, R., van Oldenborgh, G.J., Vecchi, G. & Wang, H.J. (2014) Near-term Climate Change: Projections and Predictability. *Climate Change 2013: The Physical Science Basis. Contribution of Working Group I to the Fifth Assessment Report of the Intergovernmental Panel on Climate Change* (eds T.F. Stocker, D. Qin, G.-K. Plattner, M. Tignor, S.K. Allen, J. Boschung, A. Nauels, Y. Xia, V. Bex & P.M. Midgley). Cambridge University Press, Cambridge, UK and New York, NY, USA.

Koricheva, J., Larsson, S. & Haukioja, E. (1998) Insect performance on experimentally stressed woody plants: A meta-analysis. *Annual Review of Entomology*, **43**, 195–216.

Kovats, R.S., Valentini, R., Bouwer, L.M., Georgopoulou, E., Jacob, D., Martin, E., Rounsevell, M. & Soussana, J.-F. (2014) 2014: Europe. *Climate Change 2014: Impacts, Adaptation, and Vulnerability. Part B: Regional Aspects. Contribution of Working Group II to the Fifth Assessment Report of the Intergovernmental Panel on Climate Change* (eds V.R. Barros, C.B. Field, D.J. Dokken, M.D. Mastrandrea, K.J. Mach, T.E. Bilir, M. Chatterjee, K.L. Ebi, Y.O. Estrada, R.C. Genova, B. Girma, E.S. Kissel, A.N. Levy, S. MacCracken, P.R. Mastrandrea & L.L. White), pp. 1267–1326. Cambridge University Press, Cambridge, UK and New York, NY, USA.

Kvedaras, O.L., Keeping, M.G., Goebel, F.-R. & Byrne, M.J. (2007) Water stress augments silicon-mediated resistance of susceptible sugarcane cultivars to the stalk borer *Eldana saccharina* (Lepidoptera : Pyralidae). *Bulletin of Entomological Research*, **97**, 175–183.

Lafrance, J. (1968) The seasonal movements of wireworms (Coleoptera: Elateridae) in relation to soil moisture and temperature in the organic soils of southwestern Quebec. *Canadian Entomologist*, **100**, 801–807.

Larsson, S. (1989) Stressful times for the plant stress – insect performance hypothesis. *Oikos*, **56**, 277–283.

Lawton, J.H. & Hassell, M.P. (1981) Asymmetrical competition in insects. *Nature*, **289**, 793–795.

Lu, X., Siemann, E., Wei, H., Shao, X. & Ding, J. (2015) Effects of warming and nitrogen on above- and below-ground herbivory of an exotic invasive plant and its native congener. *Biological Invasions*, **17**, 2881–2892.

Masters, G.J. & Brown, V.K. (1992) Plant-mediated interactions between two spatially separated insects. *Functional Ecology*, **6**, 175–179.

Masters, G.J. & Brown, V.K. (1997) Host-plant mediated interactions between spatially separated herbivores: effects on community structure. *Multitrophic Interactions in Terrestrial Systems – 36th Symposium of the British Ecological Society* (eds A.C. Gange & V.K. Brown), pp. 217–237. Blackwell Science, Oxford.

Masters, G.J., Brown, V.K. & Gange, A.C. (1993) Plant mediated interactions between above- and below-ground insect herbivores. *Oikos*, **66**, 148–151.

McKenzie, S.W., Hentley, W.T., Hails, R.S., Jones, T.H., Vanbergen, A.J. & Johnson, S.N. (2013a) Global climate change and aboveground-belowground insect interactions. *Frontiers in Plant Science*, **4**, 412.

McKenzie, S.W., Vanbergen, A.J., Hails, R.S., Jones, T.H. & Johnson, S.N. (2013b) Reciprocal feeding facilitation by above- and below-ground herbivores. *Biology Letters*, **9**, 20130341.

Meehan, T.D., Couture, J.J., Bennett, A.E. & Lindroth, R.L. (2014) Herbivore-mediated material fluxes in a northern deciduous forest under elevated carbon dioxide and ozone concentrations. *New Phytologist*, **204**, 397–407.

Moran, N.A. & Whitham, T.G. (1990) Interspecific competition between root-feeding and leaf-galling aphids mediated by host-plant resistance. *Ecology*, **71**, 1050–1058.

Newington, J.E., Setälä, H., Bezemer, T.M. & Jones, T.H. (2004) Potential effects of earthworms on leaf-chewer performance. *Functional Ecology*, **18**, 746–751.

Newman, J.A., Anand, M., Henry, H.A.L., Hunt, S. & Gedalof, Z. (2011) *Climate Change Biology*. CABI, Wallingford, Oxfordshire, UK.

Parker, W.E. & Howard, J.J. (2001) The biology and management of wireworms (*Agriotes* spp.) on potato with particular reference to the UK. *Agricultural and Forest Entomology*, **3**, 85–98.

Parmesan, C., Root, T.L. & Willig, M.R. (2000) Impacts of extreme weather and climate on terrestrial biota. *Bulletin of the American Meteorological Society*, **81**, 443–450.

Poveda, K., Steffan-Dewenter, I., Scheu, S. & Tscharntke, T. (2003) Effects of below- and above-ground herbivores on plant growth, flower visitation and seed set. *Oecologia*, **135**, 601–605.

Poveda, K., Steffan-Dewenter, I., Scheu, S. & Tscharntke, T. (2005) Effects of decomposers and herbivores on plant performance and aboveground plant–insect interactions. *Oikos*, **108**, 503–510.

Price, P.W. (1991) The plant vigor hypothesis and herbivore attack. *Oikos*, **62**, 244–251.

Quinn, M.A. & Hall, M.H. (1992) Compensatory response of a legume root-nodule system to nodule herbivory by *Sitona hispidulus*. *Entomologia Experimentalis et Applicata*, **64**, 167–176.

Ratzka, A., Vogel, H., Kliebenstein, D.J., Mitchell-Olds, T. & Kroymann, J. (2002) Disarming the mustard oil bomb. *Proceedings of the National Academy of Sciences of the United States of America*, **99**, 11223–11228.

Riutta, T., Slade, E.M., Bebber, D.P., Taylor, M.E., Malhi, Y., Riordan, P., Macdonald, D.W. & Morecroft, M.D. (2012) Experimental evidence for the interacting effects of forest edge, moisture and soil macrofauna on leaf litter decomposition. *Soil Biology & Biochemistry*, **49**, 124–131.

Robinson, E.A., Ryan, G.D. & Newman, J.A. (2012) A meta-analytical review of the effects of elevated CO_2 on plant–arthropod interactions highlights the importance of interacting environmental and biological variables. *New Phytologist*, **194**, 321–336.

Ryalls, J.M.W. (2016) *The impacts of climate change and belowground herbivory on aphids via primary metabolites*. PhD thesis (Appendix IV, doi: 10.6084/m9.figshare.3114637.v1), Western Sydney University, Australia.

Ryalls, J.M.W., Moore, B.D., Riegler, M., Gherlenda, A.N. & Johnson, S.N. (2015) Amino acid-mediated impacts of elevated carbon dioxide and simulated root herbivory on aphids are neutralised by increased air temperatures. *Journal of Experimental Botany*, **66**, 613–623.

Ryalls, J.M.W., Riegler, M., Moore, B.D., Lopaticki, G. & Johnson, S.N. (2013) Effects of elevated temperature and CO_2 on aboveground–belowground systems: a case study with plants, their mutualistic bacteria and root/shoot herbivores. *Frontiers in Plant Science*, **4**, 445.

Salt, D.T., Fenwick, P. & Whittaker, J.B. (1996) Interspecific herbivore interactions in a high CO_2 environment: root and shoot aphids feeding on *Cardamine*. *Oikos*, **77**, 326–330.

Scherber, C., Gladbach, D.J., Stevnbak, K., Karsten, R.J., Schmidt, I.K., Michelsen, A., Albert, K.R., Larsen, K.S., Mikkelsen, T.N., Beier, C. & Christensen, S. (2013) Multi-factor climate change effects on insect herbivore performance. *Ecology and Evolution*, **3**, 1449–1460.

Scheu, S. (2003) Effects of earthworms on plant growth: patterns and perspectives. *Pedobiologia*, **47**, 846–856.

Scheu, S., Theenhaus, A. & Jones, T.H. (1999) Links between the detritivore and the herbivore system: effects of earthworms and Collembola on plant growth and aphid development. *Oecologia*, **119**, 541–551.

Schultz, J.C., Appel, H.M., Ferrieri, A.P. & Arnold, T.M. (2013) Flexible resource allocation during plant defense responses. *Frontiers in Plant Science*, **4**, 324.

Seal, D.R., McSorley, R. & Chalfant, R.B. (1992) Seasonal abundance and spatial distribution of wireworms (Coleoptera, Elateridae) in Georgia sweet potato fields. *Journal of Economic Entomology*, **85**, 1802–1808.

Simpson, K.L.S., Jackson, G.E. & Grace, J. (2012) The response of aphids to plant water stress – the case of *Myzus persicae* and *Brassica oleracea* var. *capitata*. *Entomologia Experimentalis et Applicata*, **142**, 191–202.

Soler, R., Bezemer, T.M., Van der Putten, W.H., Vet, L.E.M. & Harvey, J.A. (2005) Root herbivore effects on above-ground herbivore, parasitoid and hyperparasitoid performance via changes in plant quality. *Journal of Animal Ecology*, **74**, 1121–1130.

Soler, R., Erb, M. & Kaplan, I. (2013) Long distance root–shoot signalling in plant–insect community interactions. *Trends in Plant Science*, **18**, 149–156.

Staley, J.T., Hodgson, C.J., Mortimer, S.R., Morecroft, M.D., Masters, G.J., Brown, V.K. & Taylor, M.E. (2007a) Effects of summer rainfall manipulations on the abundance and vertical distribution of herbivorous soil macro-invertebrates. *European Journal of Soil Biology*, **43**, 189–198.

Staley, J.T. & Johnson, S.N. (2008) Climate change impacts on root herbivores. *Root Feeders – an ecosystem perspective* (eds S.N. Johnson & P.J. Murray), pp. 192–213. CABI, Wallingford, UK.

Staley, J.T., Mortimer, S.R., Masters, G.J., Morecroft, M.D., Brown, V.K. & Taylor, M.E. (2006) Drought stress differentially affects leaf-mining species. *Ecological Entomology*, **31**, 460–469.

Staley, J.T., Mortimer, S.R. & Morecroft, M.D. (2008) Drought impacts on above–belowground interactions: do effects differ between annual and perennial host species? *Basic and Applied Ecology*, **9**, 673–681.

Staley, J.T., Mortimer, S.R., Morecroft, M.D., Brown, V.K. & Masters, G.J. (2007b) Summer drought alters plant-mediated competition between foliar- and root-feeding insects. *Global Change Biology*, **13**, 866–877.

Stevnbak, K., Scherber, C., Gladbach, D.J., Beier, C., Mikkelsen, T.N. & Christensen, S. (2012) Interactions between above- and belowground organisms modified in climate change experiments. *Nature Climate Change*, **2**, 805–808.

Tariq, M., Rossiter, J.T., Wright, D.J. & Staley, J.T. (2013a) Drought alters interactions between root and foliar herbivores. *Oecologia*, **172**, 1095–1104.

Tariq, M., Wright, D.J., Bruce, T.J.A. & Staley, J.T. (2013b) Drought and root herbivory interact to alter the response of above-ground parasitoids to aphid infested plants and associated plant volatile signals. *PLOS ONE*, **8**, e69013.

Tariq, M., Wright, D.J., Rossiter, J.T. & Staley, J.T. (2012) Aphids in a changing world: testing the plant stress, plant vigour and pulsed stress hypotheses. *Agricultural and Forest Entomology*, **14**, 177–185.

Thompson, L., Thomas, C.D., Radley, J.M.A., Williamson, S. & Lawton, J.H. (1993) The effect of earthworms and snails in a simple plant community. *Oecologia*, **95**, 171–178.

van Dam, N.M., Harvey, J.A., Wäckers, F.L., Bezemer, T.M., van der Putten, W.H. & Vet, L.E.M. (2003) Interactions between aboveground and belowground induced responses against phytophages. *Basic and Applied Ecology*, **4**, 63–77.

van Dam, N.M. & Heil, M. (2010) Multitrophic interactions below and above ground: *en route* to the next level. *Journal of Ecology*, **99**, 77–88.

van der Putten, W.H., Vet, L.E.M., Harvey, J.A. & Wäckers, F.L. (2001) Linking above- and belowground multitrophic interactions of plants, herbivores, pathogens, and their antagonists. *Trends in Ecology and Evolution*, **16**, 547–554.

Vandegehuchte, M.L., de la Peña, E. & Bonte, D. (2010) Interactions between root and shoot herbivores of *Ammophila arenaria* in the laboratory do not translate into correlated abundances in the field. *Oikos*, **119**, 1011–1019.

Wand, S.J.E., Midgley, G.F., Jones, M.H. & Curtis, P.S. (1999) Responses of wild C_4 and C_3 grass (Poaceae) species to elevated atmospheric CO_2 concentration: a meta-analytic test of current theories and perceptions. *Global Change Biology*, **5**, 723–741.

Wardle, D.A., Bardgett, R.D., Klironomos, J.N., Setälä, H., van der Putten, W.H. & Wall, D.H. (2004) Ecological linkages between aboveground and belowground biota. *Science*, **304**, 1629–1633.

Waring, C.L. & Cobb, N.S. (1992) The impact of plant stress on herbivore population dynamics. *Insect-Plant Interactions* (ed E.A. Bernays). CRC-Press, Boca Raton.

White, T.C.R. (1969) An index to measure weather-induced stress of trees associated with outbreaks of psyllids in Australia. *Ecology*, **50**, 905–909.

White, T.C.R. (1984) The abundance of invertebrate herbivores in relation to the availability of nitrogen in stressed food plants. *Oecologia*, **63**, 90–105.

Wurst, S., Dugassa-Gobena, D., Langel, R., Bonkowski, M. & Scheu, S. (2004a) Combined effects of earthworms and vesicular-arbuscular mycorrhizas on plant and aphid performance. *New Phytologist*, **163**, 169–176.

Wurst, S., Dugassa-Gobena, D. & Scheu, S. (2004b) Earthworms and litter distribution affect plant-defensive chemistry. *Journal of Chemical Ecology*, **30**, 691–701.

Wurst, S. & Jones, T.H. (2003) Indirect effects of earthworms (*Aporrectodea caliginosa*) on an above-ground tritrophic interaction. *Pedobiologia*, **47**, 91–97.

Wurst, S. & Rillig, M.C. (2011) Additive effects of functionally dissimilar above- and belowground organisms on a grassland plant community. *Journal of Plant Ecology*, **4**, 221–227.

Zaller, J.G., Parth, M., Szunyogh, I., Semmelrock, I., Sochurek, S., Pinheiro, M., Frank, T. & Drapela, T. (2013) Herbivory of an invasive slug is affected by earthworms and the composition of plant communities. *BMC Ecology*, **13**, 20.

Zvereva, E.L. & Kozlov, M.V. (2012) Sources of variation in plant responses to belowground insect herbivory: a meta-analysis. *Oecologia*, **169**, 441–452.

13

Forest Invertebrate Communities and Atmospheric Change

Sarah L. Facey and Andrew N. Gherlenda

Hawkesbury Institute for the Environment, Western Sydney University, NSW 2751, Australia

Summary

Predicting the responses of invertebrate species, and the communities they form, to global change is one of the great challenges facing modern ecology. Invertebrates play vitally important roles in forests, underpinning fundamental ecosystem processes like nutrient cycling and pollination. Changes in the composition of our atmosphere, associated with increased levels of carbon dioxide (CO_2) and ozone (O_3), have the potential to affect the abundance, diversity and structure of invertebrate communities and the ecosystems they support. This chapter reviews the findings from the body of work looking at the responses of invertebrates to changes in CO_2 and O_3 concentrations with a special focus on the results from Free-Air Enrichment studies. The most consistent finding across the studies we review is the idiosyncratic nature of the responses of invertebrate species to the elevation of CO_2 and/or O_3. This finding can be explained to some extent by bottom-up and top-down processes. These include the species- and genotype-specific responses of host plant chemistry and differences in the abilities of individual insect species to physiologically and behaviourally overcome changes in resource quality. Although evidence is clearly mixed, certain general conclusions can be made regarding the influence of CO_2 and/or O_3 on invertebrates. Forest invertebrate herbivores tend to respond negatively to elevated concentrations of CO_2. This response is likely due to diminished food-plant quality. Conversely, predators and parasitoids may benefit under enriched-CO_2 conditions as prey susceptibility increases. Elevated O_3 concentrations generally have opposing effects: herbivores show a tendency to consume more and develop faster while higher trophic levels experience declines in performance. Therefore, simultaneous elevation of both gases, such as is found in reality, may moderate the effects of either gas in isolation. There also appears to be some capacity for invertebrate communities to rebound over time, as evidenced by long-term studies. From the few community-level studies available, the current conclusion is that the structure of invertebrate communities will not be strongly disrupted by increases in CO_2 and O_3. This suggests that the ecosystem processes underpinned by these communities may be maintained under future atmospheres in these systems, though more work is needed.

Looking forward, we emphasize the critical need for long-term studies of invertebrate responses at the population and community-level within natural systems. Such studies will be particularly important in tropical regions where no such information currently exists. Studies incorporating multiple climatic and atmospheric factors will also be of great value, such as those looking at the

Global Climate Change and Terrestrial Invertebrates, First Edition. Edited by Scott N. Johnson and T. Hefin Jones.
© 2017 John Wiley & Sons, Ltd. Published 2017 by John Wiley & Sons, Ltd.

combined effects of atmospheric change and alterations in water availability. These studies will allow us to better predict the effects of future climates on these fundamental ecological systems.

13.1 Why Are Forest Invertebrate Communities Important?

Invertebrates form the foundation of terrestrial ecosystems, far outnumbering their vertebrate counterparts in terms of abundance, biomass and diversity (Wilson, 1987). In addition to comprising unrivalled levels of biodiversity – such as that found in the tropics – forest arthropod communities have multi-faceted roles in ecosystem functioning. Herbivorous arthropods play a central part in nutrient cycling through consuming plant material and returning nutrients to the forest floor as frass (Hunter, 2001; Meehan et al., 2014; Gherlenda et al., 2016a), with others decomposing leaf litter and non-living materials (Speight et al., 1999). The feeding and burrowing of these detritivorous invertebrates contributes to soil formation, fertility and health which, in turn, determine the capacity of the system to sustain primary production (Schowalter, 2006). Many insects are also involved in the pollination of a wide range of plant species, further supporting plant growth and diversity (Potts et al., 2010). At higher trophic levels, arthropods form the diet of insectivorous vertebrates such as birds and reptiles (see Thomas et al., this volume), as well as invertebrate predators and parasitoids. In addition to underpinning and contributing to the biodiversity and functioning of forest ecosystems, forest invertebrates play roles in other systems. Arthropod natural enemies, such as parasitoid wasps for instance, can have beneficial functions as pest-suppressors in agricultural systems and forests may act as reservoirs of these insects at the landscape-scale (Bianchi et al., 2006; Bianchi et al., 2008).

Whilst the majority of invertebrates arguably have supporting, stabilising roles in the maintenance of diverse forest ecosystems (Schowalter, 2006), there is the potential for invertebrates to become pests, particularly in species-poor plantations (reviewed in Hartley, 2002), when they consume abnormally high amounts of plant material. Further, natural and managed systems alike are at increasing risk from non-native, invasive arthropods (Tobin et al., 2014); a recent study estimated that wood-boring invasive insects cost 1.7 billion USD annually in the United States, in government expenditure alone (Aukema et al., 2011). Thus, forest invertebrates have strong ecological and economic importance.

Clearly, forest ecosystems – and the invertebrate communities which they encompass – represent invaluable global resources from several different viewpoints. Forests are vital from an ecological perspective, providing clean air and other ecosystem services. In addition, forests often harbour great biodiversity – indeed, 15 of the 25 global biodiversity hot spots are areas of tropical forest (Myers et al., 2000). On the other hand, forests can be viewed as economic sources of capital from forestry and other related industries. Perhaps most importantly in the context of this chapter, the world's major forests represent vast carbon stores and the largest terrestrial sink at a time when there is increasing pressure to reduce emissions of carbon dioxide (CO_2). Tropical forests, for instance, comprise 50% of the carbon stored in terrestrial biomass (Ainsworth & Long, 2005). Further, it has been estimated that, globally, forests are responsible for sequestering up to 60% of the carbon emitted from burning fossil fuels (Pan et al., 2011), highlighting the potential for the world's forests to slow increases in atmospheric CO_2 concentrations.

13.2 Atmospheric Change and Invertebrates

Changes in the composition of the Earth's atmosphere, associated with emissions of CO_2 and other greenhouse gases, have the potential to affect invertebrates and the ecosystems they underpin.

Predicting the responses of species, and the communities they comprise, to global change is one of the great challenges facing modern ecology (Gilman et al., 2010; Andrew et al., 2013; Facey et al., 2014). Many studies have looked at the implications of altered atmospheres for natural and farmed systems, with most measuring plant responses (Coviella & Trumble, 1999). So far, the majority of studies considering the effects of elevated greenhouse gas concentrations on plants and the animals they support have been concerned with the direct and indirect effects of carbon dioxide (CO_2) and tropospheric ozone (O_3).

As for other less-studied atmospheric gases, mono-nitrogen oxides (NO_x) and sulphur dioxide (SO_2) are two indirect greenhouse gases known to be detrimental to plant growth (Allen, 1992), through wet and dry deposition and the resulting acidification of ecosystems. Continued emissions of these gases are likely to have negative indirect consequences for heterotrophic food webs, the specifics of which have been reviewed elsewhere (Greaver et al., 2012; Park, 2014). In addition, atmospheric NO_x also play a role in the formation of tropospheric ozone, by reacting with other gases in the presence of sunlight, further exacerbating the negative effects of NO_x gases.

The effects of the two direct greenhouse gases nitrous oxide (N_2O) and methane (CH_4) on invertebrates will be indirect through their general warming effects on average global temperatures. The consequences of global warming will be profound for ectothermic arthropods (reviewed in Bale et al., 2002), but are beyond the scope of this chapter. This chapter reviews invertebrate community responses to changes in CO_2 and O_3 concentrations with a special focus on the results from Free-Air Enrichment studies (Fig. 13.1). Where possible, we also compare findings with experiments from more controlled settings, as well as those from non-forest ecosystems.

13.3 Responses of Forest Invertebrates to Elevated Carbon Dioxide Concentrations

13.3.1 Herbivores

Whilst instances of direct effects of elevated CO_2 on insect herbivores are scarce in the literature (but see Awmack et al., 1997; Stange, 1997; Mondor et al., 2004), the indirect consequences, mediated by physiological and morphological changes in plants, can be extensive. Elevated atmospheric concentrations of CO_2 have been shown to stimulate plant growth across a range of systems, through so-called 'carbon fertilisation' effects (Lamarche et al., 1984). These larger plants often have changed primary and secondary chemistry as a result of physiological changes occurring in the plant in response to increased CO_2 availability; indeed, a recent meta-analysis reported an average increase in plant tissue C:N ratios of 19% when grown under elevated CO_2 conditions (Robinson et al., 2012). Given the reliance of arthropods on their food sources to acquire nitrogen for growth (Mattson, 1980), changes occurring in plants in response to elevated CO_2 conditions often lead to negative effects for herbivorous invertebrates. Numerous studies in multiple systems have shown decreases in insect herbivore development rates, conversion efficiency, pupal mass and survival under elevated CO_2, which may or may not be mitigated by increases in consumption rates – so called 'compensatory feeding' (Bezemer & Jones, 1998; Buse et al., 1998; Zvereva & Kozlov, 2006; Stiling & Cornelissen, 2007; Robinson et al., 2012). In addition to compensatory feeding, insect herbivores may display other behavioural or physiological responses to altered foliar nutrition in an attempt to minimise impacts on their growth and development resulting from nutritionally sub-optimal food. Generalist herbivores may choose to preferentially feed on tree species or genotypes which are minimally impacted by changes in CO_2 concentrations (Agrell et al., 2005). Further, some insect herbivores can

Figure 13.1 A typical FACE set-up in a forest system (*Eucalyptus* 'EucFACE', Sydney, Australia). Studies from FACE sites have benefits over those in more controlled, simplified environments in that they integrate the complex of biotic and abiotic factors occurring in the system

increase their nitrogen use efficiencies when feeding on leaves grown under elevated CO_2 in order to overcome nutrient reductions (Williams et al., 1998; Hättenschwiler & Schafellner, 2004).

Insects belonging to the order Lepidoptera are the most commonly studied within forested FACE and open-top chamber sites (Table 13.1). In these systems, elevated CO_2 has been observed to increase lepidopteran larval developmental time and reduce larval survival, relative growth rate and pupal mass (Roth et al., 1998; Lindroth et al., 2002; Stiling et al., 2002; Knepp et al., 2007). Evidence has also been found for compensatory feeding by leaf mining Lepidoptera in the field (Stiling et al., 1999, 2002, 2003), further supporting the findings from glasshouse work. Thus, as seen in more controlled experimental systems, the overall effect of elevated CO_2 on forest invertebrate larval performance has proved to be negative. However, other field-based studies have found no effect (Kopper et al., 2001; Kopper & Lindroth, 2003) or positive effects (Hättenschwiler & Schafellner, 2004; Couture et al., 2012; Couture & Lindroth, 2012) on larval performance parameters under elevated CO_2. Such variation in the direction of change and/or the responsiveness of insect performance to elevated CO_2 may be due to differences in the responses of their food sources, that is, variation between individual tree species. For instance, Hättenschwiler and Schafellner (2004) observed a 30% decrease in the relative growth rate of gypsy moth larvae under elevated CO_2 when feeding on oak, whereas relative growth rate increased by 29% when feeding on hornbeam. In a similar study, Lindroth and Kinney (1998) reported a 30% decrease in the growth rate of gypsy moth larvae under elevated CO_2 conditions when feeding on aspen, while growth rate was not altered when larvae fed on maple. Different genotypes of the same tree species can also display this disparity in insect herbivore responses (Roth et al., 1998; Lindroth et al., 2002; Agrell et al., 2005). Experiments conducted with plants grown in greenhouse conditions have observed similar, generally negative, trends in insect performance (Stiling & Cornelissen, 2007; Robinson et al., 2012), as well as host-plant dependent differences in insect performance (Gherlenda et al., 2015), mirroring those results obtained from field experiments under elevated CO_2 conditions. Thus, studies in various settings have revealed generally negative effects of elevated CO_2 on invertebrate herbivores, though there is considerable variation between study species.

The differing responses of insects to various food-plant species and genotypes may be explained by the individualistic responses of plant chemistry to elevated CO_2. Secondary compounds present within leaves may increase in response to elevated CO_2, and the degree of this increase may be species dependent (Ryan et al., 2010). Elevated CO_2 may also alter the responsiveness of some secondary metabolite pathways, with some being more responsive than others to elevated CO_2 (Bidart-Bouzat & Imeh-Nathaniel, 2008; Lindroth, 2012). Further, differences in foliar chemistry under elevated CO_2 concentrations may be determined by the duration of CO_2 exposure or the developmental stage of the trees during CO_2 fumigation (Gunderson & Wullschleger, 1994; Curtis, 1996; Ellsworth et al., 2012; Couture et al., 2014).

Some secondary compounds are known to bind nitrogen, potentially reducing its availability to insects (Schweitzer et al., 2008), thus impacting their growth and development. However, insects have developed strategies, such as high gut pH or gut surfactants, to overcome the protein binding activity of some secondary compounds (Salminen & Karonen, 2011; Vihakas et al., 2015). On the other hand, secondary compounds may deter or prohibit insects from compensatory feeding (Roth et al., 1998), limiting their ability to mitigate CO_2-induced reductions in foliar nutritional quality. Thus, species and/or genotype-specific differences in foliar chemistry under elevated CO_2, coupled with the plasticity of herbivorous insect species in dealing with these changes in foliar chemistry, may account for the varying response in insect performance under elevated CO_2 conditions.

Table 13.1 Summary of the literature observing individual insect responses to elevated concentrations of CO_2 and O_3 conducted within forested FACE sites.

Factor(s)	Insect species	Tree species	Insect parameters	Summary outcomes	Reference
$CO_2 \times O_3$	Forest tent caterpillar *Malacosoma disstria*	Birch (*Betula papyrifera*) Aspen (*Populus tremuloides*, genotype 216 and 259)	Consumption	Increased consumption at eCO_2 and eO_3. Larvae preferred aspen over birch at eCO_2 which was opposite at eO_3. No $CO_2 \times O_3$ interaction on consumption. Aspen genotype affected consumption at elevated CO_2 and O_3.	(Agrell et al. 2005)
$CO_2 \times O_3$	Aphid *Cepegilletteα betulaefoliae*	Birch (*Betula papyrifera*)	Developmental time Adult weight RGR	No effect of CO_2 or O_3 on aphid performance.	(Awmack et al. 2004)
$CO_2 \times O_3$	Gypsy moth *Lymantria dispar*	Birch (*Betula papyrifera*) Aspen (*Populus tremuloides*)	Developmental time Survivorship Pupal weight Fecundity	Increased survival and egg production at eCO_2. eO_3 decreased survival, female pupal weight and egg production while developmental time was increased. Combination of eCO_2 and O_3 ameliorated effects on developmental time and survival. Tree species influenced fecundity.	(Couture & Lindroth 2012)
$CO_2 \times O_3$	Gypsy moth *Lymantria dispar* Forest tent caterpillar *Malacosoma disstria*	Birch (*Betula papyrifera*) Aspen (*Populus tremuloides*)	Growth Consumption ECD ECI	Growth of forest tent caterpillar increased at eCO_2. $CO_2 \times O_3$ ameliorated effects on growth and consumption. eO_3 decreased growth, ECD and ECI while consumption increased at eO_3 for both insect species. Gypsy moth leaf consumption decreased at eCO_2. eCO_2 did not affect growth, ECD and ECI for both species while leaf consumption by forest tent caterpillars was not affected.	(Couture et al. 2012)
CO_2	Gypsy moth *Lymantria dispar*	Oak (*Quercus petraea*) Hornbeam (*Carpinus betulus*) Beech (*Fagus sylvatica*)	RGR	RGR decreased on oak. RGR increased on hornbeam. RGR did not change on beech.	(Hättenschwiler & Schafellner 2004)

(Continued)

Table 13.1 (Continued)

Factor(s)	Insect species	Tree species	Insect parameters	Summary outcomes	Reference
$CO_2 \times O_3$	forest tent caterpillar *Malacosoma disstria* dipteran parasitoid *Compsilura concinnata*	Aspen (*Populus tremuloides*, genotype 216 and 259)	Developmental time Larval mass Pupal mass Survivorship (parasitoid)	*Herbivore* eCO_2 alone had little effect on herbivore performance and was reduced when CO_2 and O_3 were combined on genotype 216. *Parasitoid* Survivorship overall decreased at O_3 and was offset in combination with CO_2. An $O_3 \times CO_2 \times$ genotype interaction was detected for survival. O_3 alone improved herbivore performance.	(Holton et al. 2003)
CO_2	*Antheraea polyphemus*	White oak (*Quercus alba*) Black oak (*Quercus velutina*)	Developmental time Survival RGR ECD AD ECI	Reduced performance at elevated CO_2. Larvae increased ECI at elevated CO_2 for white oak only.	(Knepp et al. 2007)
$CO_2 \times O_3$	Whitemarked Tussock Moth *Orgyia leucostigma*	Birch (*Betula papyrifera*)	Developmental time Pupal mass	eCO_2 and eO_3 had no effect on insect performance.	(Kopper et al. 2001)
$CO_2 \times O_3$	Forest tent caterpillar *Malacosoma disstria*	Aspen (*Populus tremuloides*, genotype 216 and 259)	Developmental time Consumption Pupal mass	eCO_2 had minimal effect on insect performance. $CO_2 \times O_3$ interaction observed in which CO_2 negated the effect of eO_3. Significant $CO_2 \times O_3 \times$ genotype observed. Larval performance improved at eO_3.	(Kopper & Lindroth 2003)
CO_2	Whitemarked Tussock Moth *Orgyia leucostigma*	Aspen (*Populus tremuloides*, genotype 216 and 259)	Developmental time Pupal mass	eCO_2 reduced insect performance. $CO_2 \times$ genotype interaction observed.	(Lindroth et al. 2002)
$CO_2 \times O_3$	Forest tent caterpillar *Malacosoma disstria* Aphid *Chaitophorus stevensis*	Aspen (*Populus tremuloides*)	Pupal mass Abundance	Forest tent caterpillar pupal mass reduced at eCO_2 which was magnified in combination with eO_3. eO_3 alone increased pupal mass. Aphid abundance unaffected by CO_2 or O_3.	(Percy et al. 2002)

AD - approximate digestibility; ECD - efficiency of conversion of digested food; ECI - efficiency of conversion of ingested food; eCO_2 - elevated CO_2 concentration; eO_3 - elevated O_3 concentration; RGR - relative growth rate.

13.3.2 Natural Enemies

Work from glasshouse experiments has shown that for invertebrate predators and parasitoids, elevated CO_2 could lead to improved fitness through increases in the susceptibility of their prey/hosts to attack, as mediated by changes in host-plant quality (e.g., Chen et al., 2005; Coll & Hughes, 2008). Parasitoids and predators have generally received little attention in forested FACE sites owing to the difficulties in monitoring their performance. One study by Stiling et al. (2002) conducted in open top chambers found that parasitoid attack rates on leaf mining Lepidoptera increased by 80% under elevated CO_2 in a scrub oak forest. This was potentially due to prolonged larval development times which led to increased susceptibility to parasitoid attack; this finding is in line with expectations from previous work in controlled settings. Increased parasitism rates, in combination with decreased herbivore performance under elevated CO_2, may suggest that insect herbivores will experience population reductions in forest systems as CO_2 concentrations continue to rise. On the other hand, predatory and parasitic organisms will also be subjected to lower quality food sources in terms of the abundance, size and chemistry of their prey/hosts (reviewed in Facey et al., 2014). A study by Percy et al. (2002) found that although the density of aphid natural enemies was increased under future levels of CO_2, the synchrony between the two trophic levels was disrupted, leading to aphid population release and thus reductions in pest regulation. Conversely, in a separate study by Holton et al. (2003), parasitoid fitness (survival, developmental time and adult mass) was not affected by elevated CO_2. Clearly, as with herbivorous insects, the responses of higher trophic levels to altered CO_2 concentrations appear to be idiosyncratic.

13.3.3 Community-Level Responses

Given the idiosyncratic nature of the responses of all levels of the system (i.e., plants, herbivores and higher trophic levels) to elevated concentrations of CO_2, making informed predictions about the fate of forest ecosystems under future atmospheres becomes increasingly daunting (e.g., Hillstrom et al., 2014). It is problematic to scale up and extrapolate the responses of one individual species seen under glasshouse conditions to the world at large around us, given the likelihood that the response of the species in question does not adequately represent those of the multiple components in natural systems. Instead, these types of studies have greater value from a mechanistic perspective and allow the formation of ideas and hypotheses for testing in the field (discussed further in Lindroth & Raffa, this volume). In reality, organisms exist in nature as parts of intricately interwoven communities, linked to one another by trophic and other biotic interactions, with each subject to change in the abiotic environment. Studies from FACE sites integrate the complex biotic and abiotic factors occurring in the system (Hillstrom & Lindroth, 2008), though these too, are not without limitations inherent to plot-level studies of this type (discussed in Lindroth & Raffa, this volume; Moise & Henry, 2010). Community-level assessments of forest invertebrate responses to elevated CO_2 have been conducted in various systems and will be the focus of this section.

As with experiments under more controlled and simplified settings, the results of community level field studies are mixed, though some patterns can be gleaned from the literature. Community and population-level studies have, on the whole, confirmed the negative effects of elevated CO_2 concentrations on forest invertebrate herbivores (Table 13.2). Other studies have shown alterations in the species richness of certain groups under future CO_2 levels, leading to increases in species dominance and changes in feeding guild composition (Altermatt, 2003; Sanders et al., 2004). In a *Eucalyptus* FACE system, Facey et al. (2016) found declines in the abundance of several orders and feeding guilds of invertebrates under elevated conditions, including chewing herbivores and omnivores. This effect was also seen for *Eucalyptus*-feeding psyllids at the same site (Gherlenda et al., 2016b). As for the

Table 13.2 Summary of the literature considering the effects of elevated CO_2 and/or O_3 concentrations on multiple species of invertebrates in forest/woodland ecosystems. Results show the effects of elevated concentrations of the given gas, that is, a reported decrease represents a decline under elevated compared with ambient conditions. FACE = Free Air Carbon dioxide Enrichment; OTC = Open Top Chamber.

Factor(s)	Ecosystem	Organism	Method	Summary of outcomes	Reference
CO_2	Loblolly pine plantation (Duke FACE)	Arthropods	Sticky traps	Decrease in herbivore abundance, Increase in carnivorous orders.	(Hamilton et al. 2012)
CO_2	Loblolly pine plantation (Duke FACE)	Micro-arthropods	Soil samples	Reduced micro-arthropod abundance.	(Hansen et al. 2001)
CO_2	Sweetgum *Liquidambar styraciflua* (Oak Ridge FACE)	Arthropods	Pitfall traps and sweep-netting	No effect on herbivory, total abundance or richness.	(Sanders et al. 2004)
CO_2	Scrub-oak community (Kennedy Space Centre OTC)	Leafmining insects	Census	Reduced abundance of leafminers. Increased mortality and leaf consumption.	(Stiling et al. 1999)
CO_2	Scrub-oak community (Kennedy Space Centre OTC)	Insect herbivores	Leaf damage, census of leafminers	Reduced herbivory and survival of leafminers. Increased mortality by parasitoids.	(Stiling et al. 2002)
CO_2	Scrub-oak community (Kennedy Space Centre OTC)	Insect herbivores	Leaf damage, census of leafminers	Reduced leaf damage by leafminers and chewers.	(Stiling et al. 2003)
CO_2	Scrub-oak community (Kennedy Space Centre OTC)	Leafmining insects	Census	Reduced abundance.	(Stiling & Cornelissen 2007)
CO_2	Scrub-oak community (Kennedy Space Centre OTC)	Arthropods	Pitfall traps and Sticky traps	Increased herbivore abundance. No effect on abundance of other arthropods.	(Stiling et al. 2010)
CO_2	Mature mixed forest (WebFACE)	Soil Collembola	Soil samples	Reduced abundance. No effect on community richness.	(Xu et al. 2013)
CO_2	Mature mixed forest (WebFACE)	Arthropods	Canopy beating	Reduced arthropod diversity.	(Altermatt 2003)
$CO_2 \times O_3$	Mixed Aspen plantation (Aspen FACE)	Soil Invertebrates	Soil samples	Reduced abundance under single exposure to CO_2 or O_3. No effect on abundance when CO_2 and O_3 are combined.	(Loranger et al. 2004)

Treatment	Study system	Insect group	Sampling method	Findings	Reference
$CO_2 \times O_3$	Trembling aspen *Populus tremuloides* (Aspen FACE)	Forest tent caterpillar, aphids and their predators	Census	Reduced pupal mass under elevated CO_2 (forest tent caterpillar). Increased aphid predator abundance. Increased pupal mass under elevated O_3 (forest tent caterpillar). O_3 increased aphid abundance and reduced aphid predator abundances. No effect on aphid abundance when CO_2 and O_3 are combined.	(Percy et al. 2002)
$CO_2 \times O_3$	Mixed Aspen plantation (Aspen FACE)	Insect herbivores	Census	Weak, idiosyncratic effects – unlikely to influence composition or abundance.	(Hillstrom et al. 2014)
$CO_2 \times O_3$	Mixed Aspen plantation (Aspen FACE)	Arthropods	Pan traps	Elevated O_3 reduced total abundance. Altered community composition. Elevated CO_2 had no effect on abundance. No effect on species richness.	(Hillstrom & Lindroth 2008)
CO_2	Loblolly pine plantation (Duke FACE)	Leaf chewing insects	Leaf damage	Reduced herbivory at elevated CO_2.	(Knepp et al. 2005)
CO_2	Loblolly pine plantation (Duke FACE)	Herbivorous insects	Leaf damage	CO_2 by plant species interaction, either reduced or no effect on herbivory.	(Hamilton et al. 2004)
$CO_2 \times O_3$	Mixed Aspen plantation (Aspen FACE)	Herbivorous insects	Leaf damage	Increased herbivory at elevated CO_2. Reduced herbivory at elevated O_3.	(Meehan et al. 2014; Couture et al. 2015)
CO_2	*Eucalyptus* woodland (EucFACE)	Leaf chewing insects	Frass collections	No effect on frass deposition.	(Gherlenda et al. 2016a)
CO_2	*Eucalyptus* woodland (EucFACE)	Arthropods	Pitfall traps, Sticky traps and Vortis sampling	Reduced total arthropod abundance, reduced abundance of certain trophic groups including herbivores, omnivores, scavengers and parasitoid wasps. No change in community structure.	(Facey et al. 2016)
CO_2	*Eucalyptus* woodland (EucFACE)	Leaf chewing insects	Leaf Damage	No effect of CO_2 on leaf damage.	(Gherlenda 2016)
CO_2	*Eucalyptus* woodland (EucFACE)	Psyllid community	Census	Reduced abundance of psyllid community at elevated CO_2.	(Gherlenda et al. 2016b)

mechanisms behind the patterns we see, the effects of elevated CO_2 concentrations on populations of herbivorous invertebrates may be dependent on underlying chemical changes occurring in food plants (e.g., Stiling et al., 2003), and has been shown to vary between host species (Altermatt, 2003); both of these findings are in line with expectations from previous work in the field and glasshouses. However, evidence for this mechanism in the most commonly used plant quality measure, C:N ratio, is still somewhat elusive, with some studies showing reductions in herbivore performance with no concurrent expected increase in this metric (Hamilton et al., 2004, 2012; Facey et al., 2016).

The assessment of CO_2-induced changes in the abundances of leaf chewing insects in forested FACE sites has often been determined by the measurement of leaf damage. Based on evidence from glasshouse experiments, two contrasting hypotheses could be drawn; (i) leaf damage will decrease as a result of reduced survival and performance of leaf chewers under elevated CO_2, or (ii) leaf damage and frass inputs will increase as leaf chewing insects display compensatory feeding under elevated CO_2. Declines in leaf damage have been observed in forested FACE sites, indicating reduced abundances of leaf chewing insects (Hamilton et al., 2004; Knepp et al., 2005). Conversely, increased leaf damage has also been observed, indicating either an increase in leaf-chewer abundance or increased compensatory feeding behaviour by comparable numbers of leaf-chewers (Hillstrom et al., 2010; Meehan et al., 2014; Couture et al., 2015). Further complicating matters, no difference in leaf damage or frass production from leaf chewing insects has been observed in an evergreen *Eucalyptus* forest under elevated CO_2 (Gherlenda, 2016; Gherlenda et al., 2016a). *Eucalyptus* trees generally grow in nutrient-poor soils and are often water limited (Duursma et al., 2015; Crous et al., 2015) which may pose challenges for insect leaf chewers. Under these conditions, water availability has been shown to be a stronger driver than elevated CO_2 in altering the abundance of leaf chewing insects (Gherlenda, 2016; Gherlenda et al., 2016a).

For higher trophic levels, evidence from community-level studies supporting the findings from population and species-level experiments is somewhat less apparent. In support of the predicted increases in natural enemy performance, increases in the abundance of the carnivorous orders Araneae and Hymenoptera (Hamilton et al., 2012) and parasitism rates (Stiling et al., 1999, 2002, 2003) have been reported in myrtle oak and loblolly pine forests, coupled with declines in herbivorous groups. However, predator abundance and richness declined in a sweetgum plantation under future concentrations of CO_2 (Sanders et al., 2004). In a *Eucalyptus* woodland, the abundance of parasitic wasps declined significantly under elevated CO_2, but the abundance of more generalist predators like spiders was unresponsive (Facey et al., 2016). Another study in a scrub oak forest found no effects of elevated CO_2 on predatory group abundance (Stiling et al., 2010). Hence, the evidence concerning the responses of natural enemies to elevated CO_2 remains mixed.

Studies looking at soil fauna under elevated CO_2 conditions have consistently found reductions in the abundance of micro-arthropods like springtails and mites (Hansen et al., 2001; Loranger et al., 2004; Xu et al., 2013), perhaps as a result of microbial changes occurring in the soil (Hansen et al., 2001). In a microcosm study, Meehan et al. (2010) found that earthworms and springtails reared on leaf litter from elevated CO_2-grown aspen trees had lower individual and population growth rates, respectively. These declines in performance were attributed to increased tannin concentrations in the leaf litter, coupled with reduced nitrogen concentrations. Soil invertebrates play important roles in litter decomposition and nutrient cycling in forest ecosystems (Hansen et al., 2001); reductions in their population densities could thus feasibly have ramifications for soil health and primary production.

Given that most studies to date report changes in the abundances of different functional groups in one direction or another, one might feasibly expect to see alterations in overall invertebrate community composition occurring as a result of elevated CO_2. Very few studies have considered the effects of future CO_2 levels on composition, with only one study reporting significant effects of CO_2 fumigation on community composition which varied over time (Hillstrom & Lindroth, 2008).

Of the other three studies looking at community composition, none reported significant changes occurring in response to CO_2 elevation (Sanders et al., 2004; Hillstrom et al., 2014; Facey et al., 2016). The same lack of CO_2 effect was true for a similar field-based experiment in a grassland ecosystem (Villalpando et al., 2009). Thus, there is limited evidence for the idea that elevated CO_2 will have strong effects on invertebrate community structure.

The most emergent trend in the literature is the idiosyncratic nature of the responses of forest invertebrates to elevated CO_2, at all levels of organisation from individual species to communities. Indeed, it is this finding which could explain the general lack of community shifts under elevated CO_2, as the many components within natural systems respond in opposing ways, effectively 'cancelling each other out' (Hamilton et al., 2012). Documented responses are individualistic both in terms of being specific to the organisms involved as well as the time point sampled; community shifts have been reported to vary from month to month and year to year in several studies (e.g., Hillstrom & Lindroth, 2008; Gherlenda, 2016; Gherlenda et al., 2016a). Looking forward, there is the potential for any emerging responses to adjust to increasing levels of CO_2. Indeed, it could be expected that the higher levels of parasitism and predator abundance reported in some studies (e.g., Hamilton et al., 2012) may not be supported in the long term if populations of herbivores continue to decline, as food resources become limiting for them. On the other hand, observed negative trends could recover – for instance, the reduced abundances of soil micro–arthropods may be expected to rebound over time as litter inputs increase in the system (Hansen et al., 2001). Additionally, concurrent studies in a scrub oak ecosystem by Stiling et al., have found that although herbivore (leaf miners and tiers) densities initially decreased under elevated CO_2 at the beginning of their experiment (Stiling et al., 2002), the total number of herbivores eventually showed a trend towards increasing under these conditions (Stiling et al., 2009, 2010). This was due to boosted plant growth, supporting higher numbers of insects at reduced densities per leaf (Stiling et al., 2009). Indeed, this supports findings from more recent work in aspen plantations documenting increases in herbivore-mediated nutrient flux (Meehan et al., 2014; Couture et al., 2015). These examples, both theoretical and empirical, support the need for long-term monitoring in order to produce the most realistic insights into how forest invertebrate communities will respond to atmospheric change (Couture & Lindroth, 2013).

13.4 Responses of Forest Invertebrates to Elevated Ozone Concentrations

13.4.1 Herbivores

In contrast to the generally positive effects of CO_2 on plant growth, tropospheric O_3 causes a cascade of negative effects on plant physiology, leading to growth inhibition and accelerated senescence (reviewed in Lindroth, 2010). Plants grown in the presence of O_3 have reduced biomass, lower photosynthetic capacity and reduced leaf area; a recent meta-analysis indicated that, at current levels of O_3, plants have decreased biomass of 7% compared with those grown at pre-industrial levels (Wittig et al., 2009). As with elevated concentrations of CO_2, O_3 can cause altered plant secondary chemistry, but in generally different ways – O_3 exposure often leads to increased levels of compounds like phenolics and reduced carbohydrate concentrations (Valkama et al., 2007; Lindroth, 2010). On one hand, this changed plant chemistry under O_3 fumigation can lead to reduced insect herbivore performance, sometimes accompanied by compensatory feeding behaviour and thus higher levels of herbivory (Coleman & Jones, 1988a; Jones & Coleman, 1988; Peltonen et al., 2010). On the other hand, some herbivorous insects achieve higher pupal masses and develop faster under O_3 fumigation (Manninen et al., 2000; Percy et al., 2002; Kopper & Lindroth, 2003; Holton et al., 2003; reviewed in Valkama et al., 2007). Hence, as with CO_2, there is variation in the strength and/or direction of the responses of insect herbivores to elevated O_3. In addition to positive and negative responses in herbivore performance, herbivorous insects have also been observed to show no response to elevated O_3 in a forested (Awmack et al., 2004) and a soybean (Hamilton et al., 2005; Dermody et al., 2008) FACE site. In those systems where herbivore performance is sensitive to O_3 fumigation, changes in the abundance and success of these groups may lead to alterations in competitive dynamics with other species, highlighting the need for studies integrating the biotic interactions occurring within systems (Coleman & Jones, 1988b).

In common with elevated CO_2, the effects of O_3 fumigation can vary between plant species, and this has been shown to lead to changes in the feeding preferences of herbivorous insects (Agrell et al., 2005). Likewise, there is a need to consider the effects of O_3 fumigation over longer timescales as systems may adapt to changes in O_3 levels and fumigation may have variable effects at different stages of invertebrate life cycles. For instance, willow leaf beetles showed a preference for ozone treated cotton-wood foliage and consumed higher levels of leaf material, suggesting that they may benefit from O_3 exposure; however over the longer term, these beetles were less fecund on O_3 treated plants and showed a preference for ovipositing on non-fumigated controls (Coleman & Jones, 1988a; Jones & Coleman, 1988). Hence, while studies have generally revealed positive effects of elevated O_3 on invertebrate herbivores, there is considerable variation between study species, as seen with studies looking at elevated CO_2.

13.4.2 Natural Enemies

For higher trophic levels, increased levels of O_3 may have negative consequences as a result of the O_3-catalysed breakdown of biogenic volatile organic compounds (BVOCs) which play a role in host location (Butler et al., 2009; Himanen et al., 2009; Pinto et al., 2010). Whilst studies have indeed shown that ozone can reduce parasitoid larval survivorship and searching efficiency (Gate et al., 1995; Holton et al., 2003), others in crop systems have shown that ozone fumigation does not significantly affect BVOC-dependent predatory mite or parasitoid search behaviour and success (Vuorinen et al., 2004; Pinto et al., 2007, 2008). These studies suggest that the foraging abilities of some invertebrates may be conserved under a higher O_3 atmosphere, particularly those sensitive to the presence of BVOCs which are resistant to O_3-breakdown (Pinto et al., 2007). Studies looking at parasitoid foraging efficiency and performance under elevated O_3 conditions in forest systems are currently lacking (with the exception of Holton et al., 2003), though based on these studies from other systems it could be expected that any responses of individual parasitoid species will be idiosyncratic in nature.

Predatory and parasitic invertebrate responses to elevated O_3 environments will depend on the behavioural and physiological responses of their prey (Lindroth, 2010) which will also be variable. One might expect that elevated O_3 may tend to benefit predators and parasitoids in those instances where prey/hosts achieve higher pupal masses. However, Holton et al. (2003) showed that while forest tent caterpillars growing under elevated O_3 were larger, parasitoid larval survivorship decreased despite access to larger food resources. This was potentially a result of chemical limitations in the host relative to the parasitoids' requirements (Holton et al., 2003). Additional causes for the reduction in parasitoid survival under elevated O_3 could be the result of improved host immunological function (Hättenschwiler & Schafellner, 1999) and faster host developmental times (Karowe & Schoonhoven, 1992). In terms of host behaviour, Mondor et al. (2004) found that aphids on trembling aspen had improved predatory escape behaviour under O_3 fumigation as they became more sensitive to alarm pheromones released by conspecifics indicating the presence of danger. This could lead to higher population densities of these insects and potential pest outbreaks as the concentration of O_3 continues to increase. Overall, there is consistent evidence for negative effects of O_3 fumigation on forest invertebrate natural enemies.

13.4.3 Community-Level Studies

Few studies exist looking at the responses of forest system invertebrates at the population and community level to elevated O_3 concentrations, with those that do often also looking at the simultaneous elevation of CO_2 (the focus of the next section). The results from such studies have provided support for the negative effects of O_3 elevation on higher trophic levels in some cases. For instance, Percy et al. (2002) found that aphid populations at Aspen FACE were more likely to grow under O_3 fumigation as natural enemies became less abundant, pointing to a loss of effective top-down aphid regulation under future O_3 scenarios. On the other hand, in a separate study on a different aphid species at Aspen FACE, Awmack et al. (2004) showed that although aphid populations grew under O_3 fumigation, they were unaffected when predators and parasitoid populations were also present. This suggests that, for this species, effective pest-suppression will be maintained under future atmospheres.

One of the three existing community-level studies by Hillstrom and Lindroth (2008) found that O_3 fumigation resulted in declines in the abundance of parasitoid wasps whilst tending to positively influence the abundance of sucking herbivores. They found time-dependent alterations in arthropod community composition in response to O_3 elevation, as a result of these population changes. In contrast, in a more recent herbivore-only study by Hillstrom and Lindroth (2014) at the same site, the opposite pattern emerged with sucking herbivores experiencing population declines relative to chewing herbivores. In an environmental gradient study, Jones and Paine (2006) also showed that chewing herbivores became more dominant compared with sucking herbivores on oak trees, as levels of O_3 pollution increased. And so, as with CO_2, elevated O_3 may cause alterations in invertebrate population abundance to the extent that species (or group) dominance alters within the ecosystem. At the same time, there is also a general lack of evidence for community shifts occurring as a result of O_3 fumigation, though studies are very limited.

Clearly, like CO_2-related changes in invertebrate communities, the effects of O_3 will likely be idiosyncratic and difficult to predict, with changes in the direction of community responses documented through time and between systems within experimental sites. In contrast with the effects of elevated CO_2 (general reductions in herbivorous groups, some increases in predators), there is evidence for enriched O_3 environments benefitting herbivores and disadvantaging parasitoids. For soil micro-arthropods however, there is evidence for more consistent declines in the abundance of microbe-feeding mites under both CO_2 and O_3 exposure (Loranger et al., 2004).

13.5 Interactions Between Carbon Dioxide and Ozone

Given the generally opposing directions of the effects of elevated CO_2 and O_3 on insect performance (Fig. 13.2), the study of the simultaneous elevation of these gases will be important in order to gain realistic insights into the effects of future atmospheres on invertebrate communities (e.g., Valkama et al., 2007; Lindroth, 2010). Despite the logistical issues associated with studies investigating multiple global change drivers, several studies have considered the effect of the elevation of these two gases on invertebrates.

In some cases, elevated O_3 has been demonstrated to mitigate the negative effects of elevated CO_2 on insect performance. The benefits of O_3 fumigation on forest tent caterpillar performance were offset when CO_2 concentrations were also elevated (Kopper & Lindroth, 2003; Holton et al., 2003), as were the negative effects of O_3 elevation on its parasitoid (Holton et al., 2003). Conversely, other studies have found evidence for interactive effects of CO_2 and O_3. A study by Percy et al. (2002), also on the forest tent caterpillar, found that the negative effects of elevated CO_2 on female pupal mass

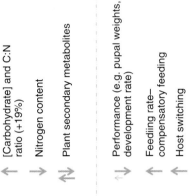

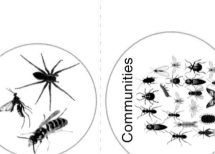

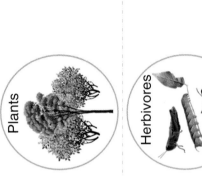

Ozone

Plants

Photosynthetic capacity

Biomass, leaf area

Plant secondary metabolites

Herbivores

Performance (e.g. pupal weights, development rate)

Feeding rate–compensatory feeding

Altered host preferences

Susceptibility to predation

Natural Enemies

Performance (e.g. parasitism rate, larval survival)

Synchrony with hosts

Abundance

Communities

Altered community structure and composition

Carbon Dioxide

Plants

[Carbohydrate] and C:N ratio (+19%)

Nitrogen content

Plant secondary metabolites

Herbivores

Performance (e.g. pupal weights, development rate)

Feeding rate–compensatory feeding

Host switching

Susceptibility to predation

Natural Enemies

Performance (e.g. parasitism rate, fecundity survival)

Synchrony with hosts

Abundance, species richness

Communities

Altered community structure and composition

Figure 13.2 Conceptual diagram summarising the main directions of the responses of invertebrates to elevated CO_2 and O_3. Arrows show the direction of responses based on the literature; the size of the arrows give some idea of our degree of confidence (i.e., the number of studies showing the same result).

were worsened under higher levels of O_3. They also found no evidence for a moderating effect of CO_2 and O_3 on aphid infestations, which were just as severe under simultaneous fumigation as when either gas was elevated singly (Percy et al., 2002). In a separate study, whitemarked tussock moths showed a non-significant trend toward reduced pupal mass under combined elevation of CO_2 and O_3, with no trends seen when either gas were elevated in isolation (Kopper et al., 2001).

At the community level, Loranger et al. (2004) found that soil micro-arthropod abundance was not significantly different compared with ambient conditions when CO_2 and O_3 were simultaneously elevated. Similarly, Hillstrom and Lindroth (2014) found that simultaneous elevated CO_2 and O_3 exposure had little effect on invertebrate community composition in aspen and birch woodland. In an earlier study (Hillstrom & Lindroth, 2008), they found generally opposing responses of invertebrate groups to O_3 and CO_2. In addition, fumigation with CO_2, O_3 and both CO_2 and O_3 caused idiosyncratic effects on invertebrate community composition which varied over time. Thus, simultaneous elevation of CO_2 and O_3, as will likely be experienced in nature, can cause invertebrates to respond in ways that are inconsistent with studies elevating either gas in isolation.

13.6 Conclusions and Future Directions

The use of forested FACE sites to assess the impacts of atmospheric change on insects is fundamental in understanding how insects will respond to future atmospheres, from the individual to the community level. Whilst the responses of invertebrates to elevated CO_2 and O_3 are highly species dependent, some general trends can be gleaned from the literature. Forest invertebrate herbivores tend to respond negatively to elevated concentrations of CO_2, whilst predators and parasitoids may benefit under enriched-CO_2 conditions. Under O_3 enrichment, on the other hand, herbivores show a tendency to consume more and develop faster, with higher trophic levels showing more consistent declines in performance. Therefore, simultaneous elevation of both gases, such as will occur in reality, could serve to moderate the effects of either gas in isolation (Kopper & Lindroth, 2003; Agrell et al., 2005; Couture & Lindroth, 2012), though evidence for this is currently mixed. This, coupled with the general lack of observed community shifts occurring under CO_2 and/or O_3 elevation, suggests that the ecosystem processes provided by invertebrate communities may not be strongly responsive to simultaneous changes in atmospheric concentrations of these gases (Hillstrom et al., 2014). However, there are still too few experiments at sufficient scales, or across the range of systems needed, to draw solid conclusions about the fate of invertebrate communities. Tropical ecosystems, for instance, remain unrepresented despite their immense significance as both biological hotspots and prominent carbon sinks (Ainsworth & Long, 2005).

As pointed out by other authors (e.g., Lindroth, 2010; Couture & Lindroth, 2013), the integration of other environmental factors beyond CO_2 and O_3 will now be required to better understand how invertebrate communities and the ecosystems they form will respond to global change. For instance, increases in temperature and changes in the timing and amount of rainfall will have profound effects on invertebrates. Effectively assessing these environmental variables will be a logistically difficult but necessary challenge at the scale of forested FACE systems. The incorporation of multiple global change variables over long time periods, coupled with greater representation across forest ecosystems, will further advance our understanding of how invertebrate communities will respond to climate change. This will allow improvements in model accuracy and more effective conservation strategy development, to ensure the sustainable management of the world's forests and the invertebrate communities that underpin them.

Acknowledgements

We are grateful to Philip Smith for proofreading this chapter.

References

Agrell, J., Kopper, B., McDonald, E.P. & Lindroth, R.L. (2005) CO_2 and O_3 effects on host plant preferences of the forest tent caterpillar (*Malacosoma disstria*). *Global Change Biology*, **11**, 588–599.

Ainsworth, E.A. & Long, S.P. (2005) What have we learned from 15 years of free-air CO_2 enrichment (FACE)? A meta-analytic review of the responses of photosynthesis, canopy properties and plant production to rising CO_2. *New Phytologist*, **165**, 351–372.

Allen, L.H. Jr. (1992) Free-air CO_2 enrichment field experiments: An historical overview. *Critical Reviews in Plant Sciences*, **11**, 121–134.

Altermatt, F. (2003) Potential negative effects of atmospheric CO_2-enrichment on insect communities in the canopy of a mature deciduous forest in Switzerland. *Mitteilungen der Schweizerischen Entomologischen Gesellschaft*, **76**, 191–199.

Andrew, N.R., Hill, S.J., Binns, M., Bahar, M.H., Ridley, E.V, Jung, M.-P., Fyfe, C., Yates, M. & Khusro, M. (2013) Assessing insect responses to climate change: What are we testing for? Where should we be heading? *PeerJ*, **1**, e11.

Aukema, J.E., Leung, B., Kovacs, K., Chivers, C., Britton, K.O., Englin, J., Frankel, S.J., Haight, R.G., Holmes, T.P., Liebhold, A.M., McCullough, D.G. & von Holle, B. (2011) Economic impacts of non-native forest insects in the continental United States. *PLOS ONE*, **6**, e24587.

Awmack, C.S., Harrington, R. & Lindroth, R.L. (2004) Aphid individual performance may not predict population responses to elevated CO_2 or O_3. *Global Change Biology*, **10**, 1414–1423.

Awmack, C., Woodcock, C.M. & Harrington, R. (1997) Climate change may increase vulnerability of aphids to natural enemies. *Ecological Entomology*, **22**, 366–368.

Bale, J.S., Masters, G.J., Hodkinson, I.D., Awmack, C., Bezemer, T.M., Brown, V.K., Butterfield, J., Buse, A., Coulson, J.C., Farrar, J., Good, J.E.G., Harrington, R., Hartley, S., Jones, T.H., Lindroth, R.L., Press, M.C., Symrnioudis, I., Watt, A.D. & Whittaker, J.B. (2002) Herbivory in global climate change research: direct effects of rising temperature on insect herbivores. *Global Change Biology*, **8**, 1–16.

Bezemer, T.M. & Jones, T.H. (1998) Plant-insect herbivore interactions in elevated atmospheric CO_2: quantitative analyses and guild effects. *Oikos*, **82**, 212–222.

Bianchi, F.J.J.A., Booij, C.J.H. & Tscharntke, T. (2006) Sustainable pest regulation in agricultural landscapes: a review on landscape composition, biodiversity and natural pest control. *Proceedings of the Royal Society B: Biological Sciences*, **273**, 1715–1727.

Bianchi, F.J.J.A., Goedhart, P.W. & Baveco, J.M. (2008) Enhanced pest control in cabbage crops near forest in The Netherlands. *Landscape Ecology*, **23**, 595–602.

Bidart-Bouzat, M.G. & Imeh-Nathaniel, A. (2008) Global change effects on plant chemical defenses against insect herbivores. *Journal of Integrative Plant Biology*, **50**, 1339–1354.

Buse, A., Good, J.E.G., Dury, S., & Perrins, C.M. (1998) Effects of elevated temperature and carbon dioxide on the nutritional quality of leaves of oak (*Quercus robur* L.) as food for the winter moth (*Operophtera brumata* L.). *Functional Ecology*, **12**, 742–749.

Butler, C.D., Beckage, N.E. & Trumble, J.T. (2009) Effects of terrestrial pollutants on insect parasitoids. *Environmental Toxicology and Chemistry*, **28**, 1111–1119.

Chen, F., Ge, F. & Parajulee, M. (2005) Impact of elevated CO_2 on tri-trophic interaction of *Gossypium hirsutum*, *Aphis gossypii*, and *Leis axyridis*. *Environmental Entomology*, **34**, 37–46.

Coleman, J.S. & Jones, C.G. (1988a) Plant stress and insect performance: cottonwood, ozone and a leaf beetle. *Oecologia*, **76**, 57–61.

Coleman, J.S. & Jones, C.G. (1988b) Acute ozone stress on Eastern Cottonwood (*Populus deltoides* Bartr.) and the pest potential of the aphid, *Chaitophorus populicola* Thomas (Homoptera: Aphididae). *Environmental Entomology*, **17**, 207–212.

Coll, M. & Hughes, L. (2008) Effects of elevated CO_2 on an insect omnivore: A test for nutritional effects mediated by host plants and prey. *Agriculture, Ecosystems & Environment*, **123**, 271–279.

Couture, J.J., Holeski, L.M., & Lindroth, R.L. (2014) Long-term exposure to elevated CO_2 and O_3 alters aspen foliar chemistry across developmental stages. *Plant, Cell & Environment*, **37**, 758–765.

Couture, J.J. & Lindroth, R.L. (2012) Atmospheric change alters performance of an invasive forest insect. *Global Change Biology*, **18**, 3543–3557.

Couture, J.J. & Lindroth, R.L. (2013) Impacts of atmospheric change on tree–arthropod interactions. *Climate Change, Air pollution and Global Challenges: Understanding and Perspectives from Forest Research*, 1st ed. (eds R. Matyssek, N. Clarke, P. Cudlin, T.N. Mikkelsen, J.-P. Tuovinen, G. Wieser & E. Paoletti), pp. 227–248. Elsevier Ltd., Oxford.

Couture, J.J., Meehan, T.D., Kruger, E.L. & Lindroth, R.L. (2015) Insect herbivory alters impact of atmospheric change on northern temperate forests. *Nature Plants*, **1**, 15016.

Couture, J.J., Meehan, T.D. & Lindroth, R.L. (2012) Atmospheric change alters foliar quality of host trees and performance of two outbreak insect species. *Oecologia*, **168**, 863–876.

Coviella, C.E. & Trumble, J.T. (1999) Effects of elevated atmospheric carbon dioxide on insect-plant interactions. *Conservation Biology*, **13**, 700–712.

Crous, K.Y., Ósvaldsson, A. & Ellsworth, D.S. (2015) Is phosphorus limiting in a mature *Eucalyptus* woodland? Phosphorus fertilisation stimulates stem growth. *Plant and Soil*, **391**, 293–305.

Curtis, P.S. (1996) A meta-analysis of leaf gas exchange and nitrogen in trees grown under elevated carbon dioxide. *Plant, Cell and Environment*, **19**, 127–137.

Dermody, O., O'Neill, B.F., Zangerl, A.R., Berenbaum, M.R. & DeLucia, E.H. (2008) Effects of elevated CO_2 and O_3 on leaf damage and insect abundance in a soybean agroecosystem. *Arthropod-Plant Interactions*, **2**, 125–135.

Duursma, R.A., Gimeno, T.E., Boer, M.M., Crous, K.Y., Tjoelker, M.G. & Ellsworth, D.S. Canopy leaf area of a mature evergreen *Eucalyptus* woodland does not respond to elevated atmospheric [CO_2] but tracks water availability. *Global Change Biology*, **22**, 1666–1676.

Ellsworth, D.S., Thomas, R., Crous, K.Y., Palmroth, S., Ward, E., Maier, C., DeLucia, E., & Oren, R. (2012) Elevated CO_2 affects photosynthetic responses in canopy pine and subcanopy deciduous trees over 10 years: a synthesis from Duke FACE. *Global Change Biology*, **18**, 223–242.

Facey, S.L., Fidler, D.B., Rowe, R.C., Bromfield, L.M., Nooten, S.S., Staley, J.T., Ellsworth, D.S. & Johnson, S.N. (2016) Atmospheric change causes declines in woodland arthropods and impacts specific trophic groups. *Agricultural and Forest Entomology*, doi: 10.111/afe.12190.

Facey, S.L., Ellsworth, D.S., Staley, J.T., Wright, D.J. & Johnson, S.N. (2014) Upsetting the order: how climate and atmospheric change affects herbivore–enemy interactions. *Current Opinion in Insect Science*, **5**, 66–74.

Gate, I.M., McNeill, S. & Ashmore, M.R. (1995) Effects of air pollution on the searching behaviour of an insect parasitoid. *Water, Air, and Soil Pollution*, **85**, 1425–1430.

Gherlenda, A.N. (2016) Effects of CO_2 and temperature on *Eucalyptus* insect herbivores from individuals to communities. PhD Thesis, Western Sydney University, Australia.

Gherlenda, A.N., Crous, K.Y., Moore, B.D., Haigh, A.M., Johnson, S.N. & Riegler, M. (2016a) Precipitation, not CO_2 enrichment, drives insect herbivore frass deposition and subsequent nutrient dynamics in a mature *Eucalyptus* woodland. *Plant and Soil*, **399**, 29–39.

Gherlenda, A.N., Esveld, J.L., Hall, A.A.G., Duursma, R.A. & Riegler, M. (2016b) Boom and bust: rapid feedback responses between insect outbreak dynamics and canopy leaf area impacted by rainfall and CO_2. *Global Change Biology*, **22**, doi: 10.1111/gcb.13334.

Gherlenda, A.N., Haigh, A.M., Moore, B.D., Johnson, S.N. & Riegler, M. (2015) Responses of leaf beetle larvae to elevated [CO_2] and temperature depend on *Eucalyptus* species. *Oecologia*, **177**, 607–617.

Gilman, S.E., Urban, M.C., Tewksbury, J., Gilchrist, G.W. & Holt, R.D. (2010) A framework for community interactions under climate change. *Trends in Ecology & Evolution*, **25**, 325–331.

Greaver, T.L., Sullivan, T.J., Herrick, J.D., Barber, M.C., Baron, J.S., Cosby, B.J., Deerhake, M.E., Dennis, R.L., Dubois, J.-J.B., Goodale, C.L., Herlihy, A.T., Lawrence, G.B., Liu, L., Lynch, J.A. & Novak, K.J. (2012) Ecological effects of nitrogen and sulfur air pollution in the US: What do we know? *Frontiers in Ecology and the Environment*, **10**, 365–372.

Gunderson, C.A. & Wullschleger, S.D. (1994) Photosynthetic acclimation in trees to rising atmospheric CO_2: a broader perspective. *Photosynthesis Research*, **39**, 369–388.

Hamilton, J.G., Dermody, O., Aldea, M., Zangerl, A.R., Rogers, A., Berenbaum, M.R. & Delucia, E.H. (2005) Anthropogenic changes in tropospheric composition increase susceptibility of soybean to insect herbivory. *Environmental Entomology*, **34**, 479–485.

Hamilton, J.G., Zangerl, A.R., Berenbaum, M.R., Pippen, J., Aldea, M. & DeLucia, E.H. (2004) Insect herbivory in an intact forest understory under experimental CO_2 enrichment. *Oecologia*, **138**, 566–573.

Hamilton, J., Zangerl, A.R., Berenbaum, M.R., Sparks, J.P., Elich, L., Eisenstein, A. & DeLucia, E.H. (2012) Elevated atmospheric CO_2 alters the arthropod community in a forest understory. *Acta Oecologica*, **43**, 80–85.

Hansen, R.A., Williams, R.S., Degenhardt, D.C. & Lincoln, D.E. (2001) Non-litter effects of elevated CO_2 on forest floor microarthropod abundances. *Plant and Soil*, **236**, 139–144.

Hartley, M.J. (2002) Rationale and methods for conserving biodiversity in plantation forests. *Forest Ecology and Management*, **155**, 81–95.

Hättenschwiler, S. & Schafellner, C. (2004) Gypsy moth feeding in the canopy of a CO_2-enriched mature forest. *Global Change Biology*, **10**, 1899–1908.

Hättenschwiler, S. & Schafellner, C. (1999) Opposing effects of elevated CO_2 and N deposition on *Lymantria monacha* larvae feeding on spruce trees. *Oecologia*, **118**, 210–217.

Hillstrom, M.L., Couture, J.J. & Lindroth, R.L. (2014) Elevated carbon dioxide and ozone have weak, idiosyncratic effects on herbivorous forest insect abundance, species richness, and community composition. *Insect Conservation and Diversity*, **7**, 553–562.

Hillstrom, M.L. & Lindroth, R.L. (2008) Elevated atmospheric carbon dioxide and ozone alter forest insect abundance and community composition. *Insect Conservation and Diversity*, **1**, 233–241.

Hillstrom, M., Meehan, T.D., Kelly, K. & Lindroth, R.L. (2010) Soil carbon and nitrogen mineralization following deposition of insect frass and greenfall from forests under elevated CO_2 and O_3. *Plant and Soil*, **336**, 75–85.

Himanen, S.J., Nerg, A.-M., Nissinen, A., Pinto, D.M., Stewart, C.N. Jr, Poppy, G.M. & Holopainen, J.K. (2009) Effects of elevated carbon dioxide and ozone on volatile terpenoid emissions and multitrophic communication of transgenic insecticidal oilseed rape (*Brassica napus*). *New Phytologist*, **181**, 174–186.

Holton, M.K., Lindroth, R.L. & Nordheim, E. V. (2003) Foliar quality influences tree-herbivore-parasitoid interactions: effects of elevated CO_2, O_3, and plant genotype. *Oecologia*, **137**, 233–244.

Hunter, M.D. (2001) Insect population dynamics meets ecosystem ecology: effects of herbivory on soil nutrient dynamics. *Agricultural and Forest Entomology*, **3**, 77–84.

Jones, C.G. & Coleman, J.S. (1988) Plant stress and insect behaviour: Cottonwood, ozone and the feeding and oviposition preference of a beetle. *Oecologia*, **76**, 51–56.

Jones, M.E. & Paine, T.D. (2006) Detecting changes in insect herbivore communities along a pollution gradient. *Environmental Pollution*, **143**, 377–387.

Karowe, D.N. & Schoonhoven, L.M. (1992) Interactions among three trophic levels: the influence of host plant on performance of *Pieris brassicae* and its parasitoid, *Cotesia glomerata. Entomologia Experimentalis et Applicata*, **62**, 241–251.

Knepp, R.G., Hamilton, J.G., Mohan, J.E., Zangerl, A.R., Berenbaum, M.R. & Delucia, E.H. (2005) Elevated CO_2 reduces leaf damage by insect herbivores in a forest community. *New Phytologist*, **167**, 207–218.

Knepp, R.G., Hamilton, J.G., Zangerl, A.R., Berenbaum, M.R. & DeLucia, E.H. (2007) Foliage of oaks grown under elevated CO_2 reduces performance of *Antheraea polyphemus* (Lepidoptera: Saturniidae). *Environmental Entomology*, **36**, 609–617.

Kopper, B.J. & Lindroth, R.L. (2003) Effects of elevated carbon dioxide and ozone on the phytochemistry of aspen and performance of an herbivore. *Oecologia*, **134**, 95–103.

Kopper, B.J., Lindroth, R.L. & Nordheim, E.V. (2001) CO_2 and O_3 Effects on Paper Birch (Betulaceae: *Betula papyrifera*) Phytochemistry and Whitemarked Tussock Moth (Lymantriidae: *Orgyia leucostigma*) Performance. *Environmental Ecology*, **30**, 1119–1126.

Lamarche, V.C. Jr, Graybill, D.A, Fritts, H.C. & Rose, M.R. (1984) Increasing atmospheric carbon dioxide: tree ring evidence for growth enhancement in natural vegetation. *Science*, **225**, 1019–1021.

Lindroth, R.L. (2010) Impacts of elevated atmospheric CO_2 and O_3 on forests: phytochemistry, trophic interactions, and ecosystem dynamics. *Journal of Chemical Ecology*, **36**, 2–21.

Lindroth, R.L. (2012) Atmospheric change, plant secondary metabolites and ecological interactions. In: *The Ecology of Plant Secondary Metabolites: From Genes to Global Processes* (eds G.R. Iason, M. Dicke & S.E. Hartley). Cambridge University Press, Cambridge.

Lindroth, R.L. & Kinney, K.K. (1998) Consequences of enriched atmospheric CO_2 and defoliation for foliar chemistry and gypsy moth performance. *Journal of Chemical Ecology*, **24**, 1677–1695.

Lindroth, R.L., Wood, S.A. & Kopper, B.J. (2002) Response of quaking aspen genotypes to enriched CO_2: Foliar chemistry and tussock moth performance. *Agricultural and Forest Entomology*, **4**, 315–323.

Loranger, G.I., Pregitzer, K.S. & King, J.S. (2004) Elevated CO_2 and O_3 concentrations differentially affect selected groups of the fauna in temperate forest soils. *Soil Biology and Biochemistry*, **36**, 1521–1524.

Manninen, A.-M., Holopainen, T., Lyytikäinen-Saarenmaa, P., & Holopainen, J.K. (2000) The role of low-level ozone exposure and mycorrhizas in chemical quality and herbivore performance on Scots pine seedlings. *Global Change Biology*, **6**, 111–121.

Mattson, W.J. Jr, (1980) Herbivory in relation to plant nitrogen content. *Annual Review of Ecology and Systematics*, **11**, 119–161.

Meehan, T.D., Crossley, M.S. & Lindroth, R.L. (2010) Impacts of elevated CO_2 and O_3 on aspen leaf litter chemistry and earthworm and springtail productivity. *Soil Biology & Biochemistry*, **42**, 1132–1137.

Meehan, T.D., Couture, J.J., Bennett, A.E. & Lindroth, R.L. (2014) Herbivore-mediated material fluxes in a northern deciduous forest under elevated carbon dioxide and ozone concentrations. *New Phytologist*, **204**, 397–407.

Moise, E.R.D. & Henry, H.A.L. (2010) Like moths to a street lamp: exaggerated animal densities in plot-level global change field experiments. *Oikos*, **119**, 791–795.

Mondor, E.B., Tremblay, M.N., Awmack, C.S. & Lindroth, R.L. (2004) Divergent pheromone-mediated insect behaviour under global atmospheric change. *Global Change Biology*, **10**, 1820–1824.

Myers, N., Mittermeier, R.A., Mittermeier, C.G., da Fonseca, G.A.B. & Kent, J. (2000) Biodiversity hotspots for conservation priorities. *Nature*, **403**, 853–858.

Pan, Y., Birdsey, R.A., Fang, J., Houghton, R., Kauppi, P.E., Kurz, W.A., Phillips, O.L., Shvidenko, A., Lewis, S.L., Canadell, J.G., Ciais, P., Jackson, R.B., Pacala, S.W., McGuire, A.D., Piao, S., Rautiainen, A., Sitch, S. & Hayes, D. (2011) A large and persistent carbon sink in the world's forests. *Science*, **333**, 988–993.

Park, C.C. (2014) *Acid Rain: Rhetoric and Reality, Reprinted*. Routledge, Oxon.

Peltonen, P.A., Vapaavuori, E.M., Heinonen, J., Julkunen-Tiitto, R., & Holpainen, J.K. (2010) Do elevated atmospheric CO_2 and O_3 affect food quality and performance of folivorous insects on silver birch? *Global Change Biology*, **16**, 918–935.

Percy, K.E., Awmack, C.S., Lindroth, R.L., Kubiske, M.E., Kopper, B.J., Isebrands, J.G., Pregitzer, K.S., Hendrey, G.R., Dickson, R.E., Zak, D.R.,Oksanen, E., Sober, J., Harrington, R. & Karnosky, D.F. (2002) Altered performance of forest pests under atmospheres enriched by CO_2 and O_3. *Nature*, **420**, 403–407.

Pinto, D.M., Blande, J.D., Nykänen, R., Dong, W.-X., Nerg, A.-M. & Holopainen, J.K. (2007) Ozone degrades common herbivore-induced plant volatiles: does this affect herbivore prey location by predators and parasitoids? *Journal of Chemical Ecology*, **33**, 683–694.

Pinto, D.M., Blande, J.D., Souza, S.R., Nerg, A.-M. & Holopainen, J.K. (2010) Plant volatile organic compounds (VOCs) in ozone (O_3) polluted atmospheres: the ecological effects. *Journal of Chemical Ecology*, **36**, 22–34.

Pinto, D.M., Himanen, S.J., Nissinen, A., Nerg, A.-M. & Holopainen, J.K. (2008) Host location behavior of *Cotesia plutellae* Kurdjumov (Hymenoptera: Braconidae) in ambient and moderately elevated ozone in field conditions. *Environmental Pollution*, **156**, 227–231.

Potts, S.G., Biesmeijer, J.C., Kremen, C., Neumann, P., Schweiger, O. & Kunin, W.E. (2010) Global pollinator declines: Trends, impacts and drivers. *Trends in Ecology and Evolution*, **25**, 345–353.

Robinson, E.A., Ryan, G.D. & Newman, J.A. (2012) A meta-analytical review of the effects of elevated CO_2 on plant–arthropod interactions highlights the importance of interacting environmental and biological variables. *New Phytologist*, **194**, 321–336.

Roth, S., Lindroth, R.L., Volin, J.C. & Kruger, E.L. (1998) Enriched atmospheric CO_2 and defoliation: effects on tree chemistry and insect performance. *Global Change Biology*, **4**, 419–430.

Ryan, G.D., Rasmussen, S. & Newman, J.A. (2010) Global atmospheric change and trophic interactions: are there any general responses? *Plant Communication from an Ecological Perspective* (eds F. Baluška & V. Ninkovic), 179–214. Springer Berlin Heidelberg, Berlin, Heidelberg.

Salminen, J.-P. & Karonen, M. (2011) Chemical ecology of tannins and other phenolics: we need a change in approach. *Functional Ecology*, **25**, 325–338.

Sanders, N.J., Belote, R.T. & Weltzin, J.F. (2004) Multitrophic Effects of elevated atmospheric CO_2 on understory plant and arthropod communities. *Environmental Entomology*, **33**, 1609–1616.

Schowalter, T. (2006) *Insect Ecology: An Ecosystem Approach*, 2nd Edition. Academic Press, London.

Schweitzer, J.A., Madritch, M.D., Bailey, J.K., LeRoy, C.J., Fischer, D.G., Rehill, B.J., Lindroth, R.L., Hagerman, A.E., Wooley, S.C., Hart, S.C. & Whitham, T.G. (2008) From genes to ecosystems: the genetic basis of condensed tannins and their role in nutrient regulation in a *Populus* model system. *Ecosystems*, **11**, 1005–1020.

Speight, M.R., Hunter, M.D. & Watt, A.D. (1999) *The Ecology of Insects: Concepts and Applications*. Blackwell Science, Oxford.

Stange, G. (1997) Effects of changes in atmospheric carbon dioxide on the location of hosts by the moth, *Cactoblastis cactorum*. *Oecologia*, **110**, 539–545.

Stiling, P., Cattell, M., Moon, D.C., Rossi, A., Hungate, B.A., Hymus, G. & Drake, B. (2002) Elevated atmospheric CO_2 lowers herbivore abundance, but increases leaf abscission rates. *Global Change Biology*, **8**, 658–667.

Stiling, P. & Cornelissen, T. (2007) How does elevated carbon dioxide (CO_2) affect plant–herbivore interactions? A field experiment and meta-analysis of CO_2-mediated changes on plant chemistry and herbivore performance. *Global Change Biology*, **13**, 1823–1842.

Stiling, P., Forkner, R. & Drake, B. (2010) Long-term exposure to elevated CO_2 in a Florida scrub-oak forest increases herbivore densities but has no effect on other arthropod guilds. *Insect Conservation and Diversity*, **3**, 152–156.

Stiling, P., Moon, D.C., Hunter, M.D., Colson, J., Rossi, A.M., Hymus, G.J. & Drake, B.G. (2003) Elevated CO_2 lowers relative and absolute herbivore density across all species of a scrub-oak forest. *Oecologia*, **134**, 82–87.

Stiling, P., Moon, D., Rossi, A., Hungate, B.A. & Drake, B. (2009) Seeing the forest for the trees: Long-term exposure to elevated CO_2 increases some herbivore densities. *Global Change Biology*, **15**, 1895–1902.

Stiling, P., Rossi, A., Hungate, B., Dukstra, P., Hinkle, C.R., Knott, W.M. III, & Drake, B. (1999) Decreased leaf-miner abundance in elevated CO_2: reduced leaf quality and increased parasitoid attack. *Ecological Applications*, **9**, 240–244.

Tobin, P.C., Parry, D. & Aukema, B.H. (2014) The influence of climate change on insect invasions in temperate forest ecosystems. *Challenges and Opportunities for the World's Forests in the 21st Century*. (ed. T. Fenning), pp. 267–293. Springer Netherlands, Dordrecht.

Valkama, E., Koricheva, J. & Oksanen, E. (2007) Effects of elevated O_3, alone and in combination with elevated CO_2, on tree leaf chemistry and insect herbivore performance: a meta-analysis. *Global Change Biology*, **13**, 184–201.

Vihakas, M., Gómez, I., Karonen, M., Tähtinen, P., Sääksjärvi, I. & Salminen, J.-P. (2015) Phenolic compounds and their fates in tropical lepidopteran larvae: modifications in alkaline conditions. *Journal of Chemical Ecology*, **41**, 822–836.

Villalpando, S.N., Williams, R.S. & Norby, R.J. (2009) Elevated air temperature alters an old-field insect community in a multifactor climate change experiment. *Global Change Biology*, **15**, 930–942.

Vuorinen, T., Nerg, A.-M. & Holopainen, J.K. (2004) Ozone exposure triggers the emission of herbivore-induced plant volatiles, but does not disturb tritrophic signalling. *Environmental Pollution*, **131**, 305–311.

Williams, R.S., Lincoln, D.E. & Norby, R.J. (1998) Leaf age effects of elevated CO_2-grown white oak leaves on spring-feeding lepidopterans. *Global Change Biology*, **4**, 235–246.

Wilson, E.O. (1987) The little things that run the world (The importance and conservation of invertebrates). *Conservation Biology*, **1**, 344–346.

Wittig, V.E., Ainsworth, E.A., Naidu, S.L., Karnosky, D.F. & Long, S.P. (2009) Quantifying the impact of current and future tropospheric ozone on tree biomass, growth, physiology and biochemistry: A quantitative meta-analysis. *Global Change Biology*, **15**, 396–424.

Xu, G.-L., Fu, S.-L., Schleppi, P. & Li, M.-H. (2013) Responses of soil Collembola to long-term atmospheric CO_2 enrichment in a mature temperate forest. *Environmental Pollution*, **173**, 23–28.

Zvereva, E.L. & Kozlov, M. V. (2006) Consequences of simultaneous elevation of carbon dioxide and temperature for plant–herbivore interactions: a metaanalysis. *Global Change Biology*, **12**, 27–41.

14

Climate Change and Freshwater Invertebrates: Their Role in Reciprocal Freshwater–Terrestrial Resource Fluxes

Micael Jonsson[1] and Cristina Canhoto[2]

[1] Department of Ecology and Environmental Science, Umeå University, SE 901 87 Umeå, Sweden
[2] Center of Functional Ecology, University of Coimbra, 3000-456 Coimbra, Portugal

Summary

Freshwater systems are strongly influenced by their terrestrial surrounding via inputs of terrestrial matter acting as a resource that may, however, also limit freshwater primary production. Climate change impacts on terrestrial environments will therefore influence freshwater communities, and these impacts will, themselves, most likely be at least as strong as those caused directly by climate change. Here, we present an overview of how climate change-induced alterations of terrestrial environments may affect freshwater habitats and their invertebrates, and discuss possible scenarios as to how these effects can, via emergent freshwater insects, feed-back to impact terrestrial systems. To obtain a more complete understanding of how climate change impacts the environment, it is important to consider the role of freshwater invertebrates as recipients as well as drivers of resource fluxes across the freshwater–terrestrial boundary.

14.1 Introduction

Global climate change, mainly due to anthropogenic emissions of greenhouse gases, is expected to increase air temperatures by 2–5 °C by the end of the twenty-first century, with consequences for a wide range of organisms, ecosystem processes and human welfare (IPCC, 2007, 2014). As air temperatures increase so will those of water (Morrill et al., 2005; Webb & Nobilis, 2007) with impacts on both freshwater organisms and the ecological processes they carry out (IPCC, 2008). Diverse, and often locally abundant, communities of freshwater invertebrates are found globally and at all trophic levels; hence, they are crucial drivers of key ecosystem processes such as organic-matter decomposition, grazing, and predator–prey interactions (Merritt et al., 1984; Covich et al., 1999). Altered freshwater invertebrate community structure in response to a changing climate will, therefore, impact the overall functioning of freshwater systems everywhere.

Global Climate Change and Terrestrial Invertebrates, First Edition. Edited by Scott N. Johnson and T. Hefin Jones.
© 2017 John Wiley & Sons, Ltd. Published 2017 by John Wiley & Sons, Ltd.

The characteristics of freshwater systems (Firth & Fisher, 1992; Meyer et al., 1999; Allan et al., 2005) and freshwater invertebrate communities (Sweeney et al., 1992; Woodward et al., 2010; Jonsson et al., 2015) can change in response to increased water temperatures; the indirect effects of climate change on freshwater invertebrate communities have received less attention. Freshwater systems are often strongly influenced by their terrestrial surroundings, and are, to varying degrees, dependent on terrestrial-to-freshwater resource fluxes (Vannote et al., 1980; Benfield, 1997). For example, inputs of plant litter (Wallace et al., 1997, 2015) and terrestrial runoff (Carpenter et al., 1998) can have considerable influence on freshwater environments. These inputs are regulated by characteristics of the riparian canopy and by the level of terrestrial net primary productivity (NPP), both of which are, in turn, regulated by climate. Indirect effects of climate change on freshwater systems, via an altered terrestrial environment, may also feed-back to influence terrestrial systems (Stenroth et al., 2015). This occurs because some freshwater invertebrates (i.e., aquatic insects) represent an important resource flux from freshwater to terrestrial environments (Ballinger & Lake, 2006; Richardson et al., 2010), and quantitative alterations in this flux, directly and/or indirectly caused by climate change, will therefore likely impact terrestrial insectivore communities (i.e., Epanchin et al., 2010; Poulin et al., 2010; Jonsson et al., 2013; Strasevicius et al., 2013), with possible cascading effects up, or down, terrestrial food chains (Henschel et al., 2001; Knight et al., 2005).

In this chapter, we present an overview of how climate change might alter resource fluxes between terrestrial and freshwater systems, focussing on the intermediary role of freshwater invertebrates (Fig. 14.1). This group of organisms is an integral part of these reciprocal fluxes, as they convert terrestrially derived and *in situ* resources to secondary production. This, at least in part, is then transferred to the terrestrial environment by the emergence of adult freshwater insects (Kraus & Vonesh, 2012; Scharnweber et al., 2014). We summarize how some expected climate change induced modifications of terrestrial environments can affect resource fluxes of terrestrial matter to freshwater systems (Fig. 14.1) before discussing how these changed resource fluxes, in addition to effects of warming per se, may influence invertebrate communities and productivity. Effects of climate change on freshwater invertebrates caused by interactive effects of warming and altered allochthonous (i.e., terrestrially derived) and/or autochthonous (i.e., in situ) resource availability and utilization are also discussed. Finally, we summarize how the above impacts may translate into altered resource fluxes from freshwater to terrestrial environments, via quantitative, qualitative and temporal changes in freshwater insect emergence, and what consequences such changes may have for terrestrial ecosystems (Fig. 14.1).

14.2   Climate-Change Effects on Riparian and Shoreline Vegetation

Global climate change will affect terrestrial vegetation in multiple but different ways at a regional level (Shaver et al., 2000; Walther et al., 2002). Because mean annual temperature and global terrestrial net primary productivity (NPP) are positively correlated (Huston & Wolverton, 2009), and as northern regions (i.e., the boreal and Arctic) are expected to experience the greatest increase in mean annual temperature (IPCC, 2007, 2014), increases in NPP will likely be more pronounced at high latitudes and at high elevations (e.g., Gao et al., 2013). Lower latitudes may experience reduced NPP due to increasingly dry conditions (Walther et al., 2002; IPCC, 2007, 2014). Similarly, precipitation may be augmented in some regions, favouring terrestrial NPP, while other areas, such as the Mediterranean, are expected to experience more severe and prolonged droughts with adverse effects on NPP (Walther et al., 2002; IPCC, 2007, 2014). The most apparent changes in vegetation cover (i.e., NPP) are predicted to occur where cold and unproductive conditions (i.e., above tree lines and in

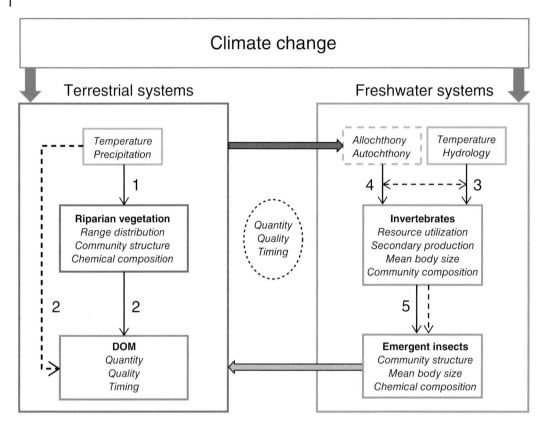

Figure 14.1 Conceptual figure of climate-change effects on terrestrial and aquatic systems. On the left, (1) represents the direct (solid arrow) effects of climate change on terrestrial (riparian and shoreline) vegetation and (2) subsequent indirect (dashed arrow) and direct effects on runoff of dissolved organic matter (DOM). On the right, (3) represents the direct effects of climate change (temperature and hydrology) and (4) the indirect effects (dashed box) of changed relative availability of basal resources, and interactions between (3) and (4), on several freshwater invertebrate parameters. (5) represents both direct and indirect influences of altered freshwater communities on emergent freshwater insect parameters. The flows of terrestrial matter to freshwater systems and aquatic matter (i.e. emergent freshwater insects) to terrestrial systems, and alterations in quantity, quality, and timing caused by climate change, are depicted in the middle.

the Arctic) currently do not allow growth and establishment of trees, or in regions where present day conditions are borderline too dry to support tree growth and survival. In these areas, warming should allow trees to expand northward and to higher elevations, or – in regions where warming induces prolonged droughts – create conditions that are too severe for trees to persist (Walther et al., 2002; Chen et al., 2011). Empirical evidence for such range expansions and contractions is, to date, however, inconclusive (Bertrand et al., 2011; Zhu et al., 2012).

Leaf litter production is tightly coupled with terrestrial NPP (Wardle et al., 2003), and decomposition of litter deposited on land and in water constitutes a major flow of energy in many ecosystems (Polis & Strong, 1996; Polis et al., 1997). The gradual changes expected in local terrestrial plant community composition in response to warming (e.g., conifers being replaced by broad-leaved species) and changed precipitation patterns (e.g., Walther et al., 2002) will change plant functional trait composition, and, hence, NPP as well as the quantity and quality of litter produced during leaf senescence

(Kominoski et al., 2013). Litter quality is based on certain traits, such as relative concentrations of carbon (C) and nitrogen (N), and the concentrations of secondary compounds (e.g., tannins, lignin, and phenolics), as these all influence palatability to consumers and, hence, rates of litter decomposition (Berg & Meentemeyer, 2002; Berg & McClaugherty, 2003; Cornwell et al., 2008). Further, leaf fall phenology differs between tree species (Dixon, 1976; Eckstein et al., 1999), and phenology is coupled with litter quality (i.e., Niinemets & Tamm, 2005; Campanella & Bertiller, 2008). For example, most conifers (i.e., low-quality litter) shed leaves continuously throughout the year, while broad-leaved species (i.e., higher-quality litter) shed their leaves in a short period of time during autumn. Climate change induced alterations of tree species composition will impact not only NPP, but also the timing, and intra- and inter-specific variability in the quality of deposited leaf litter (Kominoski et al., 2013).

In warm and wet environments, where resources are abundant, one plant strategy is to grow in height to escape intra- and inter-specific competition for light (Hautier et al., 2009). As this strategy requires allocation of resources to biomass production, it results in lower investment in secondary compounds (i.e., defence against herbivores) (Bazzaz et al., 1987; Coley, 1988). This reduction will inevitably increase the quality (i.e., palatability) of the litter produced. In contrast, increasing concentrations of atmospheric carbon dioxide (CO_2), and subsequent greater CO_2 uptake by vegetation, may result in poorer litter quality as the C:N ratio increases, as well as concentrations of lignin and phenolics (Norby et al., 2001; Stiling & Cornelissen, 2007). Besides general increases in temperature and changed levels of precipitation, that may have predictable effects on terrestrial vegetation, greater stochasticity in temperature and precipitation fluctuations (i.e., occurrence of 'extreme' weather events) is also expected with a changing climate (IPCC, 2007, 2014; Fischer & Knutti, 2015; Ledger & Milner, 2015; Nilsson et al., 2015). Effects of changed occurrence frequency in stochastic, extreme events on terrestrial vegetation are, however, difficult to predict, and robust predictions are lacking in the literature.

14.3 Climate-Change Effects on Runoff of Dissolved Organic Matter

As a consequence of increased terrestrial NPP and increased precipitation, higher levels of organic matter are available to the soil layer. Freshwater systems – especially in northern regions – are predicted to receive increased amounts of dissolved organic matter (DOM) from terrestrial runoff (Larsen et al., 2011; Christensen et al., 2012). Some freshwater systems have naturally high concentrations of DOM (primarily originating from wetlands); for these, increased terrestrial DOM runoff from forest soils (that generally produce higher-quality DOM) will likely increase the proportion of high-quality DOM (Berggren et al., 2010). Conversely, in southern, warmer regions, where dry conditions might have strong adverse impacts on terrestrial vegetation, inputs of DOM may become more sporadic but of higher magnitudes following rare, extreme rain episodes (Alpert et al., 2002; Nunes et al., 2009). To complicate matters, spates due to extreme rain events can, however, also result in great *exports* of organic C and nutrients from freshwater systems (Giling et al., 2015). Qualitative changes in DOM runoff may also occur, in addition to quantitative changes, as a consequence of changed forest composition and/or temperature, and following CO_2-induced changes in leaf litter chemistry and structure (i.e., litter quality; Bazzaz et al., 1987; Niinemets & Tamm, 2005). Finally, the timing of DOM runoff inputs, and their level of stochasticity, may change (Giling et al., 2015). In northern regions, large inputs of DOM typically take place during spring snowmelt (Berggren et al., 2010). With warmer winters and/or wetter summers (IPCC, 2007), DOM runoff will likely occur, albeit at lower levels, continuously throughout the year, and will not be concentrated in a short peak in spring. Hence, in response to climate change, runoff of DOM is expected to change not only quantitatively, but also qualitatively and temporally (Fig. 14.1).

14.4 Climate Change Effects on Basal Freshwater Resources Via Modified Terrestrial Inputs

Climate-change induced alterations of the terrestrial vegetation can have several implications for freshwater systems, particularly in heterotrophic forested streams where food webs are largely dependent upon the input of terrestrial matter as a basal food resource (Vannote et al., 1980; Benfield, 1997). In areas currently lacking trees, such as in dry areas, the Arctic and at high elevations, sunlight stimulates freshwater primary productivity (i.e., algae), as long as nutrients are not limiting (Seekell et al., 2015). If improved terrestrial growing conditions due to a warming climate allows for the establishment of trees, light limitation will reduce freshwater (i.e., stream) primary production, causing a shift from autotrophy to heterotrophy (Vannote et al., 1980; Benfield, 1997). The opposite situation will occur where trees are lost a result of extreme drought conditions (Walther et al., 2002; Lecerf et al., 2005; Chen et al., 2011). Between these two extremes, there will also be more gradual changes in terrestrial NPP resulting in the modification of the level of shading of some freshwaters (e.g., Kominoski et al., 2013). Shading that limits freshwater primary production may also result from enhanced light absorption due to higher DOM concentrations (i.e., 'brownification'; Carpenter et al., 1998; Carpenter et al., 2005; Karlsson et al., 2009). Such effects are most prominent in lakes and large rivers; they will be less likely to occur in shallow streams. Nonetheless, in streams shaded by riparian vegetation, it is possible that small increases in DOM are enough to reduce the incident light to a level below that which is critical for primary production (Seekell et al., 2015).

Climate change may affect the quantity and quality of terrestrial organic matter (OM) received by freshwaters (Tuchman et al., 2003). This will have likely impacts on the structure and secondary production of freshwater food webs (Carpenter et al., 1998; Carpenter et al., 2005; Karlsson et al., 2009; Berggren et al., 2010). Allochthonous nutrients may increase primary production (i.e., algae) in nutrient-limited freshwater systems, as long as light does not become limiting (Seekell et al., 2015). The C transported with the DOM can, in itself, act as an energy subsidy for freshwater heterotrophic production (Berggren et al., 2010). The threshold for light limitation may, however, occur at relatively low DOM concentrations (Seekell et al., 2015). Further, due to the recalcitrant nature (i.e., low quality) of terrestrial DOM, such subsidies rarely fully compensate for a simultaneous decrease in autotrophic production. Hence, increased allochthonous inputs typically lower freshwater secondary production (due to reduced basal resource quality) and food-web energy-transfer efficiency (Carpenter et al., 1998; Karlsson et al., 2009; Kelly et al., 2014). Similarly, the recalcitrant nature of terrestrial derived litter, in combination with the low assimilation efficiency of detritivorous freshwater invertebrates (Cummins et al., 1996), results in levels of secondary production lower than if equal quantities of in situ resources (i.e., freshwater primary production) constituted by the pool of basal resources (e.g., Friberg & Jacobsen 1999; Lecerf & Chauvet, 2008). As such, even if climate-changed induced alterations of precipitation patterns can have variable effects among regions, such changes will result in less seasonal (i.e., less predictable) control of, and more temporally stochastic extreme impacts on, primary and secondary freshwater production via effects on shading and resource availability/quality, respectively.

Shifts in tree community composition, as a consequence of climate change, may also alter the timing of litter inputs, as tree species differ in phenology of litter senescence (Kikuzawa, 2004; Campanella

& Bertiller, 2008). For example, litter inputs from deciduous species peak in autumn, while conifer-ous species drop their needles in a less seasonal fashion. Regardless of changes in tree community composition, sporadic and extreme rainfall events as a consequence of climate change, and subse-quent spates, can, at least, be as important for temporal availability of allochthonous resources, if litter is periodically washed out (Argerich et al., 2008; Giling et al., 2015). Likewise, low water flows due to dry conditions may decrease allochthonous resource availability in streams, due to increased distance to the riparian vegetation (England & Rosemond, 2004; Lake, 2011; Arroita et al., 2015; Giling et al., 2015).

14.5 Effects of Altered Terrestrial Resource Fluxes on Freshwater Invertebrates

Freshwater invertebrate secondary production and community composition are closely related with terrestrial input characteristics (Vannote et al., 1980; Lecerf et al., 2005; Karlsson et al., 2009; Kelly et al., 2014). The main reasons are that increased inputs of allochthonous resources often coincide with reduced light levels and, thus, reduced freshwater primary production (Vannote et al., 1980; Car-penter et al., 1998; Karlsson et al., 2009; Seekell et al., 2015). Moreover, as allochthonous resources are typically of lower quality to invertebrates than autochthonous resources (e.g., Friberg & Jacobsen 1999; Franken et al., 2005; but see Albariño et al., 2008), an overall reduction in resource quality leads to lower total invertebrate secondary production (Karlsson et al., 2009; Kelly et al., 2014). Conversely, a relative increase in autochthonous resources through, for example, opening up of riparian canopies should increase invertebrate secondary production (e.g., Friberg & Jacobsen, 1999; Hannesdóttir et al., 2013). Hence, shifts in the relative contribution of the two different types of basal resources to total resource availability should affect macroinvertebrate secondary production and community composition at all trophic levels (e.g., Karlsson et al., 2009; Kelly et al., 2014); the amplitude of effects will depend on by how much autochthonous productivity is reduced (i.e., level of shading) and on terrestrial input quality (Seekell et al., 2015).

As basal resources are altered, so is their stoichiometry (i.e., C:N:P ratios). Resource quality to invertebrate consumers is, therefore, reduced (e.g., Sterner & Elser, 2002). This, in itself, influences food-web energy-transfer efficiency, and thus total secondary production, as stoichiometric mis-matches between food and consumers lead to inefficient energy transfers 'up' the food chain (Frost et al., 2002; Cross et al., 2005; Frainer et al., 2016). Similarly, as a consequence of changed basal resources, in situ concentrations of polyunsaturated fatty acids (PUFA) change. These are compounds of significant physiological importance for consumers at all taxonomic levels, in both freshwater and terrestrial systems. Consumers may be able to synthesize their own PUFAs, but, because this is an energetically costly process (Brett & Müller-Navarra, 1997), they are primarily obtained through con-sumption (Gladyshev et al., 2009). Further, large amounts of PUFAs can only be synthesized de novo by algae; PUFA are generally limited in terrestrial plants (Mills et al., 2001; Wolff et al., 2001; Glady-shev et al., 2013). As consumer fitness and food-web energy-transfer efficiency strongly depend on dietary PUFA availability (Müller-Navarra et al., 2000; Brett et al., 2009), decreases in autotrophic pro-duction caused by increased shading from riparian vegetation, or from DOM inputs, can be expected to also lower freshwater invertebrate secondary production.

Leaf litter represents the main basal resource in small forested streams (Wallace et al., 1997, 2015; Abelho, 2001). Changed allochthonous resource diversity, quality and seasonality are known to affect the structure and functioning of these ecosystems through effects on detritivorous invertebrates (Canhoto & Graça, 1999; Lecerf et al., 2005; Riipinen et al., 2010) and/or through changes in fungal-mediated decomposition (e.g., Gonçalves & Canhoto, 2009; Gonçalves et al., 2014). Fungi, namely aquatic hyphomycetes, are crucial links between leaf litter and detritivorous freshwater invertebrates (Canhoto & Graça, 2008). Temperature-driven effects on fungal biomass and community composition have been reported to affect microbial-mediated litter decomposition (Fernandes et al., 2012; Gonçalves et al., 2013, 2015). Modified leaf-litter palatability, due to altered fungal colonization, may also result from climate-change induced modifications in stream-flow dynamics (England & Rosemond, 2004; Lake, 2011; Giling et al., 2015; Gonçalves et al., 2016). Independent of cause, such consequences for leaf-litter palatability may alter leaf processing by freshwater invertebrates and thus, the dominant patterns of nutrient cycling and energy flow in small streams (Canhoto et al., 2016). Leaf-eating invertebrates (i.e., shredders; Cummins, 1973) are known to prefer certain leaf (e.g., *Alnus glutinosa*) and fungal (e.g., *Tetracladium marchalianum*) species, fungal assemblages (e.g., high diversity) and degree of leaf conditioning (i.e., high fungal colonization). Species-specific combinations of the triad (i.e., leaves–fungi-shredders) are determinants of leaf consumption and of invertebrates' growth, elemental composition, survivorship and fecundity (Canhoto & Graça, 1999, 2008; Gonçalves et al., 2014; Mas-Martí et al., 2015). Accordingly, changes in leaf-litter quality caused by alterations in riparian composition and/or CO_2-mediated recalcitrance are also known to affect fungal dynamics and invertebrates' life history descriptors, such as growth and developmental rates, with impacts on secondary production (Canhoto & Graça, 1995; Dray et al., 2014; Ferreira et al., 2010, 2015; Mas-Martí et al., 2015).

In summary, climate change will, via effects on basal resources, impact freshwater invertebrates and, thus, entire freshwater food webs (e.g., Karlsson et al., 2009; Kelly et al., 2014). Given that these resources can change not only quantitatively and temporally, but also simultaneously qualitatively, such effects are currently immensely difficult to predict.

14.6 Direct Effects of Warming on Freshwater Invertebrates

Several studies have investigated effects of warming on freshwater invertebrate taxa, and the results are mixed; growth rates have been found to decrease (e.g., Sweeney & Schnack, 1977; Sweeney, 1978), increase (e.g., Rempel & Carter, 1987; Greig et al., 2012) or remain unchanged (e.g., Mas-Martí et al., 2015), or effects have been taxon-specific (Hogg & Williams, 1996). In studies that have considered entire invertebrate communities, a general pattern is that mean individual size decreases as a consequence of greater metabolic demands in response to warming (Daufresne et al., 2009; Forster et al., 2012). Small taxa, however, have higher mass-specific respiration rates than larger taxa (Huxley, 1932; Brown et al., 2004), and if increased metabolic demands cannot be compensated for by increased feeding rates, small taxa may be lost due to starvation earlier than larger taxa, resulting in an increased mean body size at community level (Jonsson et al., 2015). In any case, warming-induced increases or decreases in mean consumer size, as well as changes in body-resources allocations (Mas-Martí et al., 2015), will influence several ecosystem processes, including predator–prey dynamics, as predators are often physically limited (i.e., specialized) to a certain range of prey sizes (Rice et al., 1993; Craig et al., 2006; Barnes et al., 2010).

Besides effects on size, warming will likely affect the rate at which freshwater invertebrates reach their optimal size (i.e., will determine their growth rate) as well as their morphological developmental rates (McKie et al., 2004; Forster et al., 2011). Growth and developmental rates are important life-history traits that have evolved in response to environmental conditions such as temperature, resource availability and predation risk (McPeek, 2004; Kollberg et al., 2013). Freshwater inverte-brates do, however, often exhibit some level of plasticity in traits such as growth and developmental rates, in immediate response to changed environmental conditions (Stoks et al., 2014). In the event of warming, freshwater insects may be able to grow faster and, hence, emerge earlier as adults (Rempel & Carter, 1987; Greig et al., 2012; Culler et al., 2015; McCauley et al., 2015). High predation risk might select against increased foraging (i.e., increased activity)in prey species and thereby limit such warming effects (Culler et al., 2014). If, however, warming promotes growth rates more in prey than in predators, prey can emerge earlier and at higher quantities as the time of exposure to *in situ* predation is reduced (Culler et al., 2015).

Net effects of warming on metabolism and subsequently on growth, size, morphological development and survival in freshwater invertebrates are, besides being body-size dependent (Huxley, 1932; Brown et al., 2004), the product of the interaction between warming and food resource availability (e.g., Villanueva et al., 2011; Jonsson et al., 2015). If food resources increase with warming, higher metabolic demands can be compensated for by increased resource consumption and/or assimilation efficiency (e.g., Villanueva et al., 2011), while, if food resources become limiting, an increase in temperature and subsequent increase in metabolic demand leads to starvation (on an individual level) or reduced body sizes (on the population and/or community level) (Daufresne et al., 2009; Forster et al., 2012; but see Jonsson et al., 2015). Both the quantity and quality of food resources are, however, important for meeting metabolic demands (e.g., Flores et al., 2014); effects of warming on macroinvertebrate communities cannot be understood unless simultaneous effects on food resources are also considered. For example, if warming occurs simultaneously to an increase in allochthonous resources relative to autochthonous resources, an overall decline in quantity or changed seasonality of resources, and/or a drop in terrestrial-derived resource quality, it will be difficult for freshwater invertebrates to compensate for increased metabolic demands via enhanced consumption rates. Hence, reduced resource availability or quality will magnify the direct adverse impact of warming on the invertebrates (Villanueva et al., 2011; Flores et al., 2014). Further, although warming should stimulate freshwater primary productivity (e.g., Petchey et al., 1999; Hannesdóttir et al., 2013), allowing consumers to compensate for warming-induced increases in metabolic rates (via increased consumption rates of high-quality resources), reduced mixing of the water column in lakes due to warming (O'Reilly et al., 2003) or brownification due to greater DOM inputs (Karlsson et al., 2009; Kelly et al., 2014), may counteract such warming effects. Nevertheless, in systems where the net effect of climate change is increased resource quantity or quality, or for taxa that are able to meet their higher metabolic demands by switching to more readily available resources, warming might have no net effect on total secondary production (Jonsson et al., 2015).

In summary, care must be taken when considering changed resource availability and quality to understand net effects of water warming, as the relationships between resources and consumers are modulated by several biotic (e.g. Gonçalves et al., 2013) and abiotic (e.g., Lagrue et al., 2011) factors. For example, a higher fungal biomass may compensate for the CO_2-mediated decreases in leaf-litter quality (Dray et al., 2014), and a more intense litter leaching of soluble compounds can be expected in warmer waters (Tuchman et al., 2002). Both these influences are relevant and may override more direct effects of warming (e.g., via metabolic activity) on detritivorous invertebrates. As such, warming can indeed be expected to affect freshwater macroinvertebrate communities, but the consequences will differ among taxa and different types of freshwater systems, depending on invertebrate body size, preferred diet, and relative levels of high- versus low-quality terrestrial and in situ resources.

14.7 Impacts of Altered Freshwater Invertebrate Emergence on Terrestrial Ecosystems

Because freshwater invertebrates are crucially important for practically all freshwater ecosystem processes, such as decomposition, filtering, grazing, and predation (Merritt et al., 1984; Covich et al., 1999), it is of considerable interest to understand how they respond to climate change. Their importance as terrestrial adults (i.e., emergent freshwater insects; Ballinger & Lake, 2006; Richardson et al., 2010), and how this importance may be modified by a changing climate (Greig et al., 2012; Jonsson et al., 2015), must, however, also be considered. At present, it is difficult to make general predictions, but changes in emergence of freshwater insects will, in turn, propagate to influence terrestrial ecosystems (Stenroth et al., 2015).

Emerging aquatic insects are ubiquitous and, both locally and on the landscape scale (Bartrons et al., 2013), an important resource for a wide variety of terrestrial insectivores such as spiders, birds, and bats, that live adjacent to lotic (Henschel et al., 2001; Murakami & Nakano, 2002; Sabo & Power, 2002a,b; Fukui et al., 2006) and lentic (Gratton et al., 2008; Jonsson & Wardle, 2009) environments. The amount of emergent freshwater insects deposited on land can be substantial (up to 1500 kg ha^{-1} yr^{-1}; Gratton et al., 2008) and sufficient to not only influence terrestrial insectivore populations but also, given that approximately 10% of the deposited insect biomass is N (Evans-White et al., 2005), to fertilize terrestrial soils and plants (Dreyer et al., 2012, 2015; Bultman et al., 2014). For this reason, emerging freshwater insects represent an important energy flux from water to land, although their importance for elemental cycling across the freshwater–terrestrial interface in some respects still is poorly understood (Jonsson & Stenroth, 2016).

Despite the complex impacts of climate change on freshwater invertebrate communities, it is only a limited number of scenarios that explore how this can be expected to alter emergence fluxes to terrestrial systems, quantitatively, temporally, and/or qualitatively. Quantitative changes in freshwater insect emergence should occur as a consequence of climate-induced alterations to freshwater basal resources. That is, lower resource availability results in lower invertebrate secondary production and, hence, lower insect emergence, and vice versa (Gratton & Vander Zanden, 2009). Several climate change scenarios predict reduced freshwater secondary production (e.g., Karlsson et al., 2009; Kelly et al., 2014) but there are also instances where augmented freshwater primary production, or shifts toward higher-quality resources, is predicted to increase secondary production (e.g., Petchey et al., 1999; Hannesdóttir et al., 2013). It is, however, important to note that increases in secondary production in some taxa might be counteracted by simultaneous diminution or loss of others (Jonsson et al., 2015; Stenroth et al., 2015).

Changed quantities of emergent insects will affect terrestrial insectivorous populations that typically respond numerically to prey fluxes (Murakami & Nakano, 2002; Sabo & Power, 2002a,b; Fukui et al., 2006; Jonsson & Wardle 2009). In turn, this numerical response can result in changed predation pressures on terrestrial prey, and if these preys are herbivores, it will alter rates of terrestrial herbivory (Henschel et al., 2001). Emergent freshwater insects are the most abundant at the freshwater–terrestrial interface; their abundance declines exponentially with distance from the freshwater system (Briers et al., 2005; Jonsson & Wardle, 2009; Muehlbauer et al., 2014). In other words, the spatial (i.e., lateral) extent of the freshwater flux, and subsequent effects on terrestrial systems, is determined by the productivity of the freshwater system (Gratton & Vander Zanden, 2009). Quantitative changes in emergent aquatic insects will, therefore, most likely alter the spatial scale at which aquatic systems influence a wide range of terrestrial consumers, terrestrial plants and terrestrial food-web dynamics (Bultman et al., 2014; Dreyer et al., 2015).

Even if secondary production of emergent freshwater insects remains the same, temporal patterns (i.e., emergence phenology) in resource flux can be altered by warming-induced changes in growth rates (e.g., Culler et al., 2015), slightly altered community composition (Jonsson et al., 2015) or loss of taxa (Strasevicius et al., 2013). The timing of emergence can be at least as important as emergence quantity, as terrestrial consumers time their activity and reproduction with seasonal fluxes in prey availability (Murakami & Nakano, 2002; Carey, 2009). For example, in northern regions, peaks in freshwater insect emergence occur before peaks in terrestrial secondary production, and emergent freshwater insects are therefore important prey for terrestrial insectivores (e.g., birds and spiders) that reproduce early in summer (Murakami & Nakano, 2002). The timing, or early occurrence, of freshwater resource fluxes is therefore critical; a temporal mismatch between consumer reproduction and prey availability can severely reduce reproductive success (Carey, 2009; Strasevicius et al., 2013). In addition, synchrony between peaks in freshwater insect production (emergence) and terrestrial prey production determine consumer responses to freshwater insect fluxes, in terms of numerical response and contribution in diet, and, thus, the impact of aquatic fluxes on terrestrial environments (Marczak et al., 2007). Hence, net effects of temporal (and quantitative) changes in freshwater insect emergence will also depend on how terrestrial resources (prey communities) respond to climate change.

From a consumer perspective, qualitative changes in emergent freshwater insects caused by climate change can be in the form of altered size structure (e.g., prey size), stoichiometry, or PUFA content. Altered size structure can be a consequence of increased or decreased growth rates, especially if effects of warming on growth rates are taxon-specific, or due to gradual changes in community composition (Jonsson et al., 2015). A changed prey-size structure will most likely impact terrestrial insectivore populations, despite no quantitative change in freshwater insect emergence (Stenroth et al., 2015). For example, larger insectivores, such as birds, prefer large over small-bodied insects (Beintema et al., 1991; Pearce-Higgins & Yalden, 2004; Britschgi et al., 2006; Buchanan et al., 2006), while small spiders (Linyphiidae) prefer small prey (Nentwig, 1980; Dreyer et al., 2012). Hence, climate-induced shifts in freshwater invertebrate community composition, and subsequent changes in size structure of emergent freshwater insects, will influence how and by what this resource flux is used (Stenroth et al., 2015). More specifically, warming-induced decreases in mean prey size (e.g., Daufresne et al., 2009) will favour small terrestrial predators, while large predators, or predators preferring large prey, will be adversely affected.

Alterations in freshwater basal resources, as a consequence of climate change, may also alter the elemental composition of emergent insects (Sterner & Elser, 2002). Although it is well known that stoichiometric mismatch, or match, is important for food-web energy-transfer efficiency (Frost et al., 2002; Cross et al., 2005), it is entirely unknown if the elemental composition of freshwater insect fluxes matters for consumers in recipient terrestrial systems. Likewise, PUFA content in emergent freshwater insects will change, especially in instances of climate-change induced reductions in freshwater primary production. As dietary PUFA is important for consumer fitness and food-web energy-transfer efficiency (Müller-Navarra et al., 2000; Brett et al., 2009), such alterations will affect the dietary value of freshwater insects to terrestrial consumers, with possible effects on terrestrial communities and food webs. This link remains not investigated (Lau et al., 2012, 2013). Further, changes in the relative contribution of freshwater and terrestrial resources in the diet, and thus to biomass of emergent insects, will influence material cycling across the freshwater–terrestrial interface (Scharnweber et al., 2014; Jonsson & Stenroth, 2016). Fluxes of freshwater insects from water to land can differ in terms of allochthony and autochthony, depending on insect community composition, insect diets, and available resources (Jonsson & Stenroth, 2016). Knowledge of what determines the allochthony-autochthony ratio in emergent freshwater insects, and how this may be altered by extrinsic influences, such as climate change, needs to be obtained for a general understanding of how resources are cycled on the large scale in the environment.

In summary, although it is possible to envision three distinct ways in which climate change might alter freshwater insect fluxes to terrestrial environments, and thus the consequences for terrestrial systems, different combinations of these three ways are the most likely scenarios. Further, consequences of the resulting alterations will depend on the characteristics of terrestrial recipient systems (Marczak et al., 2007) and how they respond to climate change. Nevertheless, an understanding of direct or indirect (via terrestrial systems) effects on freshwater systems is crucial for obtaining a holistic understanding of how global climate change will alter our ecosystems and movements of resources across ecosystem boundaries.

14.8   Conclusions and Research Directions

Freshwater invertebrates will be important mediators of future climate-change effects on the ecosystem scale, not only for freshwaters but also for terrestrial systems (Fig. 14.1). In particular, effects of changed quantity, quality and timing of allochthonous inputs to freshwater systems will, to a large extent, depend on how freshwater invertebrate communities respond to changes in basal resources. This will in turn be reflected in resource fluxes back to the terrestrial environment. While it seems immensely complicated to predict net effects of climate change on freshwater invertebrates – effects are most likely highly context dependent – important information regarding climate-change effects can be obtained by studying how cross-boundary resource fluxes are altered (quantitatively, qualitatively, and temporally), and what consequences this has for freshwater invertebrate communities and for feedbacks to terrestrial systems in the form of emergent freshwater insects (Fig. 14.1).

Based on existing knowledge gaps highlighted in this chapter, we propose a series of future research directions regarding climate-change effects on freshwater invertebrates, with special focus on the connectedness between terrestrial and freshwater environments, as mediated by resource consumption by freshwater invertebrates and lateral movements of adult freshwater insects (Fig. 14.2).

14.8.1   Effects of Simultaneous Changes in Resource Quality and Temperature on Freshwater Invertebrate Secondary Production

More detailed studies on the relative importance of altered resource quality and how this interacts with warming per se to influence freshwater invertebrate secondary production are needed.

14.8.2   Effects of Changed Resource Quality and Temperature on the Size Structure of Freshwater Invertebrate Communities

Previous studies on direct effects of warming on freshwater invertebrates have produced mixed results; still, few studies have considered how warming interacts with effects of changed resource quality in freshwater systems.

14.8.3   Effects of Changed Resource Quality on Elemental Composition (i.e., Stoichiometry, Autochthony versus Allochthony, and PUFA Content) of Freshwater Invertebrates

It is important to understand how elemental composition of freshwater invertebrates responds to specific changes in resource quality as a consequence of climate change in much more detail.

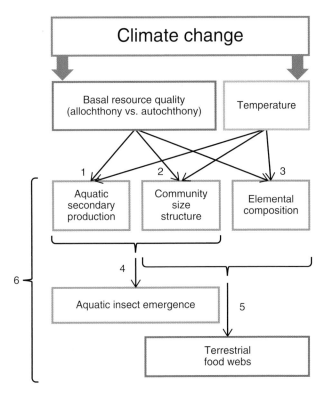

Figure 14.2 Illustration of the interactions and logical sequence underlying the proposed research questions (1-6). Suggestions intend to narrow the prevailing knowledge gaps on how climate-change driven alterations in basal resource quality, together with warming, may influence aquatic invertebrate secondary production (1), community size structure (2), and elemental composition (3); how this can influence aquatic insect emergence (4), what consequences such changes have for terrestrial food webs (5), and how climate change influences resource cycling across the aquatic–terrestrial boundary on the large scale (6).

14.8.4 Effects of Changed Freshwater Invertebrate Community Composition and Secondary Production on Freshwater Insect Emergence

Total freshwater invertebrate secondary production and production of emergent freshwater insects should be positively correlated, but warming and/or altered basal resources might modify this relationship, with potential reductions in connectivity between freshwater and terrestrial systems.

14.8.5 Effects of Changed Quality (i.e., Size Structure and Elemental Composition) of Emergent Freshwater Insects on Terrestrial Food Webs

It is not known whether the quality of the freshwater flux matters for the utilization of this resource by terrestrial consumers, responses in terrestrial consumer populations or community composition, or for general effects on terrestrial food webs.

14.8.6 Effects of Climate Change on Landscape-Scale Cycling of Matter Across the Freshwater–Terrestrial Interface

Reciprocal resource fluxes across the freshwater–terrestrial interface represent an important, though yet not well characterized, cycling of elements in the environment. How climate change will impact this cycling is practically unknown.

Investigating these proposed questions (Fig. 14.2) will not only render a better understanding of how climate change will affect freshwater invertebrates, but also give insights into the more general consequences of climate change for ecosystem processes and functioning of freshwater and terrestrial systems, and freshwater–terrestrial interactions, where freshwater invertebrates are crucial components.

Acknowledgements

We are grateful to Philip Smith for proofreading this chapter.

References

Abelho, M. (2001) From litterfall to breakdown in streams: a review. *TheScientificWorldJOURNAL*, **1**, 656–680.

Argerich, A., Martí, E., Sabater, F., Ribot, M., von Schiller, D. & Rivera, J.L. (2008) Combined effects of leaf litter inputs and a flood on nutrient retention in a Mediterranean mountain stream during fall. *Limnology and Oceanography*, **53**, 631–641.

Albariño, R., Villanueva, V.D. & Canhoto, C. (2008) The effect of sunlight on leaf litter quality reduces growth of the shredder *Klapopteryx kuscheli*. *Freshwater Biology*, **53**, 1881–1889.

Allan, J.D., Palmer, M.A. & Poff, N.L. (2005) Climate change and freshwater ecosystems. *Climate Change and Biodiversity* (eds T.E. Lovejoy & L. Hannah), pp. 272–290. Yale University Press, New Haven CT.

Alpert, P., Ben-Gai, T., Baharad, A., Benjamini, Y., Yekutieli, D., Colacino, M., Diodato, L., Ramis, C., Homar, V., Romero, R., Michaelides, S. & Manes, A. (2002) The paradoxical increase of Mediterranean extreme daily rainfall in spite of decrease in total values. *Geophysical Research Letters*, **29**, 1536.

Arroita, M., Aristi, I., Díez, J., Martinez, M., Oyarzun, G. & Elosegi, A. (2015) Impact of water abstraction on storage and breakdown of coarse organic matter in mountain streams. *Science of The Total Environment*, **503–504**, 233–240.

Ballinger, A. & Lake, P.S. (2006) Energy and nutrient fluxes from rivers and streams into terrestrial food webs. *Marine and Freshwater Research*, **57**, 15–28.

Barnes, C., Maxwell, D., Reuman, D.C. & Jennings, S. (2010) Global patterns in predator-prey size relationships reveal size dependency of trophic transfer efficiency. *Ecology*, **91**, 222–232.

Bartrons, M., Papeş, M., Diebel, M.W., Gratton, C. & Vander Zanden, M.J. (2013) Regional-level inputs of emergent aquatic insects from water to land. *Ecosystems*, **16**, 1353–1363.

Bazzaz, F.A., Chiariello, N.R., Coley, P.D. & Pitelka, L.F. (1987) Allocating resources to reproduction and defense. *BioScience*, **37**, 58–67.

Beintema A.J., Thissen, J.B., Tensen, D. & Visser, G.H. (1991) Feeding ecology of Charadiiform chicks in agricultural grassland. *Ardea*, **79**, 31–43.

Benfield, E.F. (1997) Comparisons of litterfall input to streams. *Journal of the North American Benthological Society*, **16**, 104–108.

Berg, B. & McClaugherty, C. (2003) *Plant Litter: Decomposition, Humus Formation, Carbon Sequestration*. Springer, UK.

Berg, B. & Meentemeyer, V. (2002) Litter quality in a north European transect versus carbon storage potential. *Plant and Soil*, **242**, 83–92.

Berggren, M., Laudon, H., Haei, M., Ström, L. & Jansson, M. (2010) Efficient aquatic bacterial metabolism of dissolved low-molecular-weight compounds from terrestrial sources. *The ISME Journal*, **4**, 408–416.

Bertrand, R., Lenoir, J., Piedallu, C., Riofrío-Dillon, G., de Ruffray, P., Vidal, C., Pierrat, J.-C. & Gégout, J.-C. (2011) Changes in plant community composition lag behind climate warming in lowland forests. *Nature*, **479**, 517–520.

Brett, M.T., Kainz, M.J., Taipale, S.J. & Seshan, H. (2009) Phytoplankton, not allochthonous carbon, sustains herbivorous zooplankton production. *Proceedings of the National Academy of Sciences of the United States of America*, **106**, 21197–21201.

Brett, M. & Müller-Navarra, D. (1997) The role of highly unsaturated fatty acids in aquatic food web processes. *Freshwater Biology*, **38**, 483–499.

Briers, R.A., Cariss, H.M., Geoghegan, R. & Gee, J.H.R. (2005) The lateral extent of the subsidy from an upland stream to riparian lycosid spiders. *Ecography*, **28**, 165–170.

Britschgi A., Spaar, R. & Arlettaz, R. (2006) Impact of grassland farming intensification on the breeding ecology of an indicator insectivorous passerine, the Whinchat *Saxicola rubetra*: lessons for overall Alpine meadowland management. *Biological Conservation*, **130**, 193–205.

Brown, J.H., Gillooly, J.F., Allen, A.P., Savage, V.M. & West, G.B. (2004) Toward a metabolic theory of ecology. *Ecology*, **85**, 1771–1789.

Buchanan G.M., Grant, M.C., Sanderson, R.A. & Pearce-Higgins, J.W. (2006) The contribution of invertebrate taxa to moorland bird diets and the potential implications of land-use management. *Ibis*, **148**, 615–628.

Bultman, H., Hoekman, D., Dreyer, J. & Gratton, C. (2014) Terrestrial deposition of aquatic insects increases plant quality for insect herbivores and herbivore density. *Ecological Entomology*, **39**, 419–426.

Campanella, M.V. & Bertiller, M.B. (2008) Plant phenology, leaf traits and leaf litterfall of contrasting life forms in the arid Patagonian Monte, Argentina. *Journal of Vegetation Science*, **19**, 75–85.

Canhoto, C., Gonçalves, A.L. & Bärlocher, F. (2016) Biology and ecological functions of aquatic hyphomycetes in a warming climate. *Fungal Ecology*, **19**, 201–218.

Canhoto, C. & Graça, M.A.S. (1995) Food value of introduced eucalypt leaves for a Mediterranean stream detritivore: *Tipula lateralis*. *Freshwater Biology*, **34**, 209–214.

Canhoto, C. & Graça, M.A.S. (1999) Leaf barriers to fungal colonization and shredders (*Tipula lateralis*) consumption of decomposing *Eucalyptus globulus*. *Microbial Ecology*, **37**, 163–172.

Canhoto, C. & Graça, M.A.S. (2008) Interactions between fungi and stream invertebrates: back to the future. *Novel Techniques and Ideas in Mycology* (eds K.R. Sridhar, F. Bärlocher & K.D. Hyde), pp. 1–22. Fungal Diversity Research Series **20**, 305–325.

Carey, C. (2009) The impacts of climate change on the annual cycles of birds. *Philosophical Transactions of the Royal Society B*, **364**, 3321–3330.

Carpenter, S.R., Caraco, N.F., Correll, D.L., Howarth, R.W., Sharpley, A.N. & Smith, V.H. (1998) Nonpoint pollution of surface waters with phosphorus and nitrogen. *Ecological Applications*, **8**, 559–568.

Carpenter, S.R., Cole, J.J., Pace, M.L., Van de Bogert, M., Bade, D.L., Bastviken, D., Gille, C.M., Hodgson, J.R., Kitchell, J.F. & Kritzberg, E.S. (2005) Ecosystem subsidies: terrestrial support of aquatic food webs from ^{13}C addition to contrasting lakes. *Ecology*, **86**, 2737–2750.

Chen, I.-C., Hill, J.K., Ohlemüller, R., Roy, D.B. & Thomas, C.D. (2011) Rapid range shifts of species associated with high levels of climate warming. *Science*, **333**, 1024–1026.

Christensen, J.H., Räisänen, J., Iversen, T., Bjørge, D., Christensen, O.B. & Rummukainen, M. (2001) A synthesis of regional climate change simulations – a Scandinavian perspective. *Geophysical Research Letters*, **28**, 1003–1006.

Coley, P.D. (1988) Effects of plant growth rate and leaf lifetime on the amount and type of anti-herbivore defense. *Oecologia*, **74**, 531–536.

Cornwell, W.K., Cornelissen, J.H.C., Amatangelo, K., Dorrepaal, E., Eviner, V.T., Godoy, O., Hobbie, S.E., Hoorens, B., Kurokawa, H., Pérez-Harguindeguy, N., Quested, H.M., Santiago, L.S., Wardle, D.A., Wright, I.J., Aerts, R., Allison, S.D., van Bodegom, P., Brovkin, V., Chatain, A., Callaghan, T.V., Díaz, S., Garnier, E., Gurvich, D.E., Kazakou, E., Klein, J.A., Read, J., Reich, P.B., Soudzilovskaia, N.A., Vaieretti, M.V. & Westoby, M. (2008) Plant species traits are the predominant control on litter decomposition rates within biomes worldwide. *Ecology Letters*, **11**, 1065–1071.

Covich, A.P., Palmer, M.A. & Crowl, T.A. (1999) The role of benthic invertebrate species in freshwater ecosystems: zoobenthic species influence energy flows and nutrient cycling. *BioScience*, **49**, 119–127.

Craig, J.K., Burke, B.J., Crowder, L.B. & Rice, J.A. (2006) Prey growth and size-dependent predation in juvenile estuarine fishes: experimental and model analyses. *Ecology*, **87**, 2366–2377.

Cross, W.F., Benstead, J.P., Frost, P.C. & Thomas, S.A. (2005) Ecological stoichiometry in freshwater benthic systems: recent progress and perspectives. *Freshwater Biology*, **50**, 1895–1912.

Culler, L.E., Ayres, M.P. & Virginia, R.A. (2015) In a warmer Arctic, mosquitoes avoid increased mortality from predators by growing faster. *Proceedings of the Royal Society of London B*, **282**, 20151549.

Culler, L.E., McPeek, M.A. & Ayres, M.P. (2014) Predation risk shapes thermal physiology of a predaceous damselfly. *Oecologia*, **176**, 653–650.

Cummings. K.W. (1973) Trophic relations of aquatic insects. *Annual Review of Entomology*, **18**, 183–206.

Cummins, K.W., Merritt, R.W. & Berg, M.B. (1996) Ecology and distribution of aquatic insects. *An Introduction to the Aquatic Insects of North America* (eds R.W. Merritt, K.W. Cummins & M.D. Berg), pp. 105–122. Kendall/Hunt Publishing Company, Dubuque, IA, USA.

Daufresne, M., Lengfellner, K. & Sommer, U. (2009) Global warming benefits the small in aquatic systems. *Proceedings of the National Academy of Sciences of the United States of America*, **106**, 12788–12793.

Dixon, K.R. (1976) Analysis of seasonal leaf fall in north temperate deciduous forests. *Oikos*, **27**, 300–306.

Dreyer, J., Hoekman, D. & Gratton, C. (2012) Lake-derived midges increase abundance of shoreline terrestrial arthropods via multiple trophic pathways. *Oikos*, **121**, 252–258.

Dray, M.W., Crowther, T.W., Thomas, S.M., A'Bear, A.D., Godbold, D.L., Ormerod, S.J., Hartley, S.E. & Jones, T.H. (2014) Effects of elevated CO_2 on litter chemistry and subsequent invertebrate detritivore feeding responses. *PLOs ONE*, **9**, e86246.

Dreyer, J., Townsend, P.A., Hook, J.C. III, Hoekman, D., Vander Zanden, M.J. & Gratton, C. (2015) Quantifying aquatic insect deposition from lake to land. *Ecology*, **96**, 499–509.

Eckstein, R.L., Karlsson, P.S. & Weih, M. (1999) Leaf life span and nutrient resorption as determinants of plant nutrient conservation in temperate-arctic regions. *New Phytologist*, **143**, 177–189.

England, L.E. & Rosemond, A.D. (2004) Small reductions in forest cover weaken terrestrial-aquatic linkages in headwater streams. *Freshwater Biology*, **49**, 721–734.

Epanchin, P.N., Knapp, R.A. & Lawler, S.P. (2010) Nonnative trout impact an alpine-nesting bird by altering aquatic-insect subsidies. *Ecology*, **91**, 2406–2415.

Evans-White, M.A., Stelzer, R.S. & Lamberti, G.A. (2005) Taxonomic and regional patterns in benthic macroinvertebrate elemental composition in streams. *Freshwater Biology*, **50**, 1786–1799.

Fernandes, I., Pascoal, C., Guimarães, H., Pinto, R., Sousa, I. & Cássio, F. (2012) Higher temperature reduces the effects of litter quality on decomposition by aquatic fungi. *Freshwater Biology*, **57**, 2306–2317.

Ferreira, V., Gonçalves, A.L., Godbold, D.L. & Canhoto, C. (2010) Effects of increased atmospheric CO_2 on the performance of an aquatic detritivore through changes in water temperature and litter quality. *Global Change Biology*, **16**, 3284–3296.

Ferreira, V., Chauvet, E. & Canhoto, C. (2015) Effects of experimental warming, litter species, and presence of macroinvertebrates on litter decomposition and associated decomposers in a temperate mountain stream. *Canadian Journal of Fisheries and Aquatic Sciences* **72**, 206–216.

Fischer, E.M. & Knutti, R. (2015) Anthropogenic contribution to global occurrence of heavy-precipitation and high-temperature extremes. *Nature Climate Change*, **5**, 560–564.

Firth, P. & Fisher, S.G. (1992) *Global Climate Change and Freshwater Ecosystems*. Springer-Verlag, New York Inc., New York, USA.

Flores, L., Larrañaga, A. & Elosegi, A. (2014) Compensatory feeding of a stream detritivore alleviates the effects of poor food quality when enough food is supplied. *Freshwater Science* **33**, 134–141.

Forster, J., Hirst, A.G. & Atkinson, D. (2011) How do organisms change size with changing temperature? The importance of reproductive method and ontogenetic timing. *Functional Ecology*, **25**, 1024–1031.

Forster, J., Hirst, A.G. & Atkinson, D. (2012) Warming-induced reductions in body size are greater in aquatic than terrestrial species. *Proceedings of the National Academy of Sciences of the United States of America*, **109**, 19310–19314.

Frainer, A., Jabiol, J., Gessner, M.O., Bruder, A., Chauvet, E. & McKie, B.G. (2016) Stoichiometric imbalances between detritus and detritivores are related to shifts in ecosystem functioning. *Oikos*, **125**, 861–871.

Franken, R.J.M., Waluto, B., Peeters, E.T.H.M., Gardeniers, J.J.P., Beijer, J.A.J. & Scheffer, M. (2005) Growth of shredders on leaf litter biofilms: the effect of light intensity. *Freshwater Biology* **50**, 459–466.

Friberg, N. & Jacobsen, D. (1999) Variation in growth of the detritivore-shredder *Sericostoma personatum* (Trichoptera). *Freshwater Biology*, **42**, 625–635.

Frost, P.C., Stelzer, R.S., Lamberti, G.A. & Elser, J.J. (2002) Ecological stoichiometry of trophic interactions in the benthos: understanding the role of C:N:P ratios in lentic and lotic habitats. *Journal of the North American Benthological Society*, **21**, 515–528.

Fukui, D., Murakami, M., Nakano, S. & Aoi, T. (2006) Effect of emergent aquatic insects on bat foraging in a riparian forest. *Journal of Animal Ecology*, **75**, 1252–1258.

Gao, Y., Zhou, X., Wang, Q., Wang, C., Zhan, Z., Chen, L., Yan, J. & Qu, R. (2013) Vegetation net primary productivity and its response to climate change during 2001–2008 in the Tibetan Plateau. *Science of the Total Environment*, **444**, 356–362.

Giling, D.P., Mac Nally, R. & Thompson, R.M. (2015) How might cross-system subsidies in riverine networks be affected by altered flow variability? *Ecosystems*, **18**, 1151–1164.

Gladyshev, M.I., Arts, M.T. & Sushchik, N.N. (2009) Preliminary estimates of the export of Omega-3 highly unsaturated fatty acids (EPA + DHA) from aquatic to terrestrial systems. *Lipids in Aquatic Ecosystems* (eds M.T. Arts, M.T. Brett & M.J. Kainz), pp. 179–209. Springer Science, New York, USA.

Gladyshev, M.I., Sushchik, N.N. & Makhutova, O.N. (2013) Production of EPA and DHA in aquatic ecosystems and their transfer to the land. *Prostaglandins and other Lipid Mediators*, **107**, 117–126.

Gonçalves, A.L. & Canhoto, C. (2009) Decomposition of eucalypt and alder mixtures: responses to variation in evenness. *Fundamental and Applied Limnology / Archive für Hydrobiologie*, **173**, 293–303.

Gonçalves, A.L., Lírio, A.V., Graça, M.A.S. & Canhoto, C. (2016) Fungal species diversity affects leaf decomposition after drought? *International Review of Hydrobiology*, **101**, 78–86.

Gonçalves, A.L., Chauvet, E., Bärlocher, F., Graça, M.A.S. & Canhoto, C. (2014) Top-down and bottom-up control of litter decomposers in streams. *Freshwater Biology*, **59**, 2172–2182.

Gonçalves, A.L., Graça, M.A.S. & Canhoto, C. (2013) The effect of temperature on leaf decomposition and diversity of associated aquatic hyphomycetes depends on the substrate. *Fungal Ecology*, **6**, 546–553.

Gonçalves, A.L., Graça, M.A.S. & Canhoto, C. (2015) Is diversity a buffer against environmental temperature fluctuations? – A decomposition experiment with aquatic fungi. *Fungal Ecology*, **17**, 96–102.

Gratton, C., Donaldson, J. & Vander Zanden, M.J. (2008) Ecosystem linkages between lakes and the surrounding terrestrial landscape in northeast Iceland. *Ecosystems*, **11**, 764–774.

Gratton, C. & Vander Zanden, M.J. (2009) Flux of aquatic insect productivity to land: comparison of lentic and lotic ecosystems. *Ecology*, **90**, 2689–2699.

Greig, H.S., Kratina, P., Thompson, P.L., Palen, W.J., Richardson, J.S. & Shurin, J.B. (2012) Warming, eutrophication, and predator loss amplify subsidies between aquatic and terrestrial ecosystems. *Global Change Biology*, **18**, 504–514.

Hannesdóttir, E.R., Gíslason, G.M., Ólafsson, J.S., Ólafsson, Ó.P. & O'Gorman, E.J. (2013) Increased stream productivity with warming supports higher trophic levels. *Advances in Ecological Research*, **48**, 285–342.

Hautier, Y., Niklaus, P.A. & Hector, A. (2009) Competition for light causes plant biodiversity loss after eutrophication. *Science*, **324**, 636–638.

Henschel, J.R., Mahsberg, D. & Stumpf, H. (2001) Allochthonous aquatic insects increase predation and decrease herbivory in river shore food webs. *Oikos*, **93**, 429–438.

Hogg, I.D. & Williams, D.D. (1996) Response of stream invertebrates to a global-warming thermal regime: an ecosystem-level manipulation. *Ecology*, **77**, 395–407.

Huston, M.A. & Wolverton, S. (2009) The global distribution of net primary production: resolving the paradox. *Ecological Monographs*, **79**, 343–377.

Huxley, J.S. (1932) *Problems of Relative Growth*. Methuen, London, UK.

IPCC (2007) Climate change 2007: the physical science basis. *Contributions of Working Group I to the Fourth Assessment Report of the Intergovernmental Panel on Climate Change* (eds S. Solomon, D. Qin, M. Manning, Z. Chen, M. Marquis, K.B. Averyt, M. Tignor and H.L. Miller). Cambridge University Press, Cambridge, U.K. and New York, NY, U.S.A, pp. 996.

IPCC (2008) *Climate Change and Water. Technical Paper of the Intergovernmental Panel on Climate Change* (eds B. Bates, Z.W. Kundzewicz, S. Wu & J. Palutikof). IPCC Secretariat, Geneva.

IPCC (2014) Summary for Policymakers. Climate Change 2014: Impacts, Adaptation, and Vulnerability. Part A: Global and Sectoral Aspects. *Contribution of Working Group II to the Fifth Assessment Report of the Intergovernmental Panel on Climate Change* (eds C.B. Field, V.R. Barros, D.J. Dokken, K.J. Mach, M.D. Mastrandrea, T.E. Bilir, M. Chatterjee, K.L. Ebi, Y.O. Estrada, R.C. Genova, B. Girma, E.S. Kissel, A.N. Levy, S. MacCracken, P.R. Mastrandrea, & L.L. White), pp. 1–32. Cambridge University Press, Cambridge, United Kingdom and New York, NY, USA.

Jonsson, M., Deleu, P. & Malmqvist, B. (2013) Persisting effects of river regulation on emergent aquatic insects and terrestrial invertebrates in upland forests. *River Research and Applications*, **29**, 537–547.

Jonsson, M., Hedström, P., Stenroth, K., Hotchkiss, E.R., Rivera Vasconcelos, F., Karlsson, J. & Byström, P. (2015) Climate change modifies the size structure of assemblages of emerging aquatic insects. *Freshwater Biology* **60**, 78–88.

Jonsson, M. & Wardle, D.A. (2009) The influence of freshwater-lake subsidies on invertebrates occupying terrestrial vegetation. *Acta Oecologica*, **35**, 698–704.

Jonsson, M. & Stenroth, K. (2016) True autochthony and allochthony in aquatic-terrestrial resource fluxes along a landuse gradient. *Freshwater Science*, **35**, 882–894.

Karlsson, J., Byström, P., Ask, J., Ask, P., Persson, L. & Jansson, M. (2009) Light limitation of nutrient-poor lake ecosystems. *Nature*, **460**, 506–509.

Kelly, P.T., Solomon, C.T., Weidel, B.C. & Jones, S.E. (2014) Terrestrial carbon is a resource, but not a subsidy, for lake zooplankton. *Ecology*, **95**, 1236–1242.

Kikuzawa, K. (2004) Ecology of leaf senescence. *Plant Cell Death Processes* (ed. L.D. Noodén), pp. 363–370. Elsevier Academic Press, San Diego, CA, US.

Knight, T.M., McCoy, M.W., Chase, J.M., McCoy, K.A. & Holt, R.D. (2005) Trophic cascades across ecosystems. *Nature*, **437**, 880–883.

Kollberg, I., Bylund, H., Schmidt, A., Gershenzon, J. & Björkman, C. (2013) Multiple effects of temperature, photoperiod and food quality on the performance of a pine sawfly. *Ecological Entomology*, **38**, 201–208.

Kominoski, J.S., Follstad Shah, J.J., Canhoto, C., Fischer, D.G., Giling, D.P., González, E., Griffiths, N.A., Larrañaga, A., LeRoy, C.J., Mineau, M.M., McElarney, Y.R., Shirley, S.M., Swan, C.M. & Tiegs, S.D. (2013) Forecasting functional implications of global changes in riparian plant communities. *Frontiers in Ecology and the Environment*, **11**, 423–432.

Kraus, J.M. & Vonesh, J.R. (2012) Fluxes of terrestrial and aquatic carbon by emergent mosquitoes: a test of controls and implications for cross-ecosystem linkages. *Oecologia*, **170**, 1111–1122.

Lagrue, C., Kominoski, J.S., Danger, M., Baudoin, J.-M., Lamothe, S., Lambrigot, D. & Lecerf, A. (2011) Experimental shading alters leaf litter breakdown in streams of contrasting riparian canopy cover. *Freshwater Biology*, **56**, 2059–2069.

Lake, P.S. (2011) *Drought and Aquatic Ecosystems: Effects and Responses.* Wiley-Blackwell, New York, USA.

Larsen, S., Andersen, T. & Hessen, D.O. (2011) Climate change predicted to cause severe increase of organic carbon in lakes. *Global Change Biology*, **17**, 1186–1192.

Lau, D.C.P., Goedkoop, W. & Vrede, T. (2013) Cross-ecosystem differences in lipid composition and growth limitation of a benthic generalist consumer. *Limnology and Oceanography*, **58**, 1149–1164.

Lau, D.C.P., Vrede, T., Pickova, J. & Goedkoop, W. (2012) Fatty acid composition of consumers in boreal lakes – variation across species, space and time. *Freshwater Biology*, **57**, 24–38.

Lecerf, A. & Chauvet, E. (2008) Intraspecific variability in leaf traits strongly affects alder leaf decomposition in a stream. *Basic and Applied Ecology*, **9**, 598–605.

Lecerf, A., Dobson, M., Dang, C.K. & Chauvet, E. (2005) Riparian plant species loss alters trophic dynamics in detritus based stream ecosystems. *Oecologia*, **146**, 432–442.

Ledger, M.E. & Milner, A.M. (2015) Extreme events in running waters. *Freshwater Biology*, **60**, 2455–2460.

Marczak, L.B., Thompson, R.M. & Richardson, J.S. (2007) Meta-analysis: trophic level, habitat, and productivity shape the food web of resource subsidies. *Ecology*, **88**, 140–148.

Mas-Martí, E., Muñoz, I., Oliva, F. & Canhoto, C. (2015) Effects of increased water temperature on leaf litter quality and detritivore performance: a whole-reach manipulative experiment. *Freshwater Biology*, **60**, 184–197.

McCauley, S.J., Hammond, J.I., Frances, D.N. & Mabry, K.E. (2015) Effects of experimental warming on survival, phenology, and morphology of an aquatic insect (Odonata). *Ecological Entomology*, **40**, 211–220.

McKie, B.G., Cranston, P.S. & Pearson, R.G. (2004) Gondwanan mesotherms and cosmopolitan eurytherms: effects of temperature on the development and survival of Australian Chironomidae (Diptera) from tropical and temperate populations. *Marine and Freshwater Research*, **55**, 759–768.

McPeek, M.A. (2004) The growth/predation risk trade-off: so what is the mechanism? *American Naturalist*, **163**, E88–E111.

Merritt, R.W., Cummins, K.W. & Burton, T.M. (1984) The role of aquatic insects in the processing and cycling of nutrients. *The Ecology of Aquatic Insects* (eds V.H. Resh and D.N. Rosenberg), pp. 134–163. Praeger Scientific, New York, USA.

Meyer, J.D., Sale, M.J., Mulholland, P.J. & Poff, N.L. (1999) Impacts of climate change on aquatic ecosystem functioning and health. *Journal of the American Water Resources Association*, **35**, 1373–1386.

Mills, G.L., McArthur, J.V., Wolfe, C., Aho, J.M. & Rader, R.B. (2001) Changes in fatty acid and hydrocarbon composition of leaves during decomposition in a southeastern blackwater stream. *Archiv für Hydrobiologie*, **152**, 315–328.

Morrill, J., Bales, R. & Conklin, M. (2005) Estimating stream temperature from air temperature: implications for future water quality. *Journal of Environmental Engineering*, **131**, 139–146.

Muehlbauer, J.D., Collins, S.F., Doyle, M.W. & Tockner, K. (2014) How wide is a stream? Spatial extent of the potential "stream signature" in terrestrial food webs using meta-analysis. *Ecology*, **95**, 44–55.

Müller-Navarra, D.C., Brett, M.T., Liston, A.M. & Goldman, C.R. (2000) A highly unsaturated fatty acid predicts carbon transfer between primary producers and consumers. *Nature*, **403**, 74–77.

Murakami, M. & Nakano, S. (2002) Indirect effect of aquatic insect emergence on a terrestrial insect population through by birds predation. *Ecology Letters*, **5**, 333–337.

Nentwig, W. (1980) The selective prey of linyphiid-like spiders and of their space webs. *Oecologia*, **45**, 236–243.

Niinemets, Ü. & Tamm, Ü. (2005) Species differences in timing of leaf fall and foliage chemistry modify nutrient resorption efficiency in deciduous temperate forest stands. *Tree Physiology*, **25**, 1001–1014.

Nilsson, C., Polvi, L.E. & Lind, L. (2015) Extreme events in streams and rivers in arctic and subarctic regions in an uncertain future. *Freshwater Biology*, **60**, 2535–2546.

Norby R.J., Cotrufo, M.F., Ineson, P., O'Neill, E.G. & Canadell, J.G. (2001) Elevated CO_2, litter chemistry, and decomposition: A synthesis. *Oecologia*, **127**, 153–165.

Nunes, J.P., Seixas, J., Keizer, J.J. & Ferreira, A.J.D. (2009) Sensitivity of runoff and soil erosion to climate change in two Mediterranean watersheds. Part I: model parameterization and evaluation. *Hydrological Processes*, **23**, 1202–1211.

O'Reilly, C.M., Alin, S.R., Plisnier, P.-D., Cohen, A.S. & McKee, B.A. (2003) Climate change decreases aquatic ecosystem productivity of Lake Tanganyika, Africa. *Nature*, **424**, 766–768.

Pearce-Higgins J.W. & Yalden, D.W. (2004) Habitat selection, diet, arthropod availability and growth of a moorland wader: the ecology of European Golden Plover *Pluvialis apricaria* chicks. *Ibis*, **146**, 335–346.

Petchey, O.L., McPhearson, P.T., Casey, T.M. & Morin, P.J. (1999) Environmental warming alters food-web structure and ecosystem function. *Nature*, **402**, 69–72.

Polis, G.A., Anderson, W.B. & Holt, R.D. (1997) Toward an integration of landscape and food web ecology: the dynamics of spatially subsidized food webs. *Annual Review in Ecology and Systematics*, **28**, 289–316.

Polis, G.A. & Strong, D.R. (1996) Food web complexity and community dynamics. *American Naturalist*, **147**, 813–846.

Poulin, B., Lefebvre, G. & Paz, L. (2010) Red flag for green spray: adverse trophic effects of *Bti* on breeding birds. *Journal of Applied Ecology*, **47**, 884–889.

Rempel, R.S. & Carter, J.C.H. (1987) Temperature influences on adult size, development, and reproductive potential of aquatic Diptera. *Canadian Journal of Fisheries and Aquatic Sciences*, **44**, 1743–1752.

Rice, J.A., Miller, T.J., Rose, K.A., Crowder, L.B., Marschall, E.A., Trebitz, A.S. & DeAngelis, D.L. (1993) Growth rate variation and larval survival: inferences from an individual-based size-dependent predation model. *Canadian Journal of Fisheries and Aquatic Sciences*, **50**, 133–142.

Richardson, J.S., Zhang, Y. & Marczak, L.B. (2010) Resource subsidies across the land–freshwater interface and responses in recipient communities. *River Research and Applications*, **26**, 55–66.

Riipinen, M.P., Fleituch, T., Hladyz, S., Woodward, G., Giller, P. & Dobson, M. (2010) Invertebrate community structure and ecosystem functioning in European conifer plantation streams. *Freshwater Biology*, **55**, 346–359.

Sabo, J.L. & Power, M.E. (2002a) Numerical response of lizards to aquatic insects and short-term consequences for terrestrial prey. *Ecology*, **83**, 3023–3036.

Sabo, J.L. & Power, M.E. (2002b) River-watershed exchange: effects of riverine subsidies on riparian lizards and their terrestrial prey. *Ecology*, **83**, 1860–1869.

Scharnweber, K., Vanni, M.J., Hilt, S., Syväranta, J. & Mehner, T. (2014) Boomerang ecosystem fluxes: organic carbon inputs from land to lakes are returned to terrestrial food webs via aquatic insects. *Oikos*, **123**, 1439–1448.

Seekell, D.A., Lapierre, J.-F., Ask, J., Bergström, A.-K., Deininger, A., Rodríguez, P. & Karlsson, J. (2015) The influence of dissolved organic carbon on primary production in northern lakes. *Limnology and Oceanography*, **60**, 1276–1285.

Shaver, G.R., Canadell, J., Chapin, F.S. III, Gurevitch, J., Harte, J., Henry, G., Ineson, P., Jonasson, S., Melillo, J., Pitelka, L. & Rustad, L. (2000) Global warming and terrestrial ecosystems: a conceptual framework for analysis. *BioScience*, **50**, 871–882.

Stenroth, K., Polvi, L.E., Fältström, E. & Jonsson, M. (2015) Land-use effects on terrestrial consumers through changed size structure of aquatic insects. *Freshwater Biology*, **60**, 136–149.

Sterner, R.W. & Elser, J.J. (2002) *Ecological Stoichiometry: The Biology of Elements from Molecules to the Biosphere*. Princeton University Press, Princeton, NJ, USA.

Stiling, P. & Cornelissen, T. (2007) How does elevated carbon dioxide (CO_2) affect plant–herbivore interactions? A field experiment and meta-analysis of CO_2-mediated changes on plant chemistry and herbivore performance. *Global Change Biology*, **13**, 1823–1842.

Stoks, R., Geerts, A.N. & De Meester, L. (2014) Evolutionary and plastic responses of freshwater invertebrates to climate change: realized patterns and future potential. *Evolutionary Applications*, **7**, 42–55.

Strasevicius, D., Jonsson, M., Nyholm, N.E.I. & Malmqvist, B. (2013) Reduced breeding success of Pied Flycatchers *Ficedula hypoleuca* along regulated rivers. *Ibis*, **155**, 348–356.

Sweeney, B.W. (1978) Bioenergetic and developmental response of a mayfly to thermal variation. *Limnology and Oceanography*, **23**, 461–477.

Sweeney, B.W., Jackson, J.K., Newbold, J.D. & Funk, D.H. (1992) Climate change and the life histories and biogeography of aquatic insects in eastern North America. *Global Climate Change and Freshwater Ecosystems* (eds P. Firth & S.G. Fisher), pp. 143–176. Springer-Verlag, New York Inc., New York, USA.

Sweeney, B.W. & Schnack, J.A. (1977) Egg development, growth, and metabolism of *Sigara alternata* (Say) (Hemiptera: Corixidae) in fluctuating thermal environments. *Ecology*, **56**, 265–277.

Tuchman N.C., Wahtera, K.A., Wetzel, R.G. & Teeri, J.A. (2003) Elevated atmospheric CO_2 alters leaf litter quality for stream ecosystems: An *in situ* leaf decomposition study. *Hydrobiologia*, **495**, 203–211.

Tuchman, N.C., Wetzel, R.G., Rier, S.T., Wahtera, K.A. & Teeri, J.A. (2002) Elevated atmospheric CO_2 lowers leaf litter nutritional quality for stream ecosystem food webs. *Global Change Biology*, **8**, 163–170.

Vannote, R.L., Minshall, G.W., Cummins, K.W., Sedell, J.R. & Cushing, C.E. (1980) The river continuum concept. *Canadian Journal of Fisheries and Aquatic Sciences*, **37**, 130–137.

Villanueva V.D., Albariño, R. & Canhoto, C. (2011) Detritivores feeding on poor quality food are more sensitive to increased temperatures. *Hydrobiologia*, **678**, 155–165.

Wallace, J.B., Eggert, S.L., Meyer, J.L. & Webster, J.R. (1997) Multiple trophic levels of a forest stream linked to terrestrial litter inputs. *Science*, **277**, 102–104.

Wallace, J.B., Eggert, S.L., Meyer, J.L. & Webster, J.R. (2015) Stream invertebrate productivity linked to forest subsidies: 37 stream-years of reference and experimental data. *Ecology*, **96**, 1213–1228.

Walther, G.-R., Post, E., Convey, P., Menzel, A., Parmesan, C., Beebee, T.J.C., Fromentin, J.-M., Hoegh-Guldberg, O. & Bairlein, F. (2002) Ecological responses to recent climate change. *Nature*, **416**, 389–395.

Wardle, D.A., Hörnberg, G., Zackrisson, O., Kalela-Brundin, M. & Coomes, D.A. (2003) Long-term effects of wildfire on ecosystem properties across an island area gradient. *Science*, **300**, 972–975.

Webb, B.W. & Nobilis, F. (2007) Long term changes in river temperature and the influence of climatic and hydrologic factors. *Hydrological Sciences Journal*, **52**, 74–85.

Wolff, R.L., Lavialle, O., Pédrono, F., Pasquier, E., Deluc, L.G., Marpeau, A.M. & Aitzetmüller, K. (2001) Fatty acid composition of Pinaceae as taxonomic markers. *Lipids*, **36**, 439–451.

Woodward, G., Perkins, D.M. & Brown, L.E. (2010) Climate change and freshwater ecosystems: impacts across multiple levels of organization. *Philosophical Transactions of the Royal Society B*, **365**, 2093–2106.

Zhu, K., Woodall, C.W. & Clark, J.S. (2012) Failure to migrate: lack of tree range expansion in response to climate change. *Global Change Biology*, **18**, 1042–1052.

15

Climatic Impacts on Invertebrates as Food for Vertebrates

Robert J. Thomas[1], James O. Vafidis[1,2] and Renata J. Medeiros[1]

[1] Cardiff School of Biosciences, Cardiff University, Wales, UK
[2] University of Western England, Bristol, UK

Summary

Invertebrates are essential components of ecosystems, that as well as delivering a diversity of functional roles, provide a crucial trophic resource for many vertebrates. Understanding the effects of climate change on the abundance, distribution and phenology of invertebrates can therefore provide important insights into the effects of climate on vertebrates, via cascading impacts across trophic levels. This chapter addresses the fundamental mechanisms by which vertebrates can be impacted by climate-driven changes in their invertebrate food supply across terrestrial and aquatic ecosystems, and discusses the individual-to-population level outcomes of those processes. The impacts of climate-driven changes in invertebrate populations can drive fluctuations in vertebrate population size and in some cases, to population extinctions. Climate-driven changes in invertebrate abundance and distributions can also drive major shifts in the spatial distribution of vertebrate predator populations across geographic, altitudinal and depth gradients. Climate-driven changes in invertebrate phenology can have substantial implications for the timing of vertebrate reproduction and other aspects of life history. Understanding these mechanisms can help develop effective actions to conserve the biodiversity of both vertebrates and their invertebrate prey, in the face of climate change.

15.1 Introduction

Much of what we currently know about the impacts of climate change on individual animals, on animal populations and on ecosystems, has been derived from studies of vertebrates. This focus is understandable, given that considerably more research and conservation resources are typically available for vertebrate studies than for studies of invertebrates (Grodsky et al., 2015). Understanding the processes by which climate change ultimately affects vertebrates does, however, require a detailed understanding of the effects of climate on the wider ecosystems on which they depend. Animals at lower trophic levels, including invertebrates, are important components of the structure and functioning of all major ecosystems. Therefore, studies of invertebrates are of central importance for understanding, predicting and mitigating the effects of climate change on vertebrates and the ecosystems on which they depend. Such studies can also provide important insights and research opportunities for understanding the cascading effects of climate change across the components of whole ecosystems, including both bottom-up and top-down effects.

Global Climate Change and Terrestrial Invertebrates, First Edition. Edited by Scott N. Johnson and T. Hefin Jones.
© 2017 John Wiley & Sons, Ltd. Published 2017 by John Wiley & Sons, Ltd.

While vertebrate populations can be affected directly by climate (e.g., mortality or impaired reproduction caused by severe storms, flooding, drought, freezing), many of the links between climate and vertebrate populations appear to be less direct, mediated by ecological processes such as predation, parasitism, disease transmission, competition and seasonal changes in habitat productivity and structure. Invertebrates play a central role in many of these processes, as prey, parasites, disease vectors and competitors (Bale et al., 2002; Harvell et al., 2002; Poulin & Mouritsen, 2006). In this chapter, we address the role of invertebrates as food resources for vertebrates, in the context of climate impacts across trophic levels. We highlight common processes that occur across the Earth's major biomes, as well as important variations in these impacts across different ecological contexts.

Climate change directly alters the physical environment, and the effects of these physical changes can in turn alter the biotic relationships between components of an ecosystem, such as the trophic relationships between vertebrates and invertebrates. Here, we outline the major classes of responses to climate change that have been widely observed among vertebrates over recent decades: (1) changes in abundance (including local, regional and global extinctions), (2) changes in distribution, and (3) changes in the timing of biological events. Many of these patterns of change have been described in considerable detail by numerous studies, yet we still know remarkably little about the mechanisms by which such changes are mediated. In many cases, the climate-driven changes in the trophic relationships between vertebrates and invertebrates are proving to be key ecological processes underlying these patterns (Van der Putten et al., 2004).

15.2   Changes in the Abundance of Vertebrates

15.2.1   Variation in Demography and Population Size

The present volume describes many examples of climate-driven changes in invertebrate demography and abundance. Likewise, the population sizes of numerous vertebrate species have been found to vary in relation to global and local climate changes. Examples can be drawn from across a wide range of ecosystems, including marine, freshwater and terrestrial systems (e.g., Edwards & Richardson, 2004; Stevens et al., 2002; Both et al., 2009) and animal taxa, including birds (Pearce-Higgins et al., 2010), mammals (Newman & Macdonald, 2015), fish (Hoegh-Guldberg et al., 2007), reptiles (Huey et al., 2010) and amphibians (Pounds et al., 2006). In many cases, the effects of climate on vertebrate populations are mediated by climate-driven changes in food availability influencing breeding success, or survival, or both. These effects can be negative or – at least initially – positive, and can sometimes be mediated by complex underlying mechanisms.

For some vertebrate species, such effects of climate on food availability are negative. For example, in southern Europe, the breeding success of crag martins *Ptyonoprogne rupestris* is severely reduced by an increase in temperature above current mean levels. Such temperature changes reduce the availability of the flying insects with which this aerial insectivore provisions its broods (Acquarone et al., 2003). In other species, the opposite effect occurs, namely a positive effect of climate change on foraging and breeding success. For example, in southern North America the Everglade snail kite *Rostrhamus sociabilis plumbeus*, is a specialist predator of apple snails *Pomacea paludosa*, whose availability is positively associated with water temperature. If water temperatures increase with climate change, it is predicted that snail kites will encounter more favourable foraging conditions, enabling them, at least initially, to advance their breeding dates and increase their breeding success (Stevens et al., 2002; Bretagnolle & Gillis, 2010). If temperatures continue to increase, however, and start to exceed the thermal tolerance of the apple snails, then the kite's food availability may decline and may ultimately become limiting to the kite's population size (Byers et al., 2013).

The impacts of climate-driven changes in food supply can be particularly evident among warm-blooded vertebrates with high metabolic rates requiring a high and regular food intake. For example, shrews (*Sorex* spp.) in temperate regions operate under tight energy budgets, such that successful breeding requires near-constant hunting during their waking hours for invertebrates, particularly beetles. Climate warming may at first increase availability (abundance and activity) of invertebrate prey for shrews (Russell & Grimm, 1990; Berthe et al., 2015), thereby alleviating the constraints of food availability on their survival, growth and reproduction (Newman & Macdonald, 2015).

Even among closely related species, the impacts of climate change on invertivorous vertebrates (i.e., vertebrates that consume invertebrates) can be highly variable, depending on the species' behaviour and ecology. For example, among the old-world warblers (Aves, Passeriformes, Sylvioidea) breeding in northwest Europe, different species exhibit increasing, decreasing, or fluctuating population sizes (Fig. 15.1). Such species-differences can be attributed to interspecific variations in the mechanisms by which climate influences demographic parameters (e.g., breeding productivity, survival and recruitment) and ultimately population growth rates. The specific mechanisms may differ between species, and include diet (Davies & Green, 1976; Bibby & Thomas, 1985; King et al., 2015), foraging ecology (Bibby et al., 1976; Leisler, 1992), migratory tendency, overwintering region, and winter habitat selection (Bibby & Green, 1981; Salewski et al., 2013; Vafidis et al., 2014). Trophic resources, which for the Sylvidae warblers are primarily invertebrates (Cramp & Brooks, 1992; King et al., 2015), are implicated in each of these mechanisms.

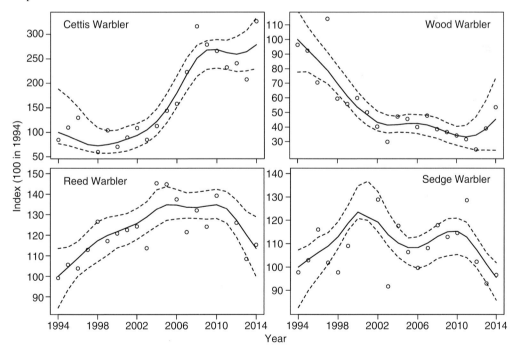

Figure 15.1 Contrasting population changes within a single taxon of insectivorous bird species (Warblers; Superfamily Sylvioidea) breeding in the UK. Dashed lines represent 95% confidence intervals around the mean population index, where a value of 100 represents the species' population size in 1994. The figure is redrawn from Breeding Bird Survey (BBS) data from the Bird Trends Report (Baillie et al., 2014) with permission from the British Trust for Ornithology (BTO). The BBS is a partnership scheme of BTO/JNCC/RSPB.

Specifically, a sustained population increase in the non-migratory Cetti's warbler *Cettia cetti* has primarily been driven by increasing winter temperatures in the birds' reedbed wintering sites, leading to higher food availability and overwinter survival of the birds (Robinson et al., 2007). A population decline in the migratory wood warbler *Phylloscopus sibilatrix*, a trans-Saharan migrant breeding in woodland, has been attributed to a decline in the quality of breeding habitat as well as changes in land-use and climate on their wintering grounds (Ockendon et al., 2012; Mallord et al., 2012; Balmer et al., 2013). Recent population fluctuations in sedge warblers *Acrocephalus schoenobaenus* have primarily been driven by the impacts of drought on the invertebrate-rich west African wetlands in which they exclusively overwinter (Peach et al., 1991; Baillie & Peach, 1992). In contrast, the congeneric reed warbler *Acrocephalus scirpaceus*, which has much more flexible overwinter habitat requirements (Vafidis et al., 2014), has shown a substantial population increase, seemingly driven by climate-driven increases in food availability on the species' reedbed breeding grounds (Vafidis et al., 2016).

Temperature is not the only aspect of climate variation that can have dramatic effects on vertebrate populations, mediated by trophic processes. An example of rainfall-driven changes in prey availability affecting predator abundances is the facultative earthworm (*Lumbricus* spp.) specialist, the European badger *Meles meles*. Badgers are particularly affected by microclimatic effects on the surfacing behaviour of earthworms, primarily *Lumbricus* spp. (Gerard, 1967; Macdonald & Newman, 2002). Population densities and survival rates of badgers tend to be highest where warm and damp soil conditions increase the availability of earthworms (Gerard, 1967; Kruuk, 1978; Johnson et al., 2002; Newman & Macdonald, 2015).

In the UK, cool May conditions and higher levels of summer rainfall prove beneficial for badger reproductive success and survival, while drought conditions have negative effects on juvenile survival. This effect of climate on survival is mediated by the low immunity of food-deprived badger cubs to endemic infection leading to widespread mortality (Macdonald & Newman, 2002; Macdonald et al., 2010). Specifically, there is an interaction between climatic conditions and the protozoan gut parasite *Eimeria melis*, which all badger cubs contract. In years of abundant food, badger cubs are able to mount an effective immune response to the infection, but in years of food limitation the badgers' immune response is weaker and the parasite causes high mortality among cubs (Newman et al., 2001; Macdonald et al., 2010).

Consequently, badger distribution and local abundance are both heavily dependent on weather patterns, particularly on rainfall, while exhibiting a relatively wide tolerance for temperature variations (Nouvellet et al., 2013). Indeed, local badger population densities can be accurately predicted in the more arid parts of their range in central Spain, using minimum rainfall measurements alone (Virgós & Casanovas, 1999). These findings suggest that the effects of future changes in rainfall patterns on badger abundance and distribution will be stronger than the effect of variation in temperature alone (Newman & Macdonald, 2015).

For some vertebrates, demographic processes may be influenced both directly by climate, and indirectly by the effects of climate variation on invertebrate food availability and foraging conditions. For example, the breeding success of Wilson's storm petrels *Oceanites oceanicus* (the smallest endothermic species breeding in the Antarctic) is dependent on the frequency of food deliveries by parents to their single nestling (Quillfeldt, 2001). This species is highly dependent on the availability of zooplankton, particularly Antarctic krill *Euphausia superba* (Quillfeldt, 2002), whose abundance during the petrel's breeding season is influenced by the extent of sea ice cover in the previous winter. In years of less extensive winter sea ice cover, food availability is scarce, leading to low food delivery rates and an increased likelihood of chick starvation (~50% of chicks starve in poor years). The frequency of food deliveries and chick survival is also related to climate variables: (1) prevailing wind conditions, which influence the transport of krill-rich surface waters towards the petrels' foraging areas, and (2) snow fall which can block the entrances to breeding burrows, physically preventing parents from delivering food to developing chicks (Quillfeldt, 2001; Büßer et al., 2004).

Similar impacts of inter-annual climate variability on sea ice extent, krill availability and breeding performance of Antarctic seabirds have been demonstrated in other bird species (e.g., Adélie penguins *Pygoscelis adeliae*, emperor penguins *Aptenodytes forsteri*, southern fulmars *Fulmarus glacialoides* and snow petrels *Pagodroma nivea*; Croxall et al., 2002; Jenouvrier et al., 2003). These examples illustrate the often complex and interacting physical and biotic mechanisms linking climate change to invertebrates and their vertebrate predators.

15.2.2 Local Extinctions

Changes in abundance at the local level can include instances of local extinction, and several such instances, across a range of ecosystems, can be attributed to climate-driven changes in trophic resources. For example, coral reefs are one of the most vulnerable ecosystems to climate change due to the widespread loss of live coral through coral bleaching caused by increases in sea surface temperatures (SST; Hoegh-Guldberg, 1999; Hughes et al., 2003; Hoegh-Guldberg et al., 2007). Climate-driven declines in the abundance of coral-building invertebrate taxa (Cnidaria) have been associated with local extinctions of coral-dwelling fishes that feed on the coral, including fishes of the genus *Gobodion* in Papua New Guinea, and butterflyfishes (family *Chaetodontidae*) of the Indian Ocean. The actual role of invertebrates in such extinctions is often unknown. For example, the disappearance of the endemic Galapagos damselfish *Azurina eupalama* coincided with the El Niño event of 1982–83 (Roberts & Hawkins, 1999), yet the mechanism by which this extinction occurred is also unknown (Monte-Luna et al., 2007).

Not all food-mediated extinctions are due to direct limitation of food availability: in a genus of insectivorous lizards (*Sceloporus*), where 12% of 200 previously documented local populations have become extinct, the likelihood of local extinction has been shown to be associated with the magnitude of recent climate warming (Sinervo et al., 2010). These local-to-regional-scale population extinctions appear to be driven by energetic shortfalls during spring, when energetic demands are high – mediated by a shifting trade-off between increased time that the lizards have to spend sheltering from high temperatures, and a consequent reduction in the time that they can spend foraging for insects. As a result of decreased foraging intensity, lizard reproductive rates are energetically limited and population growth rates are negative (Sinervo et al., 2010).

Despite these examples, the number of documented instances of local extinction mediated by climate-driven changes in invertebrate food supply is so far surprisingly small, but is likely to increase as anthropogenic climate changes continue to take hold and increasingly start to exceed the range of conditions to which species are adapted.

15.2.3 Global Extinctions

Examples of global extinction that can be directly attributed to climate change are even more scarce than examples of local or regional extinction. The golden toad *Bufo periglenes* of Costa Rica is sometimes cited as one such example (Pounds & Crump, 1994; Pounds et al., 2006), but the cause of its extinction is not known with certainty. Climate change is one of several hypotheses, via the ecological effects of El Niño (lower rainfall, higher wind speeds, causing desiccation), perhaps in interaction with *Chytrid* fungus infection, changes in food availability and other processes such as atmospheric pollution. Interactions between multiple processes such as these can result in a "deadly cocktail" of natural and anthropogenic changes, leading ultimately to extinction (Blaustein & Kiesecker, 2002).

There are no confirmed historical global extinctions of coral reef fishes (Jones et al., 2002), although the fate of the Mauritius wrasse *Anampses viridus*, and the Galapagos damselfish *Azurina eupalama*, remains uncertain, and three other species are considered to be critically endangered (Hawkins et al., 2000). Although *Gobiodon* fish species have become locally extinct in parts of Papua New Guinea (see above), they are not facing global extinction because they also exist in other regions of the western Pacific where they are less at risk (Munday et al., 1999).

The number of documented climate-driven local and global extinctions of vertebrates appears, at least to date, to be relatively small, and the number that are clearly driven by invertebrates as trophic resources is even smaller. The magnitude of the extinction crisis that is forecast under projected climate change scenarios is, however, startling (e.g., Thomas et al., 2004a). Expected extinctions by the end of the twenty-first century include 20% of lizard species (and the local extinction of ~40% of lizard sub-populations), 14% of birds, and 40% of amphibians (Barnosky et al., 2011). Clearly, understanding the role of invertebrates as foraging resources for vertebrates will be critical for the mitigation of climate impacts on vertebrate populations and the conservation of all of Earth's species, communities and ecosystems.

15.3 Changes in the Distribution of Vertebrates

Animal distributions can change in response to climate, as populations track suitable environmental conditions and resources in space and time (Hughes, 2000). Species' distributions are frequently climate-limited, both among vertebrates and invertebrates (Beaumont & Hughes, 2002; Beale et al., 2008; Pearce-Higgins & Green, 2014; Roques et al., 2015), so changing climate is frequently associated with shifts in distribution. A widely used approach to predicting future species distribution is "climate envelope" modelling which is based on tolerances to the physical environment (Box, 1981). One criticism of this approach is that it has often ignored the importance and complexity of biological processes such as trophic interactions, dispersal and evolutionary changes, in determining species distributions (e.g., Pearson & Dawson, 2003). Indeed, more recent studies have incorporated such processes into the climate envelope modelling framework (e.g., Araújo & Luoto, 2007; Preston et al., 2008; Holt & Barfield, 2009; Van der Putten et al., 2010).

Climate-driven changes in distribution can involve changes in geographical ranges, altitudinal or depth distributions, as well as more local micro-climate selection driving local changes in distributions (e.g., based on species-specific preferences for aspect, shade and substrate). The research literature describes numerous instances of range shifts associated with climate change, many of which may be mediated by trophic resources.

15.3.1 Geographical Range Shifts

The majority of climate-driven species range shifts described in the literature relate to horizontal geographical (rather than vertical altitudinal or depth) shifts, by which species track suitable climate conditions (e.g., temperature isotherms) and/or prey resources. Many of these geographical shifts are latitudinal, pole-ward shifts of mid-to high-latitude species (Pearce-Higgins & Green, 2014), whereby species move northwards in the northern hemisphere and southwards in the southern hemisphere, but range shifts are also evident along other climatic gradients such as maritime-continental axes in temperature or rainfall (VanDerWal et al., 2013).

301 Climatic Impacts on Invertebrates as Food for Vertebrates | 301

Although range shifts are widely described in well-studied taxa such as fish and birds, the underlying mechanisms often remain largely unknown. Possible examples of vertebrate range shifts occurring in association with climate-driven range shifts in their prey include teleost fish shifting their distribution in association with distribution shifts in their zooplankton prey (Poloczanska et al., 2013).

Large-scale changes in the biogeography of important zooplankton assemblages in the northeast Atlantic Ocean are associated with changing distributions of many pelagic fish. For example, Sardines *Sardina pilchardus* and anchovies *Engraulis encrasicolus* are becoming more common in seas around northern Europe (Cunningham, 1890). Although the causal mechanisms are unclear, this range shift is consistent with the northwards spread of a smaller warm-water copepod prey species *Calanus helgolandicus* in the North Sea, which has entirely replaced the larger boreal congener *C. finmarchicus* (Reid et al., 2003; Beare et al., 2004; Helaouët & Beaugrand, 2007). Likewise, other species of fish such as red mullet *Mullus surmuletus* and bass *Dicentrarchus labrax* have extended their ranges northwards in the North Atlantic, in association with increases in SST and range shifts of their invertebrate prey (Brander et al., 2003).

Range shifts of vertebrates do not necessarily track shifting prey populations sufficiently rapidly to prevent population declines. For example, populations of Atlantic salmon *Salmo salar* and Atlantic cod *Gadus morhua* in the northeast Atlantic are in decline, seemingly in response to changes in the size of copepod prey available during a critical period for juvenile survival, due to the shifts in geographical ranges of the prey species that the vertebrate predators have failed to track (Beaugrand & Reid, 2003; Beaugrand et al., 2003).

In cases of small insectivorous birds such as the Cetti's warbler, which have advanced their range rapidly northwards, the new range has been shown primarily to be limited by overwinter temperatures – although whether this is an effect of food limitation under low temperatures, or direct mortality due to freezing, is yet to be determined. There is a need for a clearer understanding of the role of invertebrate availability in mediating range shifts of their predators, but current studies are starting to address this (Vafidis et al., 2016).

15.3.2 Altitudinal Range Shifts

Terrestrial animals may track appropriate climatic conditions by moving to higher altitudes, though this is only possible if higher ground is available. An example of an altitudinal shift of a vertebrate species driven by climatic impacts on their invertebrate prey, is the case of insectivorous wading birds in the European uplands. Golden plover *Pluvialis apricaria* breeding success depends on the abundant and synchronized emergence of craneflies (Tipulidae) during the plovers' breeding season. The timing and synchronicity of cranefly emergence relies on low temperatures during the previous August. Craneflies have become less abundant at lower altitudes where climate warming in late summer has reduced emergence rates (Pearce-Higgins et al., 2010). This has had the effect of pushing golden plovers to breed at higher altitudes, where August temperatures are still low enough to cause synchronized cranefly emergence in the following year (Bale et al., 2002; Pearce-Higgins et al., 2005, 2010).

Likewise, a recent rapid shift in the distribution of the pine processionary moth *Thaumetopoea pityocampa* to higher altitudes in southern Europe (Battisti et al., 2006), may have important consequences for their avian and mammal predators (Barbaro & Battisti, 2011). Predators with different foraging strategies feed on different life-history stages of the moth, but the differential consequences of the prey's altitudinal shift on these different classes of predators remains unknown.

15.3.3 Depth Range Shifts

As climate change generates higher temperatures and lower precipitation, soil becomes drier and less suitable for foraging invertebrates such as earthworms. In response, soil-dwelling invertebrates may bury themselves deeper in the soil in order to reach wetter conditions. Lower levels of food availability induced by this effect, for surface- or shallow-feeding vertebrate predators of soil-dwelling invertebrates has consequences for the predators' body condition and breeding success, for example as described above for Eurasian badgers (Section 15.1a), as well as for birds such as song thrush *Turdus philomelos* (Peach et al., 2004), northern lapwing *Vanellus vanellus* (Beintema & Visser, 1989; Beintema et al., 1991) and common snipe *Gallinago gallinago* (Green, 1988).

The marine analogues of such shifts in terrestrial systems are bathymetric (depth) changes. In marine ecosystems, climate-driven shifts to deeper, cooler and less acidified waters are evident, both among invertebrates (e.g., crustacea; Linares et al., 2015; Morris et al., 2015) and vertebrates (e.g., fish; Perry et al., 2005). However, examples of vertebrate bathymetric changes being mediated by changes in their invertebrate prey appear, to date, to be lacking.

15.3.4 Food-Mediated Mechanisms and Trophic Consequences of Range Shifts

The numerous examples of climate-driven horizontal geographical, altitudinal and depth-range shifts illustrated above, raise important questions regarding the mechanisms underlying such distributional changes, and their consequences for predator–prey relationships. If predators and prey differ in their sensitivity to climate change and in their distributional responses, trophic relationships may change as predators become separated from their original prey and encounter new potential prey species. Even when predators and prey species shift range at the same rate, the trophic relationships between them may be altered due to changes in the type and relative abundance of competitors and alternative prey (Holt & Huxel, 2007; Holt & Barfield, 2009).

There are substantial differences between taxa in the rate at which vertebrate and invertebrate distributions are changing in response to climate, due to species differences in exposure and sensitivity to climate change, as well as differences in dispersal ability (Palmer et al., 2015). These differential responses of animal taxa drive changes in ecological communities (Lovejoy & Hannah, 2005), leading to changing trophic relationships as individuals encounter novel species of potential prey, predators, parasites and competitors (Tylianakis et al., 2008; Van der Putten et al., 2010). Dietary breadth (diversity of prey types) and dietary wariness (willingness to incorporate new prey types into the diet) may differ substantially even between individuals within a vertebrate population (Thomas et al., 2004b, 2010). Under changed climatic conditions predators will face changing availability of different taxa of potential prey. There is likely to be strong selection pressure on these aspects of foraging behaviour at expanding range margins, in favour of incorporating newly available prey types into the diet under the changed conditions.

Range expansion occurs by the dispersal of individuals beyond the historic range margin, whereas range contraction can occur by either of two processes. The first process involves individuals abandoning part of the current range. The second process involves the local extinction of individuals that fail to move from those parts of the range that have become unsuitable. There is evidence that species range margins may be advancing more rapidly at their leading (advancing) boundaries than at their trailing (retreating) boundaries, leading to a temporary range expansion (Pearce-Higgins & Green, 2014). One explanation for this apparent pattern is that there is a time-lag between the environment in a location becoming unsuitable for a prey species and the predator species' eventual local extinction. For example, if prey abundance decreases, the local extinction of a predator may not be immediate. Differences in the rate of movement at the range boundary between predator and prey taxa may lead to prey species moving away from parts of their predator's range, requiring predators to switch prey and adapt to new conditions. Examples of such dietary shifts exhibited by vertebrates in relation to climate-driven changes to their invertebrate prey are, however, currently lacking.

15.4 Changes in Phenology of Vertebrates, and Their Invertebrate Prey

Among the earliest biological signals of anthropogenic climate change to be described were changes in phenology, the timing of biological events associated with climate, such as vegetation growth, animal breeding seasons, and the timing of animal migrations (Hughes, 2000; Chmielewski & Rötzer, 2001; Both & Visser, 2001). Phenological changes have been extensively studied across a broad range of plant (Linderholm, 2006; Cleland et al., 2007; Elzinga et al., 2007), fungal (Kauserud et al., 2010; Boddy et al., 2014) and animal (Bale et al., 2002; Gordo, 2007) taxa. These studies together provide compelling evidence of the cascading impacts of physical climate change across trophic levels, affecting marine and terrestrial ecosystems from the tropics to the poles. As outlined below, phenological changes are important aspects of the mechanisms by which demographic and population changes among vertebrates are mediated by climate impacts at lower trophic levels.

15.4.1 Consequences of Phenological Changes for Trophic Relationships

Differences in the extent of phenological shifts of different taxa in response to warming temperatures can lead to a mismatch between the timing of breeding in vertebrates, and the availability of their invertebrate trophic resources (Both, 2010; Ovaskainen et al., 2013). These climate-driven impacts on food availability for breeding vertebrates can limit aspects of their survival and fecundity (Rodenhouse & Holmes, 1992; Seward et al., 2014; Vafidis et al., 2016) and therefore may represent a key mechanism driving changes in vertebrate demographic parameters and ultimately population change.

15.4.2 Phenological Mismatches in Marine Ecosystems

The concept of phenological mismatches was first developed in the context of marine fisheries (Hjort, 1914; Cushing, 1969, 1990; Edwards & Richardson, 2004). Survival of the larvae of fish such as Altantic cod *Gadus morhua* and haddock *Melanogrammus aeglefinus* varies substantially between years, and the mismatch hypothesis proposes that survival is high when the abundance of zooplankton is high in the critical first few days following exhaustion of the fish larvae's yolk supply (Beaugrand et al., 2003). Zooplankton abundance at this time depends on the complex cascading effects of regional temperatures, wind speed, upwelling and ocean currents on SST, which together influence the timing of the spring bloom of phytoplankton, and in turn the abundance of zooplankton (such as *Calanus* spp. copepods) on which the young fish feed. A close match between the spring peak in zooplankton abundance and the period when fish larvae cease to rely on their yolk sacs, and begin to forage independently, allows high larval survival and recruitment (Ellertsen et al., 1989; Fortier et al., 1995; Beaugrand et al., 2003). Low abundance of zooplankton during this critical period can impair recruitment of fish both through starvation-induced mortality, and because larvae that grow slowly suffer higher predation (Durant et al., 2007).

Climate-driven trophic mismatches mediated by marine zooplankton have also been described for a range of planktivorous and piscivorous seabird species (Durant et al., 2007; Suryan et al., 2006; Hipfner, 2008). For example, breeding productivity of Cassin's auklet *Ptychoramphus aleuticus* is positively associated with the seasonal availability of a major invertebrate prey species, the copepod *Neocalanus cristatus* (Abraham & Sydeman, 2004). The availability of the copepod during the auklets' chick provisioning period is negatively associated with ocean temperature; in years of warm waters, the copepods were less abundant overall in the near-surface waters where the auklets forage (Mackas et al., 2007). In these warmer years, copepods disappeared from the diet of Cassin's auklet nestlings 2–3 weeks earlier than in years of cooler waters, leading to reduced chick survival, lower chick fledging mass and a reduction in overall breeding productivity (Hedd et al., 2002). Statistical models of the system indicated that the timing of copepod availability (rather than overall abundance) had the greatest influence on auklet productivity (Hipfner, 2008).

15.4.3 Phenological Mismatches in Terrestrial Ecosystems

Numerous studies have described phenological changes among birds, perhaps the vertebrate taxon for which such changes are most visible and culturally significant (Cocker & Mabey, 2005). These phenological changes have affected both the timing of breeding and of migration. For example, the first arrival dates of migratory birds have advanced substantially in association with increases in spring temperatures. Comparisons across studies of multiple species have indicated that the scale of this advance is associated with migration tendency (resident, short and long distance migrant), the rate of temperature change along the migration route as well as in the breeding and wintering areas (Pearce-Higgins & Green, 2014).

A particularly well-studied instance of a trophic mismatch between birds and their invertebrate prey in a terrestrial ecosystem occurs in warm temperate oak (*Quercus* spp.) woodlands. These are highly seasonal habitats, exhibiting a sharp peak in primary productivity in spring (oak leaf burst), triggering a peak in emergence of caterpillars which feed on the young oak leaves, before increasing tannin and phenolic content renders the oak leaves inedible. The abundant caterpillars in turn form the primary component of the food provisioned by songbirds to their nestlings, such as the migratory pied flycatcher *Ficedula hypoleuca* and the resident great tit *Parus major*. Geometrid moth caterpillars are a particularly important component of the diet of these nestlings during the first days after hatching, with spiders becoming important as the nestlings continue to grow (Sanz, 1998; Huhta et al., 1998; Arnold et al., 2007; Pagani-Núñez et al., 2011).

Climate fluctuations, particularly change in spring temperatures, affect the timing of caterpillar emergence, and thus the availability of prey resources for insectivorous birds breeding in these woodlands. Climate warming generally advances spring phenology in oak woodlands, but the extent of phenological shifts varies between predator (bird) and prey (caterpillar) trophic levels (Both et al., 2009; Thackeray et al., 2010). This can lead to a mismatch between the timing of breeding by birds, and the availability of the trophic resources required for successful reproduction (Fig. 15.2). Since breeding requires large quantities of food to support the energetic activities of egg production, incubation and nestling provisioning, changes in food availability can have large impacts on reproductive success (Daan et al., 1989; Rodenhouse & Holmes, 1992; Both, 2010).

Such mismatches have been implicated in the decline of migratory woodland songbirds such as the pied flycatcher, which have failed to advance the onset of their breeding season sufficiently to match the earlier emergence of caterpillars. The caterpillars in turn have advanced their emergence in response to the earlier availability of palatable young oak leaves. This leads to a growing mismatch between nestling food requirement and food availability. For example, in one study, caterpillar peak abundance advanced by 0.75 days per year between 1988 and 2005, coinciding with increasing spring temperatures and earlier leaf burst. Pied flycatcher chick hatching has, however, advanced by only 0.5 days per year over the same period, leading to a progressively increasing mismatch in phenology between bird breeding and invertebrate abundance. The resulting mismatch between the time at

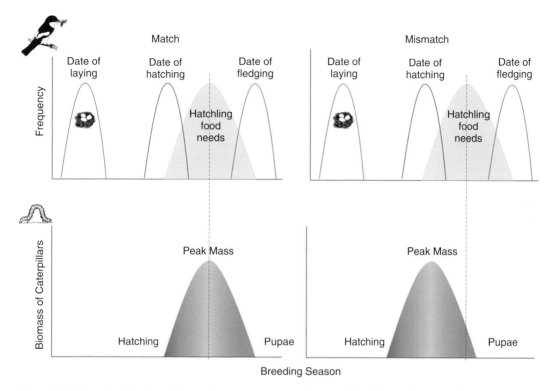

Figure 15.2 Phenological mismatches between vertebrate predators and their invertebrate trophic resources can arise when predators and prey alter their phenology at different rates. Figure adapted from Grossman, 2004.

which caterpillars are most abundant and the chick-feeding period has led to lower breeding productivity (Both et al., 2009).

The oak–caterpillar–songbird food web provides a classic case study of a phenological mismatch between the breeding season of vertebrate predators and the peak availability of their invertebrate prey. Phenological mismatches are, however, not inevitable, and indeed there is evidence that substantial mismatches are actually rather rare phenomena in terrestrial ecosystems. Whether a trophic mismatch occurs depends on two key factors: the behaviour and ecology of vertebrates and habitat differences in prey phenology.

15.4.3.1 Behaviour and Ecology of the Vertebrates

The onset of breeding is more strongly constrained for long-distance migrant birds, whose arrival on the breeding grounds is primarily influenced by the timing of departure from the wintering grounds. This may be controlled by climate-independent cues on the wintering ground (e.g., small changes in daylength), or by climatic variables along the migration route (e.g., food availability at migration stopover sites that are often hundreds or thousands of miles away from the breeding location). In either case, the timing of arrival is largely decoupled from conditions on the breeding grounds at the time of arrival. As a result of such decoupling, long-distance migrants have generally failed to advance their spring arrival dates sufficiently to track the corresponding advance in invertebrate phenology, leading to trophic mismatch in many cases. In contrast, species which remain on the breeding grounds throughout the year (e.g., great tits) may be better able to advance their breeding season, avoiding a mismatch between food availability and the nestling period, with little effect on breeding productivity (Pearce-Higgins & Green, 2014).

15.4.3.2 Habitat Differences in Prey Phenology

The extent of the mismatch varies substantially between ecosystems, habitats and specific predator–prey systems (Burger et al., 2012). A mismatch is most likely to occur in habitats where invertebrate availability exhibits a brief but intense peak (e.g., marine systems and temperate oak woodland). In these circumstances, there is a large effect on vertebrate breeding success depending on whether the offspring are produced within the brief period of peak food availability, or not. In other habitats (such as temperate wetlands), invertebrate availability may exhibit a broad peak of long duration, in which case breeding success is much less dependent on the timing of reproduction in relation to invertebrate availability (Dunn & Winkler, 2010). Even within oak woodlands, individual trees vary in their timing of leaf-burst. For example, in a sample of 36 trees studied by Crawley and Akhteruzzaman (1988), there was a range of 25 days between leaf burst in the earliest and latest individual trees. These differences in phenology are associated with significant differences in the timing of invertebrate availability. Thus, at a small scale (within woodlands) the individual heterogeneity in leaf-burst within a bird's territory will be an important determinant of the extent of any trophic mismatch. These habitat differences and local heterogeneity in the timing and extent of peak food availability lead to substantial differences in the importance of the mismatch phenomenon in different ecological contexts.

Mid-to-high latitude marine ecosystems may be particularly vulnerable to trophic mismatches because the recruitment success of higher trophic levels is highly dependent on synchronization with pulsed plankton production (Edwards & Richardson, 2004). In equatorial waters, plankton production is less seasonal and thus predator productivity is less strongly constrained by the timing of breeding. Likewise, in terrestrial ecosystems the peak of caterpillar availability is very brief in oak woodlands (~3 weeks), whereas the peak is much more protracted in other broadleaved and coniferous woodlands (Burger et al., 2012).

There is no evidence for phenological mismatch affecting songbird populations breeding in warm temperate wetlands because warming increases overall invertebrate productivity across a broad seasonal peak (Both, 2010). Macro-benthic invertebrates such as Nematocera, especially Chironomidae, emerge in huge numbers continuously throughout the summer months, providing an abundant food supply for wetland birds (Covich et al., 1999; Both, 2010; Dunn et al., 2011).

In contrast to temperate woodlands, warmer spring temperatures are predicted to advance and lengthen the window of high invertebrate availability in wetlands. This occurs through advancing emergence by direct physiological means (e.g., growth and respiration; Péry and Garric, 2006) as well as enhancing the rates of hatching success and voltinism (Eggermont & Heiri, 2012). This increases the availability of invertebrates through the breeding season, allowing insectivorous birds, such as Eurasian reed warblers, to nest earlier, which increases their breeding productivity (Halupka et al., 2008; Schaefer et al., 2006).

The situation is different for vertebrate insectivores inhabiting wetlands in the Arctic, where the effects of climate change are most severe and rapid (Gilg et al., 2009). Such effects include dramatic changes in spring temperatures but also water levels which are likely to result in substantial shifts in the timing of availability of emergent invertebrate taxa (particularly mosquitoes and Chironomids), which migratory insectivorous wading birds feed to their young (Schekkerman et al., 2003). Across these studies, it is apparent that oak woodlands and Arctic tundra are unusual among terrestrial habitats in the highly seasonal nature of their invertebrate productivity.

15.5 Conclusions

Our review has highlighted three major classes of impacts of climate change on insectivorous vertebrates that can be mediated by their invertebrate food supply. Firstly, climate-driven changes in invertebrate abundance mediates changes in vertebrate population growth rates that frequently involve population change (both increases and decreases in different contexts), and in some cases involve local population extinction. There are not yet any definitive examples of global extinctions mediated by climate-driven changes in invertebrate prey availability, but current projections indicate that such extinctions are likely to occur and increase in frequency as climate change moves beyond the current variation and the current tolerances of many taxa.

Secondly, climate impacts on invertebrates mediates changes in the distribution of vertebrates as well as their local abundance. Distributional changes can be manifested as horizontal geographical, or vertical, altitudinal and depth shifts, and invertebrate prey resources have been implicated in each of these contexts in a range of ecosystems. Invertebrate-mediated horizontal geographical shifts are widely studied, whereas altitudinal shifts and particularly depth shifts are relatively poorly documented.

Thirdly, climate impacts on invertebrate phenology are often more substantial than the corresponding impacts on vertebrate phenology, leading to substantial trophic mismatches leading to the timing of high invertebrate availability and the timing of reproduction in vertebrates which rely on such resources to provision their offspring. There are substantial differences between habitats in the magnitude of such phenological mismatches, depending on the behaviour and ecology of the predator and the phenology of the prey. Oak woodlands and Arctic tundra are highlighted as habitats where mismatches are relatively large, whereas in temperate wetlands, such mismatches are minimal.

Although there are evidently differences between taxa and habitats in the magnitude of climate impacts, it is difficult at this stage to make quantitative comparisons between ecosystems. This is due to the relatively piecemeal nature of existing studies across different ecosystems and taxonomic groups, yet such comparisons are likely to become more possible as the range and diversity of ecological studies of climate impacts continues to proliferate.

Our emerging understanding of the importance of invertebrate prey resources for vertebrate populations highlights several important applications:

i) Mitigation (protecting) against the effects of climate change, and compensating (providing replacements) for habitats or populations damaged by climate change, are the two major classes of adaptation strategies for conservation of vertebrates in the face of changing climate conditions. To be effective, both of these approaches require bottom-up management of trophic resources for vertebrates, including sufficient invertebrate prey for insectivores.

ii) As important predators of invertebrates, vertebrates can play an important role in the biological control of invertebrate pests (Civantos et al., 2012). By altering the trophic relationships between predators and prey, climate may alter the effectiveness of pest control, either positively or negatively.

iii) Vertebrates provide ecosystem services such as human nutrition (e.g., fisheries), recreation (e.g., ecotourism, hunting and fishing) and pollination (e.g., birds and bats are important pollinators in the tropics).

These applications of our emerging understanding of the role of invertebrates in mediating the impacts of climate change on vertebrates' behaviour, ecology and populations, are likely to increase in importance and urgency over the coming years and decades as changes in temperature, rainfall and other climate variables increasingly disrupt ecosystem structure and function. The challenge for ecologists and conservationists is to not only document these changes but to develop effective mitigation and adaptation strategies for a changing world. The examples described in this chapter highlight the diversity of contexts, mechanisms and impacts that need to be considered in the design of these strategies. It is our hope that the holistic understanding of climate impacts on invertebrates promoted in this book will help to facilitate the development of such applications for conserving the biodiversity both of vertebrates and their invertebrate prey, in the face of climate change.

15.6   Postscript: Beyond the Year 2100

Climate projections are typically made to the end of the present century. Climate change, however, will not end in 2100, and it is already evident that as a result of greenhouse gas emissions which have already occurred, humankind has already committed itself to a very long period of anthropogenic climate forcing that will extend long beyond the present century, regardless of any future changes in emissions. The long-term impacts of these climate changes will likely endanger many of the Earth's species (Thomas et al., 2004a), and many will inevitably be driven to local or global extinction. This chapter has identified the central role of invertebrates as trophic resources mediating these impacts on vertebrates, across trophic levels, and across entire ecosystems. Our review highlights the importance of addressing the proximate mechanisms for understanding, predicting and even mitigating these impacts of climate change. Our current understanding of these proximate mechanisms is "disturbingly limited" (Cahill et al., 2012), and a greater understanding of climate control over individuals, species (including both vertebrate and invertebrate taxa) and their trophic interactions is a clear research priority for ecologists over the coming years.

Acknowledgements

We are grateful to Philip Smith for assistance in proofreading this chapter. We thank Jeremy Smith, Richard Facey, Ian Vaughan and Hefin Jones for the discussion of ideas for this chapter.

References

Abraham, C.L. & Sydeman, W.J. (2004) Ocean climate, euphausiids and auklet nesting: inter-annual trends and variation in phenology, diet and growth of a planktivorous seabird, *Ptychoramphus aleuticus*. *Marine Ecology Progress Series*, **274**, 235–250.

Acquarone, C., Cucco, M. & Malacarne, G. (2003) Reproduction of the Crag Martin (*Ptyonoprogne rupestris*) in relation to weather and colony size. *Ornis Fennica*, **80**, 79–85.

Araújo, M.B. & Luoto, M. (2007) The importance of biotic interactions for modelling species distributions under climate change. *Global Ecology and Biogeography*, **16**, 743–753.

Arnold, K.E., Ramsay, S.L., Donaldson, C. & Adam, A. (2007) Parental prey selection affects risk-taking behaviour and spatial learning in avian offspring. *Proceedings of the Royal Society B: Biological Sciences*, **274**, 2563–2569.

Baillie, S.R. & Peach, W.J. (1992) Population limitation in Palaearctic – African migrant passerines. *Ibis*, **134**, 120–132.

Baillie, S.R., Marchant, J.H., Leech, D.I., Massimino, D., Sullivan, M.J.P., Eglington, S.M., Barimore, C., Dadam, D., Downie, I.S., Harris, S.J., Kew, A.J., Newson, S.E., Noble, D.G., Risely, K. & Robinson, R.A.

(2014) *BirdTrends 2014: Trends in Numbers, Breeding Success and Survival for UK Breeding Birds.* Research Report 662. BTO, Thetford.

Bale, J.S., Masters, G.J., Hodkinson, I.D., Awmack, C., Bezemer, T.M., Brown, V.K., Butterfield, J., Buse, A., Coulson, J.C., Farrar, J., Good, J.E.G., Harrington, R., Hartley, S., Jones, T.H., Lindroth, R.L., Press, M.C., Symrnioudis, I. Watts, A.D. & Whittaker, J.B. (2002) Herbivory in global climate change research: direct effects of rising temperature on insect herbivores. *Global Change Biology*, **8**, 1–16.

Balmer, D., Gillings, S., Caffrey, B., Swann, B., Downie, I. & Fuller, R. (2013) *Bird Atlas 2007-11: The Breeding and Wintering Birds of Britain and Ireland.* British Trust for Ornithology, Thetford.

Barbaro, L. & Battisti, A. (2011) Birds as predators of the pine processionary moth (Lepidoptera: Notodontidae). *Biological Control*, **56**, 107–114.

Barnosky, A.D., Matzke, N., Tomiya, S., Wogan, G.O.U., Swartz, B., Quental, T.B., Marshall, C., McGuire, J.L., Lindsey, E.L., Maguire, K.C., Mersey, B. & Ferrer, E.A. (2011) Has the Earth's sixth mass extinction already arrived? *Nature*, **471**, 51–57.

Battisti, A., Stastny, M., Buffo, E. & Larsson, S. (2006) A rapid altitudinal range expansion in the pine processionary moth produced by the 2003 climatic anomaly. *Global Change Biology*, **12**, 662–671.

Beale, C.M., Lennon, J.J. & Gimona, A. (2008) Opening the climate envelope reveals no macroscale associations with climate in European birds. *Proceedings of the National Academy of Sciences of the United States of America*, **105**, 14908–14912.

Beare, D.J., Burns, F., Greig, A., Jones, E.G., Peach, K., Kienzle, M., McKenzie, E. & Reid, D.G. (2004) Long-term increases in prevalence of North Sea fishes having southern biogeographic affinities. *Marine Ecology Progress Series*, **284**, 269–278.

Beaugrand, G. (2003) Long-term changes in copepod abundance and diversity in the north-east Atlantic in relation to fluctuations in the hydroclimatic environment. *Fisheries Oceanography*, **12**, 270–283.

Beaugrand, G. & Reid, P.C. (2003) Long-term changes in phytoplankton, zooplankton and salmon related to climate. *Global Change Biology*, **9**, 801–817.

Beaugrand, G., Brander, K.M., Lindley, J.A., Souissi, S. & Reid, P.C. (2003) Plankton effect on cod recruitment in the North Sea. *Nature*, **426**, 661–664.

Beaumont, L.J. & Hughes, L. (2002) Potential changes in the distributions of latitudinally restricted Australian butterfly species in response to climate change. *Global Change Biology*, **8**, 954–971.

Beintema, A.J. & Visser, G.H. (1989) The effect of weather on time budgets and development of chicks of meadow birds. *Ardea*, **77**, 181–192.

Beintema, A.J., Thissen, J.B., Tensen, D. & Visser, G.H. (1991) Feeding ecology of charadriiform chicks in agricultural grassland. *Ardea*, **79**, 31–44.

Berthe, S.C.F., Derocles, S.A.P., Lunt, D.H., Kimball, B.A. & Evans, D.M. (2015) Simulated climate-warming increases Coleoptera activity-densities and reduces community diversity in a cereal crop. *Agriculture, Ecosystems & Environment*, **210**, 11–14.

Bibby, C.J., Green, R.E., Pepler, G.R.M. & Pepler, P.A. (1976) Sedge warbler migration and reed aphids. *British Birds*, **69**, 384–399.

Bibby, C.J. & Green, R.E. (1981) Autumn migration strategies of reed and sedge warblers. *Ornis Scandinavica*, **12**, 1–12.

Bibby, C.J. & Thomas, D.K. (1985) Breeding and diets of the reed warbler at a rich and a poor site. *Bird Study*, **32**, 19–31.

Blaustein, A.R. & Kiesecker, J.M. (2002) Complexity in conservation: lessons from the global decline of amphibian populations. *Ecology Letters*, **5**, 597–608.

Boddy, L., Büntgen, U., Egli, S., Gange, A.C., Heegaard, E., Kirk, P.M., Mohammad, A. & Kauserud, H. (2014) Climate variation effects on fungal fruiting. *Fungal Ecology*, **10**, 20–33.

Both, C. & Visser, M.E. (2001) Adjustment to climate change is constrained by arrival date in a long-distance migrant bird. *Nature*, **411**, 296–298.

Both, C., Van Asch, M., Bijlsma, R.G., Van Den Burg, A.B. & Visser, M.E. (2009) Climate change and unequal phenological changes across four trophic levels: constraints or adaptations? *Journal of Animal Ecology*, **78**, 73–83.

Both, C. (2010). Food Availability, Mistiming, and Climatic Change. *Effects of Climate Change on Birds* (eds A.P. Møller, W. Fiedler and P. Berthold), pp. 129–147. Oxford University Press, Oxford.

Box, E.O. (1981) *Macroclimate and plant forms: an introduction to predictive modelling in phytogeography*. Dr W. Junk, The Hague.

Brander, K., Blom, G., Borges, M.F., Erzini, K., Henderson, G., MacKenzie, B.R., Mendes, H., Ribeiro, J., Santos, A.M.P. & Toresen, R. (2003) Changes in fish distribution in the eastern North Atlantic: are we seeing a coherent response to changing temperature? In *ICES Marine Science Symposia*, **219**, 261–270.

Bretagnolle, V. & Gillis, H. (2010) Predator–prey interactions and climate change. *Effects of Climate Change on Birds* (eds A.P. Møller, W. Fiedler and P. Berthold), pp. 227–248. Oxford University Press, Oxford.

Büßer, C., Kahles, A. & Quillfeldt, P. (2004) Breeding success and chick provisioning in Wilson's storm-petrels *Oceanites oceanicus* over seven years: frequent failures due to food shortage and entombment. *Polar Biology*, **27**, 613–622.

Burger, C., Belskii, E., Eeva, T., Laaksonen, T., Mägi, M., Mänd, R., Qvarnström, A., Slagsvold, T., Veen, T., Visser, M.E., Wiebe, K.L., Wiley C., Wright J. & Both, C. (2012) Climate change, breeding date and nestling diet: how temperature differentially affects seasonal changes in pied flycatcher diet depending on habitat variation. *Journal of Animal Ecology*, **81**, 926–936.

Byers, J.E., McDowell, W.G., Dodd, S.R., Haynie, R.S., Pintor, L.M. & Wilde, S.B. (2013) Climate and pH predict the potential range of the invasive apple snail (*Pomacea insularum*) in the Southeastern United States. *PLOS ONE*, **8**, e56812.

Cahill, E.A., Aiello-Lammens, M.E., Fisher-Reid, M.C., Hua, X., Karanewsky, C.J., Riu, H.Y., Sbeglia, G.C., Spagnolo, F., Waldron, J.B., Warsi, O. & Wiens, J.J. (2012) How does climate change cause extinction? *Proceedings of the Royal Society B: Biological Sciences*, **282**, 20121890.

Chmielewski, F.-M. & Rötzer, T. (2001) Response of tree phenology to climate change across Europe. *Agricultural and Forest Meteorology*, **108**, 101–112.

Civantos, E., Thuiller, W., Maiorano, L., Guisan, A. & Araújo, M.B. (2012) Potential impacts of climate change on ecosystem services in Europe: the case of pest control by vertebrates. *BioScience*, **62**, 658–666.

Cleland, E.E., Chuine, I., Menzel, A., Mooney, H.A. & Schwartz, M.D. (2007) Shifting plant phenology in response to global change. *Trends in Ecology & Evolution*, **22**, 357–365.

Cocker, M. & Mabey, R. 2005. *Birds Britannica*. Random House, London.

Covich, A.P., Palmer, M.A. & Crowl, T.A. (1999) The role of benthic invertebrate species in freshwater ecosystems: zoobenthic species influence energy flows and nutrient cycling. *BioScience*, **49**, 119–127.

Cramp, S. & Brooks D.J. (1992) *Handbook of the Birds of Europe, the Middle East and North Africa: The Birds of the Western Palearctic, Vol. 6 Warblers*. Oxford University Press, Oxford.

Crawley, M.J. & Akhteruzzaman, M. (1988) Individual variation in the phenology of oak trees and its consequences for herbivorous insects. *Functional Ecology*, **2**, 409–415.

Croxall, J.P., Trathan, P.N. & Murphy, E.J. (2002) Environmental change and Antarctic seabird populations. *Science*, **297**, 1510–1514.

Cunningham, J.T. (1889) Anchovies in the English Channel. *Journal of the Marine Biological Association of the United Kingdom*, **1**, 328–339.

Cushing, D.H. (1969) The regularity of the spawning season of some fishes. *Journal du Conseil / Conseil Permanent International pour l'Exploration de la Mer*, **33**, 81–92.

Cushing, D.H. (1990) Plankton production and year-class strength in fish populations: an update of the match/mismatch hypothesis. *Advances in Marine Biology*, **26**, 249–293.

Daan, S., Dijkstra, C., Drent, R. & Meijer, T. (1989) Food supply and the annual timing of avian reproduction. *Acta Congressus Internationalis Ornithologici*, **19**, 392–407.

Davies, N.B. & Green, R.E. (1976) The development and ecological significance of feeding techniques in the reed warbler (*Acrocephalus scirpaceus*). *Animal Behaviour*, **24**, 213–229.

Dunn, P.O. & Winkler, D.W. (2010) Effects of climate change on timing of breeding and reproductive success in birds. *Effects of Climate Change on Birds* (eds A.P. Møller, W. Fiedler & P. Berthold), pp. 113–128. Oxford University Press, Oxford.

Dunn, P.O. Winkler, D.W. Whittingham, L.A, Hannon, S.J. & Robertson, R.J. (2011) A test of the mismatch hypothesis: How is timing of reproduction related to food abundance in an aerial insectivore? *Ecology*, **92**, 450–461.

Durant, J.M., Hjermann, D.Ø., Ottersen, G. & Stenseth, N.C. (2007) Climate and the match or mismatch between predator requirements and resource availability. *Climate Research*, **33**, 271–283.

Edwards, M. & Richardson, A.J. (2004) Impact of climate change on marine pelagic phenology and trophic mismatch. *Nature*, **430**, 881–884.

Eggermont, H. & Heiri, O. (2012) The chironomid-temperature relationship: expression in nature and palaeoenvironmental implications. *Biological Reviews*, **87**, 430–456.

Ellertsen, B., Fossum, P., Solemdal, P. & Sundby, S. (1989) Relation between temperature and survival of eggs and first-feeding larvae of northeast Arctic cod (*Gadus morhua* L.). *Rapports et Procès-Verbaux des Réunions du Conseil International pour l'Exploration de la Mer*, **191**, 209–219.

Elzinga, J.A., Atlan, A., Biere, A., Gigord, L., Weis, A.E. & Bernasconi, G. (2007) Time after time: flowering phenology and biotic interactions. *Trends in Ecology & Evolution*, **22**, 432–439.

Fortier, L., Ponton, D. & Gilbert, M. (1995) The match/mismatch hypothesis and the feeding success of fish larvae in ice-covered southeastern Hudson Bay. *Marine Ecology Progress Series*, **120**, 11–27.

Gerard, B.M. (1967). Factors affecting earthworms in pastures. *Journal of Animal Ecology*, **36**, 235–252.

Gilg, O., Sittler, B. & Hanski, I. (2009) Climate change and cyclic predator–prey population dynamics in the high Arctic. *Global Change Biology*, **15**, 2634–2652.

Gordo, O. (2007) Why are bird migration dates shifting? A review of weather and climate effects on avian migratory phenology. *Climate Research*, **35**, 37–58.

Green, R.E. (1988) Effects of environmental factors on the timing and success of breeding of common snipe *Gallinago gallinago* (Aves: Scolopacidae). *Journal of Applied Ecology*, **25**, 79–93.

Grodsky, S.M., Iglay, R.B., Sorenson, C.E. & Moorman, C.E. (2015) Should invertebrates receive greater inclusion in wildlife research journals? *The Journal of Wildlife Management*, **79**, 529–536.

Grossman, D. (2004) Spring forward. *Scientific American*, **290**, 85–91.

Halupka, L., Dyrcz, A. & Borowiec, M. (2008) Climate change affects breeding of reed warblers *Acrocephalus scirpaceus*. *Journal of Avian Biology*, **39**, 95–100.

Harvell, C.D., Mitchell, C.E., Ward, J.R., Altizer, S., Dobson, A.P., Ostfeld, R.S. & Samuel, M.D. (2002) Climate warming and disease risks for terrestrial and marine biota. *Science*, **296**, 2158–2162.

Hawkins, J.P., Roberts, C.M. & Clark, V. (2000) The threatened status of restricted-range coral reef fish species. *Animal Conservation*, **3**, 81–88.

Hedd, A., Ryder, J.L., Cowen, L.L. & Bertram, D.F. (2002) Inter-annual variation in the diet, provisioning and growth of Cassin's auklet at Triangle Island, British Columbia: responses to variation in ocean climate. *Marine Ecology Progress Series*, **229**, 221–232.

Helaouët, P. & Beaugrand, G. (2007) Macroecology of *Calanus finmarchicus* and *C. helgolandicus* in the North Atlantic Ocean and adjacent seas. *Marine Ecology Progress Series*, **345**, 147–165.

Hipfner, J.M. (2008) Matches and mismatches: ocean climate, prey phenology and breeding success in a zooplanktivorous seabird. *Marine Ecology Progress Series*, **368**, 295–304.

Holt, R.D. & Huxel, G.R. (2007) Alternative prey and the dynamics of intraguild predation: theoretical perspectives. *Ecology*, **88**, 2706–2712.

Holt, R.D. & Barfield, M. (2009) Trophic interactions and range limits: the diverse roles of predation. *Proceedings of the Royal Society B: Biological Sciences*, **276**, 1435–1442.

Hoegh-Guldberg, O. (1999) Climate change, coral bleaching and the future of the world's coral reefs. *Marine and Freshwater Research*, **50**, 839–866.

Hoegh-Guldberg, O., Mumby, P.J., Hooten, A.J., Steneck, R.S., Greenfield, P., Gomez, E., Harvell, C.D., Sale, P.F., Edwards, A.J., Caldeira, K., Knowlton, N., Eakin, C.M., Iglesias-Prieto, R., Muthiga, N., Bradbury, R.H., Dubi, A., & Hatziolos, M.E. (2007) Coral reefs under rapid climate change and ocean acidification. *Science*, **318**, 1737–1742.

Hjort, J. (1914) Fluctuations in the great fisheries of northern Europe viewed in the light of biological research. *Rapports et Procès-Verbaux des Réunions du Conseil International pour l'Exploration de la Mer*, **20**, 1–228.

Huey, R.B., Losos, J.B. & Moritz, C. (2010) Are lizards toast? *Science*, **328**, 832–833.

Hughes, L. (2000) Biological consequences of global warming: is the signal already apparent? *Trends in Ecology & Evolution*, **15**, 56–61.

Hughes, T.P., Baird, A.H., Bellwood, D.R., Card, M., Connolly, S.R., Folke, C., Grosberg, R., Hoegh-Guldberg, O., Jackson, J.B.C., Kleypas, J., Lough, J.M., Marshall, P., Nyström, M., Palumbi, S.R., Pandolfi J.M., Rosen, B. & Roughgarden, J. (2003) Climate change, human impacts, and the resilience of coral reefs. *Science*, **301**, 929–933.

Huhta, E., Jokimakp, J. & Rahko, P. (1998) Distribution and reproductive success of the Pied Flycatcher *Ficedula hypoleuca* in relation to forest patch size and vegetation characteristics; the effect of scale. *Ibis*, **140**, 214–222.

Jenouvrier, S., Barbraud, C. & Weimerskirch, H. (2003) Effects of climate variability on the temporal population dynamics of southern fulmars. *Journal of Animal Ecology*, **72**, 576–587.

Johnson, D.D.P., Jetz, W. & Macdonald, D.W. (2002) Environmental correlates of badger social spacing across Europe. *Journal of Biogeography*, **29**, 411–425.

Jones, G.P., Munday, P.L. & Caley, M.J. (2002) Rarity in coral reef fish communities. *Coral Reef Fishes: Dynamics and Diversity in a Complex Ecosystem* (ed. P.F. Sale), pp. 81–101. Academic Press, San Diego.

Kauserud, H., Heegaard, E., Semenov, M.A., Boddy, L., Halvorsen, R., Stige, L.C., Sparks, T.H., Gange, A.C. & Stenseth, N.C. (2010) Climate change and spring-fruiting fungi. *Proceedings of the Royal Society of London B: Biological Sciences*, **277**, 1169–1177.

King, R.A., Symondson, W.O.C. & Thomas, R.J. (2015) Molecular analysis of faecal samples from birds to identify potential crop pests and useful biocontrol agents in natural areas. *Bulletin of Entomological Research*, **105**, 261–272.

Kruuk, H. (1978) Foraging and spatial organisation of the European badger, *Meles meles* L. *Behavioral Ecology and Sociobiology*, **4**, 75–89.

Leisler, B. (1992) Habitat selection and coexistence of migrants and Afrotropical residents. *Ibis*, **134**, 77–82.

Linares, C., Vidal, M., Canals, M., Kersting, D.K., Amblas, D., Aspillaga, E., Cebrián, E., Delgado-Huertas, A., Díaz, D., Garrabou, J., Hereu, B., Navarro, L., Teixidó, N. & Ballesteros, E. (2015) Persistent natural acidification drives major distribution shifts in marine benthic ecosystems. *Proceedings of the Royal Society: B Biological Sciences*, **282**, 20150587.

Linderholm, H.W. (2006) Growing season changes in the last century. *Agricultural and Forest Meteorology*, **137**, 1–14.

Lovejoy, T.E. & Hannah, L.J. (eds) (2005) *Climate Change and Biodiversity*. Yale University Press, New Haven.

Macdonald, D.W. & Newman, C. (2002) Population dynamics of badgers (*Meles meles*) in Oxfordshire, U.K.: numbers, density and cohort life histories, and a possible role of climate change in population growth. *Journal of Zoology*, **256**, 121–138.

Macdonald, D.W., Newman, C., Buesching, C.D. & Nouvellet, P. (2010) Are badgers 'Under The Weather'? Direct and indirect impacts of climate variation on European badger (*Meles meles*) population dynamics. *Global Change Biology*, **16**, 2913–2922.

Mackas, D.L., Batten, S. & Trudel, M. (2007) Effects on zooplankton of a warmer ocean: recent evidence from the Northeast Pacific. *Progress in Oceanography*, **75**, 223–252.

Mallord, J.W., Charman, E.C., Cristinacce, A. & Orsman, C.J. (2012) Habitat associations of Wood Warblers *Phylloscopus sibilatrix* breeding in Welsh oakwoods. *Bird Study*, **59**, 403–415.

Monte-Luna, P. del, Lluch-Belda, D., Serviere-Zaragoza, E., Carmona, R., Reyes-Bonilla, H., Aurioles-Gamboa, D., Castro-Aguirre, J.L., Próo, S.A.G. del, Trujillo-Millán, O. & Brook, B.W. (2007) Marine extinctions revisited. *Fish and Fisheries*, **8**, 107–122.

Morris, J.P., Thatje, S., Ravaux, J., Shillito, B., Fernando, D. & Hauton, C. (2015) Acute combined pressure and temperature exposures on a shallow-water crustacean: Novel insights into the stress response and high pressure neurological syndrome. *Comparative Biochemistry and Physiology Part A: Molecular & Integrative Physiology*, **181**, 9–17.

Munday, P.L., Harold, A.S. & Winterbottom, R. (1999) Guide to coral-dwelling gobies, genus *Gobiodon* (Gobiidae), from Papua New Guinea and the Great Barrier Reef. *Revue Française d'Aquariologie, Herpétologie*, **26**, 53–58.

Newman, C., Macdonald, D.W. & Anwar, M.A. (2001) Coccidiosis in the European Badger, *Meles meles*, in Wytham Woods: Infection and consequences for growth and survival. *Parasitology*, **123**, 133–142.

Newman, C. & Macdonald, D.W. (2015) The Implications of climate change for terrestrial UK Mammals. Terrestrial biodiversity Climate Change Impacts Report Card Technical paper. Living With Environmental Change (LWEC) Partnership.

Nouvellet, P., Newman, C., Buesching, C.D. & Macdonald, D.W. (2013) A multi-metric approach to investigate the effects of weather conditions on the demographic of a terrestrial mammal, the European badger (*Meles meles*). *PLOS ONE*, **8**, e68116.

Ockendon, N., Hewson, C.M., Johnston, A. & Atkinson, P.W. (2012) Declines in British-breeding populations of Afro-Palaearctic migrant birds are linked to bioclimatic wintering zone in Africa, possibly via constraints on arrival time advancement. *Bird Study*, **59**, 111–125.

Ovaskainen, O., Skorokhodova, S., Yakovleva, M., Sukhov, A., Kutenkov, A., Kutenkova, N., Shcherbakov, A., Meyke, E. & Del Mar Delgado, M. (2013) Community-level phenological response to climate change. *Proceedings of the National Academy of Sciences of the United States of America*, **110**, 13434–13439.

Pagani–Núñez, E., Ruiz, Í., Quesada, J., Negro, J.J. & Senar, J.C. (2011) The diet of great tit *Parus major* nestlings in a Mediterranean Iberian forest: the important role of spiders. *Animal Biodiversity and Conservation*, **34**, 355–361.

Palmer, G., Hill, J.K., Brereton, T.M., Brooks, D.R., Chapman, J.W., Fox, R., Oliver, T.H. & Thomas, C.D. (2015) Individualistic sensitivities and exposure to climate change explain variation in species' distribution and abundance changes. *Science Advances*, **1**, e1400220.

Peach, W., Baillie, S. & Underhill, L. (1991) Survival of British sedge warblers *Acrocephalus schoenobaenus* in relation to west African rainfall. *Ibis*, **133**, 300–305.

Peach, W.J., Denny, M., Cotton, P.A., Hill, I.F., Gruar, D., Barritt, D., Impey, A. & Mallord, J. (2004) Habitat selection by song thrushes in stable and declining farmland populations. *Journal of Applied Ecology*, **41**, 275–293.

Pearce-Higgins, J.W., Yalden, D.W. & Whittingham, M.J. (2005) Warmer springs advance the breeding phenology of golden plovers *Pluvialis apricaria* and their prey (Tipulidae). *Oecologia*, **143**, 470–476.

Pearce-Higgins, J.W., Dennis, P., Whittingham, M.J. & Yalden, D.W. (2010) Impacts of climate on prey abundance account for fluctuations in a population of a northern wader at the southern edge of its range. *Global Change Biology*, **16**, 12–23.

Pearce-Higgins, J.W. & Green, R.E. (2014) *Birds and Climate Change: Impacts and Conservation Responses*. Cambridge University Press, Cambridge.

Pearson, R.G. & Dawson, T.P. (2003) Predicting the impacts of climate change on the distribution of species: are bioclimate envelope models useful? *Global Ecology and Biogeography*, **12**, 361–371.

Perry, A.L., Low, P.J., Ellis, J.R. & Reynolds, J.D. (2005) Climate change and distribution shifts in marine fishes. *Science*, **308**, 1912–1915.

Péry, A.R.R. & Garric, J. (2006) Modelling effects of temperature and feeding level on the life cycle of the midge *Chironomus riparius*: an energy-based modelling approach. *Hydrobiologia*, **553**, 59–66.

Poloczanska, E.S., Brown, C.J., Sydeman, W.J., Kiessling, W., Schoeman, D.S., Moore, P.J., Brander, K., Bruno, J.F., Buckley, L.B., Burrows, M.T., Duarte, C.M., Halpern, B.S., Holding, J., Kappel,] C.V., O'Connor, M.I., Pandolfi, J.M., Parmesan, C., Schwing, F., Thompson, S.A. & Richardson, A.J. (2013) Global imprint of climate change on marine life. *Nature Climate Change*, **3**, 919–925.

Poulin, R. & Mouritsen, K.N. (2006) Climate change, parasitism and the structure of intertidal ecosystems. *Journal of Helminthology*, **80**, 183–191.

Pounds, J.A. & Crump, M.L. (1994) Amphibian declines and climate disturbance: the case of the golden toad and the harlequin frog. *Conservation Biology*, **8**, 72–85.

Pounds, J.A., Bustamante, M.R., Coloma, L.A., Consuegra, J.A., Fogden, M.P.L., Foster, P.N., La Marca, E., Masters, K.L., Merino-Viteri, A., Puschendorf, R., Ron, S.R., Sánchez-Azofeifa, G.A., Still, C.J. & Young, B.E. (2006) Widespread amphibian extinctions from epidemic disease driven by global warming. *Nature*, **439**, 161–167.

Preston K.L., Rotenberry, J.T., Redak, R.A. & Allen, M.F. (2008) Habitat shifts of endangered species under altered climate conditions: importance of biotic interactions. *Global Change Biology*, **14**, 2501–2515.

Quillfeldt, P. (2001) Variation in breeding success in Wilson's storm petrels: influence of environmental factors. *Antarctic Science*, **13**, 400–409.

Quillfeldt, P. (2002) Seasonal and annual variation in the diet of breeding and non-breeding Wilson's storm-petrels on King George Island, South Shetland Islands. *Polar Biology*, **25**, 216–221.

Roberts, C.M. & Hawkins, J.P. (1999) Extinction risk in the sea. *Trends in Ecology & Evolution*, **14**, 241–246.

Robinson, R.A., Freeman, S.N., Balmer, D.E. & Grantham, M.J. (2007) Cetti's Warbler *Cettia cetti*: analysis of an expanding population. *Bird Study*, **54**, 230–235.

Rodenhouse, N.L. & Holmes, R.T. (1992) Results of experimental and natural food reductions for breeding black-throated blue warblers. *Ecology*, **73**, 357–372.

Roques, A., Rousselet, J., Avci, M., Avtzis, D.N., Basso, A., Battisti, A., Ben Jamaa, M.L., Bensidi, A., Berardi, L., Berretima, W., Branco, M., Chakali, G., Çota, E., Dautbašić, M., Delb, H., El Alaoui El Fels, M.A., El Mercht, S., El Mokhefi, M., Forster, B., Garcia, J., Georgiev, G., Glavendekić, M.M., Goussard, F., Halbig, P., Henke, L., Hernaňdez, R., Hódar, J.A., İpekdal, K., Jurc, M., Klimetzek, D., Laparie, M., Larsson, S., Mateus, E., Matošević, D., Meier, F., Mendel, Z., Meurisse, N., Mihajlović, L., Mirchev, P., Nasceski, S., Nussbaumer, C., Paiva, M.-R., Papazova, I., Pino, J., Podlesnik, J, Poirot, J., Protasov, A.,

Rahim, N., Peña, G.S., Santos, H., Sauvard, D., Schopf, A., Simonato, M., Tsankov, G., Wagenhoff, E., Yart, A., Zamora, R., Zamoum, M. & Robinet, C. (2015) Climate warming and past and present distribution of the processionary moths (*Thaumetopoea* spp.) in Europe, Asia Minor and North Africa. *Processionary Moths and Climate Change: An Update* (ed. A. Roques), pp. 81–161. Springer Netherlands, Dordrecht.

Russell, G.W. & Grimm, E.C. (1990) Effects of global climate change on the patterns of terrestrial biological communities. *Trends in Ecology & Evolution*, **5**, 289–292.

Salewski, V., Hochachka, W.M. & Fiedler, W. (2013) Multiple weather factors affect apparent survival of European passerine birds. *PLOS ONE*, **8**, e59110.

Sanz, J.J. (1998) Effect of habitat and latitude on nestling diet of Pied Flycatchers *Ficedula hypoleuca*. *Ardea*, **86**, 81–88.

Schaefer, T., Ledebur, G., Beier, J. & Leisler, B. (2006) Reproductive responses of two related coexisting songbird species to environmental changes: global warming, competition, and population sizes. *Journal of Ornithology*, **147**, 47–56.

Schekkerman, H., Tulp, I., Piersma, T. & Visser, G.H. (2003) Mechanisms promoting higher growth rate in Arctic than in temperate shorebirds. *Oecologia*, **134**, 332–342.

Seward, A.M., Beale, C.M., Gilbert, L., Jones, T.H. & Thomas, R.J. (2014) The impact of increased food availability on reproduction in a long-distance migratory songbird: implications for environmental change? *PLOS ONE*, **9**, e111180.

Sinervo, B., Méndez-De-La-Cruz, F., Miles, D.B., Heulin, B., Bastiaans, E., Villagrán-Santa Cruz, M., Lara-Resendiz, R., Martínez-Méndez, N., Calderón-Espinosa, M.L., Meza-Lázaro, R.N., Gadsden, H., Avila, L.J., Morando, M., De la Riva, I.J., Sepulveda, P.V., Rocha, C.F.D., Ibargüengoytía, N., Puntriano, C.A., Massot, M., Lepetz, V., Oksanen, T.A., Chapple, D.G., Bauer, A.M., Branch, W.R., Clobert, J. & Sites, J.W. Jr. (2010) Erosion of lizard diversity by climate change and altered thermal niches. *Science*, **328**, 894–899.

Stevens, A.J., Welch, Z.C., Darby, P.C. & Percival, H.F. (2002) Temperature effects on Florida applesnail activity: implications for snail kite foraging success and distribution. *Wildlife Society Bulletin*, **30**, 75–81.

Suryan, R.M., Irons, D.B., Brown, E.D., Jodice, P.G.R. & Roby, D.D. (2006) Site-specific effects on productivity of an upper trophic-level marine predator: bottom-up, top-down, and mismatch effects on reproduction in a colonial seabird. *Progress in Oceanography*, **68**, 303–328.

Thackeray, S.J., Sparks, T.H., Frederiksen, M., Burthe, S., Bacon, P.J., Bell, J.R., Botham, M.S., Brereton, T.M., Bright, P.W., Carvalho, L., Clutton-Brock, T., Dawson, A., Edwards, M., Elliott, J.M., Harrington, R., Johns, D., Jones, I.D., Jones, J.T., Leech, D.I., Roy, D.B., Scott, W.A., Smith, M., Smithers, R.J., Winfield, I.J. & Wanless, S. (2010) Trophic level asynchrony in rates of phenological change for marine, freshwater and terrestrial environments. *Global Change Biology*, **16**, 3304–3313.

Thomas, C.D., Cameron, A., Green, R.E., Bakkenes, M., Beaumont, L.J., Collingham, Y.C., Erasmus, B.F.N., De Siqueira, M.F., Grainger, A., Hannah, L., Hughes, L., Huntley, B., van Jaarsveld, A.S., Midgley, G.F., Miles, L., Ortega-Huerta, M.A., Peterson, A.T., Phillips, O.L. & Williams, S.E. (2004a) Extinction risk from climate change. *Nature*, **427**,145–148.

Thomas, R.J., Bartlett, L.A., Marples, N.M., Kelly, D.J. & Cuthill, I.C. (2004b) Prey selection by wild birds can allow novel and conspicuous colour morphs to spread in prey populations. *Oikos*, **106**, 285–294.

Thomas, R.J., King, T.A., Forshaw, H.E., Marples, N.M., Speed, M.P. & Cable, J. (2010) The response of fish to novel prey: evidence that dietary conservatism is not restricted to birds. *Behavioral Ecology*, **21**, 669–675.

Tylianakis, J.M., Didham, R.K., Bascompte, J. & Wardle, D.A. (2008) Global change and species interactions in terrestrial ecosystems. *Ecology Letters*, **11**, 1351–1363.

Vafidis, J.O., Vaughan, I.P., Jones, T.H., Facey, R.J., Parry, R. & Thomas, R.J. (2014) Habitat use and body mass regulation among warblers in the Sahel Region during the non-breeding season. *PLOS ONE*, **9**, e113665.

Vafidis JO, Vaughan IP, Jones TH, Facey RJ, Thomas RJ (2016) Experimental analysis of the mechanisms underlying behavioural adaptation to climate change in a migratory songbird. *PLOS ONE, e0159933.*

Van der Putten, W.H., De Ruiter, P.C., Bezemer, T.M., Harvey, J.A., Wassen, M. & Wolters, V. (2004) Trophic interactions in a changing world. *Basic and Applied Ecology*, **5**, 487–494.

Van der Putten, W.H., Macel, M. & Visser, M.E. (2010) Predicting species distribution and abundance responses to climate change: why it is essential to include biotic interactions across trophic levels. *Philosophical Transactions of the Royal Society B: Biological Sciences*, **365**, 2025–2034.

VanDerWal, J., Murphy, H.T., Kutt, A.S., Perkins, G.C., Bateman, B.L., Perry, J.J. & Reside, A.E. (2013) Focus on poleward shifts in species' distribution underestimates the fingerprint of climate change. *Nature Climate Change*, **3**, 239–243.

Virgós, E. & Casanovas, J.G. (1999) Environmental constraints at the edge of a species distribution, the Eurasian badger (*Meles meles* L.): a biogeographic approach. *Journal of Biogeography*, **26**, 559–564.

Part IV

Evolution, Intervention and Emerging Perspectives

16

Evolutionary Responses of Invertebrates to Global Climate Change: the Role of Life-History Trade-Offs and Multidecadal Climate Shifts

Jofre Carnicer[1,2,3], Chris Wheat[4], Maria Vives[1], Andreu Ubach[1], Cristina Domingo[5], Sören Nylin[4], Constantí Stefanescu[2,6], Roger Vila[7], Christer Wiklund[4] and Josep Peñuelas[2]

[1] Department of Evolutionary Biology, Ecology and Environmental Sciences, Universitat de Barcelona, 08028 Barcelona, Spain
[2] CREAF, Cerdanyola del Vallès 08193, Spain
[3] GELIFES, Conservation Ecology Group, 9747 AG, Groningen, The Netherlands
[4] Department of Zoology, Stockholm University, Sweden
[5] Department of Geography, Autonomous University of Barcelona, 08193 Cerdanyola del Vallès, Spain
[6] Museum of Natural Sciences of Granollers, 08402 Granollers, Spain
[7] IBE, Institute of Evolutionary Biology, 08003 Barcelona, Spain

Summary

Life-history trade-offs will likely constrain the simultaneous optimisation of correlated suites of traits as invertebrates respond to global climate change. Therefore, a synthetic description of the fundamental trade-offs driving invertebrate evolutionary responses to global warming is needed. Here, five complementary trade-offs are described, including pleiotropic effects of endocrine hormonal signalling pathways, resource allocation, enzyme multi-functionality, thermal stability–kinetic efficiency, and various forms of water loss trade-offs. We also identify a gap in the study of the evolutionary responses of invertebrates to global warming: the examination of multidecadal climate dynamics and the emergence of non-analogous climates. The available evidence supports that different functional haplotypes effectively track decadal climate variability, creating new genetic landscapes that may contingently determine the adaptive capacity of species. The development of methods to extract DNA from ancient and historical samples now allows the study of historical biogeographical patterns of genetic polymorphisms (SNPs). When coupled with multidecadal information on climatic variability, it allows the study of multidecadal evolutionary responses of invertebrates to global warming and to abrupt shifts in drought regimes. This research is, in turn, facilitated by emerging modelling approaches that can effectively integrate genotype–phenotype–environment associations. Ultimately, this permits landscape studies of turnover in polymorphic candidate genes and the characterisation of non-linear threshold responses to drought and climatic variables.

16.1 Introduction

Multiple mechanisms shaping the responses of invertebrates to climate change have been described and empirically tested (Bale et al., 2002; Deutsch et al., 2008; Chown et al., 2011; Buckley & Kingsolver, 2012; Woods et al., 2015; Sunday et al., 2014; Schilthuizen & Kellermann, 2014).

Global Climate Change and Terrestrial Invertebrates, First Edition. Edited by Scott N. Johnson and T. Hefin Jones.
© 2017 John Wiley & Sons, Ltd. Published 2017 by John Wiley & Sons, Ltd.

A major research challenge remains, however, to integrate these diverse mechanisms. Here we explore whether life-history theory provides an integrative framework for the mechanisms mediating the responses of invertebrates to global warming (Stearns, 1992; Roff, 2002; Flatt & Heyland, 2011). More specifically, we suggest that life-history trade-offs are key to the integration of the implied mechanisms because they will constrain the simultaneous optimisation of correlated suites of traits. Precisely identifying and listing the different types of trade-offs specifically associated with the responses of invertebrates to global warming may consequently be useful.

This chapter will mainly focus on two major gaps that, in our opinion, feature in the study of the evolutionary responses of invertebrates to global warming.

I. A synthetic description of the fundamental trade-offs mediating invertebrate evolutionary responses to global warming.
II. The roles of multidecadal climate dynamics and drought regime shifts in driving distributional shifts, long-term population responses and evolutionary responses.

The two main sections of this chapter are devoted to analyses of these two points. Finally, the main conclusions and emerging research lines in the field are elaborated. An exhaustive review of the mechanisms driving the evolutionary responses of invertebrates to climate change is excluded due to the large number of mechanisms involved, but a brief, synthetic overview is provided in Tables 16.1 and 16.2 and Box 16.1.

Box 16.1 Exposure, Sensitivity, and Adaptive Capacity

Three components of invertebrate vulnerability to global warming have been widely discussed recently: the *exposure* or magnitude of climatic variation and impacts on the studied populations, the *sensitivity* due to the intrinsic traits of the species, plasticity and its genetic diversity, and the species' *adaptive capacity* (Williams et al., 2008; Woods et al., 2015; Pacifici et al., 2015). Adaptive capacity generally refers to the relative ability of one population to cope with the impacts of global warming. This term therefore integrates the species' demographic, plastic, and evolutionary responses to global change and its reactiveness to efforts of management and conservation (Williams et al., 2008).

The *exposure* or magnitude of climatic variation with impacts on invertebrate populations is determined by macroclimatic conditions of air temperature, relative humidity, wind, and solar radiation (Woods et al., 2015). Macroclimatic impacts and the width of climatic niches of many invertebrate groups systematically differ between latitudes and biomes (Deutsch et al., 2008; Kellermann et al., 2009). For example, higher latitudes experience greater seasonal and interannual thermal and climatic variation. As a result, terrestrial invertebrate ectotherms at higher latitudes are generally characterised by broader thermal performance curves and ranges of resistance to desiccation and cold (Deutsch et al., 2008; Kellermann et al., 2009). Complex microclimatic mosaics, however, are systematically observed in both local-scale studies and comprehensive global assessments, and invertebrates respond to this climatic heterogeneity using behavioural responses to avoid lethal or negative impacts of the meteorological conditions (Suggitt et al., 2012; Woods et al., 2015; Sunday et al., 2014; Oliver et al., 2014). Microhabitat elements can amplify or dampen macroclimatic effects, for example due to the effects of leaf transpiration or amplification by bare soil. Sensitivity and adaptability are terms used in other areas of evolutionary and population biology; for example, sensitivity also estimates the impact of an absolute change in vital rates on the rates of population growth (Benton & Grant, 1999). Similarly, adaptive capacity is a perhaps confusing term that integrates the species' demographic, plastic, and evolutionary responses to global change but also its reactiveness to efforts of management and conservation (Conrad, 1983, Williams et al., 2008). A cautionary use of these complex terms is therefore recommended.

Table 16.1 A review of the various processes and mechanisms affecting invertebrate exposure to the impacts of global warming. Exposure is ultimately determined by large-scale spatial and temporal macroclimatic gradients and a variety of local-scale effects. Species' traits and behavioural responses are also key to determining exposure, and therefore exposure and species sensitivity are ultimately linked.

Scale	Multiple hypotheses and key factors mediating invertebrate responses induced by climatic warming		Vulnerability component	References
Local	Microclimatic mosaics of temperature, relative humidity, wind, and solar radiation	Microclimatic mosaics determine responses induced by climate change and interact with behavioural responses	E	Scriber, 1996, Potter et al., 2009, Feder et al., 2010, De Frenne et al., 2013, Woods et al., 2015, Sunday et al., 2014
Local	Microsite amplification and cooling effects	Bare soils amplify environmental temperature, and leaf transpiration reduces thermal impacts on insect eggs	E	Potter et al., 2009, Pincebourde & Woods, 2012, Woods et al., 2015
Local	Behavioural buffering and thermal refugia	Invertebrate movement is key to determining experienced thermal environments	E, S, A	Feder et al., 2010, Sunday et al., 2014
Regional - Large	Altitudinal and latitudinal radiation clines	Solar radiation declines with increasing latitude and increases with increasing altitude. Organisms at high elevation can experience frequent thermal stress due to high levels of solar radiation.	E	Buckley et al., 2013
Large	Longer growing season	Climate change will increase the effective length of the growing season in mid- and high-latitude regions, and may increase population fitness	E, S	Kingsolver et al., 2013

(Continued)

Table 16.1 (Continued)

Scale	Multiple hypotheses and key factors mediating invertebrate responses induced by climatic warming	Vulnerability component	References
Large	Complex relationships between latitude and thermal variation	E, S	Addo-Bediako et al., 2000, Deutsch et al., 2008, Kingsolver et al., 2013
	Higher latitudes experience greater seasonal and interannual thermal variation. As a result, insect ectotherms are generally characterised by broader thermal performance curves at higher latitudes. Qualitatively different negative impacts of climate change are expected at low, mid-, and high latitudes		
Large	Larger impacts of short-term periods of maximum temperatures at mid-latitudes	E, S	Kingsolver et al., 2013
	Climate change could increase the frequency of high temperatures at mid-latitudes above the critical maximum temperatures		
Regional Large	Latitudinal variation in the relative importance of environmental cues	E, S	Brakefield & Zwaan, 2011, Navarro-Cano et al., 2015
	Latitudinal gradients exist for the relative importance of the photoperiodic and temperature cues determining phenological responses. Photoperiod is typically the critical variable in temperate latitudes, whereas temperature dominates in the tropics		
Large	Multidecadal droughts	E	Carnicer et al., 2011
	Shifts in drought regimes can affect the composition of insect communities, the spatial mosaics of intraspecific ecotypes, and the associated eco-evolutionary dynamics		
Large	Multi-annual, drought-induced large-scale disruption of insect food webs	E	Carnicer et al., 2011
	Extreme, multi-annual droughts can disrupt insect food webs and produce long-lasting effects on invertebrate communities		

Table 16.2 A review of the various mechanisms mediating invertebrate responses to the impacts of global warming. The vulnerability component involved is highlighted (E=exposure, S=sensitivity, A=adaptive capacity).

Multiple hypotheses and key factors mediating invertebrate responses induced by climatic warming		Vulnerability component	References
Egg-placing behaviour and host-plant selection	Egg-placing behaviour and host-plant selection modify experienced thermal environments	E, S, A	Merrill et al., 2008, Ashton et al., 2009, Bennett et al., 2015
Interactions between behavioural buffering and other ecological factors	Egg height laying behaviour interacts with thermal environment, food quality, and incidental predation by grazers	E, S, A	McBride & Singer, 2010, Bennett et al., 2015
Specificity in the effects of climate change and meteorological events on different life-cycle stages	Different life stages inhabit different microhabitats and also differ in their thermal and desiccation sensitivities	S	Roy et al., 2001, Kingsolver et al., 2011, Woods et al., 2015
Plastic larval development and developmental polymorphism (compensatory growth and catch-up growth)	Intraspecific, plastic variation in the number of larval instars, development and growth rates interacts with local weather and influences survival, life span, fecundity, and population dynamics	S, A	Esperk et al., 2007, Saastamoinen et al., 2013
Insect polyphenism	Seasonal polyphenism is associated with functionally divergent responses to temperature and desiccation stress	S	Brakefield & Zwaan, 2011, Hartfelder & Emlen, 2012
Dispersal polymorphism	Dispersal polymorphism and intraspecific variability in locomotory capacity are associated with functionally divergent responses to temperature and desiccation stress and to life history trade-offs	S	Feder et al., 2010
Photoperiod and photoperiodic-direction cues	Photoperiod and photoperiodic direction are used as environmental cues regulating winter diapause, polyphenism, and life-history adaptation to seasonal climate	S	Nylin, 1989, 1992
Photoperiod and temperature cues	Coupled additive effects of photoperiod and temperature on polyphenism	S	Nylin, 2013

(Continued)

Table 16.2 (Continued)

Multiple hypotheses and key factors mediating invertebrate responses induced by climatic warming	Vulnerability component	References
Interactions between species traits and climate-induced phenological responses	Overwintering stage, host-plant type, and other traits influence phenological responses. Species overwintering as adults or pupae and host-plant specialists tend to advance the phenology more in response to climate change.	S	Altermatt, 2010, Diamond et al., 2011, Wiklund & Friberg, 2011, Karlsson, 2014, Navarro-Cano et al., 2015
Altered biotic interactions and phenological mismatches	Phenology of host plants and hosts can be differentially affected by temperature, leading to temporal mismatches, fitness variation, and altered interaction strength	S	Tylianakis et al., 2008, Singer & Parmesan, 2010, Navarro-Cano et al., 2015, Posledovich et al., 2015
Altered function of photoperiodic cues	Higher temperatures can modify developmental rates and decouple the synchrony between critical photoperiodic and diapause-induction responses	S	Bale & Hayward, 2010
Altered endosymbiotic relationships	The ecological tolerances of obligate endosymbiotic bacteria may determine species responses to climate change	S	Ohtaka & Ishikawa, 1991, Wernegreen, 2012
Colonisation of new habitat types	Colonisation of new habitat types during range expansions induced by climate change	S, A	Thomas et al., 2001, Chen et al., 2011, Parmesan et al., 2015
Intraspecific latitudinal variation in thermal-reaction norms	Temperature-induced phenological responses are linked to intraspecific latitudinal variation in thermal-reaction norms associated with local adaptation	S, A	Posledovich et al., 2014, Stålhandske et al., 2014, Navarro-Cano et al., 2015
Negative and positive effects of winter warming on diapause and on population performance	Low winter temperatures are required for the completion of diapause in many species, and therefore increased winter temperatures often have negative effects on populations. In contrast, increased winter temperatures can have positive effects on non-diapausing, frost-sensitive species.	S, A	Pullin & Bale, 1989, Bale et al., 2002, Irwin & Lee, 2003, Crozier, 2004, Hahn & Denlinger, 2011, Williams et al., 2012, Williams et al., 2015

Effects of increased temperature on metabolic rates and water loss	Global warming has exponential effects on invertebrate metabolic rates and impacts on water-loss budgets	S, A	Dillon et al., 2010
Summer diapause and aestivation responses	Implies multiple changes such as reducing stress of dehydration by upregulating heat-shock proteins, cuticle biochemical changes, reduced metabolic rate, regulation of spiracular valves, and reduced moisture loss in waste products	S, A	Masaki, 1980, Wolda, 1988, Benoit, 2010
Interactions between winter-diapause propensity and other conditions and traits	Interactions between winter-diapause propensity, resource type (host-plant diet), sex, and environmental cues (temperature, photoperiod)	S, A	Hahn & Denlinger, 2011
Rapid trait evolution in response to climatic warming	Rapid evolution of phenology, body mass, photoperiodic day-length responses, the proportion of individuals entering diapause and the amount of shortening of day length required to trigger diapause, and desiccation resistance	A	Masaki, 1978, Pullin, 1986, Bale et al., 2002, Bradshaw & Holzapfel, 2001, Hoffman, 2010, Chown et al., 2011, van Asch et al., 2013, Higgins et al., 2014, Kopp et al., 2014
Counter-gradient variation	Involves opposing environmental and genetic variation of traits along gradients. For example, some populations at higher latitudes or altitudes develop faster to complete their life cycle within a shorter season and therefore respond differently to temperature	S, A	Ayres & Scriber, 1994, Conover & Schultz, 1995
Seasonal trade-offs in the allocation of melanisation	Trade-offs between nitrogen-costly melanisation, immune function, and reproduction	S, A	Karlsson & Wickman, 1989, Karlsson et al., 2008, Prasai & Karlsson, 2011, Diamond & Kingsolver, 2011, Hartfelder & Emlen, 2012
Interactions between season length, latitude and size	Longer season length affects the optimal development time and hence growth rate and size	E, S, A	Nylin & Svärd, 1991

(Continued)

Table 16.2 (Continued)

Multiple hypotheses and key factors mediating invertebrate responses induced by climatic warming	Vulnerability component	References
Trade-offs between heritable adult size and other traits	S, A	Bennett et al., 2015
Trade-offs between heritable adult size, advances in the phenology of adult emergence in response to warming, and fecundity		
Trade-offs between melanisation, thermal and demographic performance, immune function, and polyphenism	S, A	Prasai & Karlsson, 2011
Different investments in immune function and melanisation in directly developing and winter–diapause developmental pathways		
Significant fitness costs of phenotypic plasticity	S, A	Chaput-Bardy et al., 2014
Phenotypic plasticity in melanisation is highly heritable, differs between families, and has fitness costs		
Eco-evolutionary dynamics	S, A	Hanski, 2011
Reciprocal feedbacks between genetic polymorphisms and environmental and metapopulation geographical variability		
Climate-responsive population ecotypes	S, A	Masaki, 1967, Hill et al., 1999, Singer et al., 1995, Hanski, 2011, Singer & McBride, 2010, 2012, Posledovich et al., 2014, Stålhandske et al., 2014, Bennett et al., 2015
Local populations differ in their thermal-reaction norms, egg-placing behaviour, host-plant specialisation, body size, and wing melanism		

16.2 Fundamental Trade-Offs Mediating Invertebrate Evolutionary Responses to Global Warming

16.2.1 Background

Evidence of adaptation to local climatic conditions has been extensively documented for many invertebrate taxa for a variety of phenotypic traits, such as phenology (Hodkinson, 2005; Karl et al., 2008; Välimäki et al., 2013), diapause (Schmidt et al., 2005, 2008; Anderson et al., 2005), body size (Gilchrist et al., 2001; Taylor et al., 2015), thermal tolerance (Hoffmann et al., 2002; Kellermann et al., 2012), desiccation tolerance (Parkash & Munjal, 2000), melanisation (Munjal et al., 1997; Telonis-Scott et al., 2011), and genetic inversions associated with climate (Umina et al., 2005; Balanyà et al., 2006; Schmidt et al., 2008; Rodríguez-Trelles et al., 2013). In addition, the development of genomics and transcriptomics now allows the study of the clinal variation of genomic structure, transposable elements, and inversions along climatic and geographical gradients (Hohenlohe et al., 2010; González et al., 2010; Kolaczkowski et al., 2011; Carnicer et al., 2012). Genomic studies document pervasive spatially varying selection acting at key genes throughout multiple genetic and metabolic pathways (Kolaczkowski et al., 2011; Hohenlohe et al., 2010). These studies suggest that evolutionary responses to global warming at the genomic level will likely involve multiple, coordinated suites of genetic traits.

In sharp contrast to the widespread empirical evidence documenting local adaptation to climatic conditions, genetic evolutionary changes in response to global warming have only been properly documented for a few invertebrate groups and for some specific traits (Thomas et al., 2001; Bradshaw & Holzapfel, 2010; Merilä & Hendry, 2014; Schilthuizen & Kellermann, 2014). For example, pioneering studies have documented a genetic change in critical photoperiod associated with climate change in the pitcher plant mosquito (Bradshaw & Holzapfel, 2001) and a genetic effect on the date of egg hatching of the winter moth (van Asch et al., 2013). Similarly, suggestive evidence for a change in thermal-performance curves in response to climate change has recently been reported for *Colias* butterflies (Higgins et al., 2014) and for shell melanisation of the snail *Cepaea nemoralis* (Ozgo & Schilthuizen, 2012; Cameron & Cook, 2013). While few empirical studies have considered genetic responses to global warming in invertebrate populations, many more studies have documented significant shifts in distributions and phenology (Parmesan et al., 1999; Roy & Sparks, 2000; Warren et al., 2001; Pöyry et al., 2009; Chen et al., 2011). A wide plurality of mechanisms has been described in recent decades and the emerging view is that multiple, highly complex mechanisms will mediate the responses of invertebrates to global warming (Tables 16.1 and 16.2). These mechanisms are often characterised by the interaction of ecological and evolutionary processes (Bale et al., 2002; Deutsch et al., 2008; Chown et al., 2011; Clusella-Trullas et al., 2011; Buckley & Kingsolver, 2012; Woods et al., 2015; Sunday et al., 2014; Schilthuizen & Kellermann, 2014). A general framework allowing the integration of these diverse mechanisms, however, is currently lacking. In this chapter we argue that this framework should be rooted in recent advances of life-history theory and in the identification of the fundamental life-history trade-offs mediating the adaptive responses to global warming. More precisely, we argue that this improved framework requires consideration of the following key elements.

1. The identification of *intraspecific life-history syndromes*, describing how life-history traits vary intraspecifically and co-vary with heritable genetic variation in polymorphic candidate genes.
2. An analysis of the *pleiotropic effects* of polymorphic candidate genes on multiple metabolic paths and their relationships with highly conserved *hormonal signalling pathways*. This analysis could provide a cornerstone for the mechanistic description of the fundamental *trade-offs* operating at the intraspecific level that mediate responses to climatic warming.

3. An improved description of the diverse trade-offs specifically affecting thermal tolerance, resistance to desiccation, water-loss responses, and dispersal capacity.

4. The study of how life-history trade-offs constrain local adaptation to climatic warming and contribute to the maintenance of genetic diversity with important effects on fitness.

5. The analysis of how life-history traits and trade-offs interact with landscape features and potentially modulate the emergence of eco-evolutionary dynamics.

16.2.2 Mechanisms Underpinning Trade-Offs

The adaptive responses of invertebrate populations to global warming will ultimately be linked to the available functional genetic variation for traits affecting fitness. The mechanisms allowing the maintenance of high genetic variability in traits closely associated with fitness have been widely debated, because natural selection is expected to progressively reduce additive genetic variance (Flatt & Heyland, 2011, Kremer et al., 2012). In contrast with this theoretical expectation, accumulating evidence from diverse common-garden experiments shows that high heritabilities for life-history traits with important fitness effects are frequently observed in natural populations and in significant genotype × environment interactions in temperature treatments (Kingsolver et al., 2004; Kallionemi & Hanski, 2011; Kvist et al., 2013), suggesting the maintenance of heritable variation by active mechanisms, which could mediate population responses to climatic variability. Genomic and transcriptomic studies have progressively identified *intraspecific life-history syndromes*, in which heritable variation in key life-history traits are significantly linked to genetic polymorphisms and to transcript profiles (Watt, 1992; Haag et al., 2005; Hanski & Saccheri, 2006; Saastamoinen & Hanski, 2008; Hanski, 2011; Kvist et al., 2013; de Jong et al., 2014). Crucially, genetic polymorphism at highly pleiotropic sites could underlie fitness differences in contrasting thermal environments, acting as a key factor mediating the local adaptive responses of populations (Kvist et al., 2013; de Jong et al., 2014).

A great diversity of patterns has been reported, but the reported life-history syndromes tend to vary between *fast* and *slow* lifestyles, with faster lifestyle genotypes often being characterised by increased dispersal capacity, higher metabolic rate, early-life fecundity, reduced life span and survival, decreased immune-function capacity, and reduced resistance to desiccation and thermal stress. We hypothesise, however, that selective regimes associated with climatic warming could simultaneously favour both fast- and slow-lifestyle genotypes. For example, global warming could select for fast, highly dispersive lifestyle genotypes for colonisation and range expansion but also promote slow, desiccation-resistant lifestyle genotypes in spatially stabilised populations under increased drought stress (Feder et al., 2010). Climatic warming can reasonably be expected to create strong selective pressures on the thermal and desiccation niches of species and on the capacity of species to disperse. In scenarios of future global warming, life-history trade-offs may play a key role in constraining the simultaneous optimisation of correlated suites of traits of populations. We suggest that the precise identification of the multiple mechanisms that may allow the emergence of trade-offs associated with desiccation resistance, range-expansion capacity, and thermal tolerance is therefore essential. We have thus identified five major hypotheses that could determine the evolutionary responses of invertebrates to global warming (Fig. 16.1):

- *Endocrine hormone-signalling pathway – antagonistic pleiotropy trade-off hypothesis*
- *Thermal stability – kinetic efficiency trade-off hypothesis*
- *Resource-allocation trade-off hypothesis*
- *Enzyme-multifunctionality (moonlighting) hypothesis*
- *Respiratory water loss – total gas exchange hypothesis and water-loss trade-off hypotheses*

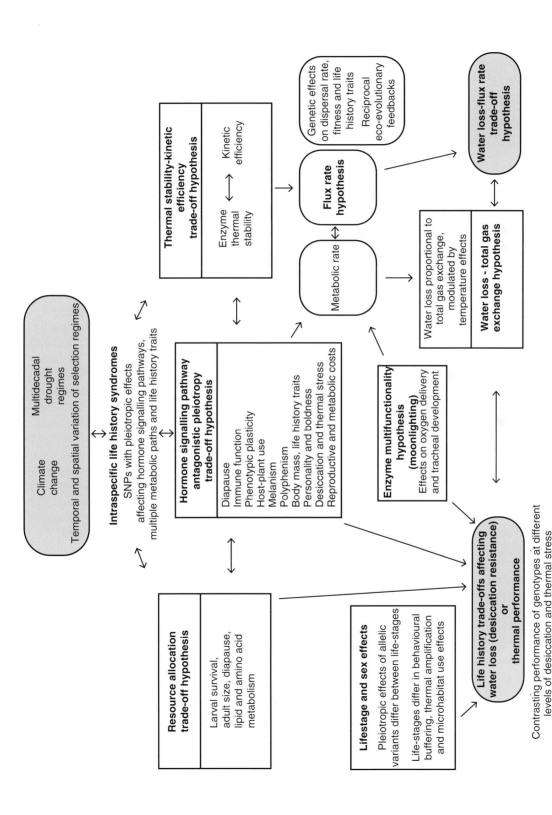

Figure 16.1 A synthesis of the major types of life-history trade-offs that could mediate the evolutionary responses of invertebrates to global warming and multidecadal drought regimes.

These five mechanisms are complementary and non-mutually exclusive. Empirical support suggests that they may be acting in natural populations. We briefly review these hypotheses in the following sections.

16.2.2.1 Endocrine Hormone-Signalling Pathway – Antagonistic Pleiotropy Trade-Off Hypothesis

Experimental research in several model organisms (e.g., *Drosophila melanogaster*, *Caenorhabditis elegans* and *Gryllus firmus*) has consistently documented life-history trade-offs that could mediate adaptive responses to climatic impacts (Feder et al., 2010; Flatt & Heyland, 2011). For example, key traits in *D. melanogaster* for climate-induced responses, such as higher resistance to desiccation and starvation, are associated with long life span, low fecundity early in life and increased adult survival (Flatt & Heyland, 2011). Similarly, the observed genetic variance in adult diapause, another trait tightly associated with climatic responses, has pleiotropic effects on stress tolerance and other life-history traits (life span, age-specific mortality, fecundity, lipid content, and developmental time) (Schmidt et al., 2005; Schmidt & Paaby, 2008; Kučerová et al., 2016). Furthermore, resistance to starvation and desiccation is reduced by immune responses to parasites in *Drosophila*, suggesting the operation of trade-offs between these two functions (Hoang, 2001). Comparative studies have concluded that these three model organisms use conserved hormonal systems to control life-history variation (reviewed in Flatt and Heyland, 2011), implying that evolutionarily conserved signalling pathways are potential key regulators of life-history traits and the associated trade-offs (Braendle et al., 2011; Kučerová et al., 2016). Among the conserved regulatory factors and signalling pathways, the insulin/insulin-like growth-signalling factor has been suggested to have a major role. In particular, this factor has pleiotropic effects on life span, growth, body size, developmental time, fecundity, longevity, stress resistance, lipid-reserve deposition, basal metabolism and increased immune function (Shingleton, 2011; Gerisch & Antebi, 2011; McKean & Lazzaro, 2011; Kučerová et al., 2016). Similarly, Juvenile hormone (JH) influences multiple functions, such as insect ecdysis and metamorphosis, longevity, egg production, diapause, growth rate, and immune function (Shingleton, 2011; Schmidt, 2011; McKean & Lazzaro, 2011). Ecdysteroids regulate polyphenism, which in turn involves differences in coordinated suites of traits, including stress resistance and physiological, morphological, behavioural, and life-history traits (Brakefield & Zwaan, 2011). For example, polyphenism in *Bicyclus anynana* is associated with differences in resistance to starvation, decreased resting metabolic rate and increased life span (Brakefield & Zwaan, 2011). In summary, there is ample evidence to support endocrine hormone-signalling pathways as key mechanisms regulating life-history trade-offs.

16.2.2.2 The Thermal Stability – Kinetic Efficiency Trade-Off Hypothesis

Temperature directly affects different enzymatic properties, such as structural stability and kinetic rate. Polymorphic enzymes of central metabolic paths are in turn frequently characterised by different thermal stabilities and kinetic rates (Watt & Dean, 2000; Kallioniemi & Hanski, 2011; Fields et al., 2015). This hypothesis states that trade-offs between the kinetic and stability properties of polymorphic enzymes could affect the evolutionary responses to global warming, selecting for different genotypes depending on the specific thermal selective regime. In this context, higher temperatures induced by global warming are expected to alter invertebrate enzymatic function via two main paths. Firstly, higher temperatures can influence the thermal stability of enzymes, because increasing temperature beyond the range to which an enzyme is adapted can denature the secondary and tertiary structures of proteins, promoting loss of enzymatic function (Fields et al., 2015). Secondly, higher temperatures associated with climate change can directly influence the rate of metabolic reactions (k), following the Arrhenius equation:

$$k = Ae^{-\frac{E_a}{RT}} \tag{16.1}$$

where A is a reaction-specific constant, T is temperature (in kelvins), R is the universal gas constant, and E_a is the Arrhenius activation energy of the reaction (Fields et al., 2015).

Genetic variation associated with enzymatic polymorphism, however, can also directly affect the rate of metabolic fluxes (k) by amino-acid changes at a few sites. These functional amino-acid changes affect the mobility of the portions of the enzyme that determine catalytic rates and conformational changes, and allow the enzyme to adapt to changes in temperature (Fields et al., 2015). Enzymes in central metabolic paths are generally under selection to balance between these two contrasting effects of temperature (thermal stability and structural flexibility) that ultimately determine the rate of catalysis and enzymatic function (Fields et al., 2015). Natural selection modifies enzymatic structure and amino-acid composition to optimise both thermal stability and structural flexibility, allowing catalysis at appropriate rates in different thermal environments. The available empirical evidence demonstrates that different enzyme genotypes can be preferentially selected in different thermal environments (Niitepold, 2010; Karl et al., 2008; 2009, Storz & Zera, 2011; Fields et al., 2015). For example, genotypes that are more thermally stable increase in frequency after regional thermal stresses (Rank & Dahlhoff, 2002; Dahlhoff et al., 2008). Furthermore, enzyme polymorphisms also affect other aspects of invertebrate performance, such as dispersal capacity, metabolic rate, reproductive capacity, and life-history traits and also significantly influence processes of range expansion (Haag et al., 2005; Hanski & Saccheri, 2006; Dahlhoff et al., 2008; Niitepold, 2010; Mitikka & Hanski, 2010; Wheat et al., 2011; Wheat & Hill, 2014). The available evidence therefore suggests that the effects of thermal stability – kinetic efficiency trade-offs can be scaled up at higher organismic and populational levels, affecting dispersal processes and potentially promoting eco-evolutionary feedbacks.

16.2.2.3 Resource-Allocation Trade-Off Hypothesis

Trade-offs have been classically described as the inevitable consequence of the allocation of a limited quantity of energy or resources between two competing organismic functions (the "Y" model). Recent advances, however, have supported a contrasting view, in which trade-offs also occur due to the pleiotropic effects of hormone-signalling pathways that are often independent of energy allocation (Flatt & Heyland, 2011). Reconciling these views is possible, because extensive empirical evidence supports the operation of the two types of trade-offs. For example, genetically based resource-allocation trade-offs have been supported in the cricket *Gryllus firmus*, in which differential flux by lipid and amino-acid metabolic paths explain trade-offs between key life-history traits (Zera & Harshman, 2011). Other studies have also empirically supported resource-allocation trade-offs in *Drosophila* and other model organisms (Bochdanovits & de Jong, 2004; St-Cyr et al., 2008). Notably, known resource-allocation trade-offs affect traits that are relevant for tolerance to temperature stress and the regulation of water loss, such as larval growth rate, body mass and cuticle investment (Bochdanovits & de Jong, 2004). For example, *Drosophila* genotypes with enhanced investments in larval protein biosynthesis, cuticle investments, growth, and body mass suffer a reduction in their rates of larval survival (Bochdanovits & de Jong, 2004). This evidence suggests that resource-allocation trade-offs may intervene in the regulation of responses of invertebrates induced by global warming.

16.2.2.4 Enzymatic-Multifunctionality (Moonlighting) Hypothesis

Genetic polymorphisms in key enzymes can alter the concentrations of intermediary metabolites of central metabolism (Storz & Zera, 2011; Marden, 2013). These intermediary metabolites in turn often act on key cell-signalling pathways and influence important life-history traits and associated trade-offs (Marden, 2013). More specifically, a key role has been proposed for the metabolites of the glycolytic pathway and the TCA cycle (Marden, 2013). Similarly, the enzyme itself can in other cases also bind to other cellular receptors, thereby regulating important organismic traits that could potentially mediate the responses of invertebrates to the impacts of climate change (Jeffery, 1999; Copley, 2003; Marden, 2013). For example, this type of effect of enzymatic multifunctionality could explain the variability of the development of oxygen-delivery networks and the associated rates of water loss and also affect other basic life-history traits such as growth rate, developmental rates, and life span (Marden, 2013; Marden et al., 2013). The life-history traits affected by the effects of moonlighting could therefore shape the evolutionary responses of invertebrates to global warming.

16.2.2.5 Respiratory Water Loss – Total Gas Exchange Hypothesis

Climate change, increased temperatures and reduced air-vapour pressures will largely affect the costs of water-vapour loss in invertebrates (respiratory and cuticular). Woods and Smith (2010) report that the loss of invertebrate respiratory water scales proportionally to gas exchange and depends on environmental temperature, ambient vapour pressure, and the dominant transport mode (convection or diffusion), following the model:

$$J' = \frac{P'(T_E) - P'_A}{\Delta P} \tau J \qquad (16.2)$$

where J' is water flux, J is gas-exchange flux, P'_A is ambient vapour pressure, and $P'(T_E)$ is the saturation vapour pressure as a function of the environmental temperature (T_E) of the external surface of the gas-exchange system of the invertebrate (Woods & Smith, 2010). τ is calculated as:

$$\tau = \frac{G'}{G} \qquad (16.3)$$

where G' and G are the relative conductances for water vapour and gas, respectively. The costs of losing respiratory water are proportional to the gas-exchange flux, which is a key element of this model, and therefore could depend on metabolic rate (Fig. 16.1). This dependence has important implications for the diverse life-history trade-offs potentially mediating the costs of water loss induced by global warming, as explained in the next section.

16.2.2.6 Water-Loss Trade-Off Hypotheses

Invertebrates respond to increased hydric stress by diverse evolutionary and phenotypically plastic responses (Gibbs et al., 2003; Dillon et al., 2010; Benoit, 2010; Chown et al., 2011). Under increased drought stress, they can reduce metabolic rates, reduce cuticular and respiratory water loss, upregulate heat-shock proteins and other stress-related proteins, increase body-water content and water absorption, regulate spiracular valves, reduce moisture loss in waste products, alter growth rates and body mass, or develop aestivation and summer-diapause responses (Benoit, 2010, Chown et al., 2011). Resistance to desiccation has a genetic basis and ample experimental evidence supports adaptive evolution in response to reduced water availability (Hoffmann & Parsons, 1989; Gibbs et al., 2003; Chown et al., 2011). Experimental studies show that the costs of invertebrate water loss are also tightly related to increased metabolic rates and environmental temperatures (Chown et al., 2011). Three key explanatory factors of water loss, that is, metabolic rate, body size and temperature, have been linked by the equation:

$$B(m, T) = b_0 m^a e^{\frac{-E}{kT}} \qquad (16.4)$$

where B is metabolic rate, T is body temperature, m is body mass, b_0 is a normalisation constant, k is the Boltzmann constant, a is the scaling exponent (usually varying between 2/3 and 3/4), and

E is the average activation energy for metabolic biochemical reactions (Dillon et al., 2010). Note that, although not explicitly indicated in equation 16.4, invertebrate body size is also largely influenced by temperature, with an increase in rearing temperature often leading to decreased adult body size (Kingsolver & Huey, 2008, Shingleton, 2011). The effects of temperature on body size are mediated by several mechanisms, such as changes in cell size and number, and can be the result of both evolutionary and phenotypically plastic responses (Shingleton, 2011). Furthermore, qualitatively different effects of temperature depending on the specific tissue or organ have been reported, as well as qualitatively different effects of temperature on different life-history traits, such as growth duration and developmental rate (Shingleton, 2011). Importantly, all these complex effects of temperature on size could potentially affect the rate of water loss and indirectly mediate the responses to global warming. Variation in body size is also regulated by hormone-signalling pathways such as the insulin/insulin-growth-factor signalling pathway, juvenile hormone, and ecdysteroids (Shingleton, 2011). Genetic polymorphisms affecting these signalling paths could therefore also determine rates of water loss and the evolutionary responses to increased drought. Finally, beyond the evolution and plasticity of body size, we suggest that the four main types of generic life-history trade-offs reviewed above can directly influence metabolic rates and the evolutionary responses associated with the costs of water loss (i.e., *endocrine hormone-signalling pathway trade-offs (antagonistic pleiotropy), thermal stability–kinetic efficiency trade-offs, resource-allocation trade-offs*, and *enzymatic multifunctionality trade-offs*) (Fig. 16.1).

16.3   The Roles of Multi-Annual Extreme Droughts and Multidecadal Shifts in Drought Regimens in Driving Large-Scale Responses of Insect Populations

Climate change is expected to increase the frequency and intensity of extreme events, including floods and heavy rains, heat waves, and multi-annual extreme droughts (IPCC, 2014; and see Meehl & Tebaldi, 2004; Carnicer et al., 2011; Peñuelas et al., 2013; Seneviratne et al., 2014). Both positive and negative effects of short-term climatic extreme events on local invertebrate population dynamics have been reported for many species (e.g., Edith's Checkerspot butterfly (*Euphydryas editha*) (Parmesan et al., 2000); *Melitaea cinxia* (Hanski & Meyke, 2005; Tack et al., 2015); Kindvall, 1995; Battisti et al., 2006; Dahlhoff et al., 2008; WallisDeVries et al., 2011). In contrast, due to limited data availability, a much more limited number of studies have quantified the decadal impacts of multi-annual and multidecadal droughts on insect communities (Shure et al., 1998; Hawkins & Holyoak, 1998; Rank & Dahlhoff, 2002; Trotter et al., 2010; Stone et al., 2010; Carnicer et al., 2011). The time span of droughts typically increases from meteorological to soil moisture and hydrological droughts, spanning from a few months to several decades.

The responses of invertebrates to multi-annual and multidecadal droughts and the interactions with global-warming trends remain poorly understood. The available studies show that droughts spanning multi-annual and multidecadal periods can synchronise the population crashes of several invertebrate species at regional and continental scales (Hawkins & Holyoak, 1998; Carnicer et al., 2011). Furthermore, large-scale multi-annual droughts can also induce long-lasting effects on the structure of invertebrate communities at the regional scale (Carnicer et al., 2011). Semiarid forests of western North America (*Pinus edulis*) were affected in 2000–2003 by a large-scale multi-annual drought that resulted in a large-scale shift in the vegetation (Breashears et al., 2005). In this system, the invertebrate communities of *P. edulis* forests have been intensively studied along gradients of soil type associated with increased drought stress (Trotter et al., 2010; Stone et al., 2010). Areas affected by chronic drought stress are characterised by very strong reductions in species richness and total abundance of invertebrates (Trotter et al., 2010; Stone et al., 2010). Notably, these effects of increased drought stress are observed at multiple trophic levels, affecting herbivores, predators, and parasitoids (Trotter et al., 2010; Stone et al., 2010). The limited available evidence generally suggests that extreme multi-annual droughts could emerge as a primary driver of the structures of invertebrate communities in the future.

Future climates and drought regimes will probably lack modern analogues (Williams & Jackson, 2007), and unprecedented dynamics of the climate are to be expected. Clusters of dry years (multi-annual droughts) may alternate with multi-annual wet periods, and decadal or multidecadal ramp-down trends may reduce water availability (Fig. 16.2A). Furthermore, the interactions of increased climatic warming and major Earth teleconnections (e.g., El Niño Southern Oscillation (ENSO), the Atlantic Multidecadal Variability (AMV), and the Pacific Decadal Oscillation (PDO)) may induce contrasting periods of dampened or increased multidecadal variability (Jackson et al., 2009). In addition, consistent spatial gradients in the type of drought dynamics are frequently observed, suggesting that future drought-induced selective regimes will be spatially structured (Fig. 16.2B) (Coll et al., 2013; Carnicer et al., 2014). Crucially, the putative genetic and evolutionary responses of invertebrate communities to these extreme scenarios of multi-annual droughts remain, to our knowledge, largely unexplored (Dahlhoff et al., 2008). Below, we analyse five main processes that could play key roles in shaping the evolutionary responses of invertebrates to changing regimes of multidecadal droughts and increased climate change.

1. *Multidecadal droughts and climate change may alter the frequency of genetic polymorphisms (single nucleotide polymorphisms, SNPs)* associated with intraspecific life-history syndromes and with the various types of trade-offs reviewed above (e.g., endocrine hormonal-signalling, resource-allocation, enzymatic-multifunctionality, thermal stability – kinetic efficiency, and water-loss efficiency trade-offs), as hypothesised in Fig. 16.3A-C. Periods of extreme multidecadal droughts could promote local extinctions of alleles, specifically those less well adapted to increased drought stress (Fig. 16.3C). Multidecadal drought could also produce peaks of population extinctions (Fig. 16.3D) and geographical movements of the trailing and leading edges of allelic distributions (Fig. 16.3E). The scarce available evidence suggests that haplotypes can effectively track decadal climatic variability (Mende & Hundsdoerfer, 2013), in support of the hypothesised scenario.

2. *Multidecadal droughts and climate change will produce different impacts on different ecotypes.* Geographically well-defined invertebrate species with relatively isolated local subpopulations often have ecotypes characterised by different genetic, phenological, and behavioural traits (e.g., Bennett et al., 2015). Crucially, different ecotypes and subpopulations are often differentially sensitive to the impacts of climate and drought (Higgins et al., 2014; Bennett et al., 2015). Ecotypes are often distributed in spatial mosaics, in which widespread and more resilient ecotypes are often intermixed with more vulnerable and specialised ecotypes (e.g., Bennett et al., 2015). Different ecotypes are expected to respond divergently to global warming, with qualitatively different distributional shifts and patterns of extinction. Multidecadal droughts and climate change, though, may also promote contacts between previously isolated ecotypes and thus promote genetic interchanges.

3. *Abrupt shifts in drought regimes will cause the loss and redistribution of genetic variation, creating new genetic landscapes* that will contingently determine the capacity of species to adapt to future impacts (Fig. 16.3E). We hypothesise that shifts in drought regimes may apply an important selective pressure that would induce significant changes in the patterns of the phenotypic and genetic variability of the local insect populations, promoting the emergence of geographical mosaics of phenotypic variability and clinal variation. These geographical mosaics could in turn act as key genetic reservoirs

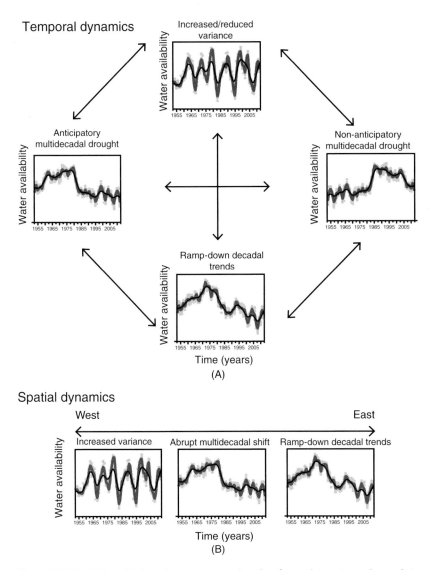

Figure 16.2 Multidecadal drought regimes may be a key factor determining the evolutionary responses of invertebrates to global change. (A) The temporal dynamics of multidecadal drought regimes may combine contrasting periods of anticipatory and non-anticipatory multidecadal trends or periods of increased variance and ramp-down trends. Anticipatory multidecadal droughts are periods of drought that anticipate the effects of future global warming. Non-anticipatory decadal droughts are mild, wet periods that temporally counteract the effects of global warming, creating contrasting selective regimes that may not coincide with future scenarios of global warming. Temporal sequences of different types of drought regimes can be expected in novel climates, with impacts on the landscapes of functional genetic variation, and creating contingent effects that may limit future evolutionary responses. (B) Drought dynamics also show spatial gradients in their trends. In this example, we illustrate three contrasting types of drought dynamics that have been observed in the Iberian Peninsula along a longitudinal transect (West-East arrow) during recent decades. Multidecadal responses of invertebrates are expected to track the spatial gradients in drought regimes and consistently differ across the spatial gradients.

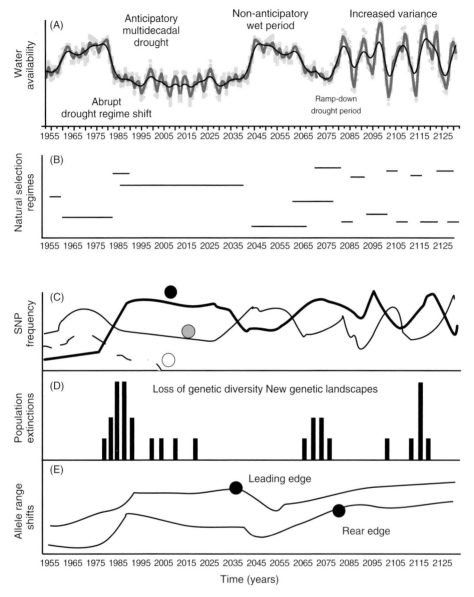

Figure 16.3 Multidecadal drought regimes and evolutionary responses of invertebrates. Changing multidecadal drought regimes (A) will alter the selective regimes experienced by invertebrates (B). This effect may produce spatiotemporal shifts in the distribution of genetic polymorphisms mediating adaptive responses to changing regimes (C). In c we illustrate three hypothetical alleles that differ in their drought sensitivity (black dot: frequency of an allele positively selected during increased drought periods; grey dot: an allele with intermediate performance during increased drought; white dot: an allele that suffers strong negative selection during increased drought and warming periods). (D) Multidecadal shifts in drought regimes may produce peaks of population extinctions, thereby creating new genetic landscapes that may determine future responses. (E) Multidecadal shifts in climate and drought regimes are expected to affect allelic distributions, causing changes in the leading and trailing edges of the distributions. The hypothetical distributional changes for a drought-resistant allele (black dot) in the leading and trailing edges are illustrated.

facilitating species evolvability in the face of impacts of future extreme climate change. Alternatively, the capacity of some species to adapt to abrupt shifts in decadal drought regimes may be limited, increasing the numbers of extinctions during abrupt shifts (Fig. 16.3D).

4. During multidecadal drought regimes, *the genetic variation associated with life-history syndromes could be maintained, in part, by eco-evolutionary dynamics and interactions with local heterogeneous landscape features.* The precise type of reciprocal eco-evolutionary feedbacks that could promote the maintenance of functional genetic variability linked to traits of desiccation resistance and thermal stress, however, remains to be explored. We hypothesise that climatic warming may favour reciprocal feedbacks in both fast- and slow-lifestyle genotypes, for example selecting for fast-lifestyle, highly dispersive genotypes with limited stress resistance during colonisation or range expansion, but also promoting a slow pace of life, by also selecting for stress-resistant genotypes in spatially stabilised populations under increased drought stress. We speculate that these divergent selective pressures could coexist in space and depend on the interaction with specific landscape features, thereby providing the opportunity for reciprocal feedback loops.

5. *Multidecadal regimes and climate change will affect specialist and generalist species differently* (Stefanescu et al., 2011a; Carnicer et al., 2013a). Generalist species will experience qualitatively different impacts, due to their greater effective population size, higher gene flow and dispersal capacity, and increased population synchrony and due to the relatively limited spatial structure of their intraspecific genetic variation (low F_{ST} values). In contrast, specialist species often have higher F_{ST} values, reduced effective population sizes, more genetic drift, more marked metapopulation structures, and susceptibility to extinction vortices. A major proportion of allelic and subpopulation extinctions is to be expected in more specialised lifestyles, consistent with recently detected trends for specialist invertebrates (Stefanescu et al., 2011b; Carnicer et al., 2013a). Similarly, *multidecadal regimes and climate change will have different impacts on the species at opposite extremes of major interspecific axes of life-history variation.* Several studies have found evidence for marked interspecific axes of covariation between phenotypic traits at the interspecific level, which could be used as a quantitative synthetic measure of life-history variation for a given taxon (i.e., interspecific-trait continua) (Carnicer et al., 2012, 2013a). For example, trait continua have been described for invertebrates and in many other groups such as plants, mammals, birds, fishes, phytoplankton, and marine bacteria (reviewed in Carnicer et al., 2012, 2013ab, 2015). A major challenge will be to quantify the differences in the functional intraspecific genetic variation that determines the responses to global warming between species at opposite extremes of interspecific life-history axes.

16.4   Conclusions and New Research Directions

A unified framework to understand the effects of global climate change on invertebrate evolutionary processes is still in its infancy due to the myriad of implied mechanisms (Tables 16.1 and 16.2). In this chapter, we suggest that the identification of the basic types of life-history trade-offs implied in the responses to global warming can provide a necessary cornerstone. We have identified five qualitative types of trade-offs that could mediate the responses of invertebrates to global climate change by constraining the optimisation of complex suites of traits. The five types of trade-offs reviewed are: endocrine hormonal signalling, resource allocation, enzymatic multifunctionality, thermal stability – kinetic efficiency, and various forms of trade-offs linked to water loss.

We have also highlighted the need for greater understanding of the evolutionary responses of invertebrates to multidecadal climate and abrupt shifts. Abrupt shifts and the emergence of multidecadal periods characterised by non-analogous climatic types are to be expected in the coming decades. The reviewed empirical evidence indicates that important evolutionary and distributional changes are to be expected in invertebrates at multidecadal timescales.

To conclude, we suggest that new methodological approaches now allow the study of the multi-decadal responses of invertebrates to climatic variation. Four main emerging research lines can be highlighted.

1. The development of methods to extract DNA from ancient and historical samples now allows the study of historical biogeographical patterns of genetic polymorphisms (SNPs) and ecotypic variants (Pääbo et al., 2004; Wandeler et al., 2007; Mende & Hundsdoerfer, 2013). When coupled with multidecadal information on climatic variability, this information would allow the study of multidecadal evolutionary responses of invertebrates to global warming and abrupt shifts in drought regimes. The multidecadal study of key polymorphic sites (SNPs) with known effects on life-history syndromes and trade-offs could provide insights to the evolution of invertebrates in changing climatic regimes.

2. New approaches to statistical modelling, such as generalised dissimilarity modelling (GDM) and gradient forests modelling, are now available to associate geographical patterns of genetic polymorphisms (SNPs) with long-term climatic records, drought indices, and phenotypic traits (Ellis et al., 2012; Fitzpatrick & Keller, 2015). GDM allows the modelling of large numbers of loci across genomes and accommodates nonlinear responses of loci to environmental gradients (Fitzpatrick & Keller, 2015). These statistical models can also be used to forecast the future spatial trends of intraspecific genetic variation, for example projecting observed gene–environment relationships across space and over time to create landscape predictions and maps of how adaptive genomic diversity (SNPs) may be disrupted by multidecadal shifts in climate (Fitzpatrick & Keller, 2015).

3. Beyond statistical approaches, recently developed theoretical models can simulate invertebrate population fitness, phenotypic plasticity, and the evolutionary responses of traits to the impacts of global change (Botero et al., 2015). A major step forward by these improved models is the inclusion of a wide range of timescales of environmental variation, varying from life-cycle timescales to much longer temporal scales that could correspond to multidecadal or multi-annual time periods. These emerging theoretical frameworks can thus model the evolutionary responses of invertebrates at timescales corresponding to a large number of life cycles. Moreover, the modelled environmental variability is characterised by different degrees of environmental predictability, allowing the modelling of the evolution of different types of strategic responses that are favoured in specific contexts. The models simulate the reversible phenotypic plasticity of the adult stage, irreversible developmental plasticity, phenotypic trait evolution allowing adaptive tracking of environmental variation, and conservative and diversifying bet-hedging strategies (Botero et al., 2015).

4. We envision a new research frontier in the comparative study of intraspecific genetic variation mediating the responses to climatic warming. We expect that species distributed along major life-history axes, characterised by contrasting lifestyles, will consistently differ in their patterns of intraspecific genetic variability and in the qualitative responses to the impacts of global warming. Preliminary evidence supports this view, reporting stronger population declines in specialist species characterised by higher F_{ST} values (Carnicer et al., 2013a). It is still unknown whether species with different lifestyles may consistently differ in the number of functional alleles maintained at key polymorphic sites and in the type of reciprocal eco-evolutionary feedbacks, phenotypically plastic responses, and life-history trade-offs mediating the responses to global warming.

Acknowledgements

We thank Philip Smith for proofreading this article. This research was supported by VENI-NWO 863.11.021, the Spanish Government grants CGL2016-78093-R, CGL2013-48074-P and CGL2013-48277-P, the Catalan Government project SGR 2014-274, and the European Research Council Synergy grant ERC-2013-SyG 610028 IMBALANCE-P.

References

Addo-Bediako, A., Chown, S.L. & Gaston, K.J. (2000) Thermal tolerance, climatic variability and latitude. *Proceedings of the Royal Society of London B: Biological Sciences*, **267**, 739–745.

Altermatt, F. (2010) Tell me what you eat and I'll tell you when you fly: Diet can predict phenological changes in response to climate change. *Ecology Letters*, **13**, 1475–1484.

Anderson, A.R., Hoffmann, A.A., McKechnie, S.W., Umina, P.A. & Weeks, A.R. (2005) The latitudinal cline in the *In(3R)Payne* inversion polymorphism has shifted in the last 20 years in Australian *Drosophila melanogaster* populations. *Molecular Ecology*, **14**, 851–858.

Ashton, S., Gutiérrez, D. & Wilson, R.J. (2009) Effects of temperature and elevation on habitat use by a rare mountain butterfly: implications for species responses to climate change. *Ecological Entomology*, **34**, 437–446.

Ayres, M.P. & Scriber, J.M. (1994) Local adaptation to regional climates in *Papilio canadensis* (Lepidoptera: Papilionidae). *Ecological Monographs*, **64**, 465–482.

Balanyà, J., Oller, J.M., Huey, R.B., Gilchrist, G.W. & Serra, L. (2006) Global genetic change tracks global climate warming in *Drosophila subobscura*. *Science*, **313**, 1773–1775.

Bale, J.S., Masters, G.J., Hodkinson, I.D., Awmack, C., Bezemer, T.M., Brown, V.K., Butterfield, J., Buse, A., Coulson, J.C., Farrar, J., Good, J.E.G., Harrington, R., Hartley, S., Jones, T.H., Lindroth, R.L., Press, M.C., Symrnioudis, I., Watt, A.D. & Whittaker, J.B. (2002) Herbivory in global climate change research: direct effects of rising temperature on insect herbivores. *Global Change Biology*, **8**, 1–16.

Bale, J.S. & Hayward, S.A.L. (2010) Insect overwintering in a changing climate. *The Journal of Experimental Biology*, **213**, 980–994.

Battisti, A., Stastny, M., Buffo, E. & Larsson, S. (2006) A rapid altitudinal range expansion in the pine processionary moth produced by the 2003 climatic anomaly. *Global Change Biology*, **12**, 662–671.

Bennett, N.L., Severns, P.M., Parmesan, C. & Singer, M.C. (2015) Geographic mosaics of phenology, host preference, adult size and microhabitat choice predict butterfly resilience to climate warming. *Oikos*, **124**, 41–53.

Benoit, J.B. (2010) Water management by dormant insects: comparisons between dehydration resistance during summer aestivation and winter diapause. *Aestivation* (ed C. Arturo Navas & J.E. Carvalho), pp. 209–229. Springer Berlin, Heidelberg.

Benton, T.G. & Grant, A. (1999) Elasticity analysis as an important tool in evolutionary and population ecology. *Trends in Ecology & Evolution*, **14**, 467–471.

Bochdanovits, Z. & de Jong, G. (2004) Antagonistic pleiotropy for life-history traits at the gene expression level. *Proceedings of the Royal Society of London B: Biological Sciences*, **271**, S75–S78.

Botero, C.A., Weissing, F.J., Wright, J. & Rubenstein, D.R. (2015) Evolutionary tipping points in the capacity to adapt to environmental change. *Proceedings of the National Academy of Sciences of the United States of America*, **112**, 184–189.

Bradshaw, W.E. & Holzapfel, C.M. (2001) Genetic shift in photoperiodic response correlated with global warming. *Proceedings of the National Academy of Sciences of the United States of America*, **98**, 14509–14511.

Bradshaw, W.E. & Holzapfel, C.M. (2010) Insects at not so low temperature: climate change in the temperate zone and its biotic consequences. *Low Temperature Biology of Insects* (eds D.L. Denlinger & R.E. Lee), pp. 242–275. Cambridge University Press, Cambridge.

Braendle, C., Heyland, A. & Flatt, T. (2011) Integrating mechanistic and evolutionary analysis of life history variation. *Mechanisms of Life History Evolution: The Genetics and Physiology of Life History Traits and Trade-offs* (eds T. Flatt & A. Heyland), pp. 3–10. Oxford University Press, Oxford.

Brakefield, P.M. & Zwaan, B.J. (2011) Seasonal polyphenisms and environmentally induced plasticity in the Lepidoptera: the coordinated evolution of many traits on multiple levels. *Mechanisms of Life History Evolution: The Genetics and Physiology of Life History Traits and Trade-offs* (eds T. Flatt & A. Heyland), pp. 243–252. Oxford University Press, Oxford.

Breshears, D.D., Cobb, N.S., Rich, P.M., Price, K.P., Allen, C. D., Balice, R. G., Romme, W.H., Kastens, J.H., Floyd, M.L., Belnap J., Anderson, J.J., Myers, O.B. & Meyer, C.W. (2005) Regional vegetation die-off in response to global-change-type drought. *Proceedings of the National Academy of Sciences of the United States of America*, **102**, 15144–15148.

Buckley, L.B. & Kingsolver, J.G. (2012) Functional and phylogenetic approaches to forecasting species' responses to climate change. *Annual Review of Ecology, Evolution, and Systematics*, **43**, 205–226.

Buckley, L.B., Miller, E.F., & Kingsolver, J.G. (2013). Ectotherm thermal stress and specialization across altitude and latitude. *Integrative and Comparative Biology*, **53**, 571–581.

Cameron, R.A.D. & Cook, L.M. (2013) Temporal morph frequency changes in sand-dune populations of *Cepaea nemoralis* (L.). *Biological Journal of the Linnean Society*, **108**, 315–322.

Carnicer, J., Coll, M., Ninyerola, M., Pons, X., Sánchez, G. & Peñuelas, J. (2011) Widespread crown condition decline, food web disruption, and amplified tree mortality with increased climate change-type drought. *Proceedings of the National Academy of Sciences of the United States of America*, **108**, 1474–1478.

Carnicer, J., Brotons, L., Stefanescu, C. & Peñuelas, J. (2012) Biogeography of species richness gradients: linking adaptive traits, demography and diversification. *Biological Reviews*, **87**, 457–479.

Carnicer, J., Stefanescu, C., Vila, R., Dincă, V., Font, X. & Peñuelas, J. (2013a) A unified framework for diversity gradients: the adaptive trait continuum. *Global Ecology and Biogeography*, **22**, 6–18.

Carnicer, J., Barbeta, A., Sperlich, D., Coll, M. & Peñuelas, J. (2013b). Contrasting trait syndromes in angiosperms and conifers are associated with different responses of tree growth to temperature on a large scale. *Frontiers in Plant Science*, **4**, 409.

Carnicer, J., Coll, M., Pons, X., Ninyerola, M., Vayreda, J., & Peñuelas, J. (2014). Large-scale recruitment limitation in Mediterranean pines: the role of *Quercus ilex* and forest successional advance as key regional drivers. *Global Ecology and Biogeography*, **23**, 371–384.

Carnicer, J., Sardans, J., Stefanescu, C., Ubach, A., Bartrons, M., Asensio, D. & Peñuelas, J. (2015) Global biodiversity, stoichiometry and ecosystem function responses to human-induced C–N–P imbalances. *Journal of Plant Physiology*, **172**, 82–91.

Chaput-Bardy, A., Ducatez, S., Legrand, D. & Baguette, M. (2014) Fitness costs of thermal reaction norms for wing melanisation in the large white butterfly (*Pieris brassicae*). *PLOS ONE*, **9**, e90026.

Chen, I.-C., Hill, J.K., Ohlemüller, R., Roy, D.B. & Thomas, C.D. (2011) Rapid range shifts of species associated with high levels of climate warming. *Science*, **333**, 1024–1026.

Chown, S.L., Sørensen, J.G. & Terblanche, J.S. (2011). Water loss in insects: an environmental change perspective. *Journal of Insect Physiology*, **57**, 1070–1084.

Clusella-Trullas, S., Blackburn, T.M. & Chown, S.L. (2011) Climatic predictors of temperature performance curve parameters in ectotherms imply complex responses to climate change. *The American Naturalist*, **177**, 738–751.

Coll, M., Peñuelas, J., Ninyerola, M., Pons, X. & Carnicer, J. (2013) Multivariate effect gradients driving forest demographic responses in the Iberian Peninsula. *Forest Ecology and Management*, **303**, 195–209.

Conrad, M. (1983) *Adaptability: The Significance of Variability from Molecule to Ecosystem*. Plenum Press, New York.

Conover, D.O. & Schultz, E.T. (1995) Phenotypic similarity and the evolutionary significance of countergradient variation. *Trends in Ecology and Evolution*, **10**, 248–252.

Copley, S.D. (2003) Enzymes with extra talents: moonlighting functions and catalytic promiscuity. *Current Opinion in Chemical Biology*, **7**, 265–272.

Crozier, L. (2004) Warmer winters drive butterfly range expansion by increasing survivorship. *Ecology*, **85**, 231–241.

Dahlhoff, E.P., Fearnley, S.L., Bruce, D.A., Gibbs, A.G., Stoneking, R., McMillan, D.M., Deiner, K., Smiley, J.T. & Rank, N.E. (2008) Effects of temperature on physiology and reproductive success of a montane leaf beetle: implications for persistence of native populations enduring climate change. *Physiological and Biochemical Zoology*, **81**, 718–732.

De Frenne, P., Rodríguez-Sánchez, F., Coomes, D. A., Baeten, L., Verstraeten, G., Vellend, M., Bernhardt-Römermanne, M., Brown, C.D., Brunet, J., Cornelis, J., Decocq, G.M., Dierschke, H., Eriksson, O., Gilliam, F.S., Hédl, R., Heinken, T., Hermy, M., Hommel, P., Jenkins, M.A., Kelly, D.L., Kirby, K.J., Mitchell, F.J.G., Naaf, T., Newman, M., Peterken, G., Petřík, P., Schultz, J., Sonnier, G., Van Calster, H., Waller, D.M., Walther, G.-R., White, P.S., Woods, K.D., Wulf, M., Graae, B.J. & Verheyen, K. (2013) Microclimate moderates plant responses to macroclimate warming. *Proceedings of the National Academy of Sciences of the United States of America*, **110**, 18561–18565.

Deutsch, C.A., Tewksbury, J.J., Huey, R.B., Sheldon, K.S., Ghalambor, C.K., Haak, D.C. & Martin, P.R. (2008) Impacts of climate warming on terrestrial ectotherms across latitude. *Proceedings of the National Academy of Sciences of the United States of America*, **105**, 6668–6672.

Diamond, S.E., Frame, A.M., Martin, R.A. & Buckley, L.B. (2011) Species' traits predict phenological responses to climate change in butterflies. *Ecology*, **92**, 1005–1012.

Diamond, S.E. & Kingsolver, J.G. (2011) Host plant quality, selection history and trade-offs shape the immune responses of *Manduca sexta*. *Proceedings of the Royal Society of London B: Biological Sciences*, **278**, 289–297.

Dillon, M.E., Wang, G. & Huey, R.B. (2010) Global metabolic impacts of recent climate warming. *Nature*, **467**, 704–706.

Ellis, N., Smith, S.J. & Pitcher, C.R. (2012) Gradient forests: calculating importance gradients on physical predictors. *Ecology*, **93**, 156–168.

Esperk, T., Tammaru, T. & Nylin, S. (2007) Intraspecific variability in number of larval instars in insects. *Journal of Economic Entomology*, **100**, 627–645.

Feder, M.E., Garland, T. Jr, Marden, J.H. & Zera, A.J. (2010) Locomotion in response to shifting climate zones: not so fast. *Annual Review of Physiology*, **72**, 167–190.

Fields, P.A., Dong, Y., Meng, X. & Somero, G.N. (2015) Adaptations of protein structure and function to temperature: there is more than one way to 'skin a cat'. *Journal of Experimental Biology*, **218**, 1801–1811.

Fitzpatrick, M.C. & Keller, S.R. (2015) Ecological genomics meets community-level modelling of biodiversity: mapping the genomic landscape of current and future environmental adaptation. *Ecology Letters*, **18**, 1–16.

Flatt, T. & Heyland, A. (2011) *Mechanisms of Life History Evolution: The Genetics and Physiology of Life History Traits and Trade-offs*. Oxford University Press, Oxford.

Gerisch, B. & Antebi, A. (2011) Molecular basis of life history regulation in C. elegans and other organisms. *Mechanisms of Life History Evolution: The Genetics and Physiology of Life History Traits and Trade-offs* (eds T. Flatt & A. Heyland) pp. 284–298. Oxford University Press, Oxford.

Gibbs, A.G., Fukuzato, F. & Matzkin, L.M. (2003) Evolution of water conservation mechanisms in *Drosophila*. *Journal of Experimental Biology*, **206**, 1183–1192.

Gilchrist, G.W., Huey, R.B. & Serra, L. (2001) Rapid evolution of wing size clines in *Drosophila subobscura*. *Genetica* **112-113**, 273–286.

González, J., Karasov, T.L., Messer, P.W. & Petrov, D.A. (2010) Genome-wide patterns of adaptation to temperate environments associated with transposable elements in Drosophila. *PLOS Genetics*, **6**, e1000905.

Haag, C.R., Saastamoinen, M., Marden, J.H. & Hanski, I. (2005) A candidate locus for variation in dispersal rate in a butterfly metapopulation. *Proceedings of the Royal Society of London B: Biological Sciences*, **272**, 2449–2456.

Hahn, D.A. & Denlinger, D.L. (2011) Energetics of insect diapause. *Annual Review of Entomology*, **56**, 103–121.

Hanski, I. & Meyke, E. (2005) Large-scale dynamics of the Glanville fritillary butterfly: landscape structure, population processes, and weather. *Annales Zoologici Fennici*, **42**, 379–395.

Hanski, I. & Saccheri, I. (2006) Molecular-level variation affects population growth in a butterfly metapopulation. *PLOS Biology*, **4**, e129.

Hanski, I.A. (2011) Eco-evolutionary spatial dynamics in the Glanville fritillary butterfly. *Proceedings of the National Academy of Sciences of the United States of America*, **108**, 14397–14404.

Hartfelder, K. & Emlen, D.J. (2012) Endocrine Control of Insect Polyphenism. *Insect Endocrinology* (ed. L.I. Gilbert), pp. 464–522, Academic Press, San Diego.

Hawkins, B.A. & Holyoak, M. (1998) Transcontinental crashes of insect populations? *The American Naturalist*, **152**, 480–484.

Hill, J.K., Thomas, C.D. & Lewis, O.T. (1999) Flight morphology in fragmented populations of a rare British butterfly, *Hesperia comma*. *Biological Conservation*, **87**, 277–283.

Higgins, J.K., MacLean, H.J., Buckley, L.B. & Kingsolver, J.G. (2014) Geographic differences and microevolutionary changes in thermal sensitivity of butterfly larvae in response to climate. *Functional Ecology*, **28**, 982–989.

Hoang, A. (2001) Immune response to parasitism reduces resistance of *Drosophila melanogaster* to desiccation and starvation. *Evolution*, **55**, 2353–2358.

Hodkinson, I.D. (2005) Terrestrial insects along elevation gradients: species and community responses to altitude. *Biological Reviews of the Cambridge Philosophical Society*, **80**, 489–513.

Hoffmann, A.A. & Parsons, P.A. (1989) Selection for increased desiccation resistance in *Drosophila melanogaster*: additive genetic control and correlated responses for other stresses. *Genetics*, **122**, 837–845.

Hoffmann, A.A., Anderson, A. & Hallas, R. (2002) Opposing clines for high and low temperature resistance in *Drosophila melanogaster*. *Ecology Letters*, **5**, 614–618.

Hoffmann, A.A. (2010) Physiological climatic limits in *Drosophila*: patterns and implications. *Journal of Experimental Biology*, **213**, 870–880.

Hohenlohe, P.A., Bassham, S., Etter, P.D., Stiffler, N., Johnson, E.A., & Cresko, W.A. (2010) Population genomics of parallel adaptation in threespine stickleback using sequenced RAD tags. *PLOS Genetics*, **6**, e1000862.

Irwin, J.T. & Lee, R.E. Jr. (2003) Cold winter microenvironments conserve energy and improve overwintering survival and potential fecundity of the goldenrod gall fly, *Eurosta solidaginis*. *Oikos*, **100**, 71–78.

IPCC 2014. Climate Change 2014: Impacts, Adaptation, and Vulnerability. Part A: Global and Sectoral Aspects. Contribution of Working Group II to the Fifth Assessment Report of the Intergovernmental Panel on Climate Change (eds V.R. Barros, C.B. Field, D.J. Dokken, M.D. Mastrandrea, K.J. Mach, T.E. Bilir, M. Chatterjee, K.L. Ebi, Y.O. Estrada, R.C. Genova, B. Girma, E.S. Kissel, A.N. Levy, S. MacCracken, P.R. Mastrandrea, & L.L. White). Cambridge University Press, Cambridge, United Kingdom and New York, NY, USA. 1132 pp.

Jackson, S.T., Betancourt, J.L., Booth, R.K. & Gray, S.T. (2009) Ecology and the ratchet of events: climate variability, niche dimensions, and species distributions. *Proceedings of the National Academy of Sciences of the United States of America*, **106**, 19685–19692.

Jeffery, C.J. (1999) Moonlighting proteins. *Trends in Biochemical Sciences*, **24**, 8–11.

de Jong, M.A., Wong, S.C., Lehtonen, R. & Hanski, I. (2014) Cytochrome P450 gene *CYP337* and heritability of fitness traits in the Glanville fritillary butterfly. *Molecular Ecology*, **23**, 1994–2005.

Kallioniemi, E. & Hanski, I. (2011) Interactive effects of *Pgi* genotype and temperature on larval growth and survival in the Glanville fritillary butterfly. *Functional Ecology*, **25**, 1032–1039.

Karl, I., Janowitz, S.A. & Fischer, K. (2008) Altitudinal life-history variation and thermal adaptation in the copper butterfly *Lycaena tityrus*. *Oikos*, **117**, 778–788.

Karl, I., Schmitt, T., & Fischer, K. (2009) Genetic differentiation between alpine and lowland populations of a butterfly is related to PGI enzyme genotype. *Ecography*, **32**, 488–496.

Karlsson, B. & Wickman, P.-O. (1989) The cost of prolonged life: an experiment on a nymphalid butterfly. *Functional Ecology*, **3**, 399–405.

Karlsson, B., Stjernholm, F. & Wiklund, C. (2008) Test of a developmental trade-off in a polyphenic butterfly: direct development favours reproductive output. *Functional Ecology*, **22**, 121–126.

Karlsson, B. (2014) Extended season for northern butterflies. *International Journal of Biometeorology*, **58**, 691–701.

Kellermann, V., van Heerwaarden, B., Sgrò, C.M. & Hoffmann, A.A. (2009) Fundamental evolutionary limits in ecological traits drive *Drosophila* species distributions. *Science*, **325**, 1244–1246.

Kellermann, V., Overgaard, J., Hoffmann, A.A., Fløjgaard, C., Svenning, J.C. & Loeschcke, V. (2012). Upper thermal limits of Drosophila are linked to species distributions and strongly constrained phylogenetically. *Proceedings of the National Academy of Sciences of the United States of America*, **109**, 16228–16233.

Kindvall, O. (1995) The impact of extreme weather on habitat preference and survival in a metapopulation of the bush cricket *Metrioptera bicolor* in Sweden. *Biological Conservation*, **73**, 51–58.

Kingsolver, J.G., Ragland, G.J. & Shlichta, J.G. (2004) Quantitative genetics of continuous reaction norms: thermal sensitivity of caterpillar growth rates. *Evolution*, **58**, 1521–1529.

Kingsolver, J.G. & Huey, R.B. (2008) Size, temperature, and fitness: three rules. *Evolutionary Ecology Research*, **10**, 251–268.

Kingsolver, J.G., Woods, H.A., Buckley, L.B., Potter, K.A., MacLean, H.J., & Higgins, J.K. (2011) Complex life cycles and the responses of insects to climate change. *Integrative and Comparative Biology*, **51**, 719–732.

Kingsolver, J.G., Diamond, S.E., & Buckley, L.B. (2013). Heat stress and the fitness consequences of climate change for terrestrial ectotherms. *Functional Ecology*, **27**, 1415–1423.

Kopp, M. & Matuszewski, S. (2014) Rapid evolution of quantitative traits: theoretical perspectives. *Evolutionary Applications*, **7**, 169–191.

Kolaczkowski, B., Kern, A.D., Holloway, A.K., & Begun, D.J. (2011) Genomic differentiation between temperate and tropical Australian populations of *Drosophila melanogaster*. *Genetics*, **187**, 245–260.

Kremer, A., Ronce, O., Robledo-Arnuncio, J.J., Guillaume, F., Bohrer, G., Nathan, R., Bridle, J.R., Gomulkiewicz, R., Klein, E.K., Ritland, K., Kuparinen, A., Gerber, S. & Schueler, S. (2012) Long-distance gene flow and adaptation of forest trees to rapid climate change. *Ecology Letters*, **15**, 378–392.

Kučerová, L., Kubrak, O.I., Bengtsson, J.M., Strnad, H., Nylin, S., Theopold, U. & Nässel, D.R. (2016) Slowed aging during reproductive dormancy is reflected in genome-wide transcriptome changes in *Drosophila melanogaster*. *BMC Genomics*, **17**, 50.

Kvist, J., Wheat, C.W., Kallioniemi, E., Saastamoinen, M., Hanski, I. & Frilander, M. J. (2013) Temperature treatments during larval development reveal extensive heritable and plastic variation in gene expression and life history traits. *Molecular Ecology*, **22**, 602–619.

Marden, J.H. (2013) Nature's inordinate fondness for metabolic enzymes: why metabolic enzyme loci are so frequently targets of selection. *Molecular Ecology*, **22**, 5743–5764.

Marden, J.H., Fescemyer, H.W., Schilder, R.J., Doerfler, W.R., Vera, J.C. & Wheat, C.W. (2013) Genetic variation in hif signaling underlies quantitative variation in physiological and life-history traits within lowland butterfly populations. *Evolution*, **67**, 1105–1115.

Masaki, S. (1967) Geographic variation and climatic adaptation in a field cricket (Orthoptera: Gryllidae). *Evolution*, **21**, 725–741.

Masaki, S. (1978) Seasonal and latitudinal adaptations in lifecycles of crickets. *Evolution of Insect Migration and Diapause* (ed. H. Dingle), pp. 72–100. Springer-Verlag, Berlin.

Masaki, S. (1980) Summer diapause. *Annual Review of Entomology*, **25**, 1–25.

McBride, C.S. & Singer, M.C. (2010) Field studies reveal strong postmating isolation between ecologically divergent butterfly populations. *PLOS Biology*, **8**, e1000529.

McKean, K.A. & Lazzaro, B. (2011) The costs of immunity and the evolution of immunological defense mechanisms. *Mechanisms of Life History Evolution: The Genetics and Physiology of Life History Traits and Trade-offs* (eds T. Flatt & A. Heyland), pp. 299–310. Oxford University Press, Oxford.

Meehl G.A. & Tebaldi, C. (2004) More intense, more frequent, and longer-lasting heat waves in the 21st century. *Science*, **305**, 994–997.

Mende, M.B. & Hundsdoerfer, A.K. (2013) Mitochondrial lineage sorting in action–historical biogeography of the *Hyles euphorbiae* complex (Sphingidae, Lepidoptera) in Italy. *BMC Evolutionary Biology*, **13**, 83.

Merilä, J. & Hendry, A.P. (2014) Climate change, adaptation, and phenotypic plasticity: the problem and the evidence. *Evolutionary Applications*, **7**, 1–14.

Merrill, R.M., Gutiérrez, D., Lewis, O.T., Gutiérrez, J., Díez, S.B. & Wilson, R.J. (2008) Combined effects of climate and biotic interactions on the elevational range of a phytophagous insect. *Journal of Animal Ecology*, **77**, 145–155

Mitikka, V. & Hanski, I. (2010) *Pgi* genotype influences flight metabolism at the expanding range margin of the European map butterfly. *Annales Zoologici Fennici*, **47**, 1–14.

Munjal, A.K., Karan, D., Gibert, P., Moreteau, B., Parkash, R. & David, J.R. (1997) Thoracic trident pigmentation in *Drosophila melanogaster*: latitudinal and altitudinal clines in Indian populations. *Genetics, Selection, Evolution*, **29**, 601–610.

Navarro-Cano, J.A., Karlsson, B., Posledovich, D., Toftegaard, T., Wiklund, C., Ehrlén, J. & Gotthard, K. (2015) Climate change, phenology, and butterfly host plant utilization. *Ambio*, **44**, S78–S88.

Niitepõld, K. (2010) Genotype by temperature interactions in the metabolic rate of the Glanville fritillary butterfly. *Journal of Experimental Biology*, **213**, 1042–1048.

Nylin, S. (1989) Effects of changing photoperiods in the life cycle regulation of the comma butterfly, *Polygonia c-album* (Nymphalidae). *Ecological Entomology*, **14**, 209–218.

Nylin, S., & Svärd, L. (1991) Latitudinal patterns in the size of European butterflies. *Holarctic Ecology*, **14**, 192–202.

Nylin, S. (1992) Seasonal plasticity in life history traits: growth and development in *Polygonia c-album* (Lepidoptera: Nymphalidae). *Biological Journal of the Linnean Society*, **47**, 301–323.

Nylin, S. (2013) Induction of diapause and seasonal morphs in butterflies and other insects: knowns, unknowns and the challenge of integration. *Physiological Entomology*, **38**, 96–104.

Ohtaka, C., & Ishikawa, H. (1991) Effects of heat treatment on the symbiotic system of an aphid mycetocyte. *Symbiosis*, **11**, 19–30.

Oliver, T.H., Stefanescu, C., Páramo, F., Brereton, T. & Roy, D.B. (2014) Latitudinal gradients in butterfly population variability are influenced by landscape heterogeneity. *Ecography*, **37**, 863–871.

Ożgo, M. & Schilthuizen, M. (2012) Evolutionary change in *Cepaea nemoralis* shell colour over 43 years. *Global Change Biology*, **18**, 74–81.

Pääbo, S., Poinar, H., Serre, D., Jaenicke-Després, V., Hebler, J., Rohland, N., Kuch, M., Krause, J., Vigilant, L., & Hofreiter, M. (2004) Genetic analyses from ancient DNA. *Annual Review of Genetics*, **38**, 645–679.

Pacifici, M., Foden, W.B., Visconti, P., Watson, J.E.M., Butchart, S.H.M., Kovacs, K.M., Scheffers, B.R., Hole, D.G., Martin, T.G., Akçakaya, H.R., Corlett, R.T., Huntley, B., Bickford, D., Carr, J.A., Hoffmann, A.A., Midgley, G.F., Pearce-Kelly, P., Pearson, R.G., Williams, S.E., Willis, S.G., Young, B. & Rondinini, C. (2015) Assessing species vulnerability to climate change. *Nature Climate Change*, **5**, 215–224.

Parkash, R. & Munjal, A.K. (2000) Evidence of independent climatic selection for desiccation and starvation tolerance in Indian tropical populations of *Drosophila melanogaster*. *Evolutionary Ecology Research*, **2**, 685–699.

Parmesan, C., Ryrholm, N., Stefanescu, C., Hill, J.K., Thomas, C.D., Descimon, H., Huntley, B., Kaila, L., Kullberg, J., Tammaru, T., Tennent, W.J., Thomas, J.A. & Warren, M. (1999) Poleward shifts in geographical ranges of butterfly species associated with regional warming. *Nature*, **399**, 579–583.

Parmesan, C., Root, T.L., & Willig, M.R. (2000) Impacts of extreme weather and climate on terrestrial biota. *Bulletin of the American Meteorological Society*, **81**, 443–450.

Parmesan, C., Williams-Anderson, A., Moskwik, M., Mikheyev, A.S. & Singer, M.C. (2015) Endangered Quino checkerspot butterfly and climate change: Short-term success but long-term vulnerability? *Journal of Insect Conservation*, **19**, 185–204.

Peñuelas, J., Sardans, J., Estiarte, M., Ogaya, R., Carnicer, J., Coll, M., Barbeta, A., Rivas-Ubach, A., Llusià, J., Garbulsky, M., Filella, I. & Jump, A.S. (2013). Evidence of current impact of climate change on life: a walk from genes to the biosphere. *Global Change Biology*, **19**, 2303–2338.

Pincebourde, S. & Woods, H.A. (2012) Climate uncertainty on leaf surfaces: the biophysics of leaf microclimates and their consequences for leaf-dwelling organisms. *Functional Ecology*, **26**, 844–853.

Pöyry, J., Luoto, M., Heikkinen, R.K., Kuussaari, M. & Saarinen, K. (2009) Species traits explain recent range shifts of Finnish butterflies. *Global Change Biology*, **15**, 732–743.

Potter, K., Davidowitz, G. & Woods, H.A. (2009) Insect eggs protected from high temperatures by limited homeothermy of plant leaves. *Journal of Experimental Biology*, **212**, 3448–3454.

Posledovich, D., Toftegaard, T., Navarro-Cano, J.A., Wiklund, C., Ehrlén, J. & Gotthard, K. (2014) Latitudinal variation in thermal reaction norms of post-winter pupal development in two butterflies differing in phenological specialization. *Biological Journal of the Linnean Society*, **113**, 981–991.

Posledovich, D., Toftegaard, T., Wiklund, C., Ehrlén, J. & Gotthard, K. (2015) The developmental race between maturing host plants and their butterfly herbivore – the influence of phenological matching and temperature. *Journal of Animal Ecology.* **84**, 1690–1699.

Prasai, K. & Karlsson, B. (2011) Variation in immune defence in relation to developmental pathway in the green-veined white butterfly, *Pieris napi*. *Evolutionary Ecology Research*, **13**, 295–305.

Pullin, A.S. (1986) Effect of photoperiod and temperature on the life-cycle of different populations of the peacock butterfly *Inachis io*. *Entomologia Experimentalis et Applicata*, **41**, 237–242.

Pullin A.S. & Bale, J.S. (1989) Effects of low temperature on diapausing *Aglais urticae* and *Inachis io* (Lepidoptera: Nymphalidae): cold hardiness and overwintering survival. *Journal of Insect Physiology*, **35**, 277–281.

Rank, N.E. & Dahlhoff, E.P. (2002) Allele frequency shifts in response to climate change and physiological consequences of allozyme variation in a montane insect. *Evolution*, **56**, 2278–2289.

Rodríguez-Trelles, F., Tarrío, R. & Santos, M. (2013) Genome-wide evolutionary response to a heat wave in *Drosophila*. *Biology Letters*, **9**, 20130228.

Roff, D.A. (2002) *Life History Evolution*. Sinauer Associates, Sunderland.

Roy, D.B. & Sparks, T.H. (2000) Phenology of British butterflies and climate change. *Global Change Biology*, **6**, 407–416.

Roy, D.B., Rothery, P., Moss, D., Pollard, E., & Thomas, J.A. (2001) Butterfly numbers and weather: predicting historical trends in abundance and the future effects of climate change. *Journal of Animal Ecology*, **70**, 201–217.

Saastamoinen, M. & Hanski, I. (2008) Genotypic and environmental effects on flight activity and oviposition in the Glanville fritillary butterfly. *The American Naturalist*, **171**, 701–712.

Saastamoinen, M., Ikonen, S., Wong, S.C., Lehtonen, R. & Hanski, I. (2013) Plastic larval development in a butterfly has complex environmental and genetic causes and consequences for population dynamics. *Journal of Animal Ecology*, **82**, 529–539.

Schilthuizen, M. & Kellermann, V. (2014) Contemporary climate change and terrestrial invertebrates: evolutionary versus plastic changes. *Evolutionary Applications*, **7**, 56–67.

Schmidt, P.S., Matzkin, L., Ippolito, M., & Eanes, W.F. (2005) Geographic variation in diapause incidence, life-history traits, and climatic adaptation in *Drosophila melanogaster*. *Evolution*, **59**, 1721–1732.

Schmidt, P.S. & Paaby, A.B. (2008) Reproductive diapause and life-history clines in North American populations of *Drosophila melanogaster*. *Evolution*, **62**, 1204–1215.

Schmidt, P.S., Zhu, C.-T., Das, J., Batavia, M., Yang, L., & Eanes, W.F. (2008) An amino acid polymorphism in the *couch potato* gene forms the basis for climatic adaptation in *Drosophila melanogaster*. *Proceedings of the National Academy of Sciences of the United States of America*, **105**, 16207–16211.

Schmidt, P. S. (2011). Evolution and mechanisms of insect reproductive diapause: a plastic and pleiotropic life history syndrome. *Mechanisms of Life History Evolution: The Genetics and Physiology of Life History Traits and Trade-offs* (eds T. Flatt & A. Heyland), pp. 221–229. Oxford University Press, Oxford.

Scriber, J.M. (1996) A new 'cold pocket' hypothesis to explain local host preference shifts in *Papilio canadensis*. *Entomologia Experimentalis et Applicata*, **80**, 315–319.

Seneviratne, S.I., Donat, M.G., Mueller, B. & Alexander, L.V. (2014) No pause in the increase of hot temperature extremes. *Nature Climate Change*, **4**, 161–163.

Shingleton, A.W. (2011) Evolution and the regulation of growth and body size. *Mechanisms of Life History Evolution: The Genetics and Physiology of Life History Traits and Trade-offs* (eds T. Flatt & A. Heyland), pp. 43–55. Oxford University Press, Oxford.

Shure, D.J., Mooreside, P.D. & Ogle, S.M. (1998) Rainfall effects on plant–herbivore processes in an upland oak forest. *Ecology*, **79**, 604–617.

Singer, M.C., White, R.R., Vasco, D.A., Thomas, C.D. & Boughton, D.A. (1995) Multi-character ecotypic variation in Edith's checkerspot butterfly. *Genecology and Ecogeographic Races* (eds A.R. Kruckeberg, R.B. Walker & A.E. Leviton), pp. 101–104. AAAS Publication, San Franscisco.

Singer, M.C. & McBride, C.S. (2010) Multitrait, host-associated divergence among sets of butterfly populations: implications for reproductive isolation and ecological speciation. *Evolution*, **64**, 921–933.

Singer, M.C. & Parmesan, C. (2010) Phenological asynchrony between herbivorous insects and their hosts: signal of climate change or pre-existing adaptive strategy? *Philosophical Transactions of the Royal Society B*, **365**, 3161–3176.

Singer, M.C. & McBride, C.S. (2012) Geographic mosaics of species' association: a definition and an example driven by plant–insect phenological synchrony. *Ecology*, **93**, 2658–2673.

Stålhandske, S., Gotthard, K., Posledovich, D. & Leimar, O. (2014) Variation in two phases of post-winter development of a butterfly. *Journal of Evolutionary Biology*, **27**, 2644–2653.

St-Cyr, J., Derome, N. & Bernatchez, L. (2008) The transcriptomics of life-history trade-offs in whitefish species pairs (*Coregonus* sp.). *Molecular Ecology*, **17**, 1850–1870.

Stearns, S.C. (1992) *The Evolution of Life Histories*. Oxford University Press, Oxford.

Stefanescu, C., Carnicer, J. & Peñuelas, J. (2011a) Determinants of species richness in generalist and specialist Mediterranean butterflies: the negative synergistic forces of climate and habitat change. *Ecography*, **34**, 353–363.

Stefanescu, C., Torre, I., Jubany, J. & Páramo, F. (2011b) Recent trends in butterfly populations from north-east Spain and Andorra in the light of habitat and climate change. *Journal of Insect Conservation*, **15**, 83–93.

Stone A.C., Gehring C.A. & Whitham T.G. (2010) Drought negatively affects communities on a foundation tree: Growth rings predict diversity. *Oecologia*, **164**, 751–761.

Storz, J.F. & Zera, A.J. (2011) Experimental approaches to evaluate the contributions of candidate protein-coding mutations to phenotypic evolution. *Molecular Methods for Evolutionary Genetics* (eds V. Orgogozo & M.V. Rockman), pp. 377–396. Humana Press, New York.

Suggitt, A.J., Stefanescu, C., Páramo, F., Oliver, T., Anderson, B.J., Hill, J.K., Roy, D.B., Brereton, T., Thomas, C.D. (2012) Habitat associations of species show consistent but weak responses to climate. *Biology Letters*, **8**, 590–593.

Sunday, J.M., Bates, A.E., Kearney, M.R., Colwell, R.K., Dulvy, N.K., Longino, J.T., & Huey, R.B. (2014) Thermal-safety margins and the necessity of thermoregulatory behavior across latitude and elevation. *Proceedings of the National Academy of Sciences of the United States of America*, **111**, 5610–5615.

Tack, A.J.M., Mononen, T. & Hanski, I. (2015) Increasing frequency of low summer precipitation synchronizes dynamics and compromises metapopulation stability in the Glanville fritillary butterfly. *Proceedings of the Royal Society of London B: Biological Sciences*, **282**, 20150173.

Taylor, M.L., Skeats, A., Wilson, A.J., Price, T.A.R. & Wedell, N. (2015). Opposite environmental and genetic influences on body size in North American *Drosophila pseudoobscura*. *BMC Evolutionary Biology*, **15**, 51.

Telonis-Scott, M., Hoffmann, A.A. & Sgrò, C.M. (2011) The molecular genetics of clinal variation: a case study of *ebony* and thoracic trident pigmentation in *Drosophila melanogaster* from eastern Australia. *Molecular Ecology*, **20**, 2100–2110.

Thomas, C.D., Bodsworth, E.J., Wilson, R.J., Simmons, A.D., Davies, Z.G., Musche, M. & Conradt, L. (2001) Ecological and evolutionary processes at expanding range margins. *Nature*, **411**, 577–581.

Trotter R.T., Cobb, N.S. & Whitham, T.G. (2010) Arthropod community and trophic structure: a comparison between extremes of plant stress. *Ecological Entomology*, **33**, 1–11.

Tylianakis, J.M., Didham, R.K., Bascompte, J. & Wardle, D.A. (2008). Global change and species interactions in terrestrial ecosystems. *Ecology Letters*, **11**, 1351–1363.

Umina, P.A., Weeks, A.R., Kearney, M.R., McKechnie, S.W. & Hoffmann, A.A. (2005) A rapid shift in a classic clinal pattern in *Drosophila* reflecting climate change. *Science*, **308**, 691–693.

Välimäki, P., Kivelä, S.M., Mäenpää, M.I. & Tammaru, T. (2013) Latitudinal clines in alternative life histories in a geometrid moth. *Journal of Evolutionary Biology*, **26**, 118–129.

van Asch, M., Salis, L., Holleman, L.J.M., van Lith, B. & Visser, M.E. (2013) Evolutionary response of the egg hatching date of a herbivorous insect under climate change. *Nature Climate Change*, **3**, 244–248.

WallisDeVries, M.F., Baxter, W. & Van Vliet, A.J.H. (2011) Beyond climate envelopes: effects of weather on regional population trends in butterflies. *Oecologia*, **167**, 559–571.

Wandeler, P., Hoeck, P.E.A. & Keller, L.F. (2007) Back to the future: museum specimens in population genetics. *Trends in Ecology & Evolution*, **22**, 634–642.

Warren, M.S., Hill, J.K., Thomas, J.A., Asher, J., Fox, R., Huntley, B., Roy, D.B., Telfer, M.G., Jeffcoate, S., Harding, P., Jeffcoate, G., Willis, S.G., Greatorex-Davies, J.N., Moss, D. & Thomas, C.D., (2001) Rapid responses of British butterflies to opposing forces of climate and habitat change. *Nature*, **414**, 65–69.

Watt, W.B. (1992) Eggs, enzymes, and evolution: natural genetic variants change insect fecundity. *Proceedings of the National Academy of Sciences of the United States of America*, **89**, 10608–10612.

Watt, W.B. & Dean, A.M. (2000) Molecular-functional studies of adaptive genetic variation in prokaryotes and eukaryotes. *Annual Review of Genetics*, **34**, 593–622.

Wernegreen, J.J. (2012) Endosymbiosis. *Current Biology*, **22**, R555–R561.

Wheat, C.W., Fescemyer, H.W., Kvist, J., Tas, E., Vera, J.C., Frilander, M.J., Hanski, I. & Marden, J.H. (2011) Functional genomics of life history variation in a butterfly metapopulation. *Molecular Ecology*, **20**, 1813–1828.

Wheat, C.W. & Hill, J. (2014) Pgi: the ongoing saga of a candidate gene. *Current Opinion in Insect Science*, **4**, 42–47.

Wiklund, C. & Friberg, M. (2011) Seasonal development and variation in abundance among four annual flight periods in a butterfly: a 20-year study of the speckled wood (*Pararge aegeria*). *Biological Journal of the Linnean Society*, **102**, 635–649.

Williams, J.W. & Jackson, S.T. (2007). Novel climates, no-analog communities, and ecological surprises. *Frontiers in Ecology and the Environment*, **5**, 475–482.

Williams, S.E., Shoo, L.P., Isaac, J.L., Hoffmann, A.A. & Langham, G. (2008) Towards an integrated framework for assessing the vulnerability of species to climate change. *PLOS Biology*, **6**, e325.

Williams, C.M., Hellmann, J. & Sinclair, B. J. (2012) Lepidopteran species differ in susceptibility to winter warming. *Climate Research*, **53**, 119–130.

Williams, C.M., Henry, H.A.L. & Sinclair, B.J. (2015) Cold truths: how winter drives responses of terrestrial organisms to climate change. *Biological Reviews*, **90**, 214–235.

Wolda, H. (1988) Insect seasonality: why? *Annual Review of Ecology, Evolution and Systematics*, **19**, 1–18.

Woods, H.A. & Smith, J.N. (2010) Universal model for water costs of gas exchange by animals and plants. *Proceedings of the National Academy of Sciences of the United States of America*, **107**, 8469–8474.

Woods, H.A., Dillon, M.E. & Pincebourde, S. (2015) The roles of microclimatic diversity and of behavior in mediating the responses of ectotherms to climate change. *Journal of Thermal Biology*, **54**, 86–97.

Zera, A.J. & Harshman, L.G. (2011) Intermediary metabolism and the biochemical-molecular basis of life history variation and trade-offs in two insect models. *Mechanisms of Life History Evolution: The Genetics and Physiology of Life History Traits and Trade-offs* (eds T. Flatt & A. Heyland), pp. 311–328. Oxford University Press, New York.

17

Conservation of Insects in the Face of Global Climate Change

Paula Arribas[1,2], Pedro Abellán[3], Josefa Velasco[4], Andrés Millán[4] and David Sánchez-Fernández[5,6]

[1] Department of Life Sciences, Natural History Museum, London SW7 5BD, UK
[2] Department of Life Sciences, Imperial College London, Ascot SL5 7PY, UK
[3] Department of Biology, Queens College, City University of New York, Flushing NY 11367, USA
[4] Department of Ecology and Hydrology, University of Murcia, Espinardo 30100, Spain
[5] Institute of Evolutionary Biology, CSIC-University Pompeu Fabra, Barcelona 08003, Spain
[6] Institute of Environmental Sciences, University of Castilla-La Mancha, Toledo 45071, Spain

Summary

Insects, which represent the vast majority of terrestrial biodiversity, are often particularly sensitive to the effects of climate change. The importance of this impact on biodiversity is now at the forefront of science, but there still remains a significant gap between predictions of species vulnerability and conservation measures in the context of global warming, particularly in the case of insects. Nowadays, the accumulated background knowledge on insect science together with the number of recent developments to study and monitor insect diversity, allows entomologists to have a positive impact on the renewal of biodiversity conservation by the development, integration and implementation of specific conservation strategies in the face of the ongoing climate change. In this chapter we address this challenge and revise some of the main areas of insect conservation in the context of the global climate change, such as the identification of the main drivers of insect species vulnerability, the adaption of insect vulnerability assessments and their translation into conservation strategies for species protection, or the role of protected areas to preserve insect diversity in the future. From the review of the most recent bibliography, some landmarks are identified in order to put insects on their place on the international conservation stage facing this impact.

17.1 Introduction

17.1.1 Insect Biodiversity

Insects are the most diverse and successful multicellular organisms on the planet, both in terms of numbers of species and abundance. They represent the vast majority of terrestrial biodiversity, around four-fifths of all metazoans, and are essential to maintain the ecological integrity of most ecosystems. Insects provide ecological services on which humans depend, such as pollination or nutrient cycling on soils, and despite the difficulties to approximate an economic value to those, it has been estimated than the ecological services provided from 'wild insects' in the United States is around \$57 billion per year (Losey & Vaughan, 2006). It is evident that the necessity for insect conservation is not only grounded on the ethical commitment of maintaining multi-scale diversity on Earth, but also on the principle of utility value and dependence for human persistence (Samways, 2005).

The inordinate magnitude of insect diversity brings about a terrible paradox for its conservation: we know that the greatest biodiversity loss in terms of number of species (but also in evolutionary history) is likely to be through the loss of insects (McKinney, 1999; Dunn, 2005), but we still have important gaps in our knowledge of insect biodiversity. We do not know how many insect species exist and how are they distributed (the so-called *Linnaean* and *Wallacean shortfalls*; Whittaker et al., 2005), or how they will respond to the changing world (Samways, 2007). Additionally, insect conservation has suffered from a general neglect (sometimes even disdain) by society, biodiversity managers and even researchers, being considered as the 'awkward kid sister' to vertebrate conservation (Dunn, 2005). But we are now in the final countdown, the ongoing biodiversity crisis, which is undeniably an insect biodiversity crisis where many species could be extinct even before they are discovered (Costello et al., 2013). Insect conservation is at a crossroads, a time of promise and peril (Spector, 2008).

17.1.2   Insect Biodiversity and Climate Change: the Research Landscape

Climate change is expected to become one of the greatest drivers of insect biodiversity loss, with impacts on species' ranges, phenology, physiology and trophic and competitive interactions already widely documented (e.g., Roy & Sparks, 2000; Pateman et al., 2012; Fox et al., 2014). Like many ectotherms, insects are often particularly sensitive to the direct effects of climate change (e.g., Thomas et al., 2004; Deutsch et al., 2008; Aragón et al., 2010). The greatest concern, however, is that climate change could be affecting insects in a synergic way with other adverse factors (Brook et al., 2008), mainly habitat loss, resulting in a 'deadly anthropogenic cocktail' for biodiversity (Travis, 2003). On the information available until now it is difficult to calibrate the total proportion of insects that might be threatened by global climate change, but the indications coming from sparse species or groups such as butterflies, moths or water beetles (e.g., Warren et al., 2001; Thomas et al., 2011; Sánchez-Fernández et al., 2013) show that some of them are in a tough situation, even on a downward trajectory toward extinction (Biesmeijer et al., 2006; Spector, 2008; Hernández-Manrique et al., 2013).

Conservation biology, and its application, needs to be revised and improved to address the emerging requirements of this global threat. Previous conservation tools and principles must to be revisited to take into account how species are able to respond to the changing scenario (Lawler, 2009). The importance of this impact is now at the forefront of science, as is reflected in the exponential number of studies in last decade estimating and evaluating the climate change effects on species (e.g., Thomas et al., 2004; Guisan & Thuiller, 2005; Pörtner & Farrell, 2009), even in the case of the insects (e.g., Deutsch et al., 2008; Gillingham et al., 2012). Nevertheless, multiple authors have stated that the translation of such information for biodiversity conservation purposes is delayed, and there still remains a significant gap between predictions of species vulnerability and conservation in the context of global warming (Heller & Zavaleta, 2009; Dawson et al., 2011). Figure 17.1 shows the number of scientific publications produced per year since 1990 as obtained from two systematic searches on the ISI Web of Knowledge combining the topics 'biodiversity', 'climate change' and 'conservation' together and also adding the string 'insect'. Such systematic searches, despite not being exhaustive, give some indication of the temporal evolution of publications addressing biodiversity and conservation in a

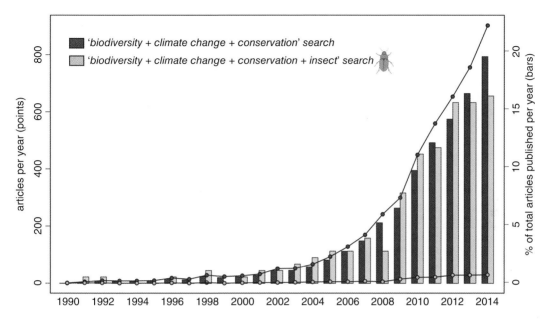

Figure 17.1 Temporal evolution of scientific publications combining topics 'Biodiversity' + 'Climate Change' + 'Conservation' and adding the string 'Insect', respectively. Points: number of publications per year for each search. Bars: proportion of the total publications of each search per year. Each search was performed on the ISI Web of Knowledge, including the Science Citation Index Expanded, Social Science Citation Index, Arts and Humanities Citation Index, and Conference Proceedings Citation Index–Science databases from 1990 to 2014. For the topic 'climate change' and 'conservation' we used a set of synonymous words in all possible combinations, that is, (climate change OR global warming OR climatic change OR climate-change OR changing climate) and (conservation OR adaption OR management OR restoration OR planning OR reserve design OR strategy OR land-use OR land use OR landscape OR protected area OR park). Please, note that the term 'biodiversity' was included in these searches to minimise the number of articles exclusively focussed on ecosystem services without an explicit mention of biodiversity.

climate change context. Moreover, from a comparative perspective, it allows us to estimate the magnitude and trend of insect diversity studies relative to the broader field that covers other groups. The number of articles matching these topics showed an astonishing exponential growth, particularly in the last five years, up to more than 900 publications in 2014. Unfortunately, those specifically linked with insects represent a much more reduced number, around a 4% (i.e., 29 publications in 2014), again evidencing the 'awkward kid sister' syndrome cited before. A positive sign is that, despite still being in the minority, papers concerning insects follow the trend for accelerated rates of publication seen for the subject more generally in recent years (see Fig. 17.1 for details). Besides, some of the key publications focusing on the development and adaption of species vulnerability categorisations and conservation strategies under climate change have been developed in insect-based systems (e.g., Kuchlein & Ellis, 1997; Warren et al., 2001; Thomas et al., 2011) emphasising their value as a model system to invigorate conservation biology.

Nowadays, the accumulated background knowledge on insect conservation together with a number of recent developments to study and monitor insect diversity allows entomologists to have a positive impact on biodiversity conservation. But such challenges should be mediated by the development, integration and implementation of insect conservation strategies in the face of the ongoing climate change. In this chapter we address this challenge and, as a complementary view to previous chapters on the different responses of invertebrates to climate change, we interpret this information in the context of conservation. Particularly, we revise some of the main areas of insect conservation in the context of the global climate change, such as the identification of the main drivers of insect species vulnerability, the adaption of insect vulnerability assessments and their translation into conservation strategies for species protection, and also the role of protected areas to preserve insect diversity in future. From the discussion of some of the recent bibliography, some landmarks are identified to try to put insects on their place on the international conservation stage facing this impact.

17.2   Vulnerability Drivers of Insect Species Under Climate Change

The vulnerability of a species to global warming will depend on its capacity: (i) to maintain present populations (species persistence or resistance) and (ii) to shift its geographical range to suitable future environments, that ultimately are related with their ecological niche breadth (and their plasticity) and dispersal capacity.

The species with broader niche breadth and/or a higher potential to respond via plastic and adaptive changes will be more able to respond to the new conditions imposed by climate change and so, their vulnerability should be lower to such environmental change (Chown & Gaston, 2008; Hoffmann, 2010). Among the multiple dimensions of species ecological niches, thermal tolerance traits are of key importance to evaluate species vulnerability to changing climate (Chown et al., 2004; Chown, 2012). This is particularly true in the case of insects where, like in other small ectotherms, temperature strongly affects individual performance (rates of locomotion, growth and feeding) and fitness (survival, reproductive rate and generation time; Chown & Nicolson, 2004). Thermal tolerance of species have shown strong phylogenetic signals in several taxonomic groups (e.g., Addo-Bediako et al., 2000; Buckley & Kingsolver, 2012; Diamond et al., 2012) potentially indicating a non-random distribution of species' vulnerability to changing temperatures across the tree of life, while a general conservationism in tolerance to heat has been reported for ectotherms (Chown et al., 2002; Araújo et al., 2013). Regarding the geographical distribution of the thermal vulnerability of insects, Deutsch et al. (2008) identified that species at higher latitudes have broader thermal tolerance and are living in climates that are currently cooler than their physiological optima, so that warming may enhance their fitness. In contrast, the warming in the tropics (despite being smaller in magnitude) is likely to have the most deleterious consequences because tropical insects are relatively sensitive to temperature change and are currently living very close to their optimal temperature.

Concerning thermal plasticity in insects, a generalized low potential for thermal acclimation has been reported, although different studies have found positive and inverse relationships among the species' upper thermal limit and acclimation capacity in invertebrates (e.g., Stillman, 2003; Calosi et al., 2008; Sánchez-Fernández et al., 2010). Further, behavioral thermoregulation has also arisen as an essential way to adapt to changing conditions which could be strongly mediated by the habitat and its heterogeneity in microclimates (Kearney et al., 2009; Gunderson & Stillman, 2015).

On the other hand, for many species of insects the primary impact of climate change could be mediated through effects on synchrony with habitat resources (Parmesan, 2006). Phenological shifts in development time, diapause, hatching, and emergence of organisms have been recorded for many species (e.g., Altermatt, 2009, 2010; Bradshaw & Holzapfel 2010), producing frequent asynchrony in predator–prey and insect–plant systems with negative consequences (Visser & Both, 2005;

Kingsolver et al., 2011; Hentley & Wade, this volume). Thus, species with a wider or more variable ecological niche (habitat requirements or diet breadth) will be less vulnerable to changes in the distribution or phenology of preys or host species (Buckley & Kingsolver, 2012).

Phenotypic and genetic variation in such physiological and biological traits provides the potential to generate selective responses to climate change (Davis et al., 2010; Somero, 2010). There are still important uncertainties about the real magnitude of adaptive evolution under rapid and ongoing climate change, but the high population size, reproductive capacity, growth rates and short generation time of many insect species could result in greater potential for adapting to new conditions for this group (Buckley & Kingsolver, 2012). Different studies have highlighted the important role of phenotypic plasticity in driving arthropod vulnerability under climate change (Chown, 2001; Chown et al., 2007). On the other hand, most of the empirical evidence for rapid adaptation to climate change comes from examples of local adaption (within species' ranges) toward higher frequencies of existing heat-tolerant genotypes (e.g., Rodríguez-Trelles & Rodríguez, 1998; Rodríguez-Trelles et al., 1998; Levitan, 2003; Bradshaw & Holzapfel, 2006).

A species' capacity to shift its geographical range to suitable future environments will depend on both its dispersal ability and habitat availability. Dispersal ability within and across habitats will determine a species' ability to track changing climate, given that greater dispersal can enable more pronounced range shifts (Pearson & Dawson, 2003), that is, the more mobile butterflies shifted their ranges further than poor dispersal species (Pöyry et al., 2009). Complete lineages of insects are more vulnerable to climate change that others via their intrinsic differences in dispersal potential (e.g., wingless insect groups). Differences can be also particularly exacerbated among the species of different habitats if dispersal potential has been constrained by the persistence and stability of the habitat (e.g., lotic species with lower dispersal ability than lentic species in the aquatic environment; Abellán et al., 2009; Arribas et al., 2012b; Sánchez-Fernández et al., 2012). Range movements can be also strongly affected by habitat fragmentation, and for instance, in many cases even good dispersers could be unable to track the changing climate (Hoegh-Guldberg et al., 2008). This is especially critical for insects confined to naturally patchy environments, such as those on inland saline waters (Sánchez-Fernández et al., 2011; Arribas et al., 2013).

17.3 Assessment of Insect Species Vulnerability to Climate Change

Classical measures of species vulnerability such as the IUCN *Red List of Threatened Species* allocate species to categories of extinction risk using quantitative rules based on sizes and decline rates of both population and distributional range areas (Akçakaya et al., 2006; Rodrigues et al., 2006). In the current context of climate change impacts on species, further extensions of these measures to include drivers of species vulnerability under climate change are needed to improve our species vulnerability estimations and categorizations (Dawson et al., 2011; Foden et al., 2013). In this framework, most of the proposed adaptations of vulnerability categorizations are based on the incorporation of the geography of climate, identifying where, how and how fast are changing the conditions which each species inhabits (Williams et al., 2007; Loarie et al., 2009). As an example on the IUCN Red List on invertebrates, Cardoso et al. (2011) propose that the incorporation of this information to the estimation of the Area of Occupancy or Extent of Occurrence criteria is of great interest to the adaption of the IUCN Red List to consider climate change. Species Distribution Models (SDMs) have been the main tool to estimate the species vulnerability to climate change and nowadays, there is an extensive literature regarding different type of SDMs that can be used, including their potential benefits and pitfalls (see Elith & Leathwick, 2009 for a review).

Despite the useful information reported by most of the SDMs on the geographical change of the climate conditions inhabited by each species, the translation of this information to species vulnerability measures is controversial because these predictions largely ignore many of the other drivers of vulnerability, which could result in significant prediction errors, that is, biological and physiological differences between species may significantly increase or reduce their vulnerability (Dawson et al., 2011; Foden et al., 2013).

Recently, integrative approaches combining SDMs with the information on multiple species traits and their plasticity have arisen to provide a more solid framework for the evaluation of species vulnerability under climate change (e.g., Williams et al., 2008; Kearney & Porter, 2009; Dawson et al., 2011; Buckley & Kingsolver, 2012; Bellard et al., 2012; Magozzi & Calosi, 2014). Publications focused on insects have highlighted the importance of these integrated vulnerability approaches as a fundamental tool to understand species responses to climate change and to adapt conservation strategies (see next section for further details). Thomas et al. (2011), using insects among other groups, proposed a framework to understand the risk of climate change on individual species, separating the threats and benefits of climate change for them. Threats were assessed by the level of climate-related decline within a species' recently occupied historical distribution, based on observed and/or projected changes. Benefits are assessed in terms of observed and/or projected increases outside the recently occupied historical range. Each species is then placed in one of six categories (high risk, medium risk, limited impact, equivalent risks and benefits, medium benefit, high benefit). The framework included consideration of questions around data adequacy and uncertainties that apply to future climatic changes and the accompanying biological responses that should help to prioritize resource allocation in measures that could facilitate the future recovery/spread of the most vulnerable species. As another example, Arribas et al. (2012a) focused on three endangered species of Iberian saline water beetles and combining measures of thermal tolerance and dispersal capability of each species with bioclimatic models, they proposed a framework to evaluate the different drivers of vulnerability under climate change for each species, with the ultimate goal of guiding conservation strategies. Results demonstrated that even species with similar habitat preferences and range size could be affected by climatic warming in very different ways, and so, the proposed conservation measures must be different (Fig. 17.2).

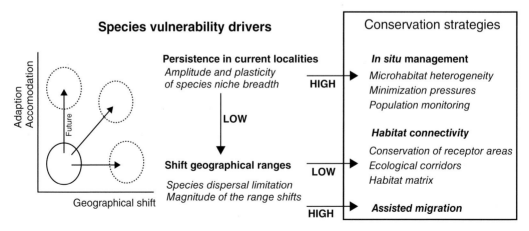

Figure 17.2 Drivers of species vulnerability under climate change and its link with each specific type of conservation strategy to mitigate the impact on species. Modified from Arribas et al. (2012a).

17.4 Management Strategies for Insect Conservation Under Climate Change

After a review of climate-change adaptation strategies for wildlife management and biodiversity conservation, Mawdsley et al. (2009) concluded that the community of wildlife and natural resource managers already possesses many of the conservation tools that are necessary to help species to adapt to climate change. For instance, the recommendations to address climate change impacts on biodiversity include a wide variety of classical strategies of conservation such as (i) the monitoring of species, (ii) the reduction of additional threats, (iii) restoring habitats, (iv) increasing connectivity between suitable habitats, (v) expanding reserve networks (see section below) and (vi) performing assisted dispersal (for a review see Heller & Zavaleta, 2009). Unfortunately, the conservation recommendations provided by most of the studies on climate impacts on species, if present, are general or ambiguous, making their implementation difficult within current management strategies (Heller & Zavaleta, 2009; Carvalho et al., 2010; Mastrandrea et al., 2010).

A few studies on insect species vulnerability under climate change have proposed that the integrated species vulnerability estimations (see previous section for extended information) could potentially play a fundamental role to bridge this gap (e.g., Thomas et al., 2011; Arribas et al., 2012a). Despite being mainly focussed on insects, the main concept is applicable to most species: a disclosed estimate of the species drivers of vulnerability under climate change allows for a more direct translation to management through the design of specific conservation strategies on those specific determinants of risk (Fig. 17.2).

Probably the main information requirement for the management of a particular species is whether it will be able to adapt to future climate conditions without the need to disperse or, if not, if the species will be able to disperse. In the first case, for species showing high capacities to deal with future climate conditions without the need to disperse, the concentration of conservation efforts in actual localities (i.e., in situ management or conservation at local scale), could be the most efficient and practical strategy. Thus, protection and conservation measures should therefore be focused on the maintenance of current populations and the minimization of other threats. In this sense, the habitat scale management in the case of insects, with measures to maintain and maximize microclimatic environments and to sustain habitat quality and extension could be fundamental to improve the persistence of populations in current localities under climatic warming (Dunn, 2005). As an example, Arribas et al. (2012a) proposed the creation of shading areas and the control of freshwater inputs as potentially essential measures to maintain current populations of the endemic water beetle *Enochrus jesusarribasi* in the saline streams of southeastern Spain. Special attention should be given to monitoring populations at high risk, or of special value, so that management interventions can occur if the risks to habitats or species increase significantly over time. For this to happen, it is essential to enhance the environmental monitoring of changes and species responses coupled with the development of local scenario-building exercises that take land (and water) use into account (see Palmer et al., 2009). Carvalho et al. (2010) proposed that the monitoring parameters for amphibians (also applicable to insects) should consider, when possible, physiological changes in thermal tolerances, phenological adjustments, behavioural thermoregulation changes, such as burrowing or adjustments in daily activity periods, quantification of dispersal rates and specially changes in population parameters.

When there is a substantial reduction or displacement in future suitable habitat with respect to present, reserves alone and/or management at local scale are unlikely to be enough to maintain insect species (Kuchlein & Ellis, 1997) and species will need to move to geographically novel and more suitable areas (Fig. 17.3A). In these cases, the potential for range shifts (i.e., species dispersal capacity)

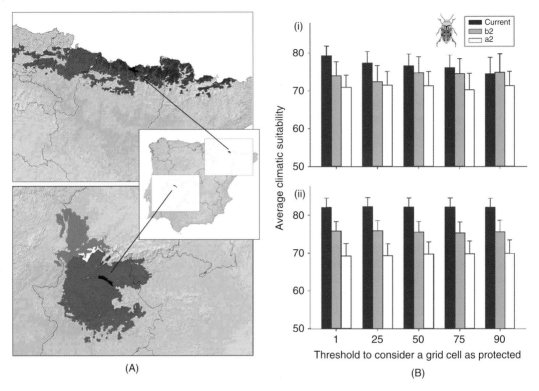

Figure 17.3 Protected areas and climate change. (A) Change in climatic conditions of two Spanish National Parks (Monfrague and Ordesa) in the future. Black areas: current extension of the National Parks. Dark grey areas: similar climatic conditions to the reserve in the present. Light grey areas: similar climatic conditions to the reserve in the future. White areas: similar climatic conditions to reserve in both present and future. Modified from Lobo (2011). B) Average climatic suitability of the potential distribution of endemic water beetles overlapping with both National Parks (i) and Natura 2000 (ii) networks in the Iberian Peninsula using different thresholds to consider a cell as protected. Bars: potential distribution as estimated for present (black bars) and future climatic conditions for the scenarios b2 (grey bars) and a2 (white bars). Modified from Sánchez-Fernández et al. (2013).

should determine which specific conservation strategies to apply (Fig. 17.2). If species are able to disperse to these new habitats, it is especially important to consider the values of sites which are not climatically suitable at present, but may be in the future, and if these can be selected and secured as a component of a current conservation plan (Bennett, 1999; Samways, 2007). Besides, as large and high-quality habitats provide source populations and locations for colonisation, the key focus for conservation should such be the retention of as much high-quality natural and semi-natural habitat as possible (Hodgson et al., 2011). However, in the case of insects, evidence suggests that it is only the more mobile generalist butterflies (Dennis, 1993; Parmesan et al., 1999), dragonflies (Aoki, 1997) or water beetles (Sánchez-Fernández et al., 2012) that are tracking climatic suitability and so special attention to this trait should be paid when evaluating the habitat connectivity for each insect group. For species with low-dispersal ability, an important increase in suitable habitat connectivity would be required through the conservation and restoration of ecological corridors (Krosby et al., 2010). In other words, it is necessary to create a network of habitats within and between present and future suitable areas, especially if they occur in highly fragmented habitats. Furthermore, the management of the spatial arrangement of habitat and matrix characteristics, together with increases in protected habitat area and habitat quality could also be effective conservation measures, as they serve to improve connectivity (Hodgson et al., 2009, 2011). Different studies have shown how insects move along corridors of remnant indigenous vegetation (e.g., Sutcliffe & Thomas, 1996; Haddad, 1999; Slade et al., 2013) and so potentially could mitigate the effects of climate change on different insect species. But the application of these strategies is not always synonymous with success. Most corridors have been designed for other (conspicuous) taxonomic groups, and since insects are small and speciose, a landscape feature that is beneficial for large mammals will not necessarily be beneficial for a particular insect species group. Indeed, Öckinger and Smith (2008) demonstrated that corridors do not always have positive effects on insect dispersal and that the effect seems to depend on the quality of the surrounding matrix, on the spatial scale in which the study is performed and on whether true dispersal or routine movements are considered.

In extreme situations (i.e., when climate change might alter habitat suitability of the areas that a species is currently occupying, and species are unable to migrate) human intervention using species translocation as a conservation strategy (Thomas, 2011), alternatively termed 'assisted migration' (McLachlan et al., 2007) or 'assisted colonisation' (Hoegh-Guldberg et al., 2008; Mueller & Hellmann, 2008), may become necessary. The 2013 IUCN guidelines define assisted colonization in broad terms as the intentional movement of an organism outside its indigenous range to avoid extinction of populations due to current or future threats (IUCN, 2013). The associated uncertainties and risks of this measure have triggered intense debate (e.g., Ricciardi & Simberloff, 2009; Schlaepfer et al., 2009; Loss et al., 2010; Hewitt et al., 2011) given the devastating ecological impact wrought by invasive species. In this sense, the 2013 IUCN guidelines place great emphasis on previous feasibility and risk analysis, including ecological and socio-economical risks, as essential components of any conservation translocation. In the same way, Hoegh-Guldberg et al. (2008) developed a decision framework that can be used to outline potential actions under a suite of possible future climate scenarios. They proposed that assisted dispersal should only be undertaken within biogeographic regions and on species with high risk, low dispersal ability or when the species range is highly fragmented. There are in the literature a number of successful cases of translocation of vertebrates, mainly birds and mammals (Seddon et al., 2014). However, well-documented examples involving other animals, such as insects, have remained scarce. There are some studies using butterflies and bioclimatic models to determine species' future habitat suitability for assisted colonization purposes. For example, Carroll et al. (2009) used bioclimatic models of the distributions of two extinct British butterflies, *Aporia crataegi* and *Polyommatus semiargus*, to investigate the potential for re-wilding in Britain. Willis et al. (2008) also used bioclimatic models to guide some of the first experimental assisted colonization of two butterfly species (*Melanargia galathea* and *Thymelicus sylvestris*) to sites ~35 and ~65 km beyond their indigenous range in northern England. The greatest success story among butterfly translocations has been the reintroduction of *Maculinea* (now *Phengaris*) *arion* in several places in Britain in the 1980s and 1990s, which had resulted in >30 local populations by 2008 (Andersen et al., 2014). Recently, Kuussaari et al. (2015) describes a successful translocation of the threatened Clouded Apollo butterfly (*Parnassius mnemosyne*) in Finland, compares this to a specific failed translocation, and presents interesting conclusions for conservation planning as specific factors contributing to the success.

17.5 Protected Areas and Climate Change

Further evaluations of vulnerability and conservation strategies focussed on particular species; the establishment of Protected Area (PA) systems represents one of the mainstays of worldwide conservation polices and is therefore central to current efforts to avoid biodiversity loss (Chape et al., 2005).

Numerous species (including insects) are highly dependent on PA for their continued persistence, occurring either entirely or largely within their bounds (Jackson & Gaston, 2008).

The locations of PAs are geographically fixed, and conservation strategies are often based on the implicit assumption of a relatively stable climate and that biological attributes are inextricably linked to place. However, the climatic conditions within PA are changing, unleashing a cascade of alterations in habitats and biological communities. Further, as species' spatial distributions track shifting environmental conditions beyond PAs, reserves may fail to retain species within their boundaries if conditions are no longer suitable for those species (Fig. 17.3A). In other words, a static notion of conserving communities and ecosystems may soon be obsolete (Hannah et al., 2002; Hannah, 2010). While a PA may lose a number of its current species, a new suite of species may shift into the protected area as individuals colonize areas of the PA, leading to species turnover in established PAs (e.g., Hannah et al., 2007; Jenkins & Joppa, 2009). In this context, it is important to anticipate how changes in climate might affect the value of existing PAs (Thomas & Gillingham, 2015), and design dynamic conservation strategies in order to accommodate these changes in climate that present evidence suggests are inevitable (Hannah et al., 2002; Hole et al., 2009; Hannah, 2010).

Contemporary changes in species ranges consistent with global warming have already been observed in a number of species, communities, and ecosystems within PAs. While such sort of information is mostly restricted to plants and vertebrates (e.g., Pounds et al., 1999; Moritz et al., 2008), there is also evidence of changes within PAs for some insects (see references below). Because PAs across the world are often disproportionately found in mountain ranges and at relatively high elevations (Joppa & Pfaff, 2009), which are (and are expected to be in the future) greatly impacted by climate change, a great proportion of insect biodiversity in these PAs is likely to be especially threatened. This is the case, for instance, of the stonefly *Zapada glacier*, endemic to alpine streams of Glacier National Park (Montana, USA), whose extremely restricted historical distribution has been further reduced over the past several decades in this park by an upstream retreat to higher, cooler sites as water temperatures increased and glacial masses decreased (Giersch et al., 2015). In the same way, Chen et al. (2009) estimated that the average altitudes of 102 montane moth species have increased over the last decades in Mount Kinabalu National Park in Borneo.

At the same time, losses from some PAs are offset by increases in others (Thomas & Gillingham, 2015). For instance, in Britain, Thomas et al. (2012) found that 256 species across eight invertebrate groups disproportionally used PAs in newly colonized areas and Gillingham et al. (2015) found that some species of butterflies and dragonflies were also significantly more abundant inside PAs in newly colonized parts of their range. On the other hand, large, mountainous PAs provide opportunities for species to take advantage of cooler conditions at higher elevations, or at different mountain aspect or slopes, without being displaced out of the reserve. An example is the silver-spotted skipper butterfly, *Hesperia comma*, which was historically restricted to sparse vegetation on south-facing hillsides in lowland Britain, the majority of which fall into PAs, and after a period of regional warming had colonized a wider range of aspects, including east-, west- and north-facing hillsides (Thomas et al., 2001).

In addition to the examples of range shifts already observed for some species within PAs, scientists attempt to anticipate how climate change might affect the value of existing reserves. Typically, these assessments are based on projections of species' future distributional shifts using SDMs (although alternative approaches may evaluate, independently of any single species, how existing PAs are represented and distributed in the current climate space of a region and how this representation and distribution may change in the future; e.g., Scott et al., 2002; Wiens et al., 2011). Maps of species ranges predicted under future climate scenarios are compared with the distribution of theoretical and/or real-world reserves, which allow evaluations of the extent to which PAs will cover the biodiversity under climate change. Recent efforts in this area have involved a wide range of taxa and geographical regions (e.g., Araújo et al., 2004, 2011; Hannah et al., 2007; Monzón et al., 2011; Thomas & Gillingham, 2015). The number of studies focused on insects is rather limited. Such a scarcity of insect-based evidence has been highlighted by Sieck et al. (2011), who reviewed the literature to determine the state of modelling of terrestrial protected areas under climate change and found limited research involving insect species. Vos et al. (2008) combined bioclimate envelope models with dispersal models on several species, including the large heath butterfly *Coenonympha tullia*, and predicted a decline in the amount of suitable habitat protected in Natura 2000 sites. More recently, other studies mainly focused on forest Lepidoptera, have achieved this task. Kharouba and Kerr (2010) working on butterfly species of Canada showed that the expected changes in species richness and composition within PA were generally the same as changes observed among random areas outside reserve boundaries, which suggest that existing protected area networks have provided little buffer against the impacts of climate change on these species. In a recent study, Ferro et al. (2014) evaluated how effective PAs in the Brazilian Atlantic Forest are in maintaining the diversity of tiger moths (Arctiinae) under climate change, showing that estimates of species turnover from current to future climate tended to be high, and most of PAs were predicted to lose more species than expected by random. Finally, Sánchez-Fernández et al. (2013) assessed the extent to which PAs in the Iberian Peninsula (mainland Spain and Portugal) cover, and will cover in the future, parts of the climatic potential distribution of water beetles close to their climatic optimums. They found that for most of them the protected networks tend to include areas with climatic conditions close to the species tolerance limit, and the expected climate change only worsened this scenario (Fig. 17.3B).

In general, evidence provided by studies assessing the extent to which PAs will cover the biodiversity in general, and the diversity of insects in particular, in the future shows a limited role of these conservation areas under climate change. Hence, it is important to modify our biodiversity protection strategies in order to mitigate the impacts of climate change. Conservation adaptation is an important part of that new focus, also concerning PAs (Mawdsley et al., 2009; Hannah, 2010). New PAs will be needed to respond to climate change and add protection to balance the projected losses in taxa, combined with dynamic spatial elements including all other areas within the landscape matrix where land use may change over time. Such a strategy should also include the design of new natural areas and restoration sites to maximize resilience (Lovejoy, 2005), the protection of refugia (areas with minimal climate impacts), movement corridors or stepping stones for wildlife dispersal and the improvement of the matrix by increasing landscape permeability to species movement (Mawdsley et al., 2009). Furthermore, management plans need to be established that cross national, regional, and continental boundaries, as the dynamics of climate change transcend national boundaries (Hannah, 2010). Finally, for a further evaluation and redesign of PAs in the face of climate change it is important to take into account all the additional drivers of species vulnerability under climate change (see previous sections) but also other factors such as the growing human population densities, intensified land-use, invasive species, etc., which are expected to dynamically influence and negatively impact the performance of PAs (Menéndez, 2007; Vos et al., 2008; Kleinbauer et al., 2010; Körner et al., 2010).

17.6   Perspectives on Insect Conservation Facing Climate Change

From the review of the literature on species conservation under climate change it is evident that insects still play a marginal role, especially when taking into account the importance of insects in terms of potential global biodiversity loss. However, some of the leading studies directly addressing the modification of vulnerability estimations to incorporate climate change threats and the use of this information to evaluate and design conservation strategies have been developed on insect groups. These studies, which were largely published in recent years, constitute a 'proof of principle' of the potential relevance of insects as model system to study the impact under climate change and its translation for conservation biodiversity, highlighting many of their positive traits for the study of this topic, such as their bioindicator value, broad diversity and adaption capacity or their potential to perform experimental approaches. It is also true that these studies are biased to a few insect groups, mainly butterflies and water beetles, from regions with a relatively broad knowledge regarding their distributions and population dynamics. As a consequence, one of the main limitations for the development of insect conservation under climate change (as in general for insect conservation) is the lack of knowledge regarding most of insect species and the inherent uncertainties that this causes. The accumulated background knowledge on insect conservation available nowadays (even including some scientific journals dedicated to this area), together with a number of recent developments (such as the application of Next Generation Sequencing to unveil the diversity of previously unapproachable insect communities), should allow us to overcome this limitation in the next few years. Entomologists can now generate more information on the status, biology and needs of insect species, and are better able to deliver it to decision makers and the broader public than ever before. Extended information about the drivers of species vulnerability for different insect species should be the key that managers need for a better link between potential response of insect species to climate change and specific management strategies (i.e., adaptation measures) that could ameliorate adverse effects of climate change on biodiversity.

Complementing this, novel management tools that promote flexible decision making such as adaptative management are also fundamentally important approaches for insect conservation. This would allow wildlife managers to incorporate scientific information (and associated uncertainties) when available, and to learn from the results of their current management activities. By using monitoring programmes to collect data on project effectiveness practitioners can then refine and adjust future management strategies on insect species. In a highly motivational paper, Spector (2008) emphasised the opportunity for insect conservation within the current biodiversity crisis. This challenge is mediated by the development, integration and implementation of insect conservation strategies in the face of the ongoing climate change. Most of the tools for this task are now on the table; the effort of researchers, biodiversity managers and policy makers will determine if in the coming years insect conservation can develop its full potential and have a positive impact and contribution to a sustainable future of biodiversity.

Acknowledgements

We thank J.M. Lobo, D.T. Bilton and the members of the Aquatic Ecology research group (Universidad de Murcia, Spain) for providing data, useful comments, discussions and help at several stages. We also express our deep gratitude to Scott Johnson and Hefin Jones for the invitation to contribute with this chapter and their useful revisions on the text. We also thank two anonymous reviewers for their revisions and Philip Smith for proofreading. Financial support was provided by a pre-doctoral grant from the Spanish Ministry of Education (FPU program) and a postdoctoral grant from the Royal Society UK (Newton International program) to P.Ar., a postdoctoral contract from the Spanish Ministry of Economy and Competitiveness (Juan de la Cierva program) and another postdoctoral contract funded by Universidad de Castilla-La Mancha and the European Social Fund (ESF) to D.S-F., as well as projects CGL2006-04159, 023/2007 (A.M.) and CGL2010- 15378 cofinanced with FEDER funds (J.V.).

References

Abellán, P., Millán, A. & Ribera, I. (2009) Parallel habitat-driven differences in the phylogeographical structure of two independent lineages of Mediterranean saline water beetles. *Molecular Ecology*, **18**, 3885–3902.

Addo-Bediako, A., Chown, S.L. & Gaston, K.J. (2000) Thermal tolerance, climatic variability and latitude. *Proceedings of the Royal Society of London B*, **267**, 739–745.

Akçakaya, H.R., Butchart, S.H.M., Mace, J.M., Stuart, S.N. & Hilton-Taylor, C. (2006) Use and misuse of the IUCN Red List Criteria in projecting climate change impacts on biodiversity. *Global Change Biology*, **12**, 2037–2043.

Altermatt, F. (2009) Climatic warming increases voltinism in European butterflies and moths. *Proceedings of the Royal Society of London B*, **277**, 1281–1287.

Altermatt, F. (2010) Tell me what you eat and I'll tell you when you fly: Diet can predict phenological changes in response to climate change. *Ecology Letters*, **13**, 1475–1484.

Andersen, A., Simcox, D.J., Thomas, J.A. & Nash, D.R. (2014) Assessing reintroduction schemes by comparing genetic diversity of reintroduced and source populations: a case study of the globally threatened large blue butterfly (*Maculinea arion*). *Biological Conservation*, **175**, 34–41.

Aoki, T. (1997) Northward expansion of *Ictinogomphus pertinax* (Selys) in eastern Shikoku and western Kinki districts, Japan (Anisoptera: Gomphidae). *Odonatologica*, **26**, 121–133.

Aragón, P., Rodríguez, M.A., Olalla-Tárraga, M.A. & Lobo, J.M. (2010) Predicted impact of climate change on threatened terrestrial vertebrates in central Spain highlights differences between endotherms and ectotherms. *Animal Conservation*, **13**, 363–373.

Araújo, M.B., Alagador, D., Cabeza, M., Nogués-Bravo, D. & Thuiller, W. (2011) Climate change threatens European conservation areas. *Ecology Letters*, **14**, 484–492.

Araújo, M.B., Cabeza, M., Thuiller, W., Hannah, L. & Williams, P.H. (2004) Would climate change drive species out of reserves? An assessment of existing reserve-selection methods. *Global Change Biology*, **10**, 1618–1626.

Araújo, M.B., Ferri-Yáñez, F., Bozinovic, F., Marquet, P.A., Valladares, F. & Chown, S.L. (2013) Heat freezes niche evolution. *Ecology Letters*, **16**, 1206–1219.

Arribas, P., Abellán, P., Velasco, J., Bilton, D.T., Millán, A. & Sánchez-Fernández, D. (2012a) Evaluating drivers of vulnerability to climate change: a guide for insect conservation strategies. *Global Change Biology*, **18**, 2135–2146.

Arribas, P., Andújar, C., Sánchez-Fernández, D., Abellán, P. & Millán, A. (2013) Integrative taxonomy and conservation of cryptic beetles in the Mediterranean region (Hydrophilidae). *Zoologica Scripta*, **42**, 182–200.

Arribas, P., Velasco, J., Abellán, P., Sánchez-Fernández, D., Andújar, C., Calosi, P., Millán, A., Ribera, I. & Bilton, D.T. (2012b) Dispersal ability rather than ecological tolerance drives differences in range size between lentic and lotic water beetles (Coleoptera: Hydrophilidae). *Journal of Biogeography*, **39**, 984–994.

Brook, B.W., Sodhi, N.S. & Bradshaw, C.J.A. (2008) Synergies among extinction drivers under global change. *Trends in Ecology and Evolution*, **23**, 453–460.

Bellard, C., Bertelsmeier, C., Leadley, P., Thuiller, W. & Courchamp, F. (2012) Impacts of climate change on the future of biodiversity. *Ecology Letters*, **15**, 365–377.

Bennett, A.F. (1999) *Linkages in the Landscape: The Role of Corridors and Connectivity in Wildlife Conservation*. UICN, Gland, Switzerland and Cambridge.

Biesmeijer, J.C., Roberts, S.P.M., Reemer, M., Ohlemüller, R., Edwards, M., Peeters, T., Schaffers, A.P., Potts, S.G., Kleukers, R., Thomas, C.D., Settele, J. & Kunin, W.E. (2006) Parallel declines in pollinators and insect-pollinated plants in Britain and the Netherlands. *Science*, **313**, 351–354.

Bradshaw, W.E. & Holzapfel, C.M. (2006). Evolutionary response to rapid climate change. *Science*, **312**, 1477–1478.

Bradshaw, W.E. & Holzapfel, C.M. (2010) Light, time, and the physiology of biotic response to rapid climate change in animals. *Annual Review of Physiology*, **72**, 147–166.

Buckley, L.B & Kingsolver, J.G. (2012) Functional and phylogenetic approaches to forecasting species' responses to climate change. *Annual Review of Ecology, Evolution and Systematics*, **43**, 205–226.

Calosi, P., Bilton, D.T. & Spicer, J.I. (2008) Thermal tolerance, acclimatory capacity and vulnerability to global climate change. *Biology Letters*, **4**, 99–102.

Cardoso, P., Borges, P.A.V., Triantis, K.A., Ferrández, M.A. & Martín, J.L. (2011) Adapting the IUCN Red List criteria for invertebrates. *Biological Conservation*, **144**, 2432–2440.

Carroll, M.J., Anderson, B.J., Brereton, T.M., Knight, S.J., Kudrna, O. & Thomas, C.D. (2009) Climate change and translocations: The potential to re-establish two regionally-extinct butterfly species in Britain. *Biological Conservation*, **142**, 2114–2121.

Carvalho, S.B., Brito, J.C., Crespo, E.J. & Possingham, H.P. (2010) From climate change predictions to actions - conserving vulnerable animal groups in hotspots at a regional scale. *Global Change Biology*, **16**, 3257–3270.

Chape, S., Harrison, J., Spalding, M. & Lysenko, I. (2005) Measuring the extent and effectiveness of protected areas as an indicator for meeting global biodiversity targets. *Philosophical Transactions of the Royal Society B*, **360**, 443–455.

Chen, I.-C., Shiu, H.-J., Benedick, S., Holloway, J.D., Chey, V.K., Barlow, H.S., Hill, J.K. & Thomas, C.D. (2009) Elevation increases in moth assemblages over 42 years on a tropical mountain. *Proceedings of the National Academy of Sciences of the United States of America*, **106**, 1479–1483.

Chown, S.L. (2001) Physiological variation in insects: hierarchical levels and implications. *Journal of Insect Physiology*, **47**, 649–660.

Chown, S.L. (2012) Trait-based approaches to conservation physiology: forecasting environmental change risks from the bottom up. *Philosophical Transactions of the Royal Society B*, **367**, 1615–1627.

Chown, S.L., Addo-Bediako, A. & Gaston, K.J. (2002) Physiological variation in insects: large-scale patterns and their implications. *Comparative Biochemistry and Physiology B: Biochemistry and Molecular Biology*, **131**, 587–602.

Chown, S.L. & Gaston, K.J. (2008) Macrophysiology for a changing world. *Proceedings of the Royal Society of London B*, **275**, 1469–1478.

Chown, S.L., Gaston, K.J. & Robinson, D. (2004) Macrophysiology: large-scale patterns in physiological traits and their ecological implications. *Functional Ecology*, **18**, 159–167.

Chown, S.L. & Nicolson, W.N. (2004) *Insect Physiological Ecology: Mechanism and Patterns*. Oxford University Press, Oxford.

Chown, S.L., Slabber, S., McGeoch, M.A., Janion, C. & Leinaas, H.P. (2007) Phenotypic plasticity mediates climate change responses among invasive and indigenous arthropods. *Proceedings of the Royal Society B*, **274**, 2531–2537.

Costello, M.J., May, R.M. & Stork, N.E. (2013) Can we name Earth's species before they go extinct? *Science*, **339**, 413–416.

Davis, C.C., Willis, C.G., Primack, R.B. & Miller-Rushing, A.J. (2010) The importance of phylogeny to the study of phenological response to global climate change. *Philosophical Transactions of the Royal Society B*, **365**, 3201–3213.

Dawson, T.P., Jackson, S.T., House, J.I., Prentice, I.C. & Mace, G.M. (2011) Beyond predictions: Biodiversity conservation in a changing climate. *Science*, **332**, 53–58.

Dennis, R.L.H. (1993) *Butterflies and Climate Change*. Manchester University Press, Manchester.

Deutsch, C.A., Tewksbury, J.J., Huey, R.B., Sheldon, K.S., Ghalambor, C.K., Haak, D.C. & Martin, P.R. (2008) Impacts of climate warming on terrestrial ectotherms across latitude. *Proceedings of the National Academy of Sciences of the United States of America*, **105**, 6668–6672.

Diamond, S.E., Sorger, D.M., Hulcr, J., Pelini, S.L., Toro, I.D., Hirsch, C., Oberg, E. & Dunn, R.R. (2012) Who likes it hot? A global analysis of the climatic, ecological, and evolutionary determinants of warming tolerance in ants. *Global Change Biology*, **18**, 448–456.

Dunn, R.R. (2005) Modern insect extinctions, the neglected majority. *Conservation Biology*, **19**, 1030–1036.

Elith, J. & Leathwick, J. R. (2009) Species distribution models: ecological explanation and prediction across space and time. *Annual Review of Ecology, Evolution and Systematics.* **40**, 677–697.

Ferro, V.G., Lemes, P., Melo, A.S. & Loyola, R. (2014) The reduced effectiveness of protected areas under climate change threatens Atlantic Forest tiger moths. *PLOS ONE*, **9**, e107792.

Foden W.B., Butchart S.H.M., Stuart S.N., Vié, J-C., Akçakaya H.R., Angulo, A., DeVantier, L.M., Gutsche, A., Turak, E., Cao, L., Donner, S.D., Katariya, V., Bernard, R., Holland, R.A., Hughes, A.F., O'Hanlon, S.E., Garnett, S.T., Şekercioğlu, Ç.H. & Mace, G.M. (2013) Identifying the world's most climate change vulnerable species: a systematic trait-based assessment of all birds, amphibians and corals. *PLOS ONE*, **8**, e65427.

Fox, R., Oliver, T.H., Harrower, C., Parsons, M.S., Thomas, C.D. & Roy, D.B. (2014) Long-term changes to the frequency of occurrence of British moths are consistent with opposing and synergistic effects of climate and land-use changes. *Journal of Applied Ecology*, **51**, 949–957.

Giersch, J.J., Jordan, S., Luikart, G., Jones, L.A., Hauer, F.R. & Muhlfeld, C.C. (2015) Climate-induced range contraction of a rare alpine aquatic invertebrate. *Freshwater Science*, **34**, 53–65.

Gillingham, P.K., Alison, J., Roy, D.B., Fox, R. & Thomas, C.D. (2015) High Abundances of species in protected areas in parts of their geographic distributions colonized during a recent period of climatic change. *Conservation Letters*, **8**, 97–106.

Gillingham, P.K., Palmer, S.C.F., Huntley, B., Kunin, W.E., Chipperfield, J.D. & Thomas, C.D. (2012) The relative importance of climate and habitat in determining the distributions of species at different spatial scales: a case study with ground beetles in Great Britain. *Ecography*, **35**, 831–838.

Guisan, A. & Thuiller, W. (2005) Predicting species distribution: offering more than simple habitat models. *Ecology Letters*, **8**, 993–1009.

Gunderson, A.R. & Stillman, J.H. (2015) Plasticity in thermal tolerance has limited potential to buffer ectotherms from global warming. *Proceedings of the Royal Society B* **282**, 20150401.

Haddad, N.M. (1999) Corridor and distance effects on interpatch movements: a landscape experiment with butterflies. *Ecological Applications*, **9**, 612–622.

Hannah, L. (2010) A Global Conservation System for Climate-Change Adaptation. *Conservation Biology*, **24**, 70–77.

Hannah, L., Midgley, G., Andelman, S., Araújo, M., Hughes, G., Martinez-Meyer, E., Pearson, R. & Williams, P. (2007) Protected area needs in a changing climate. *Frontiers in Ecology and the Environment*, **5**, 131–138.

Hannah, L., Midgley, G.F., Lovejoy, T., Bond, W.J., Bush, M., Lovett, J.C., Scott, D. & Woodward, F.I. (2002) Conservation of biodiversity in a changing climate. *Conservation Biology*, **16**, 264–268.

Heller, N.E. & Zavaleta, E.S. (2009) Biodiversity management in the face of climate change: A review of 22 years of recommendations. *Biological Conservation*, **142**, 14–32.

Hernández-Manrique, O.L., Sánchez-Fernández, D., Numa, C., Galante, E., Verdú, J.R. & Lobo, J.M. (2013) Extinction trends of threatened invertebrates in peninsular Spain. *Journal of Insect Conservation*, **17**, 235–244.

Hewitt, N., Klenk, N., Smith, A.L., Bazely, D.R., Yan, N., Wood, S., MacLellan, J.I., Lipsig-Mumme, C. & Henriques, I. (2011) Taking stock of the assisted migration debate. *Biological Conservation*, **144**, 2560–2572.

Hodgson, J.A., Moilanen, A., Wintle, B.A. & Thomas, C.D. (2011) Habitat area, quality and connectivity: striking the balance for efficient conservation. *Journal of Applied Ecology*, **48**, 148–152.

Hodgson, J.A., Thomas, C.D., Wintle, B.A. & Moilanen, A. (2009) Climate change, connectivity and conservation decision making: back to basics. *Journal of Applied Ecology*, **46**, 964–969.

Hoegh-Guldberg, O., Hughes, L., McIntyre, S., Lindenmayer, D.B., Parmesan, C., Possingham, H.P. & Thomas, C.D. (2008) Assisted colonization and rapid climate change. *Science*, **321**, 345–346.

Hoffmann, A.A. (2010) Physiological climate limits in *Drosophila*: patterns and implications. *Journal of Experimental Biology*, **213**, 870–880.

Hole, D.G., Willis, S.G., Pain, D.J., Fishpool, L.D., Butchart, S.H.M., Collingham, Y.C., Rahbek, C. & Huntley, B. (2009) Projected impacts of climate change on a continent-wide protected area network. *Ecology Letters*, **12**, 420–431.

IUCN. (2013) *Guidelines for Reintroductions and Other Conservation Translocations*. IUCN, Gland, Switzerland and Cambridge.

Jackson, S.F. & Gaston, K.J. (2008) Land use change and the dependence of national priority species on protected areas. *Global Change Biology*, **14**, 2132–2138.

Jenkins, C.N. & Joppa, L. (2009) Expansion of the global terrestrial protected area system. *Biological Conservation*, **142**, 2166–2174.

Joppa, L.N. & Pfaff, A. (2009) High and far: biases in the location of protected areas. *PLOS ONE*, **4**, e8273.

Kearney, M. & Porter, W. (2009) Mechanistic niche modelling: combining physiological and spatial data to predict species' ranges. *Ecology Letters*, **12**, 334–350.

Kearney, M., Shine, R. & Porter, W.P. (2009) The potential for behavioral thermoregulation to buffer 'cold-blooded' animals against climate warming. *Proceedings of the National Academy of Sciences of the United States of America*, **106**, 3835–3840.

Kharouba, H.M. & Kerr, J.T. (2010) Just passing through: Global change and the conservation of biodiversity in protected areas. *Biological Conservation*, **143**, 1094–1101.

Kingsolver, J.G., Woods, H.A., Buckley, L.B., Potter, K.A., MacLean, H.J. & Higgins, J.K. (2011) Complex life cycles and the responses of insects to climate change. *Integrative and Comparative Biology*, **51**, 719–732.

Kleinbauer, I., Dullinger, S., Peterseil, J. & Essl, F. (2010) Climate change might drive the invasive tree *Robinia pseudacacia* into nature reserves and endangered habitats. *Biological Conservation*, **143**, 382–390.

Körner, K., Treydte, A.C., Burkart, M. & Jeltsch, F. (2010) Simulating direct and indirect effects of climatic changes on rare perennial plant species in fragmented landscapes. *Journal of Vegetation Science*, **21**, 843–856.

Krosby, M., Tewksbury, J., Haddad, N.M. & Hoekstra, J. (2010) Ecological connectivity for a changing climate. *Conservation Biology*, **24**, 1686–1689.

Kuchlein, J.H. & Ellis, W.N. (1997) Climate-induced changes in the microlepidoptera fauna of the Netherlands and the implications for nature conservation. *Journal of Insect Conservation*, **1**, 73–80.

Kuussaari, M., Heikkinen, R.K., Heliölä, J., Luoto, M., Mayer, M., Rytteri, S. & Von Bagh, P. (2015) Successful translocation of the threatened Clouded Apollo butterfly (*Parnassius mnemosyne*) and metapopulation establishment in southern Finland. *Biological Conservation*, **190**, 51–59.

Lawler, J.J. (2009) Climate change adaptation strategies for resource management and conservation planning. *Annals of the New York Academy of Sciences*, **1162**, 79–98.

Levitan, M. (2003) Climatic factors and increased frequencies of 'southern' chromosome forms in natural populations of *Drosophila robusta*. *Evolutionary Ecology Research*, **5**, 597–604.

Loarie, S.R., Duffy, P.B., Hamilton, H., Asner, G.P., Field, C.B. & Ackerly, D.D. (2009) The velocity of climate change. *Nature*, **462**, 1052–1055.

Lobo, J.M. (2011) Vulnerabilidad de las áreas protegidas y de zonas de interés para la biodiversidad ante el cambio climático. *Biodiversidad en España. Base de la Sostenibilidad ante el Cambio Global*. pp. 360–368. Observatorio de la Sostenibilidad de España, Ministerio de Medio Ambiente y Medio Rural y Marino, Madrid.

Losey, J.E. & Vaughan, M. (2006) The economic value of ecological services provided by insects. *BioScience*, **56**, 311–323.

Loss, S.R., Terwilliger, L.A. & Peterson, A.C. (2010) Assisted colonization: Integrating conservation strategies in the face of climate change. *Biological Conservation*, **144**, 92–100.

Lovejoy, T.E. (2005) Conservation with a changing climate. *Climate Change and Biodiversity* (eds T.E. Lovejoy & L. Hannah), pp. 325–328. Yale University Press, New Haven.

Magozzi, S. & Calosi, P. (2014) Integrating metabolic performance, thermal tolerance, and plasticity enables for more accurate predictions on species vulnerability to acute and chronic effects of global warming. *Global Change Biology*, **21**, 181–194.

Mastrandrea, M.D., Heller, N.E., Root, T.L. & Schneider, S.H. (2010) Bridging the gap: linking climate-impacts research with adaptation planning and management. *Climate Change*, **100**, 87–101.

Mawdsley, J.R., O'Malley, R. & Ojima, D.S. (2009) A review of climate-change adaptation strategies for wildlife management and biodiversity conservation. *Conservation Biology*, **23**, 1080–1089.

McKinney, M.L. (1999) High rates of extinction and threat in poorly studied taxa. *Conservation Biology*, **13**, 1273–1281.

McLachlan, J.S., Hellmann, J.J. & Schwartz, M.W. (2007) A framework for debate of assisted migration in an era of climate change. *Conservation Biology*, **21**, 297–302.

Menéndez, R. (2007) How are insects responding to global warming? *Tijdschrift voor Entomologie*, **150**, 355–365.

Monzón, J., Moyer-Horner, L. & Palamar, M.B. (2011) Climate change and species range dynamics in protected areas. *BioScience*, **61**, 752–761.

Moritz, C., Patton, J.L., Conroy, C.J., Parra, J.L., White, G.C. & Beissinger, S.R. (2008) Impact of a century of climate change on small-mammal communities in Yosemite National Park, USA. *Science*, **322**, 261–264.

Mueller, J.M. & Hellmann, J.J. (2008) An assessment of invasion risk from assisted migration. *Conservation Biology*, **22**, 562–567.

Öckinger, E. & Smith, H.G. (2008) Do corridors promote dispersal in grassland butterflies and other insects? *Landscape Ecology*, **23**, 27–40.

Palmer, M.A., Lettenmaier, D.P., Poff, N.L., Postel, S.L., Richter, B. & Warner, R. (2009) Climate change and river ecosystems: protection and adaptation options. *Environmental Management*, **44**, 1053–1068.

Parmesan, C. (2006) Ecological and evolutionary responses to recent climate change. *Annual Review of Ecology, Evolution and Systematics*, **37**, 637–669.

Parmesan, C., Ryrholm, N., Stefanescu, C., Hill, J.K., Thomas, C.D., Descimon, H., Huntley, B., Kaila, L., Kullberg, J., Tammaru, T., Tennent, W.J., Thomas, J.A. & Warren, M. (1999) Poleward shifts in geographical ranges of butterfly species associated with regional warming. *Nature*, **399**, 579–583.

Pateman, R.M., Hill, J.K., Roy, D.B., Fox, R. & Thomas, C.D. (2012) Temperature-dependent alterations in host use drive rapid range expansion in a butterfly. *Science*, **336**, 1028–1030.

Pearson, R.G. & Dawson, T.P. (2003) Predicting the impacts of climate change on the distribution of species: are bioclimate envelope models useful? *Global Ecology and Biogeography*, **12**, 361–371.

Pörtner, H.O. & Farrell, A. (2009) Adapting to climate change. Response. *Science*, **323**, 876–877.

Pounds, J.A., Fogden, M.P.L. & Campbell, J.H. (1999) Biological response to climate change on a tropical mountain. *Nature*, **398**, 611–615.

Pöyry, J. Luoto, M., Heikkinen, R.K., Kuussaari, M. & Saarinen, K. (2009) Species traits explain recent range shifts of Finnish butterflies. *Global Change Biology*, **15**, 732–743.

Ricciardi, A. & Simberloff, D. (2009) Assisted colonization is not a viable conservation strategy. *Trends in Ecology and Evolution*, **24**, 248–253.

Rodrigues, A.S.L, Pilgrim, J.D., Lamoreux, J.F., Hoffmann, M., Brooks, T.M. (2006) The value of the IUCN Red List for conservation. *Trends in Ecology and Evolution*, **21**, 71–76.

Rodríguez-Trelles, F. & Rodríguez, M.A. (1998) Rapid micro-evolution and loss of chromosomal diversity in *Drosophila* in response to climate warming. *Evolutionary Ecology*, **12**, 829–838.

Rodríguez-Trelles, F., Rodríguez, M.A. & Scheiner, S.M. (1998) Tracking the genetic effects of global warming: *Drosophila* and other model systems. *Ecology and Society*, **2**, 2.

Roy, D.B. & Sparks, T.H. (2000) Phenology of British butterflies and climate change. *Global Change Biology*, **6**, 407–416.

Samways, M.J. (2005) *Insect Diversity Conservation*. Cambridge University Press, Cambridge.

Samways, M.J. (2007) Insect conservation: a synthetic management approach. *Annual Review of Entomology*, **52**, 465–487.

Sánchez-Fernández, D., Abellán, P., Picazo, F., Millán, A., Ribera, I. & Lobo, J.M. (2013) Do protected areas represent species' optimal climatic conditions? A test using Iberian water beetles. *Diversity and Distributions*, **19**, 1407–1417.

Sánchez-Fernández, D., Calosi, P., Atfield, A., Arribas, P., Velasco, J., Spicer, J.I., Millán, M. & Bilton, D.T. (2010) Reduced salinities compromise the thermal tolerance of hypersaline specialist diving beetles. *Physiological Entomology*, **35**, 265–273.

Sánchez-Fernández, D., Lobo, J.M., Abellán, P. & Millán, A. (2011) Environmental niche divergence between genetically distant lineages of an endangered water beetle. *Biological Journal of the Linnean Society*, **103**, 891–903.

Sánchez-Fernández, D., Lobo, J.M., Millán, A. & Ribera, I. (2012) Habitat type mediates time to equilibrium in the geographical distribution of Iberian diving beetles. *Global Ecology and Biogeography*, **21**, 988–997.

Schlaepfer, M.A., Helenbrook, W.D., Searing, K.B. & Shoemaker, K.T. (2009) Assisted colonization: evaluating contrasting management actions (and values) in the face of uncertainty. *Trends in Ecology and Evolution*, **24**, 471–472.

Scott, D., Malcolm, J.R. & Lemieux, C. (2002) Climate change and modelled biome representation in Canada's national park system: implications for system planning and park mandates. *Global Ecology and Biogeography*, **11**, 475–484.

Seddon, P.J. Griffiths, C.J., Soorae, P.S. & Armstrong, D.P. (2014) Reversing defaunation: restoring species in a changing world. *Science*, **345**, 406–412.

Sieck, M., Ibisch, P.L., Moloney, K.A. & Jeltsch, F. (2011) Current models broadly neglect specific needs of biodiversity conservation in protected areas under climate change. *BMC Ecology*, **11**, 12.

Slade, E.M., Merckx, T., Riutta, T., Bebber, D.P., Redhead, D., Riordan, P. & Macdonald, D.W. (2013) Life-history traits and landscape characteristics predict macro-moth responses to forest fragmentation. *Ecology*, **94**, 1519–1530.

Somero, G.N. (2010) The physiology of climate change: how potentials for acclimatization and genetic adaptation will determine 'winners' and 'losers'. *Journal Experimental Biology*, **213**, 912–920.

Spector, S. (2008) Insect conservation – a time of crisis and opportunity. *American Entomologist*, 98–100.

Stillman, J.H. (2003) Acclimation capacity underlies susceptibility to climate change. *Science*, **301**, 65.

Sutcliffe, O.L. & Thomas, C.D. (1996) Open corridors appear to facilitate dispersal of ringlet butterflies (*Aphantopus hyperantus*) between woodland clearings. *Conservation Biology*, **10**, 1359–1365.

Thomas, C.D. (2011) Translocation of species, climate change, and the end of trying to recreate past ecological communities. *Trends in Ecology and Evolution*, **26**, 216–221.

Thomas, C.D., Bodsworth, E.J., Wilson, R.J., Simmons, A.D., Davies, Z.G., Musche, M. & Conradt, L. (2001) Ecological and evolutionary processes at expanding range margins. *Nature*, **411**, 577–581.

Thomas, C.D., Cameron, A., Green, R.E., Bakkenes, M., Beaumont, L.J., Collingham, Y.C., Erasmus, B.F.N., de Siqueira, M.F., Grainger, A., Hannah, L., Hughes, L., Huntley, B., van Jaarsveld, A.S., Midgley, G.F., Miles, L., Ortega-Huerta, M.A., Peterson, A.T., Phillips, O.L. & Williams, S.E. (2004) Extinction risk from climate change. *Nature*, **427**, 145–148.

Thomas, C. D. & Gillingham, P. K. (2015) The performance of Protected Areas for biodiversity under climate change. *Biological Journal of the Linnean Society*. **115**, 718–730.

Thomas, C.D., Gillingham, P.K., Bradbury, R.B., Roy, D.B., Anderson, B.J., Baxter, J.M., Bourn, N.A.D., Crick, H.Q.P., Findon, R.A., Fox, R., Hodgson, J.A., Holt, A.R., Morecroft, M.D., O'Hanlon, N.J., Oliver, T.H., Pearce-Higgins, J.W., Procter, D.A., Thomas, J.A., Walker, K.J., Walmsley, C.A., Wilson, R.J. & Hill, J. K. (2012) Protected areas facilitate species' range expansions. *Proceedings of the National Academy of Sciences of the United States of America*, **109**, 14063–14068.

Thomas, C.D., Hill, J.K., Anderson, B.J., Bailey, S., Beale, C.M., Bradbury, R.B., Bulman, C.R., Crick, H.Q.P., Eigenbrod, F., Griffiths, H.M., Kunin, W.E., Oliver, T.H., Walmsley, C.A., Watts, K., Worsfold, N.T. & Yardley, T. (2011) A framework for assessing threats and benefits to species responding to climate change. *Methods in Ecology and Evolution*, **2**, 125–142.

Travis, J.M.J. (2003) Climate change and habitat destruction: a deadly anthropogenic cocktail. *Proceedings of the Royal Society of London B*, **270**, 467–473.

Visser, M.E. & Both, C. (2005) Shifts in phenology due to global climate change: the need for a yardstick. *Proceedings of the Royal Society of London B*, **272**, 2561–2569.

Vos, C.C., Berry, P., Opdam, P., Baveco, H., Nijhof, B., O'Hanley, J., Bell, C. & Kuipers, H. (2008) Adapting landscapes to climate change: examples of climate-proof ecosystem networks and priority adaptation zones. *Journal of Applied Ecology*, **45**, 1722–1731.

Warren, M.S., Hill, J.K., Thomas, J.A. & Asher, J., Fox, R., Huntley, B., Roy, D.B., Telfer, M.G., Jeffcoate, S., Harding, P., Jeffcoate, G., Willis, S.G., Greatorex-Davies, J.N., Moss, D. & Thomas, C.D. (2001) Rapid responses of British butterflies to opposing forces of climate and habitat change. *Nature*, **414**, 65–69.

Whittaker, R.J., Araújo, M.B., Jepson, P., Ladle, R.J., Watson, J.E.M. & Willis, K.J. (2005) Conservation Biogeography: assessment and prospect. *Diversity and Distributions*, **11**, 3–23.

Wiens, J.A., Seavy, N.E. & Jongsomjit, D. (2011) Protected areas in climate space: What will the future bring? *Biological Conservation*, **144**, 2119–2125.

Williams, J.W., Jackson, S.T. & Kutzbach, J.E. (2007) Projected distributions of novel and disappearing climates by 2100 AD. *Proceedings of the National Academy of Sciences of the United States of America*, **104**, 5738–5742.

Williams, S.E., Shoo, L.P., Isaac, J.L., Hoffmann, A.A. & Langham, G. (2008) Towards an integrated framework for assessing the vulnerability of species to climate change. *PLOS Biology*, **6**, 2621–2626.

Willis, S.G., Hill, J.K., Thomas, C.D., Roy, D.B., Fox, R., Blakeley, D.S. & Huntley, B. (2008) Assisted colonization in a changing climate: a test-study using two U.K. butterflies. *Conservation Letters*, **2**, 45–51.

18

Emerging Issues and Future Perspectives for Global Climate Change and Invertebrates

Scott N. Johnson[1] and T. Hefin Jones[2]

[1] Hawkesbury Institute for the Environment, Western Sydney, NSW 2751, Australia
[2] School of Biosciences, Cardiff University, Cardiff CF10 3AX, UK

18.1 Preamble

'Everybody complains about the weather, but nobody does anything about it' is a long standing quip of uncertain origin (often misattributed to Mark Twain). As Bloom (2010) points out, however, the second part of this sentence is now inaccurate since anthropogenic influences are affecting our climate and long-term weather patterns. As we discussed in Chapter 1, the latest IPCC report (IPCC, 2014) states that there is a clear human influence on the climate and declares that it is *extremely likely* that human influence has been the dominant cause of observed warming since 1950, with the level of confidence having increased since the Fourth IPCC Report in 2007 (IPCC, 2007). Climate change research has evolved with each IPCC report with new issues and perspectives coming to the fore, many of which pertain to invertebrate biology. In this chapter, we aim to identify emerging issues from the proceeding chapters, and put forward future perspectives in global climate change and invertebrate biology research.

18.2 Multiple Organisms, Asynchrony and Adaptation in Climate Change Studies

A recurring theme of this book is the fact that climate change might cause asynchrony in distributions and life-cycles of interacting organisms. This may take many forms ranging from reduced top-down control of pest invertebrates by natural enemies (Chapters 6 and 10) to starvation of vertebrates that depend on invertebrates as a source of food (Chapter 15). As Forrest explains in Chapter 5, asynchrony between plant and invertebrate phenologies is particularly concerning for specialist pollinators (e.g., solitary bees), if the flight period of the adults becomes asynchronous with the flowering of plants they pollinate.

Examining past records, as described by Palmer and Hill (Chapter 2), clearly shows the onset of phenological events in invertebrate lifecycles is occurring earlier due, in many instances, to warmer temperatures. The rate of this change varies between taxa. This has the potential to introduce asynchrony between interacting organisms. A famous example is the earlier emergence of winter moth (*Operophtera brumata*), which historically is becoming increasingly out-of-step with budburst in its host plant, oak (Visser & Holleman, 2001). In this species, asynchrony between egg hatching and bud-burst is a major source of mortality. Interestingly, when this was tested experimentally by increasing temperatures by 3 °C in solar domes the phenologies of winter moth and oak trees advanced at the same rate so there was no asynchrony (Buse & Good, 1996). This illustrates how historical (see Palmer & Hill, Chapter 2) and experimental (see Linroth & Raffa, Chapter 3) approaches can provide different answers to the same question.

It could be argued that where climate change introduces asynchrony between invertebrates and resources they depend on, then there is a strong selection pressure for the invertebrates to adapt. Focussing on aphids, Dixon (2003) suggested that when synchronisation between bud-burst and egg hatch is important for invertebrate fitness, insects would maintain close synchronisation overall. This seems plausible with gradual changes in the climate, such as warming and elevated atmospheric CO_2, especially when the invertebrate has fast generation times and is able to adapt rapidly to changing circumstances. Where climate change leads to greater variation in climatic conditions (e.g., extreme precipitation and temperature events), it could be argued that it would be much harder for inter-acting organisms to adapt and stay in synchrony. Indeed, this seems to be the case for parasitism of caterpillars where increasing variation in precipitation patterns impaired the ability of parasitoids to track host populations (Stireman et al., 2005).

The question of climate change induced asynchrony and adaptation clearly needs further research, most likely using approaches based on historical records and empirical studies. The extent to which invertebrates can adapt and return to synchrony with a resource they exploit will undoubtedly require longer-term studies than we currently see in the literature. It could be hypothesised, however, that invertebrates that are able to adapt and restore synchrony will be those that have rapid rates of repro-duction and where synchronising with the resource is important (i.e., there is a strong selection pressure). Moreover, adaptation and synchronisation are likely to be more evident in response to gradual changes in climate than more variable climatic changes (e.g., extreme events, see Section 18.4), which will be far harder to adapt to.

18.3   Multiple Climatic Factors in Research

As discussed in Chapter 1, global climate change is multifaceted and involves changes in greenhouse gases, which are linked to changes in air temperature and precipitation patterns. Indeed, climate change itself is just one component of global change, which include land use change, acid deposi-tion, pollution and invasive spread of species (Vitousek et al., 1997), all of which have consequences for the planet's ecosystems. Newman et al. (2011) provide extensive coverage of this topic, but it is increasingly apparent that testing multiple aspects of climate change can provide different answers than when testing each one separately. The net effect on invertebrates may be additive, whereby the effect is simply the sum of the effects of the climatic factors acting independently, or it may be inter-active. Interactions occur when the impact of one factor is either intensified or mitigated by another climatic factor.

While the number of invertebrate studies involving more than one climatic factor is increasing, these are still too few to make robust generalisations except to say that interactions between fac-tors are frequently evident (Fuhrer, 2003; Zvereva & Kozlov, 2006; DeLucia et al., 2012; Scherber et al., 2013; Facey et al., 2014). Why are there so few multi-factor climate change studies? The main answer to this question is that logistical constraints militate against them. With each level of treat-ment, experiments double in terms of experimental units, which increases the resources needed to conduct the study, and may reduce the quantity and/or quality of the responses being measured.

The biggest problem, however, is adequate replication and specifically the issue of pseudoreplication which is very common in climate change research (Newman et al., 2011; Johnson et al., 2016).

Pseudoreplication (a term first coined in Hurlbert, 1984) is a thorny issue in climate change studies that is discussed by Lindroth and Raffa (Chapter 3). It occurs frequently because when researchers apply environmental treatments this usually takes place in controlled cabinets, glasshouses, shelter or Free Air CO_2 Enrichment (FACE) rings, the unit of replication for those treatments is the cabinet, glasshouse, shelter, or ring, respectively. Subunits (e.g., individual plants) are not independently subjected to the treatment, and therefore not true replicates. This artificially inflates the degrees of freedom in statistical tests, resulting in a larger F statistic (or equivalent) and a higher P value than is strictly justified. The argument goes 'how do you know it's the treatment and not some unmeasured effect of the cabinet, glasshouse, shelter or FACE ring?' We have edited many papers where reviewers have honed in on this issue and rejected the paper without further consideration. In our experience, this seems particularly true of invertebrate studies. The issue of pseudoreplication is real, and few argue the contrary (though see Oksanen, 2001), but it is worth considering whether the actual biological inferences of a particular study are really compromised by pseudoreplication. Indeed Hurlbert, the authority on the subject, himself states 'there should be no automatic rejection of [such] experiments' (Hurlbert, 2004).

Pseudoreplication has been helped to some extent with the advent of contemporary statistical tests, such as nested designs and random/mixed effect models, which take the lack of independence between pseudoreplicates into account (Chaves, 2010; Leather et al., 2014; Davies & Gray, 2015). These tests, however, are only applicable when a treatment combination has been repeated in more than one cabinet, glasshouse, shelter or FACE ring. In terms of conducting multi-factorial experiments this is still challenging, as one of us recently discussed in Johnson et al. (2016). Let's imagine that a researcher wanted to look at the impacts of two CO_2 concentrations at three temperature regimes (i.e., six treatment combinations) on the performance of an insect herbivore. A paltry N = 3 would require 18 identical chambers. This isn't a realistic option for most researchers and, in any case, may generate type II errors (i.e., real responses are not detected statistically) with such low degrees of freedom.

Repeating the experiment several times is another option for avoiding pseudoreplication, but this is very time consuming, it locks up experimental facilities for large periods of time and is not really an option for fixed length research projects (e.g., PhD studies). One of us (Jones) faced this very problem with an experiment that needed to look at long-term plant responses to elevated atmospheric CO_2; our solution at the time was to use 'chamber switching' (Bezemer et al., 1998). Experimental units, in this case potted grasses, were moved within, and then between, chambers with corresponding changes in environmental conditions (Bezemer et al., 1998). We must be clear at this stage that this does not eliminate pseudoreplication, but it reduces its effects by equalising any unintended 'chamber effects' across all experimental units. A potential criticism of 'chamber switching' is that it might affect experimental units (e.g., plants) differently during different stages of their development (Potvin & Tardif, 1988). Researchers have overcome this by conducting experiments in stages so experimental units experience particular chambers at identical stages (e.g., of plant development as in Vuorinen et al., 2004a,b).

Johnson et al. (2016) made an, admittedly crude, comparison between three published experiments looking at how CO_2 and temperature affected lucerne growth; one was replicated in chambers, one replicated in time and one that adopted the 'chamber switching' approach. They found very similar results across all three studies, suggesting that while 'chamber swapping' doesn't eliminate the pseudoreplication problem it appears to minimise 'chamber effects' and gave similar results to fully replicated experiments – at least in the lucerne system in this experimental platform. The authors do, however, urge caution, which we repeat here, that *'researchers working in other systems also take a cautious approach with regard to careful replication until they can develop confidence that their observed effects are real and repeatable'*. Moreover, *'The statistical significance of numerical differences remain inflated, however, so it would be judicious to treat any marginally significant results with caution.'*

Investigating multiple environmental factors is widely advocated, and many would argue an absolute necessity (Fuhrer, 2003; Zvereva & Kozlov, 2006; DeLucia et al., 2012; Scherber et al., 2013; Facey et al., 2014). Given that elevated atmospheric CO_2 and temperature frequently interact and are predicted to increase in tandem (IPCC, 2014), it is hard to argue that these factors should not be investigated together. It seems paradoxical to be satisfied with an incomplete, or even misleading, biological answer because it was delivered via a robust statistical test. We recommend that researchers avail themselves of contemporary statistical tests, wherever possible, to account for the lack of independence between experimental units. If that is not possible they should take a cautious approach to interpreting results from studies involving pseudoreplication until they have evidence that observed effects are real and repeatable. We are in agreement with Jonathan Newman et al. (2011) that *'as long as authors are clear about the use of pseudoreplicates, and the readers appreciate the potential problems interpreting such results, then such studies are valuable despite their pseudoreplication'*.

18.4   Research Into Extreme Climatic Events

Climate change research involving invertebrates has come a long way and we now have a better mechanistic understanding of how, for example, herbivorous insects are affected by elevated atmospheric CO_2 (Robinson et al., 2012; Zavala et al., 2013; Ode et al., 2014). One area that perhaps deserves more attention in invertebrate climate change research is the impact of extreme events on invertebrates, which last for comparatively short periods of time, but may have greater impact on invertebrates than mean climatic changes (Jentsch et al., 2007). In particular, 2000-2011 was likely to have been the warmest decade globally for at least a millennium, with several heatwaves occurring during this period. In 2003, for example, Europe experienced its hottest summer for at least 500 years (Luterbacher et al., 2004). Examining European summer temperatures anomalies for 1500-2010 shows that the five warmest summers occurred from 2002 onwards (Fig. 18.1A). Moreover, the probability of these summer heatwaves occurring is rising sharply (Fig. 18.1B). Over the course of 90 days in 2012 and 2013, Australia experienced two 'angry summers' which broke 123 and 156 meteorological records, respectively (Johnson, 2014). At the same time, extreme rainfall events in the form of droughts, floods and storms have occurred throughout the world (IPCC, 2014).

Why do extreme events, particularly heatwaves, matter for invertebrates? As ectotherms, invertebrate metabolism and overall activity is mainly driven by the ambient temperature. High temperatures causes respiration to increase beyond rates that resources can be acquired to facilitate this, leading to a net loss of energy and eventual death (Seifert et al., 2015). Extreme events that exceed the thermal tolerances of invertebrate species therefore promote population declines and cause local extinctions. Such events also have the capacity to cause 'community closure' where they render a habitat unsuitable for invertebrates to become re-established once conditions return to normal (Lundberg et al., 2000). Whether community closure occurs depends on the extent and duration of the extreme event. Herbivorous rotifers, for example, suffered community closure at 39 °C but not at 29 °C (Seifert et al., 2015). At 39 °C, reintroduced rotifers were confronted with substantially different (presumably unsuitable) food environments which ultimately led to extinction. As discussed

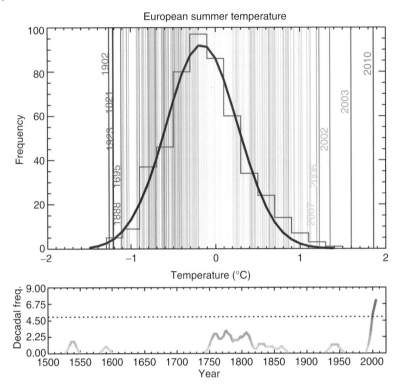

Figure 18.1 European summer temperatures for 1500–2010. (A) Statistical frequency distribution of best-guess reconstructed and instrument-based European ([35°N, 70°N], [25°W, 40°E]) summer land temperature anomalies (degrees Celsius, relative to the 1970–1999 period) for the 1500–2010 period (vertical lines). The five warmest and coldest summers are highlighted. Grey bars represent the distribution for the 1500–2002 period (Luterbacher et al., 2004), with a Gaussian fit in black. Data for the 2003–2010 period are from Hansen et al. (1999). (B) The running decadal frequency of extreme summers, defined as those with temperature above the 95th percentile of the 1500–2002 distribution. A 10-year smoothing is applied. Dotted line shows the 95th percentile of the distribution of maximum decadal values that would be expected by random chance. Reproduced from Barriopedro et al. (2011) with permission.

by Arribas and Jonsson in Chapter 17, reintroducing beneficial invertebrates to regions where they suffered local extinction may be possible, however, this is unlikely to be successful where community closure has occurred after an extreme event.

18.5 Climate change and Invertebrate Biosecurity

The term 'biosecurity' is used to describe anything to do with the protection of the economy, environment and human health from the harmful impacts of exotic pests and weeds in terms of entry, establishment and spread. It differs from 'quarantine' in that it includes organisms that do not necessarily vector disease, but cause economic or environmental damage (Gullen & Cranston, 2014). There are several reasons why exotic invertebrates flourish in novel habitats. These include an ability to outcompete natives, abundance of a resource that has not co-evolved with the invertebrate (e.g., an undefended plant) and an absence of natural enemies (Liebhold & Tobin, 2008). Most exotic invertebrates don't flourish when they arrive in new habitats because arrival in small numbers limits the probability of success and enhances chances of extinction via Allee effects (Liebhold & Tobin, 2008). Another barrier to success is that the habitat is colder than the lower thermal requirements of the invertebrates. Global climate change and specifically global warming could remove this barrier and lead to biosecurity breaches by invertebrates. Added to this, many of the species that are candidates for invasions are generalists and highly adaptable so are generally able to tolerate or take advantage of change and disturbance (Sutherst et al., 2007).

Global warming has enabled alien species to expand into regions in which they previously could not survive and reproduce. An early paper on this subject by Walther et al. (2009) gives numerous examples of how global warming has facilitated invasions of invertebrates into new regions, which they describe occur at four stages. (1) Introduction occurs because global warming opens up new areas that were previously unsuitable. (2) Colonisation (casual stage) occurs because survival increases when the temperature resembles that of their native range. (3) Establishment (naturalisation) occurs via successful reproduction at the new higher temperatures. (4) Spread (invasion stage) occurs as warming continues, extending the range available to the invader but also giving it a competitive advantage over natives that are less well adapted to the warmer temperatures. A good example of this process is the southern green stink bug (*Nezara viridula*), which was originally a sub-tropical species but has continued to spread northwards in Japan since invading in the 1960s. Colonisation was achieved through reduced mortality during milder winters and *N. viridula* now outcompetes the native *Nezara antennata* (Musolin, 2007).

Where the population dynamics of invasive invertebrates is heavily influenced by temperature, it is also possible that invasions will be more successful when they move into warmer areas, via faster generation times. In Japan, for instance, the invasive American fall webworm (*Hyphantria cunea*) completes three generations per year, compared with two in its native range (Gomi et al., 2007). While warming (both mean and extreme events) is likely to facilitate invasions of invertebrates more than other factors, other environmental factors such as altered precipitation have also facilitated the spread of invasive species such as the Argentine ant (*Linepithema humile*) into new regions of California (Heller et al., 2008). A recent analysis of range shifts for '100 of the world's worst invasive species' (26 of which are invertebrates) using both temperature and precipitation data, predicted that ranges of terrestrial and aquatic invertebrates would extend by 18% and 59%, respectively (Bellard et al., 2013). The authors noted that climate change might lead to the loss of some invasive alien species in some regions, but invertebrates were not particularly well represented in this.

After habitat loss, biological invasions are the biggest threat to the planet's biodiversity (Sandlund et al., 1999), and invasive invertebrates cause widespread economic losses and are often incredibly difficult to control. All eradication measures against red imported fire ants (*Solenopsis invicta*), for instance, have failed and annual control costs amount to several US\$ billion in the United States alone (Gullan & Cranston, 2014). Indeed, the etymology of its name reflects is pest status; *invicta* derives from the Latin for 'unconquered'. It has since invaded Australia and New Zealand (2001), and mainland China (2004), where it is equally troublesome. In addition to global climate change, invertebrate invasions are likely to be aided by a number of other factors including increasing global trade and transcontinental travel (Hill et al., 2016). Given this, increasing frequencies of invertebrate invasions as a result of climate change seem inevitable (Bellard et al., 2013), and we suggest is a more serious threat than has previously been recognised in mainstream climate change research.

18.6 Concluding Remarks

As discussed in the Introduction, with the exception of mankind, invertebrates shape Earth's ecosystems more than any other taxa. The authors of the preceding chapters make a compelling case for why we need to understand how invertebrates will respond and adapt to climate change. In some instances, invertebrates may even help us mitigate the impacts of global climate change (see Chapter 11). The climate will continue to change, even with substantial reductions in CO_2 emissions, so as well as understanding invertebrate responses, researchers should be equally concerned with how best to adapt and remediate damaging scenarios. These may range from impaired ecosystem services to increased impacts of pests. The good news is that we have never been better equipped to do this. We have a significant amount of scientific literature to make generalisations and predictions, meaningful amounts of historical records to project forwards and increasingly sophisticated climatic models for distribution and abundance of invertebrates (e.g., Elith & Leathwick, 2009). Moreover, researchers have never had better access to experimental facilities for manipulating environmental conditions – 20 years ago it would have been fairly uncommon for glasshouses to have the capacity to monitor and control CO_2 concentrations but now it is the norm. Researchers have also being innovative and shared designs for affordable experimental facilities (e.g., Godfree et al., 2011; Messerli et al., 2015), opening up research opportunities for many. At the other end of the scale, we have the engineering capacity to build large field-based experimental platforms, such as FACE, which are allowing researchers to address increasingly complex questions. A particularly encouraging trend is for climate change experiments to be linked via regional and global networks. These coordinated experimental networks, principally focusing on rainfall, temperature and atmospheric manipulation, are ideally suited for comparative studies at regional to global scales. One example is Drought-Net (www.drought-net.org), which currently involves a network of rainfall manipulation experiments at over 100 sites across the globe. Many of these networks also allow researchers to collect, contribute, store, share and integrate data across disciplines. This approach has greatly increased the capacity of the research community to advance climate change science. All of these advances will enable us to understand better which little things will run the world in the future.

References

Barriopedro, D., Fischer, E.M., Luterbacher, J., Trigo, R.M. & García-Herrera, R. (2011) The hot summer of 2010: Redrawing the temperature record map of Europe. *Science*, **332**, 220–224.

Bellard, C., Thuiller, W., Leroy, B., Genovesi, P., Bakkenes, M. & Courchamp, F. (2013) Will climate change promote future invasions? *Global Change Biology*, **19**, 3740–3748.

Bezemer, T.M., Thompson, L.J. & Jones, T.H. (1998) *Poa annua* shows inter-generational differences in response to elevated CO_2. *Global Change Biology*, **4**, 687–691.

Bloom, A.J. (2010) Global Climate Change. *Convergence of Disciplines*. Sinauer Associates, Sunderland, MA, USA.

Buse, A. & Good, J. (1996) Synchronization of larval emergence in winter moth (*Operophtera brumata* L.) and budburst in pedunculate oak (*Quercus robur* L) under simulated climate change. *Ecological Entomology*, **21**, 335–343.

Chaves, L.F. (2010) An entomologist guide to demystify pseudoreplication: data analysis of field studies with design constraints. *Journal of Medical Entomology*, **47**, 291–298.

Davies, G.M. & Gray, A. (2015) Don't let spurious accusations of pseudoreplication limit our ability to learn from natural experiments (and other messy kinds of ecological monitoring). *Ecology and Evolution*, **5**, 5295–5304.

DeLucia, E.H., Nabity, P.D., Zavala, J.A. & Berenbaum, M.R. (2012) Climage change: resetting plant–insect interactions. *Plant Physiology*, **160**, 1677–1685.

Dixon, A.F.G. (2003) Climate change and phenological asynchrony. *Ecological Entomology*, **28**, 380–381.

Elith, J. & Leathwick, J.R. (2009) Species Distribution Models: Ecological explanation and prediction across space and time. *Annual Review of Ecology Evolution and Systematics*, **40**, 677–697.

Facey, S.L., Ellsworth, D.S., Staley, J.T., Wright, D.J. & Johnson, S.N. (2014) Upsetting the order: how climate and atmospheric change affects herbivore–enemy interactions. *Current Opinion in Insect Science*, **5**, 66–74.

Fuhrer, J. (2003) Agroecosystern responses to combinations of elevated CO_2, ozone, and global climate change. *Agriculture Ecosystems & Environment*, **97**, 1–20.

Godfree, R., Robertson, B., Bolger, T., Carnegie, M. & Young, A. (2011) An improved hexagon open-top chamber system for stable diurnal and nocturnal warming and atmospheric carbon dioxide enrichment. *Global Change Biology*, **17**, 439–451.

Gomi, T., Nagasaka, M., Fukuda, T. & Hagihara, H. (2007) Shifting of the life cycle and life-history traits of the fall webworm in relation to climate change. *Entomologia Experimentalis et Applicata*, **125**, 179–184.

Gullan, P.J. & Cranston, P.S. (2014) *The Insects: An Outline of Entomology*. Wiley Blackwell, Oxford, UK.

Hansen, J., Ruedy, R., Glascoe, J. & Sato, M. (1999) GISS analysis of surface temperature change. *Journal of Geophysical Research: Atmospheres*, **104**, 30997–31022.

Heller, N.E., Sanders, N.J., Shors, J.W. & Gordon, D.M. (2008) Rainfall facilitates the spread, and time alters the impact, of the invasive Argentine ant. *Oecologia*, **155**, 385–395.

Hill, M.P., Clusella-Trullas, S., Terblanche, J.S. & Richardson, D.M. (2016) Drivers, impacts, mechanisms and adaptation in insect invasions. *Biological Invasions*, **18**, 883–891.

Hurlbert, S.H. (1984) Pseudoreplication and the design of ecological field experiments. *Ecological Monographs*, **54**, 187–211.

Hurlbert, S.H. (2004) On misinterpretations of pseudoreplication and related matters: a reply to Oksanen. *Oikos*, **104**, 591–597.

IPCC (2007) Climate Change 2007: Impacts, adaptations and vulnerability. Contribution of Working Group II to the Fourth Assessment report of the Intergovernmental Panel on Climate Change (eds M.L. Parry, O.F. Canziani, J.P. Palutikof, P.J. Van Der Linden & C.E. Hanson), pp. 391–431. Cambridge University Press, Cambridge, UK and New York, NY, USA.

IPCC (2014) Climate Change 2014 – Impacts, Adaptation and Vulnerability. Part A: Global and Sectoral Aspects. *Contribution of Working Group II to the Fifth Assessment Report of the Intergovernmental Panel on Climate Change* (eds C.B. Field, V.R. Baros, D.J. Dokken, K.J. Mach, M.D. Mastrandrea, T.E. Bilir, M. Chatterjee, K.L. Ebi, Y.O. Estrada, R.C. Genova, B. Girma, E.S. Kissel, A.N. Levy, S. MacCracken, P.R. Mastrandrea & L.L. White). Cambridge University Press, Cambridge, UK and New York, NY, USA.

Jentsch, A., Kreyling, J. & Beierkuhnlein, C. (2007) A new generation of climate-change experiments: events, not trends. *Frontiers in Ecology and the Environment*, **5**, 365–374.

Johnson, S.N. (2014) Editorial overview: Global change biology and insects: Which little things will run the world in the future? *Current Opinion in Insect Science*, **5**, vi–viii.

Johnson, S.N., Gherlenda, A.N., Frew, A. & Ryalls, J.M.W. (2016) The importance of testing multiple environmental factors in legume-insect research: replication, reviewers and rebuttal. *Frontiers in Plant Science*, **7**, 489.

Leather, S.R., Basset, Y. & Didham, R.K. (2014) How to avoid the top ten pitfalls in insect conservation and diversity research and minimise your chances of manuscript rejection. *Insect Conservation and Diversity*, **7**, 1–3.

Liebhold, A.M. & Tobin, P.C. (2008) Population ecology of insect invasions and their management. *Annual Review of Entomology*, **53**, 387–408.

Lundberg, P., Ranta, E. & Kaitala, V. (2000) Species loss leads to community closure. *Ecology Letters*, **3**, 465–468.

Luterbacher, J., Dietrich, D., Xoplaki, E., Grosjean, M. & Wanner, H. (2004) European seasonal and annual temperature variability, trends, and extremes since 1500. *Science*, **303**, 1499–1503.

Messerli, J., Bertrand, A., Bourassa, J., Belanger, G., Castonguay, Y., Tremblay, G., Baron, V. & Seguin, P. (2015) Performance of low-cost open-top chambers to study long-term effects of carbon dioxide and climate under field conditions. *Agronomy Journal*, **107**, 916–920.

Musolin, D.L. (2007) Insects in a warmer world: ecological, physiological and life-history responses of true bugs (Heteroptera) to climate change. *Global Change Biology*, **13**, 1565–1585.

Newman, J.A., Anand, M., Henry, H.A.L., Hunt, S. & Gedalof, Z. (2011) *Climate Change Biology*. CABI, Wallingford, UK.

Ode, P.J., Johnson, S.N. & Moore, B.D. (2014) Atmospheric change and induced plant secondary metabolites — are we reshaping the building blocks of multi-trophic interactions? *Current Opinion in Insect Science*, **5**, 57–65.

Oksanen, L. (2001) Logic of experiments in ecology: is pseudoreplication a pseudoissue? *Oikos*, **94**, 27–38.

Potvin, C. & Tardif, S. (1988) Sources of variability and experimental designs in growth chambers. *Functional Ecology*, **2**, 123–130.

Robinson, E.A., Ryan, G.D. & Newman, J.A. (2012) A meta-analytical review of the effects of elevated CO_2 on plant-arthropod interactions highlights the importance of interacting environmental and biological variables. *New Phytologist*, **194**, 321–336.

Sandlund, O.T., Schei, P.J. & Viken, Å. (eds) (1999) *Invasive Species and Biodiversity Management*. Springer, Berlin.

Scherber, C., Gladbach, D.J., Stevnbak, K., Karsten, R.J., Schmidt, I.K., Michelsen, A., Albert, K.R., Larsen, K.S., Mikkelsen, T.N., Beier, C. & Christensen, S. (2013) Multi-factor climate change effects on insect herbivore performance. *Ecology and Evolution*, **3**, 1449–1460.

Seifert, L.I., Weithoff, G. & Vos, M. (2015) Extreme heat changes post-heat wave community reassembly. *Ecology and Evolution*, **5**, 2140–2148.

Stireman, J.O.I., Dyer, L.A., Janzen, D.H., Singer, M.S., Lill, J.T., Marquis, R.J., Ricklefs, R.E., Gentry, G.L., Hallwachs, W., Coley, P.D., Barone, J.A., Greeney, H.F., Connahs, H., Barbosa, P., Morais, H.C. & Diniz, I.R. (2005) Climatic unpredictability and parasitism of caterpillars: Implications of global warming. *Proceedings of the National Academy of Sciences of the United States of America*, **102**, 17384–17387.

Sutherst, R., Baker, R.H.A., Coakley, S.M., Harrington, R., Kriticos, D.J. & Scherm, H. (2007) Pests under global change - meeting your future landlords? *Terrestrial Ecosystems in a Changing World* (eds J. Canadell, D.E. Pataki & L.F. Pitelka), pp. 211–226. Springer-Verlag, Berlin, Germany.

Visser, M.E. & Holleman, L.J.M. (2001) Warmer springs disrupt the synchrony of oak and winter moth phenology. *Proceedings of the Royal Society B-Biological Sciences*, **268**, 289–294.

Vitousek, P.M., Mooney, H.A., Lubchenco, J. & Melillo, J.M. (1997) Human domination of Earth's ecosystems. *Science*, **277**, 494–499.

Vuorinen, T., Nerg, A.-M. & Holopainen, J.K. (2004a) Ozone exposure triggers the emission of herbivore-induced plant volatiles, but does not disturb tritrophic signalling. *Environmental Pollution*, **131**, 305–311.

Vuorinen, T., Nerg, A.-M., Ibrahim, M.A., Reddy, G.V.P. & Holopainen, J.K. (2004b) Emission of *Plutella xylostella*-induced compounds from cabbages grown at elevated CO_2 and orientation behavior of the natural enemies. *Plant Physiology*, **135**, 1984–1992.

Walther, G.-R., Roques, A., Hulme, P.E., Sykes, M.T., Pyšek, P., Kuehn, I., Zobel, M., Bacher, S., Botta-Dukát, Z., Bugmann, H., Czúcz, B., Dauber, J., Hickler, T., Jarošik, V., Kenis, M., Klotz, S., Minchin, D., Moora, M., Nentwig, W., Ott, J., Panov, V.E., Reineking, B., Robinet, C., Semenchenko, V., Solarz, W., Thuiller, W., Vilà, M., Vohland, K. & Settele, J. (2009) Alien species in a warmer world: risks and opportunities. *Trends in Ecology & Evolution*, **24**, 686–693.

Zavala, J.A., Nabity, P.D. & DeLucia, E.H. (2013) An emerging understanding of mechanisms governing insect herbivory under elevated CO_2. *Annual Review of Entomology*, **58**, 79–97.

Zvereva, E.L. & Kozlov, M.V. (2006) Consequences of simultaneous elevation of carbon dioxide and temperature for plant–herbivore interactions: a metaanalysis. *Global Change Biology*, **12**, 27–41.

Species Index

a

Acacia
   A. falcata   52, 55, 56
   A. obtusata   53, 56
   A. parvipinnula   56
Acrocephalus
   A. schoenobaenus   298
   A. scirpaceus   298
Acyrthosiphon
   A. malvae   161
   A. pisum   102, 151, 155–158, 182–184, 239
Aedes spp.   135, 137
   A. aegypti   129–131, 133, 134, 139
   A. albopictus   126, 129–131, 133–137
   A. flavopictus   126, 135–137
   A. japonicus   126, 135–137, 139
   A. triseriatus   129
   A. vigilax   129
Agasicles hygrophila   104, 105
Agriotes spp.   190, 237
Alliaria petiolata   188
Alnus glutinosa   280
Alopecurus pratensis   99
Alternanthera
   A. philoxeroides   104, 105
   A. sessilis   105, 234
Amphorophora idaei   182, 187
Anampses viridis   300
Andrena spp.   74
   A. bicolor   78
Andropogon gayanus   96
Angophora hispida   53, 56
Anopheles spp.   129
   A. albimanus   15
   A. arabiensis   130

   A. gambiae   130, 131, 133
   A. stephensi   129
   A. walkeri   133
Antheraea polyphemus   258
Anthophora spp.   78
Aphelenchoides spp.   205
Aphidius
   A. avenae   183
   A. colemani   184, 239
   A. eadyi   182
   A. ervi   96, 103, 182–184, 187, 190, 240
   A. matricariae   182, 183
   A. picipes   182, 186
   A. rhopalosiphi   183
Aphis
   A. fabae   150, 183, 233, 236
   A. gossypii   150, 161, 182
Apis spp.   74
   A. mellifera   72, 74, 80
Aporia crataegi   50, 357
Aporrectodea caliginosa   96, 212, 213, 215
Aptenodytes forsteri   299
Arabidopsis thaliana   156, 237
Argiope bruennichi   51
Aricia agestis   51, 83, 104
Armigeres subalbatus   131
Artemisia tridentata   155
Ascosphaera spp.   82
Atalopedes campestris   50
Aulacorthum solani   161
Azurina eupalama   300

b

Babesia canis   116
Banksia spp.   55

Global Climate Change and Terrestrial Invertebrates, First Edition. Edited by Scott N. Johnson and T. Hefin Jones.
© 2017 John Wiley & Sons, Ltd. Published 2017 by John Wiley & Sons, Ltd.

Bemisia tabaci   100
Betula
   B. papyrifera   257, 258
   B. pendula   55
Boloria titania   82
Bombus spp.   72, 74, 75, 78, 80
   B. impatiens   78
   B. terrestris   74, 80
Borrelia burgdorferi s.l.   115
Brachycaudus helichrysi   152
Brassica spp.   190
   B. oleracea   185, 186, 191, 238, 241–242
Brevicoryne brassicae   150, 156, 182–185, 238–239, 241–242
Bromus hordeaceus   102
Buchnera aphidicola   151
Bufo periglenes   300

c

Caenorhabditis elegans   330
Calanus spp.   304
   C. finmarchicus   301
   C. helgolandicus   301
Callistemon pinifolius   56
Candidatus Serratia symbiotica   151
Capreolus capreolus   116
Capsella bursa-pastoris   236
Cardamine
   C. hirsute   232
   C. pratensis   211, 233
Cardiaspina densitexta   241
Carpinus betulus   257
Centaurea diffusa   101
Cepaea nemoralis   327
Cepegillettea betulaefoliae   182–184, 257
Cettia cetti   298
Chaitophorus stevensis   258
Chromatomyia syngenesiae   237
Chrysopa sinica   182
Cleora scriptaria   52
Clinopodium vulgare   237
Coccinella septempunctata   102, 151
Coenonympha tullia   359
Coleomegilla maculata   183
Colias spp.   77, 83, 327
Colletes spp.   74
   C. inequalis   74

Compsilura concinnata   182–184, 258
Costelytra zealandica   102
Cotesia
   C. astrarches   104
   C. bignellii   183
   C. marginiventris   183, 184
   C. melanoscela   182
   C. melitaearum   82
   C. plutellae   182–184, 186, 191
   C. vestalis   182, 183, 186
Cucumis sativus   232
Cucurbita moschata   232
Culex spp.   127, 131
   C. annulirostris   129
   C. inatomii   139
   C. nigripalpus   133
   C. pipiens   114, 129, 131, 133, 139
   C. quinquefasciatus   129–131, 133
   C. restuans   130, 133, 139
   C. stigmatosa   139
   C. tarsalis   129, 131
   C. tritaeniorhynchus   129
Culicoides spp.   116
   C. sonorensis   130
Cymbopogon refractus   209

d

Daviesia corymbosa   53, 56
Delia radicum   210, 238, 239
Dermacentor reticulatus   115, 116
Diaeretiella rapae   100, 182–185, 239
Dicentrarchus labrax   301
Drosophila spp.   330, 331
   D. melanogaster   330

e

Eimeria melis   299
Eisenia fetida   213
Elatobium abietinum   152
Encarsia formosa   100
Engraulis encrasicolus   301
Enochrus jesusarribasi   355
Epiblema strenuana   101–102
Epichloë spp.   100
   E. coenophiala   101
   E. festucae var. *lolii*   100

Erebia medusa 80
Erigeron speciosus 81
Erynnis propertius 50, 51, 80, 83
Eucalyptus spp. 235, 236, 255, 259, 261, 262
   E. globulus 243
Eudorylaimus spp. 205
Euphausia superba 299
Euphydryas
   E. aurinia 183
   E. editha 79, 333
   E. editha bayensis 75, 81

f
Fagus sylvatica 257
Festuca arundinacea 157
Ficedula hypoleuca 305
Filenchus spp. 205
Fulmarus glacialoides 299

g
Gadus morhua 301, 304
Gallinago gallinago 302
Glomus spp. 215
Gobodion spp. 299, 300
Grammia incorrupta 98
Gratiana boliviana 98
Gryllus firmus 330, 331
Gynaephora menyuanensis 104

h
Haemagogus spp. 129
Hakea gibbosa 56
Halotydeus destructor 61
Harmonia axyridis 102, 182, 190
Helicoverpa armigera 182
Hepialus californicus 184, 211
Hesperia comma 358
Heteronychus arator 94, 211
Heterorhabditis marelatus 184
Hippodamia convergens 182, 183
Holcus lanatus 99
Hordeum vulgare 157
Humulus lupulus 18
Hyperomyzus lactucae 160
Hyphantria cunea 373

i
Ilex aquifolium 52
Iridomyrmex purpureus 60
Ixodes ricinus 114–116

l
Lantana camara 98
Lasiommata megera 76
Lecanopsis formicarum 242
Leis axyridis 182
Leishmania peruviana 130
Leptospermum squarrosum 56
Linepithema humile 373
Liquidambar styraciflua 260
Listronotus bonariensis 101
Lolium spp. 94
   L. perenne 93, 100, 213
Longidorus elongatus 207
Lucilia
   L. cuprina 112, 117
   L. sericata 112, 113, 117, 119, 122
Lumbricus spp. 298
   L. rubellus 212, 213, 216
   L. terrestris 212
Lutzomyia
   L. longipalpis 130
   L. trapidoi 134
Lymantria dispar 37, 182, 257
Lysiphlebia japonica 182
Lysiphlebus fabarum 183

m
Macropiper excelsum 52
Macrosiphum euphorbiae 158
Maculinea arion see Phengaris arion
Malacosoma disstria 37, 40, 182–184, 257, 258,
   264
Manduca sexta 77
Medicago spp. 99
   M. sativa 155, 158, 210, 233, 239
   M. truncatula 157
Megachile rotundata 72, 77, 82
Melanargia galathea 357
Melanogrammus aeglefinus 304
Melanoplus femurrubrum 54
Meles meles 298

Melissodes rustica   78
Melitaea cinxia   82, 333
Meloidogyne incognita   234
Mesonychoteuthis hamiltoni xvii
Metopolophium dirhodum   152
Microctonus aethiopoides   105
Microlaena stipoides   209
Microplitis mediator   182
Mullus surmuletus   301
Myzus persicae   152, 156, 161, 162, 182–184, 190, 238–239, 241–242

n

Nematodirus battus   114
Neocalanus cristatus   304
Neotyphodium lolii   100 see also *Epichloë, E. festucae* var. *lolii*
Nezara
   N. antennata   373
   N. viridula   373
Nicotiana
   N. benthamiana   159
   N. tabacum   159
Nomia melanderi   72, 77
Nothofagus pumilio   54

o

Obtusicauda coweni   155
Oceanites oceanicus   299
Octotoma
   O. championi   98
   O. scabripennis   98
Oechalia schellenbergii   182
Onchocerca volvulus   130
Operophtera brumata   368
Orgyia leucostigma   258
Osmia spp.   72, 74, 78–80
   O. bicornis   79
Otiorhynchus sulcatus   209
Oulema melanopus   183

p

Pagodroma nivea   299
Papilio zelicaon   50, 51
Pararge aegeria   19, 24
Parnassius

P. mnemosyne   357
   P. smintheus   80
Parthenium hysterophorus   101–102
Parthenolecanium quercifex   183
Parus major   305
Paspalum dilatatum   94
Pemphigus populitransversus   207, 233
Pennisetum clandestinum   94
Pentalonia nigronervosa   161
Peponapis pruinosa   78
Perilitus brevicollis   183
Phalaris aquatica   239
Phenacoccus herreni   184
Phengaris arion   357
Phleum pratense   207
Phratora vulgatissima   183
Phyllopertha horticola   236
Phylloscopus sibilatrix   298
Phytomyza ilicis   52
Phytoseiulus persimilis   183
Pieris brassicae   188
Pinus
   P. edulis   333
   P. taeda   260–262
Pisaurina mira   54
Plagiodera versicolora   264
Plantago lanceolata   232
Plasmodium spp.   129
   P. chabaudi   129
   P. falciparum   129
   P. vivax   129
Plutella xylostella   182–184, 186, 240
Pluvialis apricaria   302
Poa pratensis   207
Podisus maculiventris   182
Polygonia c-album   17–19
Polygonum bistorta   82
Polyommatus semiargus   357
Pomacea paludosa   296
Populus tremuloides   257, 258, 261
Pratylenchus penetrans   207
Propylea japonica   182
Psoroptes ovis   113
Ptychoramphus aleuticus   304
Ptyonoprogne rupestris   296
Pygoscelis adeliae   299

q

Quercus spp.   83, 256, 304
   Q. alba   258
   Q. myrtifolia   262
   Q. petraea   257
   Q. velutina   258

r

Rhipicephalus sanguineus   113
Rhodnius prolixus   131
Rhopalosiphum
   R. maidis   97
   R. padi   96, 103, 150–152, 161, 162, 184, 240,
     242
Rickettsia spp.   115, 151
Rostrhamus sociabilis plumbeus   296

s

Salmo salar   301
Sardina pilchardus   301
Sceloporus spp.   300
Schedonorus phoenix   94
Sericesthis nigrolineata   209
Simulium damnosum   130
Sinapis arvensis   232
Sitobion avenae   150, 152, 161, 182, 183, 186, 188
Sitona spp.   233
   S. discoideus   105, 233, 234, 239, 240
   S. lepidus   209, 233 (*see also Sitona, S.*
     obsoletus)
   S. obsoletus   3, 104
Sminthurus viridis   104
Solanum
   S. dulcamara   158
   S. viarum   98
Solenopsis invicta   373
Sonchus spp.   237

Sorex spp.   297
Speyeria mormonia   80, 81
Spodoptera
   S. exigua   183, 184, 192
   S. frugiperda   37
Spodoptera littoralis   188
Stephensia brunnichella   190, 237

t

Taeniopoda eques   60
Telopea speciosissima   53, 56
Teratocephalus spp.   205
Tetracladium marchalianum   280
Tetranychus urticae   183
Tetrastichus julis   183
Thaumetopoea pityocampa   302
Themeda triandra   97
Therioaphis maculata   103, 156
Thymelicus sylvestris   357
Triatoma infestans   130
Trifolium repens   93, 209, 233
Triticum aestivum   232
Tropilaelaps spp.   82
Trypanosoma cruzi   127, 130
Turdus philomelos   302

u

Ulmus glabra   18
Urtica dioica   18

v

Vanellus vanellus   302
Vicia faba   157

z

Zapada glacier   358
Zea mays   153

Subject Index

a

A. philoxeroides   104

above-/ below-ground interactions

Brassicaceae case study   237–239, **238–239**, 244

C:N ratios   233, **235**, 243, 244

CO$_2$ concentration effects   233–236, **235**, 243, 244

decomposer–herbivore/precipitation alteration 240

deposition pathways   234–236, **235**, 243

detritivore–shoot herbivore   232, 244

FACE studies   235–236

frequency, precipitation effects on   239–240

glucosinolates   238, 242

induced susceptibility   231

mechanisms of   230, **230**

mixed plants, altered rainfall   239–240

multiple factor testing   243–244

multiple species testing   244

N availability   239–240

overview   6, 229–231, **230**, 245

photosynthesis rates   232

precipitation changes   236–242, **238**

relative water content   236–237

root herbivore–pollinator   232

secondary metabolites   231–232

shoot/root herbivores   231–232, 243

temperature effects   234

Acacia falcata   52, **53**, 55, **56**

Acacia obtusata   **56**

Acacia parvipinnula   **56**

Acyrthosiphon malvae   161

Aedes spp.

A. aegypti   129–131, 133–134, **134**, 139

A. albopictus   126, 129–130, 134–138, **136–137**

A. flavopictus   126, **136**, 137–138

A. japonicus   126, 134–138, **136–137**

A. vigilax   129

African black beetle (*Heteronychus arator*)   211

Agasicles hygrophila Selman & Vogt   105

Agriotes spp. (click beetles)   190, 237, 242

alfalfa leaf cutter bee (*Megachile rotundata*)   77, **77**

alkali bee (*Nomia melanderi*)   77, **77**

Alliaria petiolata   188–189

Alopecurus pratensis L.   99–100

Alternanthera philoxeroides (Mart.)   105

Alternanthera sessilis (Linn.)   105

American fall web worm (*Hyphantria cunea*)   373

amino acids as defence compound   99

Amphorophora idaei   186

anchovies (*Engraulis encrasicolus*)   301

Andrena bicolor   78

Angophora hispida   **56**

Anguinidae   205

Anise swallowtail (*Papilio zelicaon*)   51

Anopheles spp.

A. albimanus   15

A. albopictus   131, 133–134

A. arabiensis   130

A. gambiae   130, 131, 133

A. stephensi   129

A. walkeri   133

Antarctic krill (*Euphausia superba*)   299

Antheraea polyphemus   **258**

Anthophora spp.   78

Aphelenchoides   205

Aphidius colemani   239, **239**

Global Climate Change and Terrestrial Invertebrates, First Edition. Edited by Scott N. Johnson and T. Hefin Jones.
© 2017 John Wiley & Sons, Ltd. Published 2017 by John Wiley & Sons, Ltd.

Aphidius ervi   96, 103, 186, 189–190, 239
Aphidius picipes   186
aphids
    aphid–host-plant interactions   **155**, 155–158,
        188
    aphid–host-plant–virus interactions   162, **162**,
        163
    banana *(Pentalonia nigronervosa)*   161
    bird cherry–oat *(Rhopalosiphum padi)*   150,
        151, 157, 161, 162, 239, 242
    black bean *(Aphis fabae)*   150, 236
    Brassicaceae case study   237–239, **238–239**
    CO_2 concentration effects   157–158, 162, 163,
        182, **184**, 185–186, **187**, 233, **235**
    corn leaf *(Rhopalosip hummaidis* (Fitch))   97
    cotton/melon *(Aphis gossypii)*   150, 161
    currant–sow thistle *(Hyperomyzus lactucae)*
        160
    endosymbionts   151
    foxglove/glasshouse-potato *(Aulacorthum*
        solani)   161
    grain *(Sitobion avenae)*   150, 152, 161, 188
    green spruce *(Elatobium abietinum)*   152
    leaf curling *(Brachycaudus helichrysi)*   152
    mealy cabbage *(Brevicoryne brassicae)*   150,
        156, 185, **238**, 238–239, 241–242
    natural enemies   259
    N availability   157
    ozone effects on   **257**, 264, 265
    pea *(Acyrthosiphon pisum)*   103, 151, 155–157,
        233, 239–240
    peach–potato *(Myzus persicae)*   156, 161, 162,
        190, **238**, 238–239, 241–242
    potato *(Macrosiphum euphorbiae)*   158
    root-feeding *(Pemphigus populi-transversus)*
        208–209, 233
    rose grain *(Metopolophium dirhodum)*   152
    shoot-feeding *(Aphis fabae fabae)*   233
    spotted *(Therioaphis maculata* (Buckton))
        103, 156
    temperature effects on   150–152, 155–158, 161
    See also herbivore-enemy interactions; plant
        disease vectors
Aporia crataegi   357
Aporrectodea caliginosa   212, 213, 215
apple snails *(Pomacea paludosa)*   296
Arabidopsis thaliana   156, 237

Araneae   262
Argentine ant *(Linepithema humile)*   373
Aricia agestis   104
Armigeres subalbatus   131
Arrhenius equation   330–331
Artemisia tridentata   155
aspen *(Populus tremuloides)*   **257–258**, 262
Aspen FACE   40–41
Atlantic cod *(Gadus morhua)*   301, 304
Atlantic salmon *(Salmo salar)*   301

b
Babesia canis infections   116
badger *(Meles meles)*   298–299
banana bunchy top virus   161
barley yellow dwarf virus   151, 158–162
bass *(Dicentrarchus labrax)*   301
Bay checker spot *(Euphydryas editha bayensis)*
    75, 81
beech *(Fagus sylvatica)*   **257**
bees. *See* insect pollinators
beetles
    African black *(Heterony chusarator* Fabricius)
        94
    Octotoma championi Baly   98
    Octotoma scabripennis Guérin-Méneville   98
    water *(Enochrus jesusarribasi)*   355, 356
biogenic volatile organic compounds (BVOCs)
    264
biological controls   93. *See also* grassland
    ecosystems
biosecurity   372–373
birch, white *(Betula papyrifera)*   **257–258**
birch, silver *(Betula pendula)*   55
blue tongue virus   116, 130
Boloria titania   82
Bombus spp.   74–75, 78
Borrelia burgdorferi   115
branch chambers   38–39
Brassicaceae case study   237–239, **238–239**, 244
Brassica oleracea   186, 241, 242
broad beans *(Vicia faba)*   157
Bromus laevipes Shear   101
brood parasites (kleptoparasites)   82
Buchnera aphidicola   151
bumble bees *(Bombus impatiens)*   78
bumble bees *(Bombus* spp.)   74–75

bumble bees *(Bombus terrestris)* 74, 80
butterflies
    black-veined white *(Aporia crataegi)* 50
    brown argus *(Aricia agestis)* 51, 83
    Clouded Apollo *(Parnassius mnemosyne)* 357
    Erebia medusa 80
    large heath *(Coenonympha tullia)* 358
    range shifts 12, 15–19, **18**, 22–24, **23**
    silver-spotted skipper *(Hesperia comma)* 358
    subalpine *(Speyeria mormonia)* 80
butternut squash *(Cucurbita moschata)* 232

c

Calanus finmarchicus 301
Calanus helgolandicus 301
Callistemon pinifolius **56**
Candidatus Serratia symbiotica 151
Capsella bursa-pastoris 236
Carbon Nutrient Balance Hypothesis 157
Cardamine hirsuta 232
Cardamine pratensis 211, 233
Cardiaspina densitexta 241
carrot root fly *(Delia radicum)* 210
Cassin's auklet *(Ptychoramphus aleuticus)* 304
caterpillars 305
cauliflower mosaic virus 154
Centaurea diffusa Lam. 101
Cepaea nemoralis 327
Cepegillettea betulaefoliae **257**
Cephalobidae 205, 206
Cetti's warbler *(Cettia cetti)* 298, 301–302
Chagas disease 130
Chaitophorus stevensis **258**
chalkbrood *(Ascosphaera* spp.) 82
chaos theory 192–193
Chironomidae 306
chlorofluorocarbons (CFCs) 191
Chromatomyia syngenesiae 237
citizen science projects 4, 13–15
click beetles *(Agriotes)* 190, 237, 242
climate change
    asynchrony, adaptation 368–369
    biosecurity 372–373
    climate envelope 47
    extreme event research 371–372, **372**
    invertebrates, impacts on 2–4
    mechanisms for 2–4

    multiple factor research 369–371
    predictions 2
    warming rate 46
climate envelope modeling 301
Clinopodium vulgare 237
clover, white *(Trifolium repens)* 233
clover root weevil *(Sitona obsoletus)* 3
Coccinella septempunctata 102, 151–152
Colias butterflies 327
Colias spp. 77
comma butterfly *(Polygonia c-album)* 17–19, **18**
controlled chambers systems 36–38, **37**
coral reefs 299–300
Costelytra zealandica 102
Costesia plutelle 186, 191
Costesia vestalis 186
Cotesia melitaearum 82–83
Coupled Model Intercomparison Project Phase 5 (CMIP5) 2
crag martins *(Ptyonoprogne rupestris)* 296
crane flies (Tipulidae) 302
cricket *(Gryllus firmus)* 331
cucumber *(Cucumis sativus)* 232
cucumber mosaic virus 159–160, 162
Culex spp.
    C. nigripalpus 133
    C. pipiens 114, 129, 131, 133, 139
    C. quinquefasciatus 130, 131
    C. restuans 130, 133, 139
    C. stigmatosa 139
    C. tarsalis 131
    C. tritaeniorhynchus 129
Culicoides sonorensis 130
Cymbidium ringspot virus 159

d

Daviesia corymbosa **56**
Delia radicum 238
dengue fever 129
detritivores 212, 232, 244
Diaeretiella rapae 100, 185, 239, **239**

e

earthworms 6, 96, 201–203, 212–216, **213**, **216**, 232, 239, 262, 298, 302
Eastern equine encephalitis 129

ectoparasitism
    climate change impacts evidence 114–116
    farmer intervention 118
    human behaviour, husbandry impacts
        116–118
    overview 5, 111–113
    parasite–host interactions 113–114
    predictive models 118–122, **119–122**
    thermal tolerance **119**, 119–120
    transmission, relative risk of 119–122,
        119–122
Eimera melis 299
Eisenia foetida 213
enzyme multifunctionality hypothesis
    (moonlighting) **329**, 331–332
Epiblema strenuana Walker 101
Epichloë 100–101
Epichloë coenophiala 101
Erigeron speciosus 81
Eucalyptus 235–236, 262
Eucalyptus globulus 243
Eudorylaimus 205
Euphydryas editha 79
European butterfly *(Lasiommata megera)* 76
Everglade snail kite *(Rostrhamus sociabilis
    plumbeus)* 296
experimental approaches
    chamber switching 370–371
    end points 35–36
    extreme event research 371–372, **372**
    holistic 32–33
    indoor closed systems 36–38, **37**
    integrated 32–33
    outdoor closed, semi-closed systems
        38–39
    outdoor open systems **37**, 39–40
    overview 4, 30–32, **31**, 41–42
    pseudoreplication in 370
    reductionism 32–33
    scale 32–33
    statistical design 33–35
    systems 36–40, **37**
    team science 40–41
    See also FACE systems; transplant
        experiments
extreme event research 371–372, **372**

f
Fabaceae 55
FACE systems
    above-/ below-ground interactions 235–236
    design generally 39–40
    disadvantages 40
    forest communities 252, **255**, 256, **257–258**,
        259–262, **260–261**, 265, 267
    overview 6
    pseudoreplication in 370
    statistical design 33–35
fig wasp (Agaonidae) 78–79
filaree red-leaf virus 161
Filenchus 205
fire ants *(Solenopsis invicta)* 373
flavonoids as defence compound 100
folivores 209, 211
forest communities
    arthropods **260–261**, 263, 265
    C:N ratios 254
    CO_2 concentration effects 252–267, **257–258**,
        260–261, **266**
    community-level responses 259–262,
        260–261, 265, **266**
    CO_2/ozone interactions 265–267, **266**
    FACE studies 252, **255**, 256, **257–258**,
        259–262, **260–261**, 265, 267
    greenhouse gases 254
    insect herbivores 254–256, **260–261**,
        263–265, **266**
    invertebrate roles in 253
    natural enemies 259, 262–264, **266**
    overview 252–253
    ozone effects 252–254, **257–258**, **260–261**,
        263–267, **266**
    species composition 262–263
forest tent caterpillar *(Malacosoma disstria)*
    257–258, 264–267
free air CO_2 enrichment systems. *See* FACE
    systems
Frescalo method 22–23
freshwater communities
    allochthony-autochthony ratio 283–284
    body size 281–283, 285
    brownification 278
    C:N ratio 277
    CO_2 concentration effects 281

detritivorous invertebrates   279–280
dissolved organic matter   **276**, 277–278, 281
emergence alteration effects   282–285
food webs   278, 282
fungal-mediated decomposition   279–280
growth, development rates   281–282
landscape-scale cycling   285
leaf-litter palatability   280, 281
overview   6, 274–275
phenological mismatches in   304
PUFAs   279, 283
resource flux effects   278–280, 283
resource quality alteration effects   284, **285**
riparian, shoreline vegetation   275–277, **276**
species composition   278–279, 283, 285
temperature effects   280–281
fritillary *(Speyeria mormonia)*   81–83

g

Galapagos damselfish *(Azurina eupalama)*
      299–300
Global Biodiversity Information Facility (GBIF)
      13–15, **14**
Gobiodon   300
grasshopper *(Melanoplus femurrubrum)*   54
grasshopper, western horse lubber *(Taeniopoda
      eques)*   60
grassland ecosystems
   biological controls in   93
   case studies   96, 97, 103
   drought effects   103
   extreme events exposure   102–104
   food webs   **95–96**, 95–97
   fungal endophytes   100–101
   herbage productivity, quality   98
   multitrophic interactions   **95–96**, 94–96
   non-target impacts   104–105
   overview   5, 92–93, 105
   pest outbreaks   104
   plant biodiversity changes in   94
   plant defence compounds   98–100
   plant phenology changes   101–102
   precipitation changes   102–103
   range shifts   103–104
   species composition   262–263
   warming, predator behaviour and   97–98
Gratiana boliviana Spaeth   98

greenhouses   36–39, **37**
green stink bug *(Nezara viridula)*   373
Gynaephora menyuanensis Yan & Chou   104
gypsy moth *(Lymantria dispar)*   **257**

h

haddock *(Melanogrammus aeglefinus)*   304
haemorrhagic disease of deer virus   130
Hakea gibbosa   **56**
Harmonia axyridis (Pallas)   102
hawkmoth *(Manduca sexta)*   77, 78
herbivore-enemy interactions
   abiotic factors, multiple   191–192
   above–below-ground interactions   231–232,
         243, 244
   agro-ecosystems   180
   aphid–host-plant interactions   **155**, 155–158,
         188
   climate change impacts   **180–184**, 181–185
   CO_2, elevated   **182**, **184**, 185–186, **187**, 192
   drought effects on   153, 156, 188–189
   extreme weather events   190
   overview   5, 179–181, **180**, 192–193
   ozone effects   **183–184**, 190–191
   precipitation changes   **184**, 188–190
   prey location   185–186
   prey quality   186, 188, 192
   temperature effects   **182–184**, 186–188, 192
   UV radiation effects   **184**, 190–191
herbivore induced plant volatiles (HIPVs)
         185–186
Holcus lanatus L.   99–100
holly *(Ilex aquifolium)*   52
holly leaf-miner *(Phytomyza ilicis)*   52
honey bees *(Apis mellifera)*   74, 80
Hoplolaimidae   205
Hordeum vulgare   157
hormone signaling pathway antagonistic
         pleiotropy trade-off hypothesis   **329**, 330
hornbeam *(Carpinus betulus)*   **257**
Hymenoptera   262

i

infectious disease vectors
   ecology, evolution of, **138**   138–139
   extrinsic incubation period (EIP)   128–129
   life history   131–134, **132**

infectious disease vectors (*contd.*)
  Nagasaki case study   128, 134–138,
    136–137
  overview   5, 126–128
  pathogens, interactions with   128–130
  physiology, development, phenology
    130–131
  population dynamics   131–134, **132**, **134**
  species interactions   131–134, **132**
  transmission efficiency   129–130
insect conservation
  adaptive evolution   353, 359
  assisted migration/colonisation   357
  biodiversity   349–352
  dispersal potential   353
  habitat asynchrony   352–353
  management strategies   355–360, **356**
  protected areas   **356**, 357–359
  range shifts   352–353, **354**, 355–356
  scientific publications   350–351, **351**
  thermal tolerance, plasticity   352, 355
  vulnerability assessment   353–354
  vulnerability drivers   352–353, **354**
insect herbivores
  folivores   209, 211
  forest communities   254–256, **260–261**,
    263–265, **266**
  soil communities   6, 207–212, **208**, **216**, 232,
    241
insect pollinators
  above-/ below-ground interactions   232
  climate change impacts   **73**, 83–84
  diapause signal   76
  Diptera   76
  food plant interactions   81–82
  natural enemies interactions   82–83
  overview   4–5, 7, 71–72
  phenology   72–75, **73**
  population decline   75–76
  range shifts   75
  temperatures, growing-season   76–79, **77**
  winters, snow pack reductions   79–80
invertebrate range shifts. *See* range shifts

j
Japanese encephalitis virus   129
joyweed, sessile (*Alternanthera sessilis*)   234

k
Kawakawa looper (*Cleora scriptaria*)   52
Kawakawa tree *(Macropiper excelsum)*   52
kleptoparasites (brood parasites)   82

l
ladybird (*Harmonia axyridis*)   190
Lantana camara L.   98
lapwing, northern (*Vanellus vanellus*)   302
Larinus minutus   101
leaf miners   237, 256, **260–261**, 262, 263
Lecanopsis formicarum   242
Leishmania spp.   130
Lepidoptera   256, 259, 359
Leptospermum squarrosum   **56**
life-history trade-offs
  adaptability, 320   **321–326**
  aestivation responses   **325**
  behavioural buffering   **323**
  biotic interactions alterations   **324**
  candidate gene pleiotropic effects   327
  compensatory, catch-up growth   **323**
  counter-gradient variation   **325**
  developmental polymorphism   **323**
  diapause alterations   **324**, **325**, 330
  dispersal polymorphism   **323**
  drought, drought regimens   333–337, **335–336**
  eco-evolutionary dynamics   **326**, 334–337,
    335–336
  ecological factor interactions   **323**
  egg-placing behaviour   **323**
  endosymbiotic relationship alterations   **324**
  enzyme multifunctionality hypothesis
    (moonlighting)   **329**, 331–332
  exposure   320, **321–326**
  heritability   **326**, 327, **329**, 330–332
  hormonal signaling pathways   327, 330
  hormone signaling pathway antagonistic
    pleiotropy trade-off hypothesis   **329**, 330
  host-plant selection   **323**
  hypotheses, key factors   **321–322**, 327–328
  insect polyphenism   **323**
  intraspecific variation   **324**, 327, 330–333
  life stage, sex effects   **329**
  mechanisms   328–330, **329**
  melanisation allocation   **325**
  metabolic rates, water loss   **325**, 328, 332

new habitat colonisation **324**

overview 7, 319–320, **321–326**, 337–338

phenological mismatches **324**

phenotypic plasticity fitness costs **326**

photoperiod, photoperiodic-direction cues **323, 324**

plastic larval development **323**

population ecotypes, climate-responsive **326**

population performance **324**

rapid trait evolution **325**

resource allocation trade-off hypothesis **329**, 331

response mediation mechanisms **323–326**

season length, latitude, size interactions **325**, 333

sensitivity 320, **321–326**

SNPs **329**, 333, 334, 338

starvation, dessication resistance **329**, 330, 332–333

statistical modeling 338

temperature cues, effects **323, 325**

theoretical modeling 338

thermal stability-kinetic efficiency trade-off hypothesis **329**, 330–331

trait-phenology interactions **324**

water loss-flux rate trade-off hypothesis **329**, 332–333

water loss-total gas exchange hypothesis **329**, 332

winter warming effects **324**

Linyphiidae 283

lizards *(Sceloporus)* 300

Lolium perenne 213

Longidorus elongatus 207

Lucilia cuprina 101, 117

Lucilia sericata 101, 117, 119, 122

Lumbricus rubellus 212, 213

Lumbricus spp. 298

Lumbricus terrestris 212, 215

Lutzomyia longipalpis 130

Lutzomyia trapidoi **134**

Lyme borreliosis 115–116

m

Maculinea (Phengaris) 357

Malacosoma disstria **258**

malaria 127–129

mason bees *(Osmia* spp.) 78–80

Mauritius wrasse *(Anampses viridus)* 300

meat ants *(Iridomyrmex purpureus)* 60

Medicago sativa 155, 156, 158, 210, 233, 239–240

Medicago spp. 99

Medicago truncatula 157

Megachilidae 80

Melanargia galathea 357

Melissodes rustica 78

Meloidogyne incognita 234

Microctonus aethiopoides Loan 105

Microlaena stipoides 209

mite, red-legged earth 61

mite, sheep scab *(Psoroptes ovis)* 113

moths

diamondback *(Plutella xylostella)* 239

grassland *(Grammia incorrupta* Hy. Edwards) 98

tiger (Arctiinae) 359

white marked tussock *(Orgyia leucostigma)* **258**

Murray Valley encephalitis 129

Myrtaceae 55

n

National Biodiversity Network (NBN) 14

Nematocera 306

nematodes 6, 203–207, **204**, 215, **216**, 217, 234

Nematodirus battus 114

Nezara antennata 373

Nicotiana benthamiana 159

Nicotiana tabacum 159

nitrous oxide 191

o

oak *(Quercus* spp.) **257–258**, 304–305

Onchocerca volvulus 130

open-top chambers 38–39

p

Pararge aegeria 24

parasitoid interactions 82–83

Parthenium hysterophorus L. 101–102

Passeriformes 297, **297**

Phalaris aquatica 239–240

Phleum pratense 207

pied flycatcher *(Ficedula hypoleuca)* 305

Pieris brassicae 188–189
Plantago lancelota 232
plant disease vectors
    aphid–host-plant interactions **155**, 155–158, 188
    aphid–host-plant–virus interactions 162, **162**, 163
    aphids **149**, 149–152
    CO_2 concentration effects **152**, 152–153, 157–158, 162, 163
    disease pyramid 148–149, **149**
    drought effects on 153, 156, 188–189
    endosymbionts 151
    host plants **152**, 152–153
    host-plant–virus interactions **158**, 158–160
    N availability 157, 192
    overview 5, 148, 162–163
    RNA-based antiviral resistance 159
    temperature effects on 150–161
    virus–aphid interactions **160**, 160–161
    viruses 154, **154**
plant–grasshopper–spider food chains 97
plant stress hypothesis 188–189, 241–242
plant vigour hypothesis 189, 241
Plasmodium chabaudi 129
Plasmodium falciparum 129
Plasmodium vivax 129
plover, golden *(Pluvialis apricaria)* 302
Plutella xylostella 186
Poa pratensis 207
pollinators. *See* insect pollinators
Polyommatus semiargus 357
potato virus Y 159, 161, 162
Pratylenchus penetrans 207
precipitation changes
    above-/below-ground interactions 236–242, **238**
    grassland ecosystems 102–103
    herbivore-enemy interactions **184**, 188–190
    soil communities **204**, 206–207, **208**, 210–211, **213**, 214, **216**
    vertebrate interactions 298–299
Propertius dusky wing *(Erynnis propertius)* 50–51
Proteaceae 55
pseudoreplication 370
pulsed stress hypothesis 241–242

q
Qudsianematidae 206

r
rain-manipulation chambers 38–39
range shifts
    butterflies 12, 15–19, **18**, 22–24, **23**
    citizen science projects 4, 13–15
    distribution size measurement 16
    DNA barcoding 24
    fixed effort transects 23–24
    grassland ecosystems 104
    habitat associations 24
    historical datasets 12–15, **14**
    insect conservation 352–353, **354**, 355–356
    insect pollinators 75
    latitudinal shifts 15–17
    marine 15, 19–20
    meta-analyses 15–16
    multi-directional shift 17–19, **18**
    overview 4, 11–13
    range margins 17, **18**
    rates of 15, 16
    soil communities 302
    spatial, temporal biases 20–23, **22–23**
    species location measurement 16–17
    success, factors affecting 373
    taxonomic bias 19–20
    transplant experiments 50–51
    tropical areas 20–21
    vertebrate interactions 301–303
red mullet *(Mullus surmuletus)* 301
reed warbler *(Acrocephalus scirpaceus)* 298, 306
Rhabditidae 205
Rhipicephalus sanguineus 113
Rhodnius prolixus 131
Rhopalosiphum padi L. 103
roe deer *(Capreolus capreolus)* 116
Ross River virus 129
rotifers 371–372
ryegrass *(Lolium perenne)* 100

s
Sachem skipper *(Atalopedes campestris)* 50
saponins 99
sardines *(Sardina pilchardus)* 301
scarab, root-feeding Argentine *(Sericesthis nigrolineata)* 209

scarab beetle *(Phyllopertha horticola)* 236

Schmalhausen's law 139

sedge warbler *(Acrocephalus schoenobaenus)* 298

shrews *(Sorex* spp.) 297

Simulium damnosum 130

Sitona discoideus 105, 234, 239–240

Sitona obsoletus Gmelin 104

skipper *(Erynnis propertius)* 80, 83

Sminthurus viridis L. 104

snipe *(Gallinago gallinago)* 302

soil communities

    aboveground net primary production 214

    chemical ecology 217

    climate change impacts 216

    climate change moderation by 3, 211–212, 214–215

    C:N ratio 209

    CO_2 impacts on 202–205, **204**, **208**, 208–215, **213**, **216**, 217

    detritivores 212

    earthworms 6, 201–203, 212–216, **213**, **216**, 232, 239, 262, 298, 302

    ecosystem level effects 207

    experimental time scale 217

    folivores 209, 211

    food webs 215

    insect herbivores 6, 207–212, **208**, **216**, 232, 241

    multiple effect studies 217

    N availability 213

    nematodes 6, 203–207, **204**, 215, **216**, 217, 234

    new habitats colonization 215–216

    overview 6, 201–203

    phenological mismatches in 304–305

    precipitation changes **204**, 206–207, **208**, 210–211, **213**, 214, **216**

    range shifts 302

    seed transport, translocation 214

    species composition 214

    temperature impacts on **204**, 205–206, **208**, 210, **213**, 214, **216**

Solanum dulcamara 158

Solanum viarum Dunal 98

song thrush *(Turdus philomelos)* 302

southern beech *(Nothofagus pumillo)* 54

soybean dwarf virus 161

species distribution models (SDMs) 48–49, 353–354, 358–359

spiders *(Pisaurina mira)* 54

Spodoptera littoralis 188–189

squash bees *(Peponapis pruinosa)* 78

St. Louis encephalitis 129

starvation, dessication resistance **329**, 330, 332–333

Stephensia brunnichella 190, 237

stonefly *(Zapada glacier)* 358

Sylvioidea 297, **297**

t

Telopea speciosissima **56**

Teratocephalus 205

thermal stability-kinetic efficiency trade-off hypothesis **329**, 330–331

Thymelicus sylvestris 357

tick *(Ixodes ricinus)* 114–117

tick, European meadow *(Dermacentor reticulatus)* 115–116

tick-borne encephalitis 114–117

Tipulidae 210

toad, golden *(Bufo periglenes)* 300

tobacco necrosis virus 154, 159

transplant experiments

    climate change impacts **47**, 47–48

    community shifts 54–57, **56**

    event timing changes 51

    feeding guild distribution 55–57, **56**

    genotypic/phenotypic responses 54

    microclimatic variation 60

    network analysis trends 57–59, **58–59**

    overview 4, 46

    predator-prey relationships 61

    principles **49**, 49–50

    range shifts 50–51

    species interactions, spatial mismatches 52–54, **53**, 61

    temperature adaptation 50, 57–61, **58–59**

Triatoma infestans 130

Triticum aestivum 232

Tropilaelaps 82

Trypanosoma cruzi 130

turnip mosaic virus 154

u

vertebrate interactions
  abundance changes 296–300, **297–298**
  adaptation strategies 307
  behaviour, ecology 305–306
  demography, population size 296–299, **297–298**, 307
  distribution changes 300–303, 307
  ecological services 307
  extinctions 299–300
  food webs 304–305
  habitat differences 306
  overview 6, 295–296, 307–308
  pest control 307
  phenology 303–307
  precipitation changes 298–299
  range shifts 301–303
  temperature effects 296–298, 301–302
veterinary disease vectors. *See* ectoparasitism

v

warblers 297–298, **297–298**, 301–302, 306
wasp spider *(Argiope bruennichi)* 51
water loss-flux rate trade-off hypothesis **329**, 332–333

water loss-total gas exchange hypothesis **329**, 332
weevils, clover root *(Sitona lepidus)* 209–210, 233
weevils, vine *(Otiorhynchus sulcatus)* 209
Western equine encephalitis 129
West Nile virus 114, 116, 127, 129
white cabbage *(Brassica oleracea)* 191
whole-tree chambers 38–39
Wider Countryside Butterfly Survey 23–24
wild mustard *(Sinapis arvensis)* 232
Wilson's storm petrels *(Oceanites oceanicus)* 299
wood warbler *(Phylloscopus sibilatrix)* 298
worm, beet army *(Spodoptera exigua* (Hübner)) 192

w

yellow fever 129
yellow vein virus 160

x

zooplankton 304, 306
zucchini yellow mosaic potyvirus 161